VOLUME TWO

STRATIGRAPHY & TIMESCALES

Advances in Sequence Stratigraphy

VOLUME TWO

STRATIGRAPHY & TIMESCALES

Advances in Sequence Stratigraphy

Edited by

MICHAEL MONTENARI

Department of Earth Sciences and Geography
Keele University
Newcastle
United Kingdom

ELSEVIER

ACADEMIC PRESS
An imprint of Elsevier

Academic Press is an imprint of Elsevier
50 Hampshire Street, 5th Floor, Cambridge, MA 02139, United States
525 B Street, Suite 1800, San Diego, CA 92101-4495, United States
The Boulevard, Langford Lane, Kidlington, Oxford OX5 1GB, United Kingdom
125 London Wall, London, EC2Y 5AS, United Kingdom

First edition 2017

ISBN: 978-0-12-813077-3
ISSN: 2468-5178

For information on all Academic Press publications
visit our website at https://www.elsevier.com/books-and-journals

Publisher: Zoe Kruze
Acquisition Editor: Jason Mitchell
Editorial Project Manager: Joanna Collett
Production Project Manager: Vignesh Tamil
Cover Designer: Alan Studholme

Typeset by SPi Global, India

CONTENTS

CONTRIBUTORS

Thomas Aigner
University of Tübingen, Tübingen, BW, Germany

Daniel Bendias
ENI, London, United Kingdom

Octavian Catuneanu
University of Alberta, Edmonton, AB, Canada

Jed Day
Illinois State University, Normal, IL, United States

Andrei Dronov
Geological Institute, Russian Academy of Sciences, Moscow, Russia

Ashton F. Embry
Embry Corp, Calgary, AB, Canada

Annette E. Götz
Earth and Environmental Sciences, University of Portsmouth, Portsmouth, United Kingdom

Jean Guex
University of Lausanne, Lausanne, Switzerland

Francis Hirsch
University of Naruto, Naruto, Japan

Michael Holz
Universidade Federal da Bahia, Salvador, Bahia, Brazil

Erik P. Johannessen
EP Skolithos, Stavanger, Norway

Ali M. Kiliç
Balıkesir University, Balıkesir, Turkey

Andrew D. Miall
University of Toronto, Toronto, ON, Canada

Michael Montenari
Geography, Geology and the Environment, Keele University, Staffordshire, United Kingdom

Michael Obermaier
Shell Projects & Technology, The Hague, The Netherlands

Pablo Plasencia
Universidad de Valencia, Valencia, Spain

Vinicius Carneiro Santana
Universidade Federal da Bahia, Salvador, Bahia, Brazil

Markus Schlaich
University of Tübingen, Tübingen, BW, Germany

Edric Brasileiro Troccoli
Universidade Federal da Bahia, Salvador, Bahia, Brazil

Paulo Augusto Vidigal-Souza
Universidade Federal da Bahia, Salvador, Bahia, Brazil

Daniel Bono Vilas-Boas
Universidade Federal da Bahia, Salvador, Bahia, Brazil

Brian J. Witzke
University of Iowa, Iowa, City, IA, United States

PREFACE

Sequence stratigraphy is nowadays regarded to be one of the most important and powerful stratigraphic concepts. Its ability to not only interpret but also to predict the spatial and temporal distribution of sediment bodies has made it an indispensable tool for both academia and industry. Sequence stratigraphy is, and has always been, a highly integrative method, which finds parts of its scientific roots even in observations and discoveries made centuries ago.

Since its invention four decades ago, sequence stratigraphy as an important scientific method and industrial tool has undergone rapid modifications and refinements. Many of these modifications were essential and clearly aimed toward a broadening of the sequence stratigraphic paradigms in order to provide effective solutions for a wide range of stratigraphic problems. Some other adaptations, although certainly conceived in good faith, introduced misconceptions and sometimes not clearly defined new technical terms to sequence stratigraphy. This led to considerable misunderstandings among researchers and resulted in questioning the scientific value and the universal applicability of sequence stratigraphy as such. Over the past decade excellent progress has been made to standardize the sequence stratigraphic nomenclature and methodology on an international level. This continued effort has established sequence stratigraphy again at the very forefront of earth scientific research.

In this book, altogether nine contributions some of a more conceptual nature as well as case studies on *Advances in Sequence Stratigraphy* are being presented. In order to provide an overview the contributions together with their abstracts are introduced in the following. The first chapter by **O. Catuneanu** on "Sequence Stratigraphy: Guidelines for a Standard Methodology" highlights "the interplay of local and global controls on accommodation and sedimentation generates basin-specific sequence stratigraphic frameworks that record cyclicity at multiple scales. There are no temporal or physical standards for the scale of any type of sequence stratigraphic unit. Sequences, systems tracts, and depositional systems can be defined at different scales, depending on the scope of the study, the resolution of the data available, and the local conditions of accommodation and sedimentation. A scale-independent methodology and nomenclature is key to the standard application of sequence stratigraphy. Stratal stacking patterns provide the basis for the definition of all units and surfaces of sequence stratigraphy. The same types of stacking patterns may be observed at different scales, in relation to stratigraphic cycles of different magnitudes. At any scale of observation (i.e., hierarchical rank), a specific type of stacking pattern defines a systems tract, and changes in stacking pattern mark the position of sequence stratigraphic surfaces. Beyond this model-independent framework, model-dependent choices with respect to the selection of the 'sequence boundary' may be made as a function of the mappability of the different types of sequence stratigraphic surfaces that are present within the study area. The model-independent methodology, inherently simple and consistent, provides the flexible platform for a standard application of sequence stratigraphy across the entire range of geological settings, stratigraphic scales, and types of data available." This chapter is followed by the contribution of **A.D. Miall** on "Sequence Stratigraphy and Geologic Time" which outlines that "the maturation of sequence stratigraphy and the development of a refined geological timescale have provided the basis for much more detailed stratigraphic reconstructions than were possible as little as a decade ago. Earlier studies showing that measured sedimentation rates vary inversely in proportion to the timescale over which they are measured may now be placed in a context that relates rates and timescales to actual geological processes. As little as 10% of elapsed time may be represented by a vertical succession at any one location. Sediment breaks account for the remainder, and constitute a much more significant component of stratigraphic processes than is generally assumed. Research methods that assume stratigraphic continuity must take this fact into account." The third chapter by **A.F. Embry and E.P. Johannessen** on "Two Approaches to Sequence Stratigraphy" is emphasizing that "sequence stratigraphy is a stratigraphic discipline in which the defined stratigraphic surfaces represent either breaks in deposition or changes in depositional trend. Two approaches for defining sequence stratigraphic surfaces and units have evolved, with one being inductive and the other deductive. The empirical, inductive approach defines surfaces based on observable, physical characteristics. Five surfaces, subaerial unconformity (SU), unconformable shoreline ravinement (SR-U), slope onlap surface (SOS), maximum regressive surface (MRS), and maximum flooding surface (MFS), are used to define inductive sequence units and to construct a correlation framework. Two different types of sequences (depositional T–R, genetic stratigraphic) and two systems tracts (transgressive systems tract, regressive systems tract) are employed in the inductive approach. In contrast, the deductive approach defines sequence stratigraphic surfaces in terms of theoretical events on a base-level curve (e.g., start base-level fall). Three deductive surfaces (SU, MRS, and MFS) used for unit definition and correlation are the same as those employed by the inductive approach. Two surfaces, not recognized in the inductive approach, are also used to delineate units and for correlation. Notably, both are chronostratigraphic surfaces which display no diagnostic physical characteristics. These surfaces include the basal surface of forced regression (BSFR), the depositional surface at the start of base-level fall, and the correlative conformity (CC), the depositional surface at the start of base-level rise. A sequence is defined as a succession of strata deposited during a full base-level cycle and associated change in accommodation or sediment supply. Recognition of the five surfaces has allowed the definition of four types of sequences (depositional sequence type 1 and 2, T–R, genetic stratigraphic), with each type being divided into four systems tracts (lowstand, transgressive, highstand, and falling stage). The decision as to what approach to use—inductive or deductive—depends on the robustness of the available data and whether the BSFR and CC can be recognized with reasonable objectivity and consistency over the study area. Caution must be exercised to avoid the pitfall of trying to force fit data into the deductive approach by interpreting an inappropriate surface (e.g., facies change) as either the BSFR or CC. In many situations, especially when seismic data are not available, it is not possible to recognize and correlate the BSFR and/or the CC with objectivity. In these cases, the inductive approach is required for unit delineation and correlation." In the fourth contribution **M. Holz, D.B. Vilas-Boas, E.B. Troccoli, V.C.V. Santana, and P.A. Vidigal-Souza** are introducing "Conceptual Models for Sequence Stratigraphy of Continental Rift Successions" and highlight that "due to the basins asymmetry and the subsidence controlled by one major fault zone, the accommodation space within a rift basin is variable, and the popular passive margin-based sequence stratigraphic scheme, where depositional systems tracts develop following a predictable order during a complete cycle of base-level change, is not applicable. For a practical and useful stratigraphic analysis of rift basins, conceptual and methodological adaptations are required. This chapter discusses some main aspects of sequence stratigraphy of continental rift basins and the conceptual adaptation needed, and proposes an integrative sequence stratigraphic model, with two different scenarios (gradual and punctuated subsidence), defining three tectonostratigraphic phases: (1) the rift initiation phase, which is characterized by isolated and restricted faults that create incipient half-grabens, infilled with fluvial and deltaic facies; (2) the rift development phase, during which the initial rift faults tend to link and to form a larger and deeper depositional area, recording lacustrine facies with an overall retrogradational trend; and (3) the rift termination phase, when the accommodation decreases, the sedimentation regime is progradational, and the rift basin is filled with fluvial–deltaic and eolian facies. The chapter closes illustrating those models with examples from selected Brazilian rift basins." The fifth chapter by **A. Dronov** on "Ordovician Sequence Stratigraphy of the Siberian and Russian Platforms" is pointing out that "the comparative analysis of the Ordovician successions of the Siberian and North American platforms demonstrates a striking similarity in the long-term lithological development and the sea-level curve interpretation. On both platforms, the Ordovician succession commences with tropical stromatolite-bearing carbonates, which abruptly change to siliciclastic deposits and the succession terminates with cool-water carbonates. Numerous K-bentonite beds in the Upper Ordovician of the North American and Siberian platforms emphasize this similarity even further. The sea-level curve for the Ordovician of the North American Platform assumes a prominent sea-level drop at the base of the Middle Ordovician and a long-term lowstand during the Dapingian and Darriwilian. The sea-level curve for the Ordovician of the Siberian Platform looks roughly the same. On the other hand, sea-level curves for the Ordovician of the Gondwanan platforms (North Africa, Yangtze Platform, Tarim, South America, Avalonia, and Baltica) seem to share different patterns. The Middle Ordovician represents there rather a highstand interval in these reconstructions. As a result, instead of one global sea-level curve for the Ordovician, it would be probably more reasonable to suggest two semi-global curves for two major tectonic regions: one of which includes the Siberian and the North American platforms and the other combines Baltica and the Gondwanan platforms. This subdivision probably reflects the position of the main Ordovician lithospheric plates." This chapter is followed by the contribution of **J. Day and B.J. Witzke** on "Upper Devonian Biostratigraphy, Event Stratigraphy, and Late Frasnian Kellwasser Extinction Bioevents in the Iowa Basin: Western Euramerica" in which is highlighted that the "Upper Devonian (Upper Frasnian) strata in the Iowa area (central United States) provide one of the best records in North America of Late Frasnian sea-level history and the associated Kellwasser Extinction Events within a detailed conodont biostratigraphic framework. Rich conodont faunas span Frasnian Zones 11 through 13c. Diverse benthic faunas of the Lime Creek Formation provide an unparalleled paleontologic record to examine the effects of the Kellwasser extinctions of benthic faunas in the tropical seaway. These strata preserve an onshore-to-offshore transect of depositional facies ranging from shallow-water carbonate shelf to basinal deeper-water shale facies in an epeiric ramp system. Late Frasnian sedimentation was initiated above a deeply incised and karstified surface on underlying Givetian–mid Frasnian carbonates. Regional stratigraphic relations require a minimal sea-level rise of at least 65–90 m to onlap this older surface. The Lime Creek Formation includes two stratigraphic sequences correlated with the global eustatic T–R cycles IId-1 and IId-2. The Lime Creek records a major extinction of benthic faunas through the Lower Kellwasser Extinction Event (LKE) associated with T–R cycle IId-2 deepening (upper part Zone 12) and onset of a major positive $\delta^{13}C_{carb}$ isotope excursion that spans the LKE and post-LKE recovery interval. The change from brachiopod-dominated to mollusk-dominated benthic associations is evidence of significant reorganization of benthic paleocommunity structure in the post-LKE survival-and-recovery interval. The near total extinction of the Late Frasnian conodont fauna during the Upper Kellwasser Event was coincident with regional expansion of Grassy Creek Shale organic-rich dysoxic–anoxic shale facies in southeast Iowa." In the seventh chapter by **A.M. Kiliç, P. Plasencia, J. Guex, and F. Hirsch** on "Challenging Darwin: Evolution of Triassic Conodonts and Their Struggle for Life in a Changing World" attention is drawn toward the fact that "the phylogeny and distribution of Triassic conodonts reveal many aspects of their natural history. Conodonts incorporate the morphologic response to temperature as well as to eustatic cycles. Speciation, radiation, and extinction are not fortuitous, and evolution uses heterochrony (progenesis and neoteny) in response to stress-generating events. Proteromorphosis (reappearance of ancestral morphs) and paedomorphosis (retention of juvenile traits) are reactions to sublethal environmental stress. This often follows radiation of fully developed forms in the recovery stage after extinction that timely matches transgressions. Evolutionary retrogradation (neoteny) during eustatic high stands often precedes extinction. The 51-Ma-long Triassic Period consists of the Induan–Olenekian first 5 Ma with a post-Permian extinction recovery rate of 13 speciations/Ma. The next 10 Ma of the Anisian–Ladinian saw a drop in speciations to 8 S/Ma, but during the remaining Carnian–Rhaetian 36 Ma, the rate was down to below 2 S/Ma, until the end of the Triassic. Roughly a dozen Permian survivors became extinct during the Induan. Soon after, Gondolellidae saw a revival with the Neospathodus–Kashmirella lineage. The Paragondolella–Misikella lineage emerged in the Aegean (Early Anisian) and dominated the Triassic scene for the next 45 Ma until extinction. In pelagic environments Gladigondolellidae ranged from Late Spathian until Mid-Carnian. Neogondolella became extinct by the Late Ladinian, and Paragondolella was replaced by Metapolygnathus during Julian times. The evolutionary trends of a free blade, platform adornment, widening of the posterior end, and bifurcation of the basal cavity were initiated sporadically during Late Ladinian and became the norm, reaching their peak in

Early Norian Ancyrogondolella. Strongly ornamented Mazzaella and Carnepigondolella marked the Julian Pluvial Event and the Late Tuvalian LST (lowstand systems tract). The atavistic (neotenic) Norigondolella came in around the Tuvalian–Lacian boundary and a generalized posterior acumination set in with Alaunian (Middle Norian) Mockina. Catastrophes of various origin, such as global LST events and temperature rises, precipitated retrogradation with Dienerian Neospathodus, Pelsonian Nicoraella, Illyrian Pseudofurnishius, Cordevolian Mosherella, Tuvalian Neocavitella, and Sevatian–Rhaetian Misikella. Bithynian Kamuellerella–Ketinella–Gedikella in the Istanbul Zone defines a North Tethyan Prepontic faunal subprovince. Late Anisian early forms of Pseudofurnishius initiated a Ladinian Sephardic faunal province along the northern Gondwanian shelf. A stepwise decrease in speciation linked Triassic conodont evolution in unison to climatic–eustatic events." This chapter is followed by the contribution of **A.E. Götz and M. Montenari** on "A Facies-Independent Trans-European Anisian–Ladinian Marker Horizon? Significance and Impact for Sequence Stratigraphy and Intra-Tethyan Correlation" in which it is proposed that "transcontinental correlation of European carbonate systems that developed during the onset of the Mesozoic breakup of Pangaea remains difficult due to pronounced lateral facies variations and in many cases facies-dependent fossil groups used for biostratigraphic dating and correlation. In addition, radiometric dating of ash layers associated with Middle Triassic carbonate successions in Europe is limited to certain areas and times of volcanic activity. Thus, Middle Triassic correlation schemes of the Tethyan and Peri-Tethyan realm—a highly dynamic area in terms of plate reorganization and related development of early Mesozoic epeiric seas—need further refinement using new correlation tools. Here, we highlight the Europe-wide occurrence of a silica-rich horizon at the Anisian–Ladinian boundary as a potential trans-European correlation horizon. Chert layers and nodules form in different depositional settings, under different conditions, and within different facies, thus being well suited for interregional correlation. However, the processes leading to chert formation in Middle Triassic shallow-marine carbonate deposits are still under discussion and require further research in order to make this prominent feature a powerful correlation tool." In the ninth chapter by **M. Schlaich, M. Obermaier, D. Bendias, and T. Aigner** on "Integrated Stratigraphy of Triassic and Jurassic Formations in the Central Oman Mountains: A Synopsis" it is emphasized that "the Triassic Mahil Formation and the Lower Jurassic Lower Mafraq Member are part of the AP6 (Arabian Plate) tectonostratigraphic mega-sequence. The overlying Mid- to Upper Jurassic

Upper Mafraq Member, Dhruma, Tuwaiq Mountain, Hanifa and Jubaila formations form the AP7 mega-sequence. All formations were studied in outcrops and in borehole cores and cuttings from the subsurface of Oman. Biostratigraphic and chemostratigraphic analyses were carried out in order to determine the age of each formation and to identify major unconformities. Biostratigraphic analyses for the Triassic Mahil Formation revealed an Olenekian to Early Anisian to Carnian age. This paper expands on the sequence stratigraphic interpretations that have been established for each individual interval from the Lower Triassic to the Upper Jurassic. It aims to integrate these previous studies in a consistent sequence stratigraphic framework based on observations in outcrops of the Central Oman Mountains."

Michael Montenari
Keele University
United Kingdom

CHAPTER ONE

Sequence Stratigraphy: Guidelines for a Standard Methodology

Octavian Catuneanu[1]
University of Alberta, Edmonton, AB, Canada
[1]Corresponding author: e-mail address: octavian@ualberta.ca

Contents

Stratigraphy & Timescales, Volume 2
ISSN 2468-5178
http://dx.doi.org/10.1016/bs.sats.2017.07.003

Abstract

The interplay of local and global controls on accommodation and sedimentation generates basin-specific sequence stratigraphic frameworks that record cyclicity at multiple scales. There are no temporal or physical standards for the scale of any type of sequence stratigraphic unit. Sequences, systems tracts, and depositional systems can be defined at different scales, depending on the scope of the study, the resolution of the data available, and the local conditions of accommodation and sedimentation. A scale-independent methodology and nomenclature is key to the standard application of sequence stratigraphy.

Stratal stacking patterns provide the basis for the definition of all units and surfaces of sequence stratigraphy. The same types of stacking patterns may be observed at different scales, in relation to stratigraphic cycles of different magnitudes. At any scale of observation (i.e., hierarchical rank), a specific type of stacking pattern defines a systems tract, and changes in stacking pattern mark the position of sequence stratigraphic surfaces. Beyond this model-independent framework, model-dependent choices with respect to the selection of the "sequence boundary" may be made as a function of the mappability of the different types of sequence stratigraphic surface that are present within the study area. The model-independent methodology, inherently simple and consistent, provides the flexible platform for a standard application of sequence stratigraphy across the entire range of geological settings, stratigraphic scales, and types of data available.

1. INTRODUCTION

Sequence stratigraphy is a type of stratigraphy that deals with the description, interpretation, classification, and nomenclature of sedimentary rocks based on their stratal stacking patterns and their stratigraphic relations. Sequence stratigraphy integrates all other types of stratigraphy and includes seismic stratigraphy (Fig. 1). The sequence stratigraphic methodology has gained considerable popularity among practitioners with interest in different aspects of the stratigraphic record, and its applications have been expanded to all depositional, tectonic, and climatic settings, from Precambrian to Phanerozoic successions (e.g., Csato et al., 2013, 2015; De Gasperi and Catuneanu, 2014; Eriksson et al., 2005, 2013; Zecchin et al., 2015). A standard workflow assumes a set of guidelines that afford a consistent application of the method across all spectrum of geological settings, stratigraphic scales, and types of data available.

Sequence stratigraphic units are bodies of sedimentary rocks that are defined and characterized on the basis of their stratal stacking patterns and their stratigraphic relations. The bounding surfaces of sequence stratigraphic units are sequence stratigraphic surfaces, which are stratigraphic contacts that

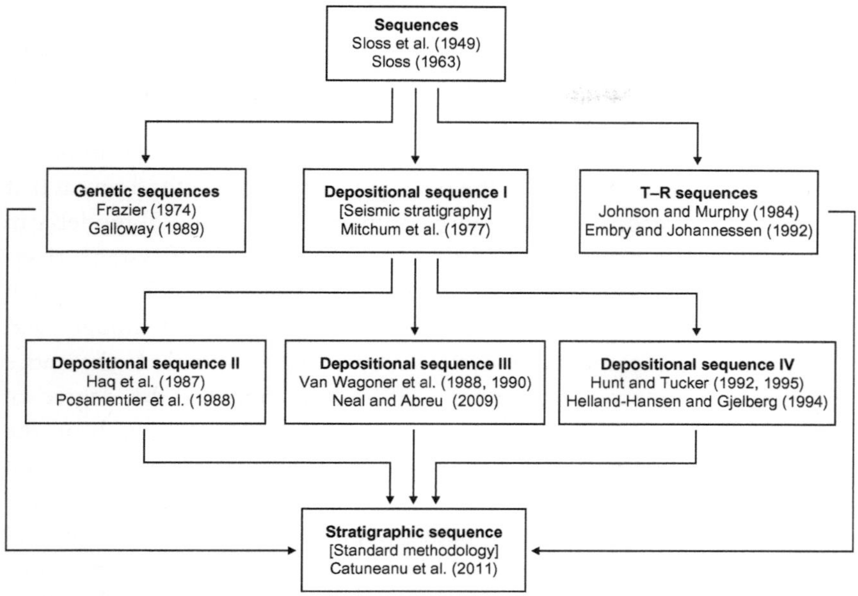

Fig. 1 Evolution of sequence stratigraphic approaches. *Modified from Catuneanu, O., Bhattacharya, J.P., Blum, M.D., Dalrymple, R.W., Eriksson, P.G., Fielding, C.R., Fisher, W.L., Galloway, W.E., Gianolla, P., Gibling, M.R., Giles, K.A., Holbrook, J.M., Jordan, R., Kendall, C.G.St.C., Macurda, B., Martinsen, O.J., Miall, A.D., Nummedal, D., Posamentier, H.W., Pratt, B.R., Shanley, K.W., Steel, R.J., Strasser, A., Tucker, M.E., 2010. Sequence stratigraphy: common ground after three decades of development. First Break 28, 21–34; Catuneanu, O., Galloway, W.E., Kendall, C.G.St.C., Miall, A.D., Posamentier, H.W., Strasser, A., Tucker, M.E., 2011. Sequence stratigraphy: methodology and nomenclature. Newsl. Stratigr. 44/3, 173–245.*

mark changes in stratal stacking pattern between the underlying and the overlying units. The sequence stratigraphic units and their bounding surfaces provide the basis for a genetic, process-based approach to stratigraphic mapping and correlation. This approach sets sequence stratigraphy apart from other correlation methods that rely on similarities of rock units in terms of lithology (i.e., lithostratigraphy), fossil assemblages (i.e., biostratigraphy), magnetic characteristics (i.e., magnetostratigraphy), geochemical signatures (i.e., chemostratigraphy), or age (chronostratigraphy).

Stratal stacking patterns provide the basis for the definition of all units and surfaces of sequence stratigraphy. At any scale of observation (i.e., hierarchical rank), a stratal stacking pattern defines a systems tract, and the surfaces that mark changes in stratal stacking pattern (i.e., systems tract boundaries) are sequence stratigraphic surfaces. A full stratigraphic cycle (i.e., which starts and ends with the same type of sequence stratigraphic surface) delineates a "sequence" (Catuneanu and Zecchin, 2013), which typically includes two or more systems tracts. This methodology transcends the difference between various approaches, as the selection of the sequence boundary takes a subordinate role in the workflow, being a function of mappability rather than an a priori model-dependent premise (Catuneanu et al., 2009, 2011). Advances in the development of the method reveal that the stratigraphic record is much more complex than theoretical models can predict; sequences may consist of variable combinations of systems tracts (e.g., Csato and Catuneanu, 2012; Zecchin and Catuneanu, 2013), which may or may not conform with the prediction of standard models; and consequently, stratigraphic frameworks may or may not include the entire spectrum of sequence stratigraphic surfaces. Additionally, the degree of mappability of each type of sequence stratigraphic surface depends on the types of data available for analysis (Catuneanu, 2006; Posamentier and Allen, 1999).

The existence of several competing approaches to the definition and classification of sequence stratigraphic units (Figs. 1 and 2) has generated considerable confusion among practitioners, with respect to the "best practice" in sequence stratigraphy. The various approaches differ in terms of (1) nomenclature of sequences, systems tracts, and sequence stratigraphic surfaces; (2) selection of surfaces which should be elevated to the rank of "sequence boundary"; (3) the approach taken to define a sequence hierarchy system; and (4) the assertions of the dominant controls on sequence development. Beyond these differences, there is a common ground of core principles that affords a unified application of sequence stratigraphy, irrespective of geological setting and types of data available. Formal recommendations on a model-independent methodology have been sanctioned by the International Subcommission on Stratigraphic Classification of the International Commission of Stratigraphy (Catuneanu et al., 2011). This work updates the core principles of sequence stratigraphy that afford a standard application of the method across the whole spectrum of geological settings, stratigraphic scales, and types of data available.

2. HISTORICAL DEVELOPMENT OF THE METHOD

Sequence stratigraphy started to emerge as a method of stratigraphic analysis ever since the recognition of unconformities in the rock record, which allowed the subdivision of the sedimentary succession into units separated by breaks in the depositional process. Following the initial definition

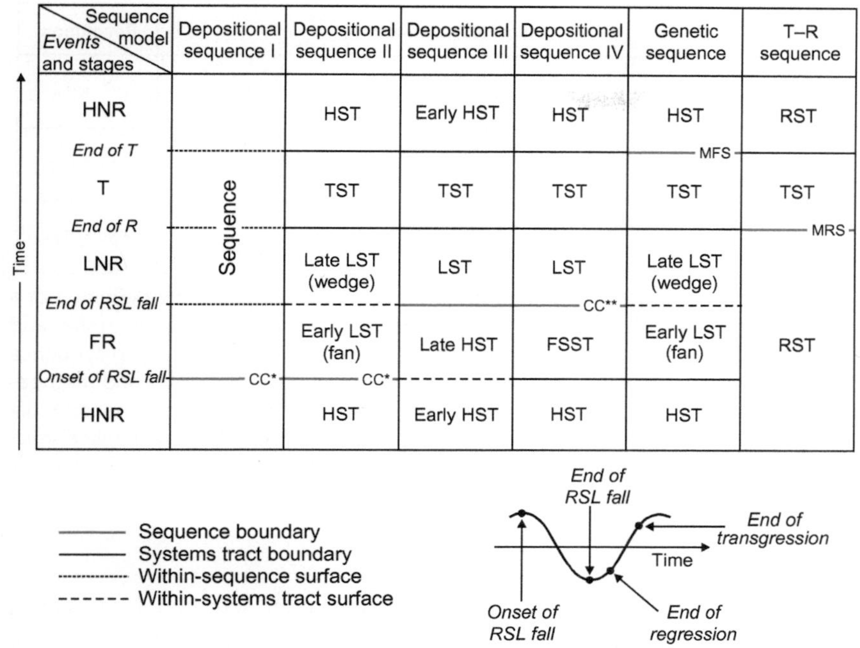

Fig. 2 Sequence stratigraphic approaches: nomenclature of systems tracts and timing of sequence boundaries. Abbreviations: *CC**, correlative conformity in the sense of Posamentier et al. (1988), referred to here as the "basal surface of forced regression"; *CC***, correlative conformity in the sense of Van Wagoner et al. (1988), referred to here as the "correlative conformity"; *FR*, forced regression; *FSST*, falling-stage systems tract; *HNR*, highstand normal regression; *HST*, highstand systems tract; *LNR*, lowstand normal regression; *LST*, lowstand systems tract; *MFS*, maximum flooding surface; *MRS*, maximum regressive surface; *R*, regression; *RSL*, relative sea level; *RST*, regressive systems tract; *T*, transgression; *T–R*, transgressive–regressive; *TST*, transgressive systems tract. References for the proponents of the various sequence stratigraphic approaches are provided in Fig. 1. *From Catuneanu, O., Galloway, W.E., Kendall, C.G.St.C., Miall, A.D., Posamentier, H.W., Strasser, A., Tucker, M.E., 2011. Sequence stratigraphy: methodology and nomenclature. Newsl. Stratigr. 44/3, 173–245.*

of a "sequence" as an unconformity-bounded unit (Longwell, 1949; Sloss et al., 1949), several sequence stratigraphic approaches have been proposed, which differ in terms of nomenclature of systems tracts and the selection of the "sequence boundary" (Fig. 2).

The concept of "sequence" evolved over time, in parallel to the trend of gradually increasing the resolution of stratigraphic analysis (Fig. 3). The development of the sequence concept started with unconformity-bounded units defined at a continental scale (Longwell, 1949; Sloss et al., 1949), by considering only interregional unconformities

Decade	Definition of "sequence"	Resolution
1940s	Rock-stratigraphic unit bounded by interregional unconformities (1)	10^2–10^3 m
1970s	A relatively conformable succession of genetically related strata bounded by unconformities or their correlative conformities (2)	10^1–10^2 m
2010s	A cycle of change in stratal stacking patterns defined by the recurrence of sequence stratigraphic surfaces in the rock record (3)	10^0–10^1 m

Fig. 3 Developments in the definition of a "sequence." Changes to the definition of a "sequence" reflect (1) the gradual increase in the resolution of stratigraphic studies and (2) the need for a more inclusive definition that accommodates all existing sequence stratigraphic approaches (Fig. 1). Following the refinements in definition and methodology, the applications of sequence stratigraphy have expanded from continental-scale correlations (1940s to 1960s) to 2D seismic-scale exploration (1970s) and subseismic-scale production development (2010s). References: (1) Longwell (1949), Sloss et al. (1949), Sloss (1963); (2) Mitchum (1977); (3) Catuneanu et al. (2011), Catuneanu and Zecchin (2013).

as sequence boundaries. As a result, the Phanerozoic sedimentary cover of North America was subdivided into only six sequences (Sloss, 1963). Subsequent work by Wheeler (1958, 1959, 1964) depicted stratigraphic cyclicity in a time domain (depth-to-time Wheeler transformation, or "Wheeler diagram" as it is known today; Qayyum et al., 2014, 2015) and recognized sequence-bounding unconformities of smaller magnitude and smaller areal extent than those considered by Sloss (1963). In doing so, Wheeler (1964) introduced the concept of "continuity surface" beyond the termination of an unconformity, which was later renamed as the "correlative conformity" in the 1970s. This was the first step toward increasing the resolution of stratigraphic studies, by decreasing the scale of a sequence. This trend continued in the 1970s, with the definition of sequences at scales compatible with the vertical seismic resolution (Mitchum et al., 1977). The trend continues today with the definition of sequences at subseismic scales (i.e., high-resolution sequence stratigraphy; Catuneanu and Zecchin, 2013; Catuneanu et al., 2011; Csato et al., 2014; Magalhaes et al., 2015; Zecchin and Catuneanu, 2013, 2015; Zecchin et al., 2015, 2017a,b).

The evolution of the concept of sequence is reflected in the revisions to the definition, which, as the scale of a "sequence" decreased through time, gradually changed the emphasis from the sequence-bounding unconformities to the stacking patterns that define the sequence itself (Fig. 3). Following this trend, the definition changed from "an unconformity-bounded unit" (1940s) to "a unit bounded by unconformities or their correlative conformities" (1970s), and eventually to "a unit bounded by any recurring surface of sequence stratigraphy" (2010s). This trend highlights the fact that as the scale of observation decreases, the magnitude and the areal extent of unconformities decrease as well, and conformable surfaces become increasingly important to delineate sequences. The result is that at high-resolution level, and also depending on the type of sequence stratigraphic surface selected as sequence boundary (Fig. 2), a sequence may no longer require unconformities at its boundaries. The latest definition of a sequence is independent of model, as it accommodates all sequence stratigraphic approaches.

3. CORE CONCEPTS

Multiple allogenic and autogenic processes interplay to generate the stratigraphic architecture of the rock record (Catuneanu and Zecchin, 2013; Muto and Steel, 1992, 1997, 2002). The nature, the intensity, and the relative importance of these controls may vary with stratigraphic age and tectonic setting, making it impossible to draw generalized conclusions with respect to the dominance of one control over others throughout geologic time and under all circumstances. However, the interplay of all controls on the architecture of the stratigraphic record always boils down to two fundamental variables, namely accommodation and sedimentation, which can be used to understand the formation of specific stratal stacking patterns that can be observed in the rock record irrespective of the dominant control(s) at *syn*-depositional time. This helps to focus the methodology on observational field criteria, irrespective of the interpreted origin of the units and bounding surfaces that are being mapped (Catuneanu and Zecchin, 2016).

3.1 Accommodation and Sedimentation

Accommodation is the space made available for sediments to fill (Jervey, 1988), primarily by basin-forming mechanisms (i.e., tectonism and sea/lake-level changes). Additional controls on accommodation include glacial isostasy, sediment loading, and compaction (Catuneanu, 2003, 2006; Miall, 2010; Posamentier and Allen, 1999). The reference horizon that marks the top of available accommodation within a sedimentary basin is represented by the eustatic sea level, in basins connected to the global ocean, and by the elevation of the outflow ridge in interior basins disconnected from the global ocean (Fig. 4). Relative to this reference horizon, accommodation

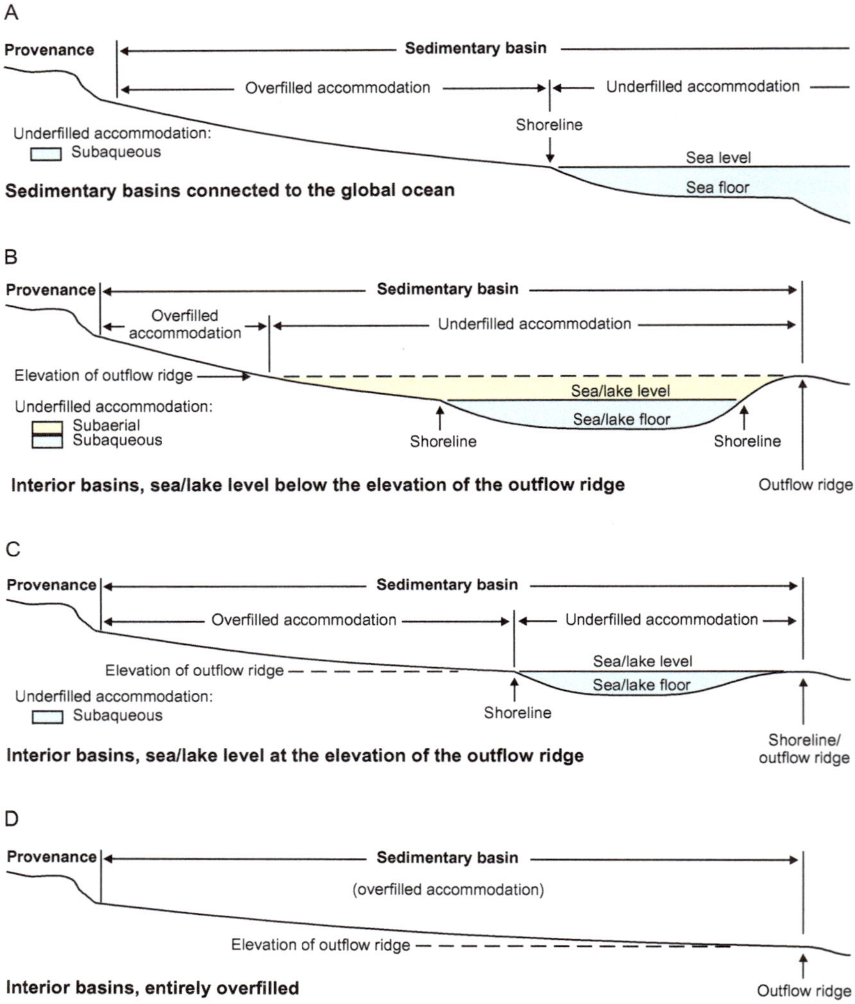

Fig. 4 Underfilled vs overfilled accommodation in sedimentary basins. Accommodation is measured up to the eustatic sea level, in basins connected to the global ocean, and up to the elevation of the outflow ridge in interior basins isolated from the global ocean. In the case of sedimentary basins connected to the global ocean (A), as well as interior basins where the sea/lake level is at the elevation of the outflow ridge (C), the shoreline marks the limit between underfilled and overfilled accommodation. In the case of interior basins where the sea/lake level is below the elevation of the outflow ridge (B), accommodation can also be underfilled in continental (e.g., fluvial, eolian) settings. Sedimentary basins become entirely overfilled where all accommodation is consumed by sedimentation (D). Note that sediment accumulation may continue in overfilled settings, driven by sediment supply that outpaces the energy of the sediment transport agents. In a long term, the sediment in excess of accommodation has a lower preservation potential where provenance areas are subject to denudation.

can be underfilled or overfilled (i.e., depositional surface below or above the reference horizon, respectively; Fig. 4). Sedimentary basins typically evolve from underfilled to overfilled stages, as accommodation is consumed by sedimentation.

Available accommodation in underfilled settings may be subaqueous (below the sea/lake level) or subaerial (in interior basins where the sea/lake level is below the elevation of the outflow ridge; Fig. 4). Subaqueous accommodation is relevant to "conventional" sequence stratigraphic frameworks, which form in relation to shoreline trajectories. Subaerial accommodation only becomes significant in interior basins devoid of subaqueous accommodation, where fluvial and/or eolian processes are isolated from any marine or lacustrine influence (e.g., the situation of a dry basin in Fig. 4B). Sequence stratigraphic frameworks may also form under overfilled accommodation conditions, as sedimentation is only in part dependent on accommodation.

Sedimentation is controlled by all processes that modify the balance between sediment supply and environmental energy, including accommodation, climate, source-area uplift, and autogenic shifts in the patterns of sediment distribution (Catuneanu, 2006). Note that sedimentation, rather than sediment supply, is the relevant variable which, along with accommodation, controls the development of stratal stacking patterns. The integration of sediment supply with the energy of sediment transport agents, as the driving force for deposition vs erosion (i.e., base-level changes), was first recognized by Barrell (1917). Base level is a surface of equilibrium between sedimentation and erosion, which can be observed at different scales in all depositional environments (Fig. 5). Changes in base level constantly shape the landscape and the seafloor profiles as the depositional surface strives to attain a state of equilibrium (Fig. 5).

The updip limit of the influence of changes in subaqueous accommodation on sedimentation provides the basis to subdivide a sedimentary basin into downstream- vs upstream-controlled settings (Fig. 5). In downstream-controlled areas, stratal stacking patterns relate to shoreline trajectories, and changes in subaqueous accommodation are described as "relative sea/lake-level changes" (Posamentier and Allen, 1999). The "conventional" systems tract nomenclature (i.e., lowstand, transgressive, highstand, falling-stage) applies specifically to downstream-controlled settings. In upstream-controlled areas, stratal stacking patterns form independently of shoreline trajectories and are described by an "unconventional" systems tract nomenclature that makes reference to the dominant depositional elements (e.g., high vs low degree of amalgamation of channels in fluvial systems).

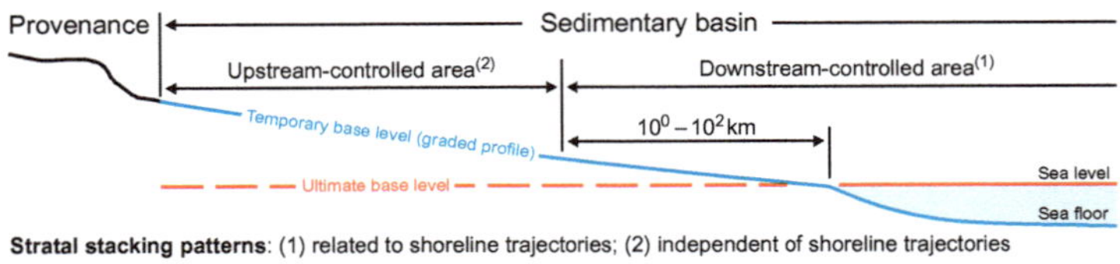

Fig. 5 Downstream-controlled vs upstream-controlled areas within a sedimentary basin. The downstream-controlled area includes continental, coastal, and marine/lacustrine systems which respond to changes in relative sea/lake level. Within a downstream-controlled area, stratal stacking patterns form in relation to shoreline trajectories. The upstream-controlled area includes continental systems beyond the influence of relative sea/lake-level changes, in which stratal stacking patterns form independently of shoreline trajectories. The same sequence stratigraphic methodology applies to both marine and lacustrine settings, whereby changes in subaqueous accommodation and shoreline trajectories control the formation and timing of "conventional" (i.e., downstream-controlled) systems tracts and bounding surfaces. Therefore, "sea level" also stands for "lake level" in terms of processes that are relevant to the construction of downstream-controlled sequence stratigraphic frameworks. The sea level is the *ultimate* base level for subaqueous deposition and continental erosion. *Temporary* base levels are also established in all depositional environments, as equilibrium profiles which the depositional surface strives to attain by means of sedimentation or erosion. The ultimate base level is linked to the concept of "accommodation," whereas the temporary base level (or "base level," as referred to in the text) is a descriptor of "sedimentation." Changes in the temporary base level lead to deposition (base-level rise) or erosion (base-level fall).

Updip from the shoreline, accommodation in continental settings (downstream- or upstream-controlled) may be either overfilled or underfilled, depending on the particular circumstances of each sedimentary basin (Fig. 4). However, the availability of subaerial accommodation takes a subordinate role in any sedimentary basin that includes subaqueous accommodation. Wherever present, the relative sea/lake level, due to its link to the shoreline, is the relevant variable that controls, along with sedimentation, the formation and timing of systems tracts and bounding surfaces in downstream-controlled settings. In such basins, subaerial accommodation also becomes irrelevant to the sedimentary processes in upstream-controlled areas. For this reason, *subaqueous* accommodation is typically inferred when reference is made to "accommodation," in any sedimentary basin that includes a marine or lacustrine depocenter, and hence, a shoreline.

The balance between the rates of sedimentation and accommodation at the shoreline is key to the formation of stratal stacking patterns that define "conventional" systems tracts in downstream-controlled settings. In underfilled basins, where accommodation is still available below the sea/lake level, the water depth represents the balance between concurrent creation of space (accommodation) and consumption of space (sedimentation). In upstream-controlled settings, whether overfilled or underfilled, stratal stacking patterns form in response to the interplay of accommodation, climate, source-area tectonism, and autogenic processes that modify the patterns of sediment distribution. Notably, accommodation is not the sole control on the formation of stratal stacking patterns, in both downstream- and upstream-controlled settings.

At any scale of observation, the balance between the rates of accommodation and sedimentation may change along a shoreline, resulting in the coeval deposition of different systems tracts along strike, and the formation of diachronous systems tract boundaries (e.g., Catuneanu, 2006; Catuneanu et al., 1998; Csato and Catuneanu, 2014; Posamentier and Allen, 1999).

3.2 Stratal Stacking Patterns

Stratal stacking patterns define the stratigraphic architecture of the sedimentary record within a sedimentary basin. Stratal stacking patterns are fundamental to the sequence stratigraphic methodology, as they provide the basis for the definition of all units and surfaces of sequence stratigraphy. Each type of stacking pattern defines a systems tract, and changes in stacking pattern (i.e., systems tract boundaries) define sequence stratigraphic surfaces. Furthermore, a full cycle of change in stratal stacking patterns, which begins and ends with the same type of sequence stratigraphic surface, defines a stratigraphic sequence (Fig. 3).

Stratal stacking patterns may be generated within the area of influence of relative sea/lake-level changes (i.e., "downstream-controlled" stacking patterns), or independently of changes in relative sea/lake level (i.e., "upstream-controlled" stacking patterns) (Fig. 5). The stacking patterns that develop in the downstream- and upstream-controlled settings are described below.

4. STRATAL STACKING PATTERNS IN DOWNSTREAM-CONTROLLED SETTINGS

Downstream-controlled settings include areas of underfilled accommodation in marine or lacustrine environments, as well as adjacent continental areas in which sedimentary processes respond to changes in the relative sea/lake level (Fig. 5). Stratal stacking patterns in downstream-controlled settings form in response to the interplay between accommodation and sedimentation *at the shoreline* (Catuneanu, 2002, 2006). The shoreline trajectory, as observed at different scales, is the key element

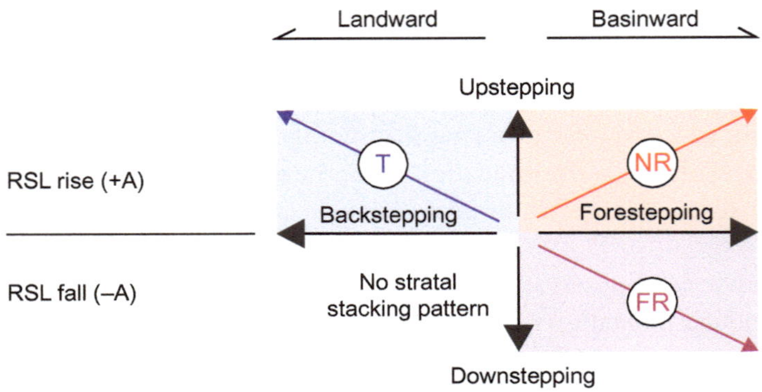

Fig. 6 Shoreline trajectories in downstream-controlled settings: normal regression (i.e., forestepping and upstepping), forced regression (i.e., forestepping and downstepping), and transgression (backstepping and upstepping). Abbreviations: *A*, accommodation; *FR*, forced regression; *NR*, normal regression; *RSL*, relative sea level; *T*, transgression.

in the definition of downstream-controlled stratal stacking patterns (Fig. 6; Catuneanu et al., 2009; Løseth and Helland-Hansen, 2001; Posamentier et al., 1992).

4.1 Normal Regression

Normal regression refers to a stratal stacking pattern defined by a combination of forestepping and upstepping of the shoreline (Figs. 7 and 8; Posamentier et al., 1992). Normal regressions occur when sediment supply outpaces the amount of accommodation generated by relative sea/lake-level rise at the shoreline. A normal regression corresponds to the "ascending regressive" shoreline trajectory of Helland-Hansen and Hampson (2009).

A normal regression that follows a forced regression of equal hierarchical rank is designated as a "lowstand" normal regression (Fig. 8). Lowstand normal regressive shorelines typically describe a concave-up trajectory, in response to the increase in the rates of creation of accommodation following the onset of relative sea/lake-level rise (Fig. 9).

A normal regression that follows a transgression of equal hierarchical rank is designated as a "highstand" normal regression (Fig. 8). Highstand normal regressive shorelines typically describe a convex-up trajectory, where the rates of creation of accommodation decrease following the maximum flooding at the end of transgression (Fig. 9).

Normal regressions are typically accompanied by the aggradation of continental topsets, with the rates of progradation being inversely proportional to the rates of topset aggradation. In turn, the rates of topset aggradation

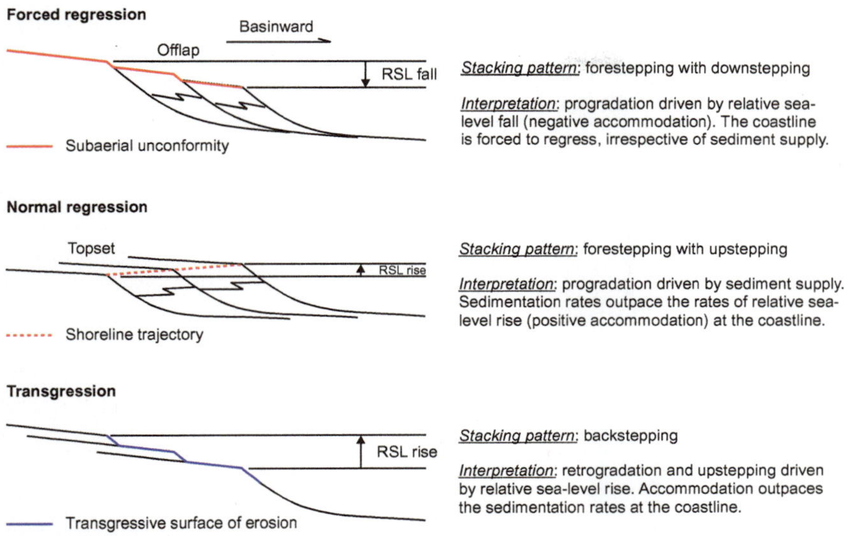

Fig. 7 Stratal stacking patterns in downstream-controlled settings: forced regression, normal regression, and transgression. The amount of upstepping of the coastline during normal regression or transgression, as well as the amount of downstepping of the coastline during forced regression, indicates the magnitude of relative sea/lake-level changes at *syn*-depositional time. This diagram illustrates common stratigraphic trends in the rock record (e.g., the development of offlap during forced regression, and the aggradation of coastal and fluvial systems during transgression). Deviations from these trends (e.g., fluvial and coastal aggradation during forced regression, and fluvial and coastal erosion during transgression) have been discussed by Catuneanu and Zecchin (2016). *From Catuneanu, O., Galloway, W.E., Kendall, C.G.St.C., Miall, A.D., Posamentier, H.W., Strasser, A., Tucker, M.E., 2011. Sequence stratigraphy: methodology and nomenclature. Newsl. Stratigr. 44/3, 173–245.*

reflect the rates of relative increase in coastal elevation (i.e., the rates of relative sea/lake-level rise). Therefore, the rates of progradation tend to decrease with time during lowstand normal regressions, and increase with time during highstand normal regressions (Fig. 8). These trends are reflected in the thickness of the beds that compose the topset units, and are particularly evident in carbonate systems where topsets include peritidal cycles (e.g., fig. 14 in Catuneanu et al., 2011).

Other contrasts between lowstand and highstand topsets are evident in fluvial systems, due to differences in gradients and energy levels between the lowstand and the highstand rivers. The patterns of change in fluvial energy within a sequence depend on the timing of the subaerial unconformity, which may form during forced regressions or transgressions (Figs. 10 and 11; see discussion in Catuneanu and Zecchin, 2016). In both cases, the gradients of the fluvial profile increase during the formation of the

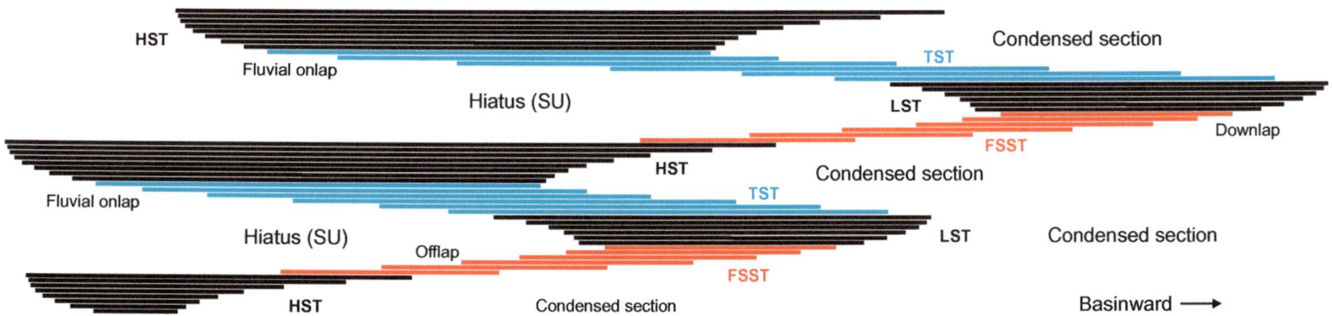

Fig. 8 Stratal stacking patterns in downstream-controlled settings, in a time domain (dip-oriented section, shelf setting). Each type of stratal stacking pattern defines a systems tract: FSST (forced regression), LST (lowstand normal regression), TST (transgression), and HST (highstand normal regression). The degree of preservation of the sedimentary record tends to increase in a downdip direction, from the continental to the marine portions of the basin. Abbreviations: *FSST*, falling-stage systems tract; *HST*, highstand systems tract; *LST*, lowstand systems tract; *SU*, subaerial unconformity; *TST*, transgressive systems tract.

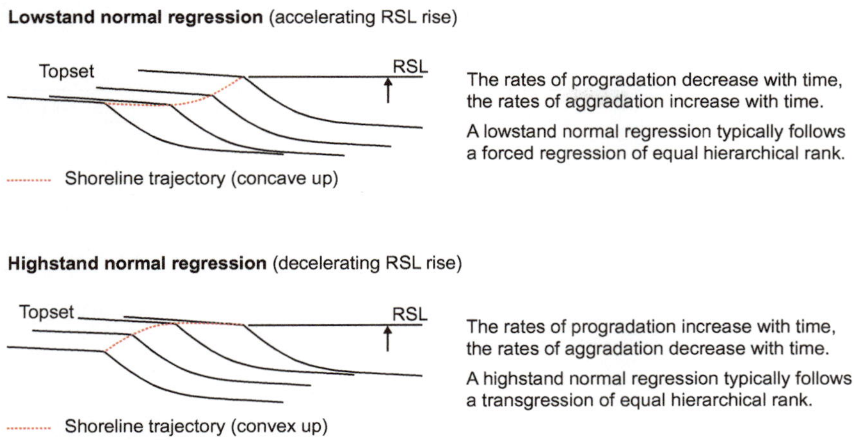

Fig. 9 Stratal stacking patterns of "lowstand" and "highstand" normal regressions. In both cases progradation is driven by sediment supply during a period of positive accommodation at the coastline (i.e., sedimentation outpaces accommodation at the coastline). A lowstand normal regression records a change in depositional trends from dominantly progradational to dominantly aggradational (concave-up shoreline trajectory). In contrast, a highstand normal regression records a change in depositional trend from dominantly aggradational to dominantly progradational (convex-up shoreline trajectory). Abbreviation: *RSL*, relative sea level. *Modified from Catuneanu, O., 2006. Principles of Sequence Stratigraphy. Elsevier, Amsterdam, 375 pp., fig. 7.20, p. 306.*

subaerial unconformity, and decrease during the rest of the stratigraphic cycle. Where the subaerial unconformity forms during forced regression (i.e., in settings where the gradient of the shoreline trajectory is steeper than the fluvial profile; Fig. 10), river systems aggrade during the deposition of the lowstand, transgressive, and highstand systems tracts (Fig. 11A). In this case, the lowstand topsets include the highest energy fluvial systems of the stratigraphic cycle. This scenario, which is most common in the stratigraphic record, is typically recorded where the provenance is far from the shoreline, allowing for the development of low-gradient fluvial profiles within the downstream-controlled settings. Where the subaerial unconformity forms during transgression (i.e., in settings where the fluvial profile is steeper than the gradient of the shoreline trajectory; Fig. 10), the aggradation of river systems occurs during the deposition of the highstand, falling-stage, and lowstand systems tracts (Fig. 11B). In this case, the highest energy fluvial systems are part of the highstand topsets. This scenario is likely where the provenance is proximal to the shoreline, leading to steep landscape gradients and the potential dominance of upstream controls all the way to the shoreline. In either case (fluvial incision during forced regression or transgression), the

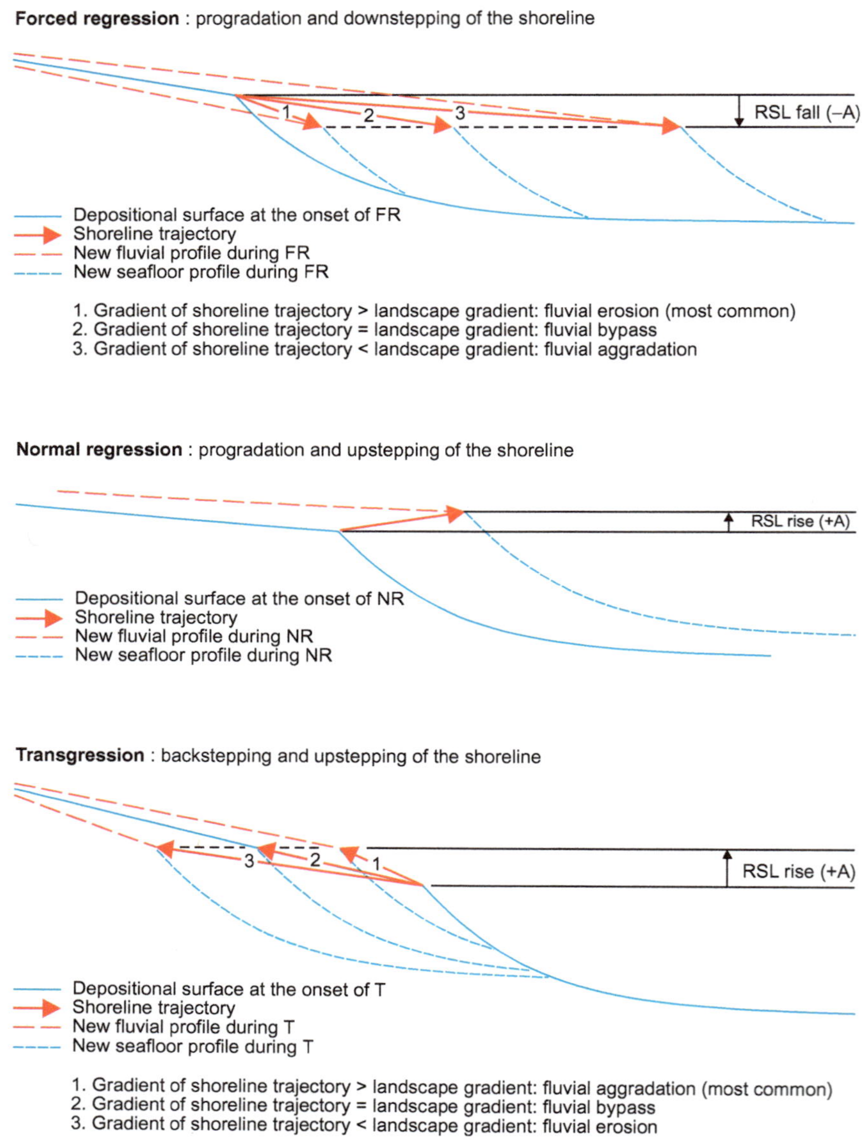

Fig. 10 Depositional trends during forced regression, normal regression, and transgression. Note that subaerial unconformities may form during either forced regression or transgression (see text for details). Abbreviations: *−A*, negative accommodation; *+A*, positive accommodation; *RSL*, relative sea level. *From Catuneanu, O., Zecchin, M., 2016. Unique vs. non-unique stratal geometries: relevance to sequence stratigraphy. Mar. Pet. Geol. 78, 184–195.*

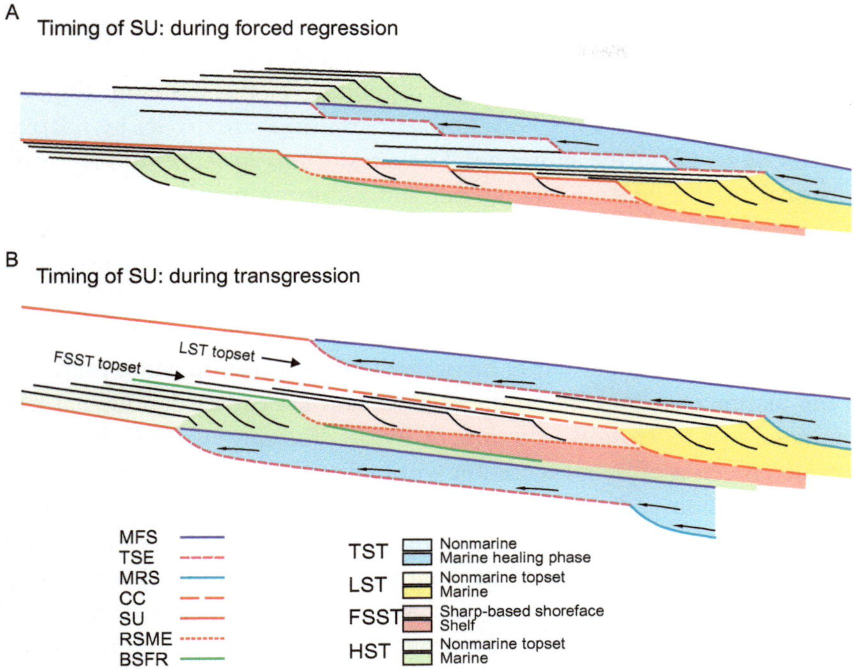

Fig. 11 Architecture of systems tracts and sequence stratigraphic surfaces in a shelf setting: (A) forced regressive and transgressive shoreline trajectories steeper than the landscape gradient, leading to the formation of the subaerial unconformity during forced regression; (B) landscape gradient steeper than the forced regressive and transgressive shoreline trajectories, leading to the formation of the subaerial unconformity during transgression. The strike variability in subsidence and sedimentation rates along the shoreline may result in the coeval development of different systems tracts between different areas of the same sedimentary basin (e.g., Catuneanu et al., 1998, 1999, 2002). Abbreviations: *BSFR*, basal surface of forced regression; *CC*, correlative conformity; *FSST*, falling-stage systems tract; *HST*, highstand systems tract; *LST*, lowstand systems tract; *MFS*, maximum flooding surface; *MRS*, maximum regressive surface; *RSME*, regressive surface of marine erosion; *SU*, subaerial unconformity; *TSE*, transgressive surface of erosion; *TST*, transgressive systems tract.

coarsest sediment is found above the subaerial unconformity, and the fluvial sequence displays a fining-upward profile that reflects the decrease in stream energy and competence with time. This trend also explains the increased likelihood of occurrence of channel amalgamation at the base of depositional sequences, in relation to the steeper gradients which promote higher energy and potentially unconfined river systems.

4.2 Forced Regression

Forced regression refers to a stratal stacking pattern defined by a combination of forestepping and downstepping of the shoreline (Figs. 7 and 8; Posamentier et al., 1992). Forced regressions are driven by relative sea/lake-level fall at the shoreline. The rates of forced regression are proportional to the rates of relative fall and sediment supply and inversely proportional to the seafloor gradients. Forced regression corresponds to the "descending regressive" shoreline trajectory of Helland-Hansen and Hampson (2009).

Forced regressions accompanied by the formation of the subaerial unconformity (Figs. 7, 8, and 11A) assume sediment accumulation primarily in the marine environment, as the downstream-controlled continental setting is subject to erosion or sediment bypass. Exceptions from this trend include processes of lateral accretion within fluvial systems, which may lead to the formation and even preservation of point bar deposits as part of a forced regressive unit. Notwithstanding this exception, a unit defined by an offlapping forced regressive stacking pattern (Figs. 7 and 8) is the only type of systems tract that consists exclusively of marine deposits. All other systems tracts in a downstream-controlled setting (i.e., defined by normal regressive or transgressive stacking patterns) typically include both continental and marine deposits. This is the case in settings where the provenance is located far from the shoreline, leading to the development of fluvial profiles with a gradient lower than the gradient of the shoreline trajectory.

Forced regressions may also be accompanied by fluvial aggradation, particularly in settings where the provenance is located close to the shoreline, leading to the development of landscape gradients steeper than the gradient of the shoreline trajectory (Catuneanu, 2006; Catuneanu and Zecchin, 2016; Posamentier and Allen, 1999; Figs. 10 and 11B). This scenario is less common in the stratigraphic record, but shows that processes of fluvial aggradation are not diagnostic of any particular systems tract. Criteria to differentiate normal regressions from atypical forced regressions with fluvial topsets have been outlined by Posamentier and Morris (2000) and discussed more recently by Catuneanu and Zecchin (2016).

4.3 Transgression

Transgression refers to a stratal stacking pattern defined by a combination of backstepping and upstepping of the shoreline (Figs. 7 and 8). Transgressions occur when sediment supply is insufficient to fill the amount of accommodation generated by relative sea/lake-level rise at the shoreline. The rates of transgression are proportional to the rates of relative rise and inversely proportional to sediment supply and the landscape gradients.

Most commonly, transgressions lead to the highest rates of fluvial aggradation, which reflect the high rates of increase in coastal elevation. Consequently, most sediment during transgression tends to be trapped within fluvial and backstepping coastal systems, leading to sediment starvation of the marine seafloors (Loutit et al., 1988). This is the case in settings where the provenance is located far from the shoreline, leading to the development of fluvial profiles with a gradient lower than the gradient of the shoreline trajectory (Fig. 11A).

Transgressions may also be accompanied by fluvial incision and the formation of subaerial unconformities, particularly in settings where the provenance is located close to the shoreline, leading to the development of landscape gradients steeper than the gradient of the shoreline trajectory (Catuneanu, 2006; Catuneanu and Zecchin, 2016; Leckie, 1994; Posamentier and Allen, 1999; Figs. 10 and 11B). This stratigraphic scenario also demonstrates that fluvial processes of aggradation or erosion are not diagnostic to the definition of systems tracts (see discussion in Catuneanu and Zecchin, 2016).

5. STRATAL STACKING PATTERNS IN UPSTREAM-CONTROLLED SETTINGS

Stratal stacking patterns in upstream-controlled settings develop beyond the influence of relative sea/lake-level changes (Fig. 5), in response to the interplay of all factors which modify the balance between sediment supply and energy flux (i.e., accommodation, climate, source-area tectonism, and autogenic controls on sediment dispersal patterns over various timescales). The degree of channel amalgamation is a key element in the definition of upstream-controlled stratal stacking patterns in fluvial systems (Fig. 12; Boyd et al., 2000; Shanley and McCabe, 1994). The ratio between channel and overbank depositional elements is the result of the interplay of three main processes in fluvial systems, namely, the rates of floodplain aggradation, the degree of channel confinement, and the frequency of avulsion (Fig. 13; Bristow and Best, 1993).

It is noteworthy that accommodation, while important, is not the sole control on the formation of upstream-controlled stratal stacking patterns

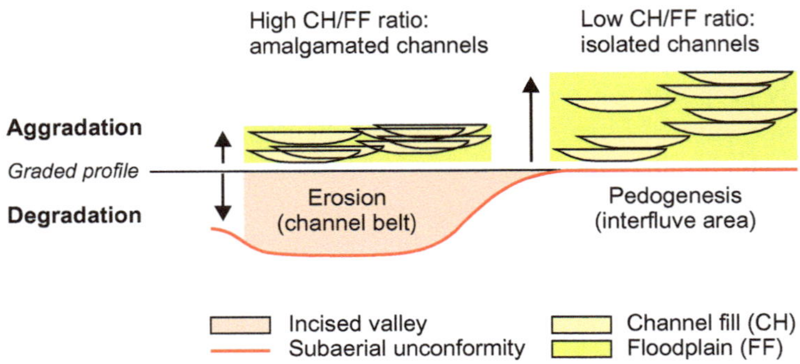

Fig. 12 Depositional trends in an upstream-controlled area. The development of "unconventional" stratal stacking patterns depends on (1) the rates of floodplain aggradation (proportional to the size of *arrows* in the diagram); (2) the ability of channels to shift laterally, which is a function of fluvial style; and (3) the frequency of avulsion. The processes and the rates of aggradation and degradation depend on all factors which control sedimentation (i.e., sediment supply vs energy flux), including accommodation, climate, source-area tectonism, and the autogenic controls on sediment dispersal patterns. The aggrading side of the diagram illustrates the seven-step evolution of a channel under identical conditions of avulsion and lateral shift. In this case, the contrast in stratigraphic architecture (i.e., amalgamated vs isolated channels) is the result of differences in the rates of floodplain aggradation.

(Bristow and Best, 1993; Miall, 2015). Early studies on the classification of stratal stacking patterns in fluvial systems have proposed a direct link between accommodation and the degree of channel amalgamation. As a result, the terminology of systems tracts made exclusive reference to accommodation conditions (i.e., low- vs high-accommodation systems tracts; e.g., Boyd et al., 2000; Leckie and Boyd, 2003). This approach is now considered as an oversimplification, as accommodation only plays a part in a much more complex interaction of controlling parameters (Miall, 2015). A more descriptive nomenclature, free of the interpretation of the underlying controls, is introduced below.

5.1 High-Amalgamation (Channel-Dominated) Stacking Pattern

The development of a high degree of channel amalgamation in fluvial systems (i.e., high channel-to-overbank ratio; Fig. 14A) is promoted by (1) low rates of floodplain aggradation; (2) unconfined channels; and (3) a high frequency of channel avulsion (Fig. 13; Bristow and Best, 1993). This fluvial architecture of amalgamated channels was previously referred to as a "low-accommodation" stacking pattern.

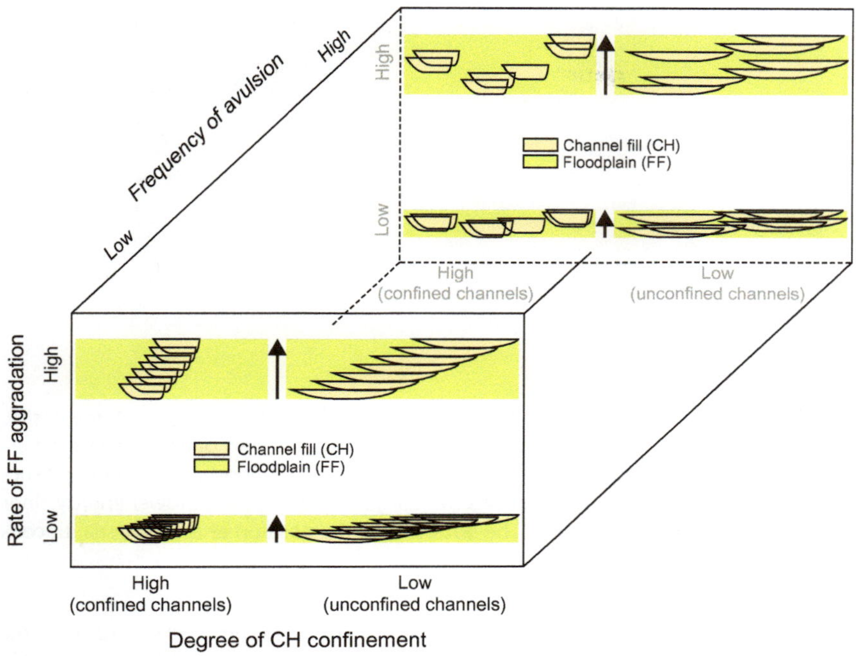

Fig. 13 Fluvial architecture under variable conditions of floodplain aggradation, channel confinement, and avulsion frequency, as illustrated by the seven-step evolution of a channel. The rates of fluvial aggradation depend on all factors which control sedimentation, including accommodation, climate, source-area tectonism, and autocyclic changes in sediment distribution. The degree of channel amalgamation is proportional to the rate of lateral channel migration (higher in unconfined rivers) and the frequency of avulsion, and inversely proportional to the rate of aggradation. *Modified from Bristow, C.S., Best, J.L., 1993. Braided rivers: perspectives and problems. In: Best, J.L., Bristow, C.S. (Eds.), Braided Rivers. Geological Society Special Publication No. 75, pp. 1–11.*

5.2 Low-Amalgamation (Floodplain-Dominated) Stacking Pattern

The development of a low degree of channel amalgamation in fluvial systems (i.e., low channel-to-overbank ratio; Fig. 14B) is promoted by (1) high rates of floodplain aggradation; (2) confined channels; and (3) a low frequency of channel avulsion (Fig. 13; Bristow and Best, 1993). This fluvial architecture, defined by isolated channels within floodplain deposits, was previously referred to as a "high-accommodation" stacking pattern.

Both types of upstream-controlled fluvial stacking patterns can be observed at different scales, in relation to stratigraphic cycles of different magnitudes (Fig. 15). The ratio between the high- and low-amalgamation

Fig. 14 Fluvial stacking patterns in an upstream-controlled setting (Triassic, Karoo Basin). (A) Channel-dominated succession: high-amalgamation stacking pattern (Molteno Formation: higher energy, coarser grained braided channels); (B) floodplain-dominated succession: low-amalgamation stacking pattern (Burgersdorp Formation: lower energy, finer grained meandering river system). The *arrows* mark the position of the subaerial unconformity that separates the two stacking patterns.

stacking patterns within lower rank sequences defines the type of higher rank (high- vs low-amalgamation) systems tract (e.g., a set of fourth-order sequences dominated by a high degree of channel amalgamation defines a third-order "high-amalgamation" systems tract; a set of fourth-order sequences dominated by floodplain deposits with isolated channels defines a third-order "low-amalgamation" systems tract; Fig. 15).

6. TYPES OF SEQUENCE STRATIGRAPHIC UNIT

Sequence stratigraphic units are defined by stratal stacking patterns and specific bounding surfaces, and not by their inferred origin, age, time span, or physical scales. The geographic extent of any type of sequence stratigraphic unit is highly variable, and determined by the development of its diagnostic stratal stacking patterns and bounding surfaces, which is a function of tectonic and depositional settings. All types of sequence stratigraphic unit consist of strata that are "genetically related"; i.e., strata that belong to the same cycle of accommodation or sediment supply (Catuneanu et al., 2009). Genetically related successions may be defined at different scales of observation (i.e., hierarchical ranks), in relation to stratigraphic cycles of different magnitudes. Therefore, there are no standards for the scale of any type of sequence stratigraphic unit.

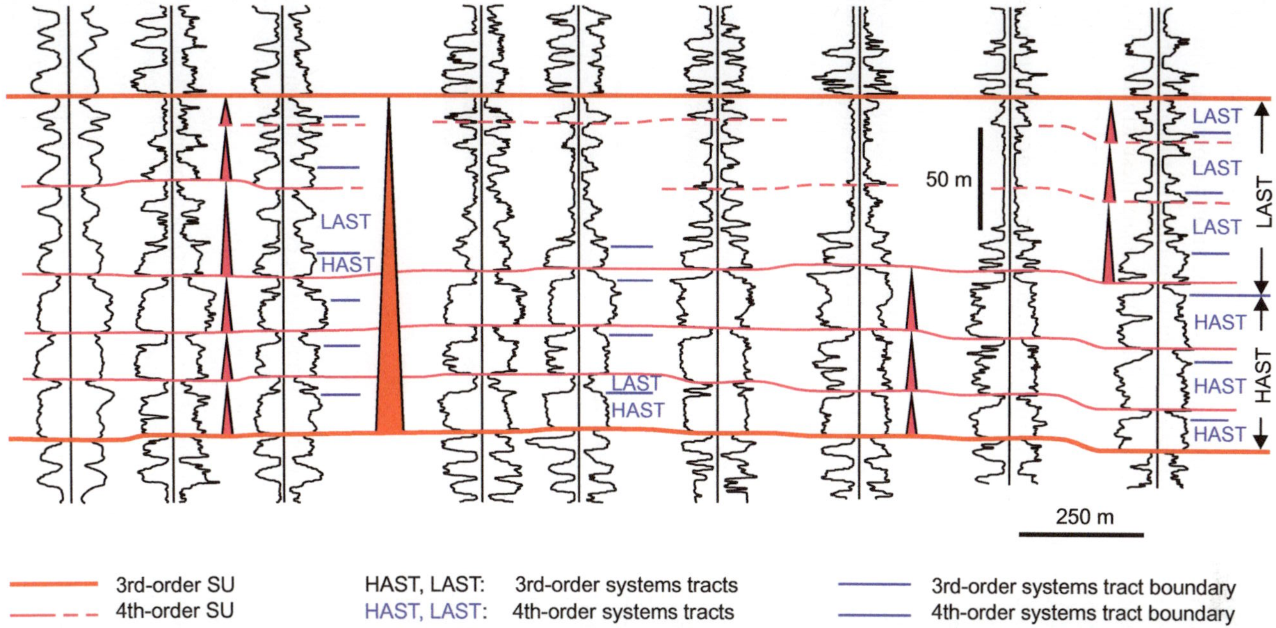

Fig. 15 Stratigraphic architecture of a fluvial succession in an upstream-controlled setting (Upper Cretaceous, Golfo San Jorge Basin). Well logs: spontaneous potential (*left*) and resistivity (*right*). The stratigraphic correlation is calibrated with production pressure data. The third-order sequence consists of nested fourth-order sequences. At each scale of observation (i.e., hierarchical level), depositional sequences display fining-upward trends (decline in fluvial energy with time) and can be subdivided into systems tracts based on the dominant fluvial stacking patterns (channel- vs floodplain-dominated successions). The ratio between the high- and low-amalgamation stacking patterns within the lower rank sequences defines the type of higher rank (high- vs low-amalgamation) systems tract. Abbreviations: *HAST*, high-amalgamation systems tract; *LAST*, low-amalgamation systems tract; *SU*, subaerial unconformity. *Data courtesy of YPF Argentina.*

The fundamental unit of sequence stratigraphy is the stratigraphic sequence. Sequences consist of component systems tracts, and systems tracts consist of depositional systems that accumulate during the development of particular stratal stacking patterns. This stratigraphic architecture can be observed at different scales, depending on the purpose of the study and the resolution of the data available. At the smallest stratigraphic scales, depositional systems consist of beds and bedsets (i.e., sedimentological cycles; Catuneanu and Zecchin, 2013). At any larger scales, systems tracts and depositional systems consist of higher frequency sequences (Figs. 15 and 16).

The scale-independent nature of sequences and systems tracts is consistent with the fact that depositional systems can also be observed at different scales (Fig. 16). The seismic stratigraphy of the 1970s imposed, by default, a minimum scale to the concepts of sequence, systems tract, and depositional system, which had to exceed the vertical seismic resolution. In reality, all these types of unit can also be defined and observed at subseismic scales

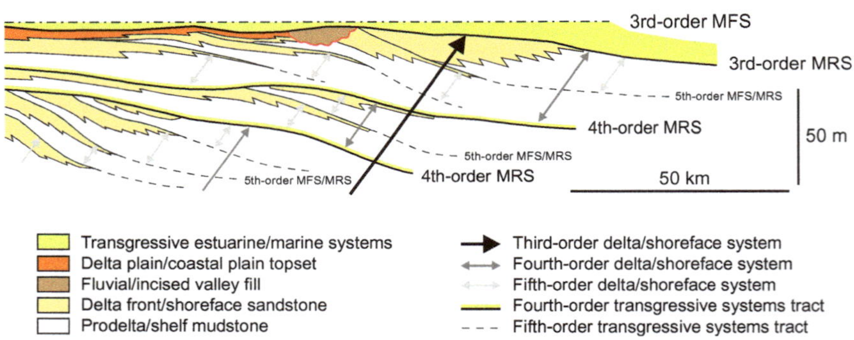

Fig. 16 Stratigraphic architecture of a prograding system in a downstream-controlled setting (Upper Cretaceous Dunvegan Formation, Western Canada Sedimentary Basin). Sequences, systems tracts, and depositional systems can be defined at different scales of observation (i.e., hierarchical levels). In this example, the third-order "delta" includes several different depositional systems that can be defined at the fourth- and fifth-order scales of observation. At all hierarchical levels, depositional systems have paleogeographic significance and correspond to specific environments of deposition. Abbreviations: *MFS*, maximum flooding surface; *MRS*, maximum regressive surface, potentially reworked in part by the transgressive surface of erosion. *Modified from Bhattacharya, J.P., 1993. The expression and interpretation of marine flooding surfaces and erosional surfaces in core; examples from the Upper Cretaceous Dunvegan Formation in the Alberta foreland basin. In: Summerhayes, C.P., Posamentier, H.W. (Eds.), Sequence Stratigraphy and Facies Associations. International Association of Sedimentologists Special Publication No. 18, pp. 125–160.*

in higher resolution studies (i.e., high-resolution sequence stratigraphy; e.g., Zecchin and Catuneanu, 2013).

Depositional systems form when the defining subenvironments and related geomorphic elements are established as dominant (but not necessarily exclusive) sediment fairways. For example, a "delta" observed at a seismic scale is defined by the dominant pattern of sediment progradation, even though, on shorter timescales, this dominant pattern is interrupted by stages of transgression (e.g., flooding of delta plain and the formation of estuaries) of lower hierarchical ranks (Fig. 16). Therefore, the dominance of a particular sediment dispersal pattern can be observed at different scales, depending on the resolution of the stratigraphic study (Fig. 16). Indeed, the definition of sequences, systems tracts, and depositional systems makes no reference to temporal or physical scales. All these types of unit are assumed to be "relatively conformable," but this conformable character is only relative to the resolution of the data available. In reality, unconformities exist at all scales, both below and above the resolution of the data available.

Despite the scale-independent and nested nature of sequence stratigraphic units, the stratigraphic architecture is not necessarily and truly "fractal," because sequences of different scales may have different controls and internal makeup (e.g., different combinations and/or relative development of component systems tracts).

6.1 Stratigraphic Sequence

A stratigraphic sequence corresponds to a cycle of change in stratal stacking patterns, defined by the recurrence of the same type of sequence stratigraphic surface in the rock record (Catuneanu and Zecchin, 2013; Catuneanu et al., 2011). A stratigraphic sequence is bounded at the base and at the top by the same type of sequence stratigraphic surface; e.g., from a maximum flooding surface to the next maximum flooding surface in the stratigraphic succession, on condition that the strata between the sequence boundaries belong to one cycle of change in accommodation or sediment supply at the selected scale of observation (Catuneanu et al., 2009).

Stratigraphic cycles may be symmetrical or asymmetrical, and the corresponding sequences may include a variable number of *distinct* systems tracts (i.e., a systems tract cannot be repeated within a sequence), up to four in the case of ideal "conventional" sequences that develop and preserve all systems tracts. Therefore, a sequence is not defined by its internal makeup, but by the recurrence of the sequence stratigraphic surface that marks its boundaries. Not all types of stratal stacking patterns (i.e., systems tracts) and bounding sequence stratigraphic surfaces may occur in the succession under analysis. For example, stratigraphic cyclicity may be defined by the repetition of transgressions and highstand normal regressions, without intervening stages of forced regression and lowstand normal regression; or, by the repetition of forced regressions and lowstand normal regressions, without intervening stages of transgression and highstand normal regression. Therefore, the types of recurring stacking patterns and bounding surfaces that define stratigraphic cyclicity may vary with the case study, which underlines the need for a model-independent approach to the sequence stratigraphic analysis.

The definition of a "sequence" was gradually revised from the 1940s to the present day, in response to the need for (1) an increase in the resolution of stratigraphic studies and (2) a more inclusive definition that accommodates all existing sequence stratigraphic approaches (Fig. 3). The current definition provides the flexibility to apply the sequence stratigraphic methodology in a manner that is independent of model, and at any scale afforded by the data available.

Different kinds of stratigraphic sequence may be defined as a function of the specific type of recurring sequence stratigraphic surface that is selected as the boundary of the cycle of change in stratal stacking pattern (Figs. 1 and 2).

6.1.1 Depositional Sequence

The depositional sequence is a stratigraphic sequence bounded by subaerial unconformities or their correlative conformities (Mitchum, 1977). Different types of "depositional sequence" have been defined (Fig. 1), which fall into two groups, as a function of the timing of the marine portion of the sequence boundary (i.e., the "correlative conformity"; Fig. 2): one group considers the correlative conformity as the paleo-seafloor at the onset of forced regression (i.e., the correlative conformity sensu Posamentier et al., 1988; herein referred to as the "basal surface of forced regression"); and another group considers the correlative conformity as the paleo-seafloor at the end of forced regression (i.e., the correlative conformity sensu Van Wagoner et al., 1988; herein referred to as the "correlative conformity").

All types of "depositional sequence," as originally defined, assume full cycles of change in accommodation and relate the sequence boundary to stages of negative accommodation. As a note of caution, however, subaerial unconformities may also form during stages of relative sea/lake-level rise and transgression (Fig. 10). Therefore, the interpretation of the underlying controls responsible for the formation of depositional sequences needs to be separated from the observation of stratal stacking patterns and be performed on a case-by-case basis (Catuneanu and Zecchin, 2016). The concept of depositional sequence applies to both downstream- and upstream-controlled settings, where subaerial unconformities may form.

6.1.2 Genetic Stratigraphic Sequence

The genetic stratigraphic sequence is a stratigraphic sequence bounded by maximum flooding surfaces (Galloway, 1989). As maximum flooding surfaces form during stages of positive accommodation, the formation of genetic stratigraphic sequences does not require stages of negative accommodation. At any scale of observation, a genetic stratigraphic sequence corresponds to a regressive–transgressive cycle, which may occur during a full cycle of change in accommodation or during a stage of positive accommodation. In the latter case, the genetic stratigraphic sequence does not include falling-stage and lowstand systems tracts, nor any sequence stratigraphic surfaces that are exclusively associated with forced regression (i.e., the basal surface of forced regression and the regressive surface of marine erosion).

The genetic stratigraphic sequence approach does not rely on the development and recognition of subaerial unconformities and correlative conformities. Instead, the physical record of transgression provides "readily recognized regionally correlative, easily and accurately datable, and robust sequence boundaries" (Galloway, 1989). The concept of genetic stratigraphic sequence applies to downstream-controlled settings, where maximum flooding surfaces may form.

6.1.3 Transgressive–Regressive Sequence

The transgressive–regressive (T–R) sequence is a stratigraphic sequence bounded by maximum regressive surfaces (Johnson and Murphy, 1984). In its original definition, the formation of T–R sequences does not require stages of negative accommodation. As maximum regressive surfaces of any hierarchical rank may form during stages of positive accommodation, the T–R sequences may be generated either during full cycles of positive–negative accommodation or during periods of positive accommodation.

A proposal to modify the definition of the T–R sequence was made to include the subaerial unconformity as the continental portion of the sequence boundary (Embry and Johannessen, 1992). However, as the maximum regressive surface is most commonly younger than the subaerial unconformity, the marine portion of the maximum regressive surface may not meet with the basinward termination of the subaerial unconformity (Embry and Johannessen, 1992). Therefore, the original definition of Johnson and Murphy (1984) is still recommended. The concept of T–R sequence applies to downstream-controlled settings, where maximum regressive surfaces may form.

6.2 Systems Tract

A systems tract is a linkage of contemporaneous depositional systems, forming the subdivision of a sequence (Brown and Fisher, 1977). Systems tracts are interpreted on the basis of stratal stacking patterns, stratigraphic relations, and types of bounding surfaces. The definition of a systems tract is independent of physical and temporal scales; systems tracts can be observed at different scales, depending on the resolution of the data available. The internal architecture of a systems tract may vary greatly from a succession of facies (beds and bedsets: sedimentological cycles within the lowest rank depositional systems) to a set of higher frequency sequences of lower hierarchical rank (Figs. 15 and 16). Systems tracts may be shoreline related, in downstream-controlled settings, or shoreline independent, in upstream-controlled settings (Fig. 5).

In downstream-controlled settings, systems tracts form in relation to specific types of shoreline trajectory (Fig. 6) and build a stratigraphic framework that may include the entire array of sequence stratigraphic surfaces and depositional systems (Fig. 11). At any scale of observation (i.e., hierarchical level), sequences may consist of different combinations of systems tracts, depending on the local conditions of accommodation and sedimentation at the time of deposition (Csato and Catuneanu, 2012). As such, systems tract successions are basin specific and may not conform with the predictions of idealized models. For this reason, the construction of the sequence stratigraphic framework needs to be performed on a case-by-case basis, with the emphasis on local data rather than model assumptions.

6.2.1 Falling-Stage Systems Tract

The falling-stage systems tract is defined by a forced regressive stratal stacking pattern (Figs. 6–8). Where the subaerial unconformity forms during forced regression, the falling-stage systems tract is bounded at the base by a marine basal surface of forced regression (i.e., the paleo-seafloor at the onset of forced regression), and at the top by the subaerial unconformity and the correlative conformity (Fig. 11A). In this case, with the exception of fluvial sediments that accumulate by processes of lateral accretion, the falling-stage systems tract consists solely of marine deposits.

Less commonly, forced regression may be accompanied by fluvial aggradation (Fig. 11B). In this case, the falling-stage systems tract is bounded at the base by a basal surface of forced regression with both continental and marine portions, and at the top by a conformity that marks the change in stacking pattern to the overlying lowstand systems tract (Fig. 11B; see Catuneanu and Zecchin, 2016 for a discussion of the field criteria that afford the distinction between a normal regression and an atypical forced regression with fluvial aggradation). The conformity at the top of the falling-stage systems tract may be reworked in part by erosional processes associated with subsequent transgression. In this case, the systems tract boundary becomes a composite surface whose precise nature at any location needs to be determined on a case-by-case basis. Where two or more sequence stratigraphic surfaces are superimposed, due to nondeposition or erosion, the name of the younger surface, which leaves the last imprint on the preserved contact, is typically used (Catuneanu, 2006).

6.2.2 Lowstand Systems Tract

The lowstand systems tract is defined by a normal regressive stacking pattern (Figs. 6–8) which follows a forced regression of the same hierarchical rank (Figs. 8 and 11). The lowstand systems tract is bounded at the base by the subaerial unconformity and/or the correlative conformity (Fig. 11). Where the lowstand systems tract is followed by transgression, the upper boundary is represented by the maximum regressive surface reworked in part by the transgressive surface of erosion (transgression accompanied by fluvial aggradation; Fig. 11A), or by a composite surface which includes the marine portion of the maximum regressive surface, the transgressive surface of erosion, and the subaerial unconformity (transgression accompanied by fluvial erosion; Fig. 11B). Where the lowstand systems tract is followed by forced regression, the upper boundary is represented by the subaerial unconformity and/or the basal surface of forced regression. Lowstand systems tracts typically include a continental topset and a marine foreset and bottomset, and tend to display a concave-up shoreline trajectory (Fig. 9).

6.2.3 Transgressive Systems Tract

The transgressive systems tract is defined by a retrogradational stratal stacking pattern (Figs. 6–8). The transgressive systems tract is bounded at the base by the maximum regressive surface reworked in part by the transgressive surface of erosion, and at the top by the maximum flooding surface (Fig. 11). Where transgression is accompanied by fluvial aggradation, the fluvial portion of the transgressive systems tract may also rest directly on top of the subaerial unconformity, beyond the updip termination of the lowstand topset (Fig. 11A). Most commonly, transgressive systems tracts record the highest rates of fluvial aggradation, as accommodation is generated rapidly during transgression, which leads to the lowest rates of siliciclastic sediment supply to the marine environment. This results in the sediment starvation of the seafloor and the formation of marine condensed sections (Loutit et al., 1988).

6.2.4 Highstand Systems Tract

The highstand systems tract is defined by a normal regressive stacking pattern (Figs. 6–8) which follows a transgression of the same hierarchical rank (Figs. 8 and 11). The highstand systems tract is bounded at the base by a maximum flooding surface with both continental and marine portions (transgression accompanied by fluvial aggradation; Fig. 11A), or by a marine maximum flooding surface and a subaerial unconformity (transgression accompanied by fluvial erosion; Fig. 11B). Where the highstand systems tract is followed by forced regression, the upper boundary is represented by the subaerial unconformity and/or the basal surface of forced regression (Fig. 11). Where the highstand systems tract is followed by transgression, the upper boundary is represented by the maximum regressive surface reworked in part by the transgressive surface of erosion (transgression accompanied by fluvial aggradation), or by a composite surface which includes the marine portion of the maximum regressive surface, the transgressive surface of erosion, and the subaerial unconformity (transgression accompanied by fluvial erosion). Highstand systems tracts typically include a continental topset and a marine foreset and bottomset, and tend to display a convex-up shoreline trajectory (Fig. 9).

In upstream-controlled settings, systems tracts form independently of relative sea/lake-level changes and shoreline shifts (Fig. 5), and reflect the combined influence of accommodation, climate, source-area uplift, and autocyclicity on depositional processes. Upstream-controlled systems tracts may also be observed at different scales (Fig. 15) and are interpreted on the basis of stratal stacking patterns defined by the dominant depositional elements. In fluvial settings, the upstream-controlled stacking patterns are defined by the high vs low channel-to-overbank ratio (Figs. 12–15).

6.2.5 High-Amalgamation Systems Tract

The high-amalgamation systems tract (formerly termed "low-accommodation" systems tract) is defined by a stacking pattern dominated by a high degree of channel amalgamation (i.e., high channel-to-overbank ratio; Figs. 14A and 15). The formation of high-amalgamation systems tracts is favored by (1) low rates of floodplain aggradation; (2) unconfined fluvial channels; and (3) a high frequency of channel avulsion (Fig. 13; Bristow and Best, 1993).

6.2.6 Low-Amalgamation Systems Tract

The low-amalgamation systems tract (formerly termed "high-accommodation" systems tract) is defined by a stacking pattern dominated by floodplain deposits, within which channels occur as isolated depositional elements (i.e., low channel-to-overbank ratio; Figs. 14B and 15). The formation of low-amalgamation systems tracts is favored by (1) high rates of floodplain aggradation; (2) confined fluvial channels; and (3) a low frequency of channel avulsion (Fig. 13; Bristow and Best, 1993).

Notably, accommodation generated by basin subsidence is, in many if not most cases, already overfilled during the development of upstream-controlled depositional sequences and component systems tracts. While accommodation (i.e., rates of subsidence) may still play an important role, fluvial processes, including the rates of floodplain aggradation, are also influenced by all other controls that modify the balance between sediment supply and energy flux at any location (i.e., climate, source-area uplift, and autocyclicity). In a long term, the rates of fluvial aggradation exceed the rates of subsidence, leading eventually to entirely overfilled basins (Fig. 4D).

The high- and low-amalgamation systems tracts were designated as "low-accommodation" and "high-accommodation" systems tracts, respectively, based on the assumption that accommodation is the main control on the degree of channel amalgamation (e.g., Boyd et al., 2000; Leckie and Boyd, 2003). However, it is becoming apparent that accommodation alone cannot always explain the development of fluvial stacking patterns, particularly where the rates of accumulation of depositional elements do not match the rates of creation of accommodation (Miall, 2015). Therefore, the revised systems tract terminology emphasizes the observation of stratal stacking patterns rather than the interpretation of the underlying controls.

6.3 Parasequence

The parasequence is a succession of genetically related beds or bedsets bounded by flooding surfaces (Van Wagoner et al., 1988, 1990). A flooding surface is a facies contact that marks an abrupt increase in water depth and, consequently, an abrupt shift to relatively more distal facies on top. The concept of parasequence applies to coastal and shallow-water settings, where flooding surfaces may form (Posamentier and Allen, 1999). Several issues with the parasequence concept led to the conclusion that other types of unit (i.e., bedsets in the case of sedimentological cycles, and high-frequency sequences in the case of stratigraphic cycles; discussion below) provide better alternatives to define sedimentary cyclicity at parasequence scales (Catuneanu, 2006; Posamentier and Allen, 1999; Zecchin and Catuneanu, 2013).

Historically, the parasequence was introduced as the building block of seismic-scale systems tracts in the low-resolution era of seismic stratigraphy. The concept of parasequence triggered confusion and controversy in sequence stratigraphy, due to its traits that are unlike the features of a sequence stratigraphic unit, including (1) restricted applicability to coastal and shallow-water settings, and (2) the allostratigraphic rather than sequence stratigraphic nature of its bounding surfaces. Another issue is the potential confusion of flooding surfaces with similar-looking facies contacts generated by processes unrelated to water-depth changes, such as the abandonment of deltaic lobes in the process of autocyclic delta-lobe switching.

Parasequence boundaries form during transgression, when "flooding" occurs. These abrupt water-deepening episodes are typically short-lived "events" that punctuate longer term trends of coastal progradation or retrogradation. The result is stepped progradation, marked by a set of forestepping parasequences (e.g., Amorosi et al., 2005, 2017), or stepped retrogradation, marked by a set of backstepping parasequences (e.g., Bruno et al., 2017).

In the case of stepped progradation, the regressive trend is interrupted by higher frequency transgressions that lead to short-term changes in coastal depositional environments during the formation of flooding surfaces (e.g., short-term estuaries that interrupt temporarily the longer term deltaic progradation; Fig. 16). The architecture of stepped progradation is commonly defined by asymmetrical forestepping parasequences, dominated by regressive deposits. Each high-frequency transgression *may* be accompanied by the formation of a flooding surface (if the diagnostic lithological discontinuity develops), but it always ends with a maximum flooding surface. For this reason, genetic stratigraphic sequences of the same hierarchical rank with parasequences provide a more reliable alternative for correlation, both within and beyond the confines of coastal and shallow-water systems (Catuneanu et al., 2009, 2011). The stepped progradation of the Po coastal plain during the Middle to Late Holocene (Amorosi et al., 2005, 2017) documents forestepping parasequences of 10^0 m and 1000 years scales, which indicate regressive–transgressive cycles, and therefore changes in coastal depositional systems, on millennial and submillennial (i.e., centennial or even smaller) timescales. In this case, sequences, systems tracts and depositional systems can also be defined at parasequence scale.

In the case of stepped retrogradation, the transgressive trend proceeds with variable rates, higher during the formation of flooding surfaces and lower during the intervening stages. The episodes of abrupt water deepening may result in the formation of several flooding surfaces during transgression, which can be used to subdivide the transgressive systems tract into parasequences (e.g., Bruno et al., 2017). In this case, there is no change in coastal depositional environments during transgression, but only episodic shifts of subenvironments in an updip direction (e.g., stepwise retreat of bayhead deltas within wave-dominated estuaries). The architecture of such stepped retrogradation is defined by backstepping parasequences dominated (or exclusively built) by transgressive deposits. The stepped retrogradation of the Po coastal plain during the Early Holocene (Bruno et al., 2017) provides an example of backstepping parasequences of 10^0 m and 1000 years scales within a millennial-scale transgressive systems tract. Without changes in coastal depositional environments during transgression, this systems tract cannot be subdivided into lower rank sequences. In this case, parasequences are bedsets within the lowest rank transgressive systems tract and component depositional systems.

The distinction between parasequences and sequences is not based on scale or internal architecture, but on the nature of their bounding surfaces. Both types of unit may include normal regressive, transgressive, and forced regressive deposits (Catuneanu et al., 2011). Furthermore, parasequences and sequences may form at overlapping scales (Catuneanu et al., 2011; Csato et al., 2014; Fielding et al., 2008; Strasser et al., 1999; Tucker et al., 2009), in which case they offer different alternatives for correlation and the definition of stratigraphic units. A stratigraphic framework of parasequences requires lithological discontinuities (i.e., flooding surfaces), whereas a stratigraphic framework of sequences relies on sequence stratigraphic surfaces which are independent of lithology (e.g., maximum flooding surfaces). Where stratigraphic cycles include both flooding surfaces and sequence stratigraphic surfaces of equal hierarchical rank (i.e., sequences developed at parasequence scales; the case of stepped progradation above), the use of high-frequency sequences is recommended (Zecchin and Catuneanu, 2013).

Flooding surfaces do not mark a change in stratal stacking pattern, unless they happen to coincide with a sequence stratigraphic surface (e.g., maximum regressive surface, transgressive surface of erosion, or maximum flooding surface). Where flooding surfaces are within-trend facies contacts (i.e., retrogradational stacking pattern below and above the contact), they do not carry the significance of a sequence stratigraphic surface. In such cases, parasequences and flooding surfaces are more appropriate for allostratigraphic rather than sequence stratigraphic studies, even though they may provide useful subdivisions for the lowest rank transgressive systems tracts that form during stepped retrogradation (e.g., Bruno et al., 2017). For all other types of conventional systems tracts, which involve shoreline regression, high-frequency genetic stratigraphic sequences that form at the scale of parasequences provide a better alternative for stratigraphic mapping and correlation. The fact that flooding surfaces do not require a change in stratal stacking pattern to form (i.e., they may or may not coincide with systems tract boundaries) is the reason why they are not included on the list of sequence stratigraphic surfaces.

7. SEQUENCE STRATIGRAPHIC SURFACES

A sequence stratigraphic surface is a type of stratigraphic contact that serves, at least in part, as a systems tract boundary (Catuneanu et al., 2011). As systems tract boundaries, sequence stratigraphic surfaces mark changes in stratal stacking pattern between the units below and above the contact (Fig. 11). This defining attribute separates a sequence stratigraphic surface from any other type of stratigraphic contact. Seven sequence stratigraphic surfaces have been defined.

7.1 Subaerial Unconformity

The subaerial unconformity (Sloss et al., 1949) is an unconformity that forms under subaerial conditions as a result of fluvial erosion or bypass, pedogenesis, wind degradation, or karstification. Subaerial unconformities may form in both downstream- and upstream-controlled settings, most commonly during periods of negative accommodation (Catuneanu et al., 2011; Fig. 11A).

Under particular circumstances (e.g., where the landscape gradient is steeper than the gradient of the shoreline trajectory), the subaerial unconformity may also form during periods of positive accommodation and transgression (Figs. 10 and 11B). Alternative terms include "lowstand unconformity" (Schlager, 1992) and "regressive surface of fluvial erosion" (Plint and Nummedal, 2000). However, the term "subaerial unconformity" is preferred because it does not link, nor restrict, the formation of this surface to stages of lowstand or regression.

The identification of a "subaerial unconformity" in the rock record requires the preservation of continental deposits (fluvial, eolian) on top. Notably, the facies preserved below a subaerial unconformity may range from continental to marine, and therefore, they are not diagnostic to the identification of this surface. Subaerial unconformities may be subsequently reworked and replaced by younger erosional surfaces, in which case the composite unconformity takes the name of the younger surface (e.g., a transgressive surface of erosion may replace an older subaerial unconformity).

7.2 Basal Surface of Forced Regression

The basal surface of forced regression (Hunt and Tucker, 1992) is a sequence stratigraphic surface that marks a change in stratal stacking pattern from normal regression (below) to forced regression (above). Most commonly, the underlying normal regression is highstand, in the case of sequences that include all systems tracts (Fig. 11), but it can also be lowstand in the case of incomplete sequences which only include falling-stage and lowstand systems tracts. In either case, the basal surface of forced regression marks the onset of forced regression.

Where the subaerial unconformity forms during forced regression (Fig. 11A), the basal surface of forced regression is a marine surface (i.e., the paleo-seafloor at the onset of forced regression) truncated at the top by the subaerial unconformity. Where the subaerial unconformity forms during transgression (Fig. 11B), the basal surface of forced regression includes both marine and continental portions. The basal surface of forced regression is also known as the "correlative conformity" in the sense of Posamentier et al. (1988).

7.3 Correlative Conformity

The correlative conformity (Hunt and Tucker, 1992; Van Wagoner et al., 1988) is a sequence stratigraphic surface that marks a change in stratal stacking pattern from forced regression (below) to lowstand normal regression (above) (Fig. 11). Where the subaerial unconformity forms during forced regression (Fig. 11A), the correlative conformity is a marine surface (i.e., the paleo-seafloor at the end of forced regression) which connects physically with the downdip termination of the subaerial unconformity at the location of the shoreline at the end of forced regression (hence, the name "correlative" conformity; Fig. 11A). This is the context in which the "correlative" conformity has been defined, and which is most common in the stratigraphic record. Where the subaerial unconformity forms during transgression (Fig. 11B), the "correlative conformity" becomes a conformable surface with both marine and continental portions, without a physical and temporal relationship with the subaerial unconformity. In either case, the correlative conformity marks the end of forced regression and the beginning of subsequent lowstand normal regression (Fig. 11); therefore, the criteria that afford the distinction between forced and normal regressive deposits (e.g., Catuneanu and Zecchin, 2016; Posamentier and Morris, 2000) are critical to the identification of this surface.

Both the basal surface of forced regression and the correlative conformity have physical expression that can be observed on various types of data, from seismic to core and outcrop. Field criteria that can be used to identify these surfaces in outcrop and subsurface have been discussed and exemplified in several publications, including Posamentier and Allen (1999), Catuneanu (2006), Catuneanu et al. (2011), and MacEachern et al. (2012).

7.4 Maximum Regressive Surface

The maximum regressive surface (Helland-Hansen and Martinsen, 1996) is a stratigraphic surface that marks the change from regression to subsequent transgression. Most commonly, this change is expressed as a shift in stratal stacking pattern from normal regression (below) to transgression (above) (Fig. 11). The underlying normal regression may be "lowstand," in the case of sequences that include stages of negative accommodation (Fig. 11), or "highstand," in the case of sequences that form during periods of positive accommodation. In the latter case, sequences consist only of transgressive and highstand systems tracts which repeat in the stratigraphic succession without intervening stages of negative accommodation (Csato and Catuneanu, 2012, 2014). Where sequences record both stages of positive and negative accommodation, but the lowstand normal regression is missing, the maximum regressive surface coincides with the correlative conformity at the end of forced regression (e.g., in extensional settings where accommodation is generated rapidly by the reactivation of faults; Martins-Neto and Catuneanu, 2010). Therefore, maximum regressive surfaces may top any type of regressive deposit (i.e., lowstand normal regressive, highstand normal regressive, or forced regressive), and the precise stratigraphic reality of each case study needs to be determined on a case-by-case basis. There are also sequences that may not include surfaces associated with transgression, at the scale of observation that matches the hierarchical rank of the sequence, where the stratigraphic cyclicity is defined by the repetition of falling-stage and lowstand systems tracts (i.e., transgressions suppressed by high sediment supply, at that particular scale of observation).

The maximum regressive surface is the paleo-seafloor at the end of regression, and its correlative surface within the nonmarine setting. At least part of the continental portion of the maximum regressive surface is reworked and replaced by the transgressive surface of erosion during subsequent transgression (Fig. 11). The marine portion of the maximum regressive surface has a better preservation potential, as it is typically onlapped by the transgressive marine "healing-phase" deposits (Posamentier and Allen, 1999).

An alternative term is "transgressive surface" (Posamentier and Vail, 1988). The term "maximum regressive surface" is recommended where emphasis is placed on the end of regression; the term "transgressive surface" is recommended where emphasis is placed on the onset of transgression.

7.5 Maximum Flooding Surface

The maximum flooding surface (Frazier, 1974; Galloway, 1989; Posamentier et al., 1988; Van Wagoner et al., 1988) is a stratigraphic surface that marks a change in stratal stacking pattern from transgression (below) to highstand normal regression (above) (Fig. 11). It is the paleo-seafloor at the end of transgression, and its correlative surface within the nonmarine setting where transgression is accompanied by fluvial aggradation (Fig. 11A). The continental portion of the maximum flooding surface is commonly associated with the highest water table relative to the topographic profile, and therefore, it may be marked by the development of regional coal seams (e.g., Bohacs and Suter, 1997; Fanti and Catuneanu, 2010; Gastaldo et al., 1993; Hamilton and Tadros, 1994; Holz et al., 2002; Shanley and McCabe, 1994; Wright and Marriott, 1993). Within marine sections, the maximum flooding surface is often "cryptic" from a lithological standpoint, at the heart of condensed sections (e.g., Carter et al., 1998; Catuneanu, 2006; Posamentier and

Allen, 1999). However, the identification of maximum flooding surfaces is still possible and aided by the integration of independent methods that may involve seismic data (e.g., observation of downlap reflection terminations on 2D seismic lines and/or paleogeographic reconstructions using 3D seismic horizon slices), well logs calibrated with lithologs (e.g., highest radioactivity in fine-grained sediments), and biostratigraphic data (e.g., highest abundance and diversity of microfossils).

Alternative terms include "final transgressive surface" (Nummedal et al., 1993), "surface of maximum transgression" (Helland-Hansen and Gjelberg, 1994), and "maximum transgressive surface" (Helland-Hansen and Martinsen, 1996). The term "maximum flooding surface" is strongly entrenched in the literature, and it is recommended for historical reasons. It is important to avoid confusion between maximum flooding surfaces and flooding surfaces. The distinction is not a matter of scale (e.g., major vs minor transgressions), but a matter of definition: a maximum flooding surface marks a change in stratal stacking pattern (i.e., it is a sequence stratigraphic surface), and it may or may not be associated with a lithological contrast; a flooding surface is a lithological discontinuity (i.e., an allostratigraphic surface), which may or may not mark a change in stratal stacking pattern. Maximum flooding surfaces and flooding surfaces may form at the same scale of observation, in relation to the same transgressions.

7.6 Transgressive Surface of Erosion

The transgressive surface of erosion (Posamentier and Vail, 1988) is an erosional surface that forms during transgression by means of wave scouring (i.e., "wave-ravinement surface"; Swift, 1975) or tidal scouring (i.e., "tidal-ravinement surface"; Allen and Posamentier, 1993) in coastal to shallow-water environments. Both types of transgressive ravinement surfaces young toward the basin margin (Nummedal and Swift, 1987) and invariably rework part of the maximum regressive surface, thus becoming systems tract boundaries (Catuneanu et al., 2011; Fig. 11). The amount of erosion associated with transgressive ravinement processes is proportional to the energy of waves and tides along the transgressive coastline (i.e., the depth of the wave base and the magnitude of the tidal range). In most cases, an average of 10–20 m of section is removed by ravinement processes during transgression (Abbott, 1998; Demarest and Kraft, 1987). This amount can increase significantly in areas of exceptionally high energy (e.g., 40 m along the Canterbury Plains shoreline in New Zealand; Leckie, 1994).

Diagnostic to the transgressive surface of erosion is the presence of backstepping estuary-mouth complex (in the case of tidal-ravinement surfaces) or shallow-water "healing-phase" (in the case of wave-ravinement surfaces) deposits on top (Catuneanu, 2006; Posamentier and Allen, 1999). The presence of a transgressive lag and/or a concentration of onlapping shell beds immediately above the contact are also common (Kidwell, 1991; Zecchin and Catuneanu, 2013). Notably, the facies preserved below a transgressive surface of erosion may range from continental to marine, and therefore are not diagnostic to the identification of this surface.

7.7 Regressive Surface of Marine Erosion

The regressive surface of marine erosion (Plint, 1988) is an erosional surface that forms during forced regression by means of wave scouring triggered by the lowering of the wave base in the process of relative sea/lake-level fall. The regressive surface of marine erosion youngs basinward and invariably reworks part of the basal surface of forced regression, thus becoming a systems tract boundary (Catuneanu, 2006; Fig. 11). The degree of preservation of the basal surface of forced regression on the shelf depends on the gradient of the forced regressive shoreline trajectory relative to the seafloor gradient. Where the shoreline trajectory is steeper than the seafloor gradient (unlike the situation depicted in Fig. 11), the basal surface of forced regression may be entirely reworked and replaced by the regressive surface of marine erosion within the shelf setting. The amount of scouring that accompanies the formation of the regressive surface of marine erosion is proportional to the magnitude of relative sea/lake-level fall. Alternative terms include "regressive ravinement surface" (Galloway, 2001) and "regressive wave ravinement" (Galloway, 2004).

Diagnostic to the regressive surface of marine erosion is the presence of "sharp-based" prograding shoreface deposits on top (Catuneanu, 2006; Plint, 1988). Below this surface, there is typically an abrupt shift to more distal and finer grained facies (Fig. 11).

8. SCALE IN SEQUENCE STRATIGRAPHY

The development of stratigraphic stacking patterns that define systems tracts in both downstream- and upstream-controlled settings requires, typically, timescales of minimum 10^2–10^3 years (e.g., Amorosi et al., 2005; Bridge and Leeder, 1979; Miall, 2015; Nanson et al., 2013; Nixon et al.,

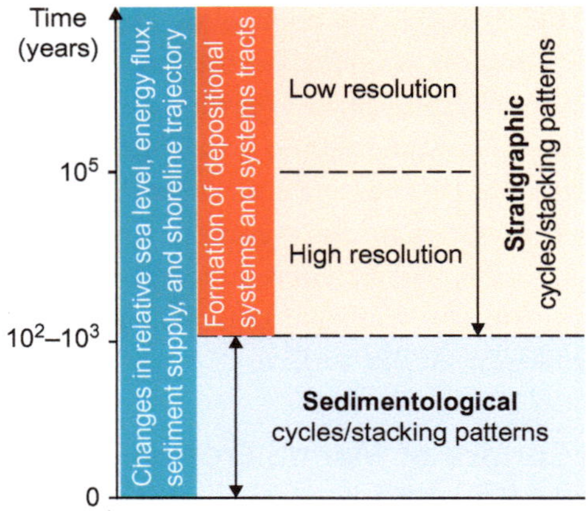

Fig. 17 Timescales of sedimentological vs stratigraphic cycles. Sedimentological cycles (i.e., beds and bedsets) are the building blocks of the lowest rank systems tracts and component depositional systems. In contrast, stratigraphic cycles (i.e., sequences) involve changes in systems tracts and component depositional systems. Both types of sedimentary cycle involve changes in stratal stacking pattern. Sedimentological stacking patterns refer to a stratal architecture that describes the internal organization of a depositional system, without changes in systems tract (e.g., cycles of change in the degree of amalgamation of storm beds; Zecchin et al., 2017b). Stratigraphic stacking patterns refer to a stratal architecture that involves changes in systems tracts and component depositional systems. The minimum timescales required to form a depositional system are commonly within a range of 10^2–10^3 years, as indicated by simulation models and high-resolution studies (e.g., Amorosi et al., 2005; Bridge and Leeder, 1979; Miall, 2015; Nanson et al., 2013; Nixon et al., 2014). Changes in relative sea level, energy flux, sediment supply, and shoreline trajectory occur at all scales, starting with the scale of tidal and fairweather-storm cycles, and may accompany the formation of both sedimentological and stratigraphic cycles. Criteria to discriminate between sedimentological and stratigraphic units and bounding surfaces have been summarized by Zecchin et al. (2017a,b).

2014; Fig. 17). These timescales afford the formation of depositional systems, whereby the defining subenvironments and related geomorphic elements are established as dominant sediment fairways in a paleogeographic context. The minimum scale of depositional systems defines the scale of the lowest rank systems tracts and stratigraphic sequences. Depositional systems can be observed at different scales, depending on the purpose of the study and the resolution of the data available (Fig. 16). Therefore, systems tracts and sequences can also be observed at different scales.

High-frequency sequences (and component systems tracts and depositional systems) are commonly observed at scales of 10^0–10^1 m and 10^2–10^5 years (e.g., Ainsworth et al., 2017; Amorosi et al., 2005; Lobo et al.,

2004; Magalhaes et al., 2015; Miall, 2015; Nanson et al., 2013; Nixon et al., 2014; Tesson et al., 1990, 2000; Zecchin et al., 2017a,b), which defines the scope of high-resolution sequence stratigraphy (Fig. 17). The stacking pattern of high-frequency sequences defines systems tracts, and component depositional systems, of higher hierarchical ranks in lower resolution studies (Figs. 15 and 16).

The sequence stratigraphic framework is scale invariant in the sense that the same stratal stacking patterns can be observed at different scales (Schlager, 2004, 2010). However, the stratigraphic architecture is not truly fractal, because sequences of different scales may have different controls and internal makeup in terms of component systems tracts. The internal architecture of sequence stratigraphic units becomes increasingly complex with the increase in the scale of observation. For example, the internal makeup of a systems tract may vary widely from a succession of beds and bedsets (i.e., sedimentological cycles) to a set of higher frequency (lower rank) sequences (i.e., stratigraphic cycles).

The scale of sequence stratigraphic units is highly variable in terms of duration, thickness, and geographic extent. There are no standards for the scale of any type of unit that contributes to the sequence stratigraphic framework, from depositional systems to systems tracts and sequences. In the context of seismic stratigraphy, the "genetically linked" processes that operate within the confines of seismic-scale depositional systems and systems tracts are interrupted by changes in depositional environment and sediment dispersal patterns that occur at smaller scales, some of which may fall below the seismic resolution (Fig. 16). Therefore, with the exception of the smallest stratigraphic scales, whereby depositional systems consist only of sedimentological cycles (e.g., 5th-order scale in Fig. 16), the stacking patterns that define systems tracts reflect *dominant trends* rather than an uninterrupted manifestation of genetically linked processes within stable environments (e.g., 4th- and 3rd-order scales in Fig. 16). Changes in paleogeography and depositional environments are recorded at different scales, with magnitudes that are proportional to their hierarchical rank (Fig. 16). The scale at which changes in the stratigraphic staking pattern can be demonstrated is a function of data resolution, which defines the stratigraphic resolution that can be achieved in any particular case study.

The notion that a sequence is a "relatively conformable" succession (i.e., with negligible internal unconformities; Mitchum, 1977; Fig. 3) cannot be used as a standard for the scale of a sequence, because the "relatively conformable" character itself is *relative* to the scale of observation and/or the resolution of the data available. In seismic stratigraphy, the scale of a sequence exceeds, by default, the vertical seismic resolution, which in turn varies with the seismic data set. Outcrop, core, and well-log data provide better resolution, and afford the recognition of sequences and unconformities at smaller scales. Any change in the type and/or the resolution of the data available results in changes to the scale of what can be perceived as a "relatively conformable" succession, and hence in inconsistencies with respect to the scale of a sequence. For this reason, the definition of a sequence must remain independent of scale and variables that change with the case study (e.g., types and resolution of data available). Sequences may be defined at different scales; at any particular scale (hierarchical rank), sequences may include internal unconformities of equal and/or lower hierarchical ranks. Moreover, depending on the selection of the sequence boundary (Fig. 2), sequences may or may not be "relatively conformable." However, sequences of any hierarchical rank consist of "genetically related" strata in the sense that they belong to the same cycle of accommodation and/or sediment supply (Catuneanu et al., 2009, 2011).

The trend in the development of sequence stratigraphy was to gradually improve the resolution of stratigraphic studies, by applying the method to increasingly smaller scales of observation. This implies refinements to the definition of a sequence and the recognition of sequences at smaller scales. Following this trend, the resolution of sequence stratigraphy improved from 10^2–10^3 m in the 1940s–1960s (i.e., scales relevant to continent-wide correlations; Sloss, 1963; Sloss et al., 1949), to 10^1–10^2 m in the 1970s (i.e., scales relevant to petroleum exploration in seismic stratigraphy; Payton, 1977), and eventually to 10^0–10^1 m with the advent of high-resolution sequence stratigraphy (i.e., scales relevant to reservoir compartmentalization and petroleum production development; e.g., Magalhaes et al., 2015; Zecchin and Catuneanu, 2015; Fig. 3). A resolution below 10^0 m is also possible, especially in carbonate systems which are more sensitive to small changes in environmental conditions (e.g., Mawson and Tucker, 2009).

With the increase in stratigraphic resolution, which is now approaching the scale of sedimentology, it becomes important to define criteria to separate between sedimentological and stratigraphic units which may form at similar scales (e.g., Zecchin et al., 2017a,b). The development of both types of unit may involve an interplay of allogenic and autogenic controls, and may be accompanied by changes in accommodation, energy flux, sediment supply, and shoreline trajectory (Fig. 17). The difference between the two types of sedimentary cycle is that the depositional environment does not change during the formation of sedimentological units (i.e., beds and bedsets within the lowest rank depositional systems), whereas stratigraphic sequences involve changes in depositional systems and systems tracts, which can be observed at different scales (Figs. 16 and 17). At any scale of observation, sequences can be subdivided into systems tracts and component depositional systems which describe the dominant architectural and depositional trends at that particular hierarchical level.

9. SEQUENCE STRATIGRAPHIC HIERARCHY

Stratigraphic cyclicity within a sedimentary succession can be observed at different scales. The classification of sequence stratigraphic units and bounding surfaces based on their relative scale and stratigraphic significance defines the concept of sequence stratigraphic hierarchy. Several sequence stratigraphic hierarchy systems have been proposed since the 1970s, based on criteria that highlight different attributes of sequences, including their temporal scales, their physical scales, or their "relatively conformable" nature. None of these hierarchy systems has received universal acceptance or validation by data in all depositional or tectonic settings.

One group of models asserts that hierarchical orders provide an adequate solution to the classification of sequences of different scales (i.e., sequences of first-order, second-order, third-order, fourth-order, etc.; Embry, 1995; Posamentier and Allen, 1999; Vail et al., 1977, 1991), although the criteria for the definition of different orders of cyclicity are subject to debate (e.g., temporal vs physical criteria). An alternative approach is to use a different nomenclature for units that develop at different scales, starting with the "relatively conformable" sequence as a reference for the hierarchy system (e.g., sequences, sequence sets, composite sequence sets; Mitchum and Van Wagoner, 1991; Sprague et al., 2003; Van Wagoner et al., 1990).

Statistical surveys show that sequences which can be observed in sedimentary basins worldwide are not organized into discrete classes of temporal or physical scales, but are rather part of a stratigraphic continuum (Carter et al., 1991; Drummond and Wilkinson, 1996). This variability of stratigraphic sequences in terms of time spans and physical dimensions is the result of the complex interplay of multiple local and global controls on accommodation and sedimentation. Indeed, the interplay of various processes within a sedimentary basin can alter or override the orderly patterns that may be expected from the natural periodicities of any specific controls on accommodation or sedimentation. Therefore, the classification of stratigraphic sequences based on their temporal or physical scales becomes arbitrary and difficult to generalize for all depositional and tectonic settings.

Attempts to define hierarchical ranks on the basis of the internal characteristics of stratigraphic cycles have also met with shortcomings and criticism. The notion that a sequence is a "relatively conformable" unit (Mitchum, 1977) fails to provide an objective standard for the scale of a sequence, as "relatively conformable" successions can be observed at different scales, depending on the resolution of the data available. Therefore, a "relatively conformable" sequence does not provide an objective reference for the construction of a hierarchy system that promotes a scale-dependent nomenclature. The definition of hierarchical ranks based on the systems tract composition of stratigraphic cycles (e.g., including or excluding lowstand systems tracts; Duval et al., 1998) is also unrealistic, because the systems tract architecture of sequences is independent of scale. For example, sequences of different scales may consist of the same combinations of systems tracts, and sequences of similar scales may consist of different combinations of systems tracts, depending on the *syn*-depositional conditions of accommodation and sedimentation (Catuneanu et al., 2011; Csato and Catuneanu, 2012, 2014).

The hierarchy systems which predict orderly patterns in the stratigraphic record are underlain by the assumption that "sequences" (i.e., the classic third-order cycles of seismic stratigraphy, which include lowstand systems tracts) are defined by full cycles of accommodation, whereas any lower rank cycles form during periods of positive accommodation, therefore missing the lowstand systems tracts (Duval et al., 1998). In reality, accommodation cycles are recorded at all scales, starting from the scale of tidal cycles, and exposure surfaces are as common as flooding surfaces in the rock record (Schlager, 2010; Vail et al., 1991). The scale of the smallest accommodation cycles that can be recognized in a particular case study remains a matter of resolution of the data available. Therefore, the classification of stratigraphic cycles based on their internal makeup is also artificial and cannot be generalized as a reproducible hierarchy system.

Stratigraphic cyclicity is basin specific, reflecting the importance of local controls on accommodation and sedimentation. The stratigraphic frameworks of individual basins reflect the unique evolution of those basins, and may differ from the stratigraphic frameworks of other sedimentary basins in terms of timing and duration of cycles, as well as the geometry of sequences and their controlling mechanisms. Moreover, stratigraphic frameworks may also differ between subbasins of the same sedimentary basin, as a result of changing accommodation and sedimentation conditions across sub-basin boundaries (e.g., Catuneanu et al., 1999, 2002; Miall et al., 2008).

The natural variability of the stratigraphic record indicates that the classification of sequences and bounding surfaces (i.e., the definition of a hierarchy system) is best approached on a basin-specific basis, rather than using any global standards or reference cycle charts. The stratigraphic reality of each sedimentary basin needs to be described on the basis of local data rather than information extrapolated from other basins. Within the context of each sedimentary basin, the relative stratigraphic significance of sequences and bounding surfaces may be measured by the degree of facies shifts associated with their formation, irrespective of the interpreted origin of their controlling mechanisms. In downstream-controlled settings, facies shifts reflect the magnitude of the associated shoreline shifts. In upstream-controlled settings, the magnitude of facies shifts is measured by the degree of change in depositional system (e.g., changes in fluvial style) or dominant depositional elements.

Hierarchical orders may be assigned within the context of each sedimentary basin, starting with the basin fill as a reference. A first-order sequence is defined as the entire sedimentary basin fill that accumulated within a specific tectonic setting (i.e., with accommodation controlled by a related set of subsidence mechanisms). In the case of "polyphase" basins, first-order sequence boundaries mark changes in the tectonic setting, which are the most significant changes that can be observed within a sedimentary succession. First-order sequences can be subdivided into lower rank sequences whose hierarchical orders are defined by the relative magnitude of facies shifts that are associated with their formation. In this approach, sequences of different hierarchical orders have no time or thickness connotations, but only a relative stratigraphic significance in relation to each other.

The possibility that stratigraphic frameworks may correlate from one basin to another in response to global controls on accommodation may not excluded, particularly in the case of tectonically "passive" settings and/or icehouse periods, but also, it cannot be generalized. The importance of local controls on accommodation, particularly in the case of tectonically active basins, cannot be underestimated and should be considered as a safe norm. For this reason, hierarchical orders are only meaningful within the context of the sedimentary basin in which they are defined, and their meaning may change from one basin to another. For example, "third-order" sequences may occur in any sedimentary basin, but they will differ from one basin to another in terms of duration, thickness, geographic extent, and controlling mechanisms. The "third-order" connotation is only meaningful with the context of the sedimentary basin in which it was defined, *relative* to the lower and higher rank cycles that occur within the same basin.

10. DISCUSSION AND CONCLUSIONS: METHODOLOGY AND NOMENCLATURE

The sequence stratigraphic framework consists of sequences and component systems tracts, which may be observed at different scales. Systems tracts may be further subdivided into sequence stratigraphic cycles (higher frequency sequences), allostratigraphic cycles (parasequences), or sedimentological cycles (beds and bedsets). The scope of the sequence stratigraphic methodology is to identify the different types of sequence stratigraphic units and bounding surfaces that develop at different scales, based on the observation of stratal stacking patterns. The existence of several competing approaches (Figs. 1 and 2) hindered, for decades, the definition of a standard methodology and the inclusion of sequence stratigraphy in international stratigraphic codes. These competing approaches differ in terms of nomenclature of sequence stratigraphic units and bounding surfaces (Figs. 1 and 2), the selection of the sequence boundary (Fig. 2), the criteria used to define a hierarchy system (e.g., temporal vs physical scales), and the assertions of the dominant controls on sequence development (e.g., global eustasy: Haq et al., 1987; Vail et al., 1977, 1991; tectonism: Embry, 1995; accommodation: Neal and Abreu, 2009; interplay of accommodation and sediment supply: Catuneanu, 2006; Catuneanu et al., 2009; Schlager, 1993; interplay of allogenic and autogenic controls: Catuneanu and Zecchin, 2013).

The standard methodology transcends the differences between the various schools, and relies on the common ground (i.e., set of core principles) that underlies all competing approaches. The model-independent guidelines are simpler than the workflow of any particular model, thus promoting greater flexibility in the application of the method. Significant progress has been made in outlining the common ground in sequence stratigraphy (Catuneanu et al., 2009, 2010, 2011), which led to the publication of formal recommendations by the International Subcommission on Stratigraphic Classification of the International Commission of Stratigraphy (Catuneanu et al., 2011): "The definition of the common ground in sequence stratigraphy should promote flexibility with respect to the choice of approach that is best suited to a specific set of conditions as defined by tectonic setting, depositional setting, data available, and scale of observation" (Catuneanu et al., 2011, p. 176); "A standard methodology can be defined based on the common ground between the different approaches, with emphasis on the observation of stratal stacking patterns in the rock record" (Catuneanu et al., 2011, p. 233). It has become clear that none of the models in Fig. 2 provides the "best practice" under all circumstances, as defined by different geological settings and types of data available.

The fundamental common ground protocol is the acquisition of local data (i.e., derived from the basin under analysis), followed by the observation of all features that help indentify stratal stacking patterns at scales selected by the practitioner or afforded by the resolution of the data available (Fig. 18). The same types of stratal stacking patterns may form at different scales, in relation to stratigraphic cycles of different magnitudes. At each scale of observation, stratigraphic cycles (i.e., sequences) may include internal unconformities of equal and/or lower hierarchical ranks. Therefore, "relatively conformable" successions, whether sequences or systems tracts, can be observed at different scales, depending on the scope of the study and the resolution of the data available (e.g., seismic-scale systems tracts typically include outcrop-scale sequences). Sequence stratigraphic frameworks are basin specific in terms of the temporal and physical scales of their units and bounding surfaces, reflecting the interplay of local and global controls on accommodation and sedimentation. Therefore, the methodology and

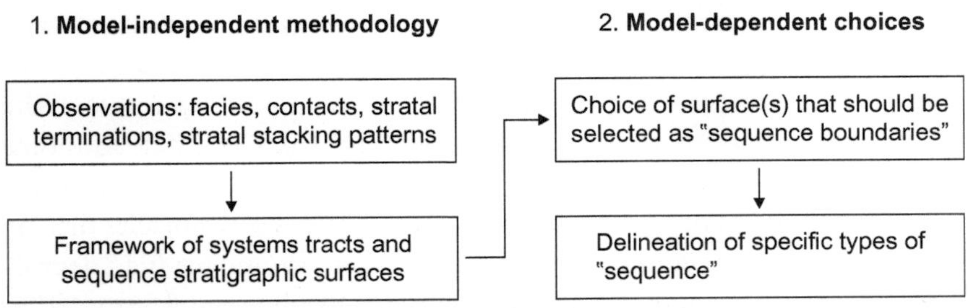

Fig. 18 Model-independent methodology vs model-dependent choices in sequence stratigraphy. The model-independent methodology is based on the observation of data and results in the construction of a stratigraphic framework of systems tracts and bounding surfaces at scales selected by the practitioner and/or afforded by the data available. Further model-dependent choices may be made with respect to the selection of surfaces that should be elevated to the status of "sequence boundary." This selection is often guided by the mappability of the different types of surface with the data available. *Modified from Catuneanu, O., Galloway, W.E., Kendall, C.G.St.C., Miall, A.D., Posamentier, H.W., Strasser, A., Tucker, M.E., 2011. Sequence stratigraphy: methodology and nomenclature. Newsl. Stratigr. 44/3, 173–245.*

nomenclature must remain independent of scale, and of the resolution of the data available, for a consistent and objective application of sequence stratigraphy.

The definition of all types of sequence stratigraphic units and bounding surfaces is independent of temporal and physical scales, and of the mechanism(s) of formation (Catuneanu et al., 2011, p. 175). The definition of a "sequence" has been revised and refined over time, to reflect conceptual developments and improvements in stratigraphic resolution (Fig. 3). With the decrease in the scale of observation (i.e., increase in stratigraphic resolution), as well as with the definition of different types of "sequence" (Fig. 1), the role of conformable surfaces as part or whole of the "sequence boundary" has become increasingly important. Within a common ground approach, a stratigraphic sequence is a cycle of change in stratal stacking patterns, defined by the recurrence of the same type of sequence stratigraphic surface (i.e., the "sequence boundary") in the rock record (Fig. 3). This definition is independent of scale, and independent of model (i.e., it accommodates all types of "sequence"). Sequences may form at different scales, and their relative stratigraphic significance is indicated by hierarchical orders. Sequences may or may not be "relatively conformable," depending in part on the selection of the sequence boundary, but always include sediments that are genetically related (i.e., which belong to the same cycle of accommodation or sedimentation; Catuneanu et al., 2009, 2011).

Stratal stacking patterns provide the basis for the definition of all units and surfaces of sequence stratigraphy. Sequences, systems tracts, and depositional systems can be observed at different scales, depending on the purpose of the study and the resolution of the data available (Fig. 16). Coastal depositional systems play a key role in defining the scales of systems tracts in downstream-controlled settings. The three-dimensional assemblages of lithofacies that are genetically linked by sedimentary processes and environments (i.e., depositional systems; Fisher and McGowen, 1967) can be observed from $<10^3$ to $>10^6$ years, depending on the resolution of the stratigraphic study (Fig. 16). The highest frequency changes in coastal environments (e.g., deltas vs estuaries in river-mouth settings, or prograding strandplains vs backstepping lagoon—barrier island systems in open coastline settings) define the lowest rank depositional systems, which consist truly and solely of genetically linked lithofacies accumulated in specific environments. These lowest rank depositional systems may form at scales of $\leq 10^0$ m and $\leq 10^3$ years, with an internal architecture defined by sedimentological cycles (beds or bedsets). At any larger scales (higher hierarchical ranks, including seismic exploration scales), depositional systems reflect dominant depositional trends rather than uninterrupted sedimentary processes in stable depositional environments, and may be subdivided into lower rank stratigraphic cycles (Figs. 16 and 17). Similarly, depositional systems in upstream-controlled settings may also undergo cyclic changes in fluvial styles and dominant depositional elements, at different scales (Fig. 15).

The observation of depositional systems at different scales affords the definition of systems tracts at different hierarchical levels. At each scale of observation (i.e., hierarchical level), systems tracts are defined by specific stacking patterns, and changes in the stacking pattern mark the position of sequence stratigraphic surfaces (e.g., a transgressive systems tract is defined by a retrogradational stacking pattern associated with shoreline transgression, and a maximum flooding surface is defined by a change from transgression to highstand normal regression; both transgressions and maximum flooding surfaces can be observed at different scales). The construction of a framework of systems tracts and bounding surfaces, at scales selected by the practitioner or afforded by the resolution of the data available, fulfills the practical purpose of sequence stratigraphy (Fig. 18). Within this framework, the practitioner can explain and predict the patterns of sediment distribution between the different depositional environments within a sedimentary basin.

Beyond the construction of a model-independent framework of systems tracts and bounding surfaces, the practitioner can make model-dependent choices with respect to the selection of the "sequence boundary" (Fig. 18). Such choices are often guided by the mappability of the different types of surface that are present within the studied section, which depends on the data available (e.g., well logs vs seismic lines). The flexibility of this model-independent workflow frees the practitioner from the rigid guidelines of any specific approach (e.g., the need to find a particular type of surface as "sequence boundary"), and from the expectations to fulfill the predictions of any particular model (e.g., the need to find orderly patterns in the stratigraphic record, or ideal successions of systems tracts). This promotes an objective construction of sequence stratigraphic frameworks, which may consist of variable successions and combinations of systems tracts (e.g., Csato and Catuneanu, 2012), at scales controlled by the geological setting (i.e., local conditions of accommodation and sedimentation), the resolution of the data available (e.g., seismic vs well-log data), and/or the scope of the study (e.g., petroleum exploration vs production development). Furthermore, the observation of stratal stacking patterns (i.e., methodology) needs to be separated from the subsequent interpretation of underlying controls (i.e., modeling) (Catuneanu and Zecchin, 2016).

Despite the scale invariance of stratal stacking patterns (i.e., same types observed at different scales), the stratigraphic architecture is not truly fractal, as sequences of different hierarchical ranks may have different controls and internal makeup in terms of component systems tracts. The internal makeup of systems tracts also becomes increasingly complex with the increase in the scale of observation, from a succession of beds and bedsets (i.e., sedimentological cycles) to a set of higher frequency sequences (i.e., stratigraphic cycles) of lower hierarchical rank. Within a framework of nested stratigraphic cycles, "relatively conformable" successions can be defined at different scales, depending on the resolution of the data available and/or the scope of the study, and therefore, they do not provide an objective reference for scale. The classification of nested sequences is basin specific, starting with the basin fill as a reference, with scales controlled by local conditions of accommodation and sedimentation.

It can be concluded that a scale-independent methodology is key to the standard application of sequence stratigraphy. The definition of sequence stratigraphic units and bounding surfaces must be independent of any variables that change with the case study (e.g., depositional and tectonic settings, and the types and resolution of the data available), in order to remain objective and consistent. The observation of stratal stacking patterns, at scales selected by the practitioner or imposed by the resolution of the data available, takes precedence over any model assumptions in terms of constructing a sequence stratigraphic framework. The model-independent methodology, inherently simple and consistent, provides the flexible platform for a standard application of sequence stratigraphy across the full spectrum of geological settings, stratigraphic scales, and types of data available.

ACKNOWLEDGMENTS

This work builds on the joint effort that led to the publication of formal recommendations on sequence stratigraphic methodology and nomenclature by the International Subcommission on Stratigraphic Classification (ISSC) of the International Commission on Stratigraphy (Catuneanu et al., 2011). I thank Bill Galloway, Chris Kendall, Andrew Miall, Henry Posamentier, Andre Strasser, and Maurice Tucker for their contributions and support as members of the ISSC working group on sequence stratigraphy. Additional insights were provided by many experts over the years, including V. Abreu, J.P. Bhattacharya, M.D. Blum, I. Csato, R.W. Dalrymple, A.F. Embry, P.G. Eriksson, C.R. Fielding, W.L. Fisher, P. Gianolla, M.R. Gibling, K.A. Giles, J.M. Holbrook, R. Jordan, B. Macurda, O.J. Martinsen, J.E. Neal, D. Nummedal, L. Pomar, B.R. Pratt, J.F.

Sarg, K.W. Shanley, R.J. Steel, C. Winker, and M. Zecchin. Feedback from Michael Montenari, Brian Pratt, and Massimo Zecchin helped to improve earlier drafts of this manuscript.

REFERENCES

Abbott, S.T., 1998. Transgressive systems tracts and onlap shellbeds from Mid-Pleistocene sequences, Wanganui Basin, New Zealand. J. Sediment. Res., v. 68, p. 253–268.

Ainsworth, R.B., Vakarelov, B.K., MacEachern, J.A., Rarity, F., Lane, T.I., Nanson, R.A., 2017. Anatomy of a shoreline regression: implications for the high-resolution stratigraphic architecture of deltas. J. Sediment. Res. 87, 1–35.

Allen, G.P. and Posamentier, H.W., 1993. Sequence stratigraphy and facies model of an incised valley fill: the Gironde Estuary, France. J. Sediment. Petrol., v. 63, no. 3, p. 378–391.

Amorosi, A., Centineo, M.C., Colalongo, M.L., Fiorini, F., 2005. Millennial-scale depositional cycles from the Holocene of the Po Plain, Italy. Mar. Geol., v. 222–223, p. 7–18.

Amorosi, A., Bruno, L., Campo, B., Morelli, A., Rossi, V., Scarponi, D., Hong, W., Bohacs, K.M., Drexler, T.M., 2017. Global Sea-Level Control on Local Parasequence Architecture from the Holocene Record of the Po Plain. Italy, Marine and Petroleum Geology (in press).

Barrell, J., 1917. Rhythms and the measurements of geological time. Geol. Soc. Am. Bull., v. 28, p. 745–904.

Bohacs, K. and Suter, J., 1997. Sequence stratigraphic distribution of coaly rocks: fundamental controls and paralic examples. Am. Assoc. Pet. Geol. Bull., v. 81, no. 10, p. 1612–1639.

Boyd, R., Diessel, C.F.K., Wadsworth, J., Leckie, D., Zaitlin, B.A., 2000. Organization of non marine stratigraphy. In: Boyd, R., Diessel, C.F.K., Francis, S. (Eds.), Advances in the Study of the Sydney Basin. Proceedings of the 34th Newcastle Symposium, University of Newcastle, Callaghan, New South Wales, Australia, pp. 1–14.

Bridge, J.S. and Leeder, M.R., 1979. A simulation model of alluvial stratigraphy. Sedimentology, v. 26, p. 617–644.

Bristow, C.S., Best, J.L., 1993. Braided rivers: perspectives and problems. In: Best, J.L., Bristow, C.S. (Eds.), Braided Rivers, pp. 1–11. Geological Society Special Publication No. 75.

Brown Jr., L.F., Fisher, W.L., 1977. Seismic stratigraphic interpretation of depositional systems: examples from Brazilian rift and pull apart basins. In: Payton, C.E. (Ed.), Seismic Stratigraphy—Applications to Hydrocarbon Exploration, pp. 213–248. American Association of Petroleum Geologists Memoir 26.

Bruno, L., Bohacs, K.M., Campo, B., Drexler, T.M., Rossi, V., Sammartino, I., Scarponi, D., Hong, W., Amorosi, A., 2017. Early Holocene Transgressive Paleogeography in the Po Coastal Plain (Northern Italy). Sedimentology (in press).

Carter, R.M., Abbott, S.T., Fulthorpe, C.S., Haywick, D.W., Henderson, R.A., 1991. Application of global sea-level and sequence-stratigraphic models in Southern Hemisphere Neogene strata. In: Macdonald, D.I.M. (Ed.), Sedimentation, Tectonics and Eustasy: Sea-Level Changes at Active Margins, pp. 41–65. International Association of Sedimentologists Special Publication 12.

Carter, R.M., Fulthorpe, C.S. and Naish, T.R., 1998. Sequence concepts at seismic and outcrop scale: the distinction between physical and conceptual stratigraphic surfaces. Sediment. Geol., v. 122, p. 165–179.

Catuneanu, O., 2002. Sequence stratigraphy of clastic systems: concepts, merits, and pitfalls. J. Afr. Earth Sci. 35, 1–43.

Catuneanu, O., 2003. Sequence Stratigraphy of Clastic Systems. Short Course Notes, vol. 16. Geological Association of Canada. 248 pp.

Catuneanu, O., 2006. Principles of Sequence Stratigraphy. Elsevier, Amsterdam. 375 pp.

Catuneanu, O. and Zecchin, M., 2013. High-resolution sequence stratigraphy of clastic shelves II: controls on sequence development. Mar. Pet. Geol., v. 39, p. 26–38.

Catuneanu, O. and Zecchin, M., 2016. Unique vs. non-unique stratal geometries: relevance to sequence stratigraphy. Mar. Pet. Geol., v. 78, p. 184–195.

Catuneanu, O., Willis, A.J., Miall, A.D., 1998. Temporal significance of sequence boundaries. Sediment. Geol. 121 (3–4), 157–178.

Catuneanu, O., Sweet, A.R., Miall, A.D., 1999. Concept and styles of reciprocal stratigraphies: Western Canada foreland basin. Terra Nova 11, 1–8.

Catuneanu, O., Hancox, P.J., Cairncross, B., Rubidge, B.S., 2002. Foredeep submarine fans and forebulge deltas: orogenic off-loading in the underfilled Karoo Basin. J. Afr. Earth Sci. 35 (4), 489–502.

Catuneanu, O., Abreu, V., Bhattacharya, J.P., Blum, M.D., Dalrymple, R.W., Eriksson, P.G., Fielding, C.R., Fisher, W.L., Galloway, W.E., Gibling, M.R., Giles, K.A., Holbrook, J.M., Jordan, R., Kendall, C.G.S.C., Macurda, B., Martinsen, O.J., Miall, A.D., Neal, J.E., Nummedal, D., Pomar, L., Posamentier, H.W., Pratt, B.R., Sarg, J.F., Shanley, K.W., Steel, R.J., Strasser, A., Tucker, M.E., Winker, C., 2009. Towards the standardization of sequence stratigraphy. Earth Sci. Rev. 92, 1–33.

Catuneanu, O., Bhattacharya, J.P., Blum, M.D., Dalrymple, R.W., Eriksson, P.G., Fielding, C.R., Fisher, W.L., Galloway, W.E., Gianolla, P., Gibling, M.R., Giles, K.A., Holbrook, J.M., Jordan, R., Kendall, C.G.S.C., Macurda, B., Martinsen, O.J., Miall, A.D., Nummedal, D., Posamentier, H.W., Pratt, B.R., Shanley, K.W., Steel, R.J., Strasser, A., Tucker, M.E., 2010. Sequence stratigraphy: common ground after three decades of development. First Break 28, 21–34.

Catuneanu, O., Galloway, W.E., Kendall, C.G.St.C., Miall, A.D., Posamentier, H.W., Strasser, A. and Tucker, M.E., 2011. Sequence stratigraphy: methodology and nomenclature. Newsl. Stratigr., v. 44/3, p. 173-245.

Csato, I. and Catuneanu, O., 2012. Systems tract successions under variable climatic and tectonic regimes: a quantitative approach. Stratigraphy, v. 9, no. 2, p. 109–130.

Csato, I. and Catuneanu, O., 2014. Quantitative conditions for the development of systems tracts. Stratigraphy, v. 11, no. 1, p. 39–59.

Csato, I, Granjeon, D., Catuneanu, O. and Baum, G.R., 2013. A three-dimensional stratigraphic model for the Messinian crisis in the Pannonian Basin, eastern Hungary. Basin Res., v. 25, p. 121–148.

Csato, I., Catuneanu, O., and Granjeon, D., 2014. Millennial-scale sequence stratigraphy: numerical simulation with Dionisos. J. Sediment. Res., v. 84, p. 394–406.

Csato, I., Toth, S., Catuneanu, O. and Granjeon, D., 2015. A sequence stratigraphic model for the Upper Miocene-Pliocene basin fill of the Pannonian Basin, eastern Hungary. Mar. Pet. Geol., v. 66, p. 117–134.

De Gasperi, A. and Catuneanu, O., 2014. Sequence stratigraphy of the Eocene turbidite reservoirs in Albacora field, Campos Basin, offshore Brazil. Am. Assoc. Pet. Geol. Bull., v. 98, no. 2, p. 279–313.

Demarest, J.M., Kraft, J.C., 1987. Stratigraphic record of Quaternary sea levels: implications for more ancient strata. In: Nummedal, D., Pilkey, O.H., Howard, J.D. (Eds.), Sea Level Fluctuation and Coastal Evolution , pp. 223–239. SEPM Special Publication 41.

Drummond, C.N. and Wilkinson, B.H., 1996. Stratal thickness frequencies and the prevalence of orderedness in stratigraphic sequences. J. Geol., v. 104, p. 1–18.

Duval, B., Cramez, C., Vail, P.R., 1998. Stratigraphic cycles and major marine source rocks. In: De Graciansky, P.C., Hardenbol, J., Jacquin, T., Vail, P.R. (Eds.), Mesozoic and Cenozoic Sequence Stratigraphy of European Basins, pp. 43–51. Society for Sedimentary Geology, Special Publication 60.

Embry, A.F., 1995. Sequence boundaries and sequence hierarchies: problems and proposals. In: Steel, R.J., Felt, V.L., Johannessen, E.P., Mathieu, C. (Eds.), Sequence Stratigraphy on the Northwest European Margin, pp. 1–11. Norwegian Petroleum Society (NPF), Special Publication 5.

Embry, A.F., Johannessen, E.P., 1992. T–R sequence stratigraphy, facies analysis and reservoir distribution in the uppermost Triassic-Lower Jurassic succession, western Sverdrup Basin, Arctic Canada. In: Vorren, T.O., Bergsager, E., Dahl-Stamnes, O.A., Holter, E., Johansen, B., Lie, E., Lund, T.B. (Eds.), In: Arctic Geology and Petroleum Potential, vol. 2 (Special Publication), pp. 121–146. Norwegian Petroleum Society (NPF).

Eriksson, P.G., Catuneanu, O., Nelson, D. and Popa, M., 2005. Controls on Precambrian sea-level change and sedimentary cyclicity. Sediment. Geol., v. 176, 1–2, p. 43–65.

Eriksson, P.G., Banerjee, S., Catuneanu, O., Corcoran, P.L., Eriksson, K.A., Hiatt, E.E., Laflamme, M., Lenhardt, N., Long, D.G.F., Miall, A.D., Mints, M.V., Pufahl, P.K., Sarkar, S., Simpson, E.L., Williams, G.E., 2013. Secular changes in sedimentation systems and sequence stratigraphy. Gondw. Res., v. 24, p. 468–489.

Fanti, F. and Catuneanu, O., 2010. Fluvial sequence stratigraphy: the Wapiti Formation, west-central Alberta, Canada. J. Sediment. Res., v. 80, p. 320–338.

Fielding, C.R., Frank, T.D., Birgenheier, L.P., Rygel, M.C., Jones, A.T., Roberts, J., 2008. Stratigraphic record and facies associations of the late Paleozoic ice age in eastern Australia (New South Wales and Queensland). In: Fielding, C.R., Frank, T.D., Isbell, J.L. (Eds.), Resolving the Late Paleozoic Ice Age in Time and Space, pp. 41–57. Geological Society of America Special Paper 441.

Fisher, W.L., McGowen, J.H., 1967. Depositional systems in the Wilcox group of Texas and their relationship to occurrence of oil and gas. Trans. Gulf Coast Assoc. Geol. Soc. 17, 105–125.

Frazier, D.E., 1974. Depositional Episodes: Their Relationship to the Quaternary Stratigraphic Framework in the Northwestern Portion of the Gulf Basin. University of Texas at Austin Bureau of Economic Geology Geological Circular 74–1. 28 pp.

Galloway, W.E., 1989. Genetic stratigraphic sequences in basin analysis I: architecture and genesis of flooding-surface bounded depositional units. Am. Assoc. Pet. Geol. Bull., v. 73, p. 125–142.

Galloway, W.E., 2001. In: The many faces of submarine erosion: theory meets reality in selection of sequence boundaries. A.A.P.G. Hedberg Research Conference on "Sequence Stratigraphic and Allostratigraphic Principles and Concepts," Dallas, August 26–29, Program and Abstracts Volume, pp. 28–29.

Galloway, W.E., 2004. Accommodation and the sequence stratigraphic paradigm. Reservoir Can. Soc. Pet. Geol., v. 31, Issue 5, p. 9–10.

Gastaldo, R., Denko, T. and Liu, Y., 1993. Application of sequence and genetic stratigraphic concepts to carboniferous coal-bearing strata: an example from the Black Warrior Basin, USA. Geogr. Rundsch., v. 82, p. 212–226.

Hamilton, D.S., Tadros, N.Z., 1994. Utility of coal seams as genetic stratigraphic sequence boundaries in non-marine basins: an example from the Gunnedah basin, Australia. Am. Assoc. Pet. Geol. Bull. 78, 267–286.

Haq, B.U., Hardenbol, J. and Vail, P.R., 1987. Chronology of fluctuating sea levels since the Triassic (250 million years ago to present). Science, v. 235, p. 1156–1166.

Helland-Hansen, W. and Gjelberg, J.G., 1994. Conceptual basis and variability in sequence stratigraphy: a different perspective. Sediment. Geol., v. 92, p. 31–52.

Helland-Hansen, W. and Hampson, G.J., 2009. Trajectory analysis: concepts and applications. Basin Res., v. 21, p. 454–483.

Helland-Hansen, W. and Martinsen, O.J., 1996. Shoreline trajectories and sequences: description of variable depositional-dip scenarios. J. Sediment. Res., v. 66, no. 4, p. 670–688.

Holz, M., Kalkreuth, W., and Banerjee, I., 2002. Sequence stratigraphy of paralic coal-bearing strata: an overview. Int. J. Coal Geol., v. 48, p. 147–179.
Hunt, D., Tucker, M.E., 1992. Stranded parasequences and the forced regressive wedge systems tract: deposition during base-level fall. Sediment. Geol. 81, 1–9.
Jervey, M.T., 1988. Quantitative geological modeling of siliciclastic rock sequences and their seismic expression. In: Wilgus, C.K., Hastings, B.S., Kendall, C.G.S.C., Posamentier, H.W., Ross, C.A., Van Wagoner, J.C. (Eds.), Sea Level Changes—An Integrated Approach, pp. 47–69. Society of Economic Paleontologists and Mineralogists (SEPM) Special Publication 42.
Johnson, J.G., Murphy, M.A., 1984. Time-rock model for Siluro-Devonian continental shelf, western United States. Geol. Soc. Am. Bull. 95, 1349–1359.
Kidwell, S.M., 1991. Condensed deposits in siliciclastic sequences: expected and observed features. In: Einsele, G., Ricken, W., Seilacher, A. (Eds.), Cycles and Events in Stratigraphy. Springer-Verlag, Berlin, pp. 682–695.
Leckie, D.A., 1994. Canterbury Plains, New Zealand—implications for sequence stratigraphic models. AAPG Bull. 78, 1240–1256.
Leckie, D.A. and Boyd, R., 2003. Towards a nonmarine sequence stratigraphic model, American Association of Petroleum Geologists Annual Convention, Salt Lake City, 11–14 May 2003, Official Program, vol. 12, p. A101.
Lobo, F.J., Tesson, M., Gensous, B., 2004. Stratal architectures of late Quaternary regressive-transgressive cycles in the Roussillon Shelf (SW Gulf of Lions, France). Mar. Pet. Geol., v. 21, p. 1181–1203.
Longwell, C.R., 1949. Sedimentary Facies in Geologic History. Geological Society of America Memoir 39. 171 pp.
Løseth, T.M., Helland-Hansen, W., 2001. Predicting the pinchout distance of shoreline tongues. Terra Nova 13, 241–248.
Loutit, T.S., Hardenbol, J., Vail, P.R., Baum, G.R., 1988. Condensed sections: the key to age-dating and correlation of continental margin sequences. In: Wilgus, C.K., Hastings, B.S., Kendall, C.G.S.C., Posamentier, H.W., Ross, C.A., Van Wagoner, J.C. (Eds.), Sea Level Changes—An Integrated Approach, pp. 183–213. SEPM Special Publication 42.
MacEachern, J.A., Dashtgard, S.E., Knaust, D., Catuneanu, O., Bann, K.L., Pemberton, S.G., 2012. Sequence stratigraphy. In: Knaust, D., Bromley, R.G. (Eds.), Trace Fossils as Indicators of Sedimentary Environments. In: Developments in Sedimentology, vol. 64. Elsevier, pp. 157–194.
Magalhaes, A., Raja Gabaglia, G., Scherer, C., Ballico, M., Guadagnin, F., Bento Freire, E., Silva Born, L., Catuneanu, O., 2015. Sequence hierarchy in the Mesoproterozoic Tombador Formation, Chapada Diamantina, Brazil. Basin Res., 1–40.
Martins-Neto, M.A. and Catuneanu, O., 2010. Rift sequence stratigraphy. Mar. Pet. Geol., v. 27, p. 247–253.
Mawson, M., Tucker, M.E., 2009. High-frequency cyclicity (millennial-scale and Milankovitch) in slope carbonates (Zechstein) Permian, NE England. Sedimentology 56, 1905–1936.
Miall, A.D., 2010. The Geology of Stratigraphic Sequences, second ed. Springer-Verlag, Berlin. 522 pp.
Miall, A.D., 2015. Updating uniformitarianism: stratigraphy as just a set of "frozen accidents" In: Smith, D.G., Bailey, R.J., Burgess, P., Fraser, A. (Eds.), Strata and Time, pp. 11–36. Geological Society, London, Special Publications, vol. 404.
Miall, A.D., Catuneanu, O., Vakarelov, B.K., Post, R., 2008. The Western Interior Basin. In: Miall, A.D. (Ed.), The Sedimentary Basins of the United States and Canada. Sedimentary Basins of the World. Elsevier, Amsterdam, pp. 329–362. Hsu, K.J. (Series Editor).

Mitchum Jr., R.M., 1977. Seismic stratigraphy and global changes of sea level, part 11: glossary of terms used in seismic stratigraphy. In: Payton, C.E. (Ed.), Seismic Stratigraphy—Applications to Hydrocarbon Exploration, pp. 205–212. American Association of Petroleum Geologists Memoir 26.

Mitchum, R.M., Jr. and Van Wagoner, J.C., 1991. High-frequency sequences and their stacking patterns: sequence stratigraphic evidence of high-frequency eustatic cycles. Sediment. Geol., v. 70, p. 131–160.

Mitchum Jr., R.M., Vail, P.R., Thompson III, S., 1977. Seismic stratigraphy and global changes of sea-level, part 2: the depositional sequence as a basic unit for stratigraphic analysis. In: Payton, C.E. (Ed.), Seismic Stratigraphy—Applications to Hydrocarbon Exploration, pp. 53–62. American Association of Petroleum Geologists Memoir 26.

Muto, T., Steel, R.J., 1997. Principles of regression and transgression: the nature of the interplay between accommodation and sediment supply. J. Sediment. Res. 67, 994–1000.

Muto, T., Steel, R.J., 1992. Retreat of the front in a prograding delta. Geology 20, 967–970.

Muto, T., Steel, R.J., 2002. Role of autoretreat and A/S changes in the understanding of deltaic shoreline trajectory: a semi-quantitative approach. Basin Res., v. 14, p. 303–318.

Nanson, R.A., Vakarelov, B.K., Ainsworth, R.B., Williams, F.M., and Price, D.M., 2013. Evolution of a Holocene, mixed-process, forced regressive shoreline: the Mitchell River delta, Queensland, Australia. Mar. Geol., v. 339, p. 22–43.

Neal, J. and Abreu, V., 2009. Sequence stratigraphy hierarchy and the accommodation succession model. Geology, v. 37, no. 9, p. 779–782.

Nixon, F.C., England, J.H., Lajeunesse, P., and Hanson, M.A., 2014. Deciphering patterns of postglacial sea level at the junction of the Laurentide and Innuitian Ice Sheets, western Canadian High Arctic. Quat. Sci. Rev., v. 91, p. 165–183.

Nummedal, D., Swift, D.J.P., 1987. Transgressive stratigraphy at sequence-bounding unconformities: some principles derived from Holocene and Cretaceous examples. In: Nummedal, D., Pilkey, O.H., Howard, J.D. (Eds.), Sea-Level Fluctuation and Coastal Evolution, pp. 241–260. Society of Economic Paleontologists and Mineralogists (SEPM) Special Publication 41.

Nummedal, D., Riley, G.W., Templet, P.L., 1993. High-resolution sequence architecture: a chronostratigraphic model based on equilibrium profile studies. In: Posamentier, H.W., Summerhayes, C.P., Haq, B.U., Allen, G.P. (Eds.), Sequence Stratigraphy and Facies Associations, pp. 55–68International Association of Sedimentologists Special Publication 18.

Payton, C.E. (Ed.), 1977. Seismic Stratigraphy—Applications to Hydrocarbon Exploration. American Association of Petroleum Geologists Memoir 26, 516 pp.

Plint, A.G., 1988. Sharp-based shoreface sequences and "offshore bars" in the Cardium Formation of Alberta; their relationship to relative changes in sea level. In: Wilgus, C.K., Hastings, B.S., Kendall, C.G.S.C., Posamentier, H.W., Ross, C.A., Van Wagoner, J.C. (Eds.), Sea Level Changes—An Integrated Approach, pp. 357–370. SEPM Special Publication 42.

Plint, A.G., Nummedal, D., 2000. The falling stage systems tract: recognition and importance in sequence stratigraphic analysis. In: Hunt, D., Gawthorpe, R.L. (Eds.), Sedimentary Response to Forced Regression, vol. 172, pp. 1–17. The Geological Society of London, Special Publication.

Posamentier, H.W., Allen, G.P., 1999. Siliciclastic Sequence Stratigraphy: Concepts and Applications. SEPM Concepts in Sedimentology and Paleontology No. 7, 210 pp.

Posamentier, H.W., Morris, W.R., 2000. Aspects of the stratal architecture of forced regressive deposits. In: Hunt, D., Gawthorpe, R.L. (Eds.), Sedimentary Responses to Forced Regressions, pp. 19–46. Geological Society Special Publication 172.

Posamentier, H.W., Vail, P.R., 1988. Eustatic controls on clastic deposition. II. Sequence and systems tract models. In: Wilgus, C.K., Hastings, B.S., Kendall, C.G.S.C.,

Posamentier, H.W., Ross, C.A., Van Wagoner, J.C. (Eds.), In: Sea Level Changes—An Integrated Approach, vol. 42. SEPM Special Publication, pp. 125–154.
Posamentier, H.W., Jervey, M.T., Vail, P.R., 1988. Eustatic controls on clastic deposition. I. Conceptual framework. In: Wilgus, C.K., Hastings, B.S., Kendall, C.G.S.C., Posamentier, H.W., Ross, C.A., Van Wagoner, J.C. (Eds.), In: Sea Level Changes—An Integrated Approach, vol. 42. SEPM Special Publication, pp. 110–124.
Posamentier, H.W., Allen, G.P., James, D.P., Tesson, M., 1992. Forced regressions in a sequence stratigraphic framework: concepts, examples, and exploration significance. Am. Assoc. Pet. Geol. Bull. 76, 1687–1709.
Qayyum, F., de Groot, P., Hemstra, N., Catuneanu, O., 2014. 4D Wheeler diagrams: concept and applications. In: Smith, D.G., Bailey, R.J., Burgess, P.M., Fraser, A.J. (Eds.), Strata and Time: Probing the Gaps in Our Understanding. The Geological Society of London, Special Publication 404, 10 pp.
Qayyum, F., Catuneanu, O. and de Groot, P., 2015. Historical developments in Wheeler diagrams and future directions. Basin Res., v. 27, p. 336–350.
Schlager, W., 1992. Sedimentology and sequence stratigraphy of reefs and carbonate platforms. In: Continuing Education Course Note Series #34. American Association of Petroleum Geologists. 71 pp.
Schlager, W., 1993. Accommodation and supply—a dual control on stratigraphic sequences. In: Cloetingh, S., Sassi, W., Horvath, F., Puigdefabregas, C. (Eds.), Basin Analysis and Dynamics of Sedimentary Basin Evolution. Sedimentary Geology, vol. 86, pp. 111–136.
Schlager, W., 2004. Fractal nature of stratigraphic sequences. Geology 32, 185–188.
Schlager, W., 2010. Ordered hierarchy versus scale invariance in sequence stratigraphy. Int. J. Earth Sci., v. 99, p. S139-S151.
Shanley, K. W. and McCabe, P. J., 1994. Perspectives on the sequence stratigraphy of continental strata: Am. Assoc. Pet. Geol. Bull., v. 78, p. 544–568.
Sloss, L.L., 1963. Sequences in the cratonic interior of North America. Geol. Soc. Am. Bull., v. 74, p. 93–114.
Sloss, L.L., Krumbein, W.C., Dapples, E.C., 1949. Integrated facies analysis. In: Longwell, C.R. (Ed.), Sedimentary Facies in Geologic History, pp. 91–124. Geological Society of America Memoir 39.
Sprague, A. R., Patterson, P.E., Sullivan, M.D., Campion, K.M., Jones, C.R., Garfield, T.R., Sickafoose, D.K., Jennette, D.C., Jensen, G.N., Beaubouef, R.T., Goulding, F.J., Van Wagoner, J.C., Wellner, R.W., Larue, D.K., Rossen, C., Hill, R.E., Geslin, J.K., Feldman, H.R., Demko, T.M., Abreu, V., Zelt, F.B., Ardill, J. and Porter, M.L., 2003. Physical stratigraphy of clastic strata: a hierarchical approach to the analysis of genetically related stratigraphic elements for improved reservoir prediction. S. Tex. Geol. Soc. Bull., v. 44, p. 7.
Strasser, A., Pittet, B., Hillgärtner, H. and Pasquier, J.-B., 1999. Depositional sequences in shallow carbonate-dominated sedimentary systems: concepts for a high-resolution analysis. Sediment. Geol., v. 128, p. 201–221.
Swift, D.J.P., 1975. Barrier-island genesis: evidence from the central Atlantic shelf, eastern U.S.A. Sediment. Geol., v. 14, p. 1–43.
Tesson, M., Gensous, B., Allen, G.P., and Ravenne, C., 1990. Late Quaternary deltaic lowstand wedges on the Rhone continental shelf, France. Mar. Geol., v. 91, p. 325–332.
Tesson, M., Posamentier, H.W., and Gensous, B., 2000. Stratigraphic organization of Late Pleistocene deposits of the western part of the Rhone shelf (Languedoc shelf) from high resolution seismic and core data. Am. Assoc. Pet. Geol. Bull., v. 84, p. 119–150.
Tucker, M.E., Gallagher, J. and Leng, M., 2009. Are beds millennial-scale cycles? An example from the Carboniferous of NE England. Sediment. Geol., v. 214, p. 19–34.

Vail, P.R., Mitchum Jr., R.M., Thompson III, S., 1977. Seismic Stratigraphy and Global Changes of Sea Level, Part Four: Global Cycles of Relative Changes of Sea Level. American Association of Petroleum Geologists Memoir 26, pp. 83–98.

Vail, P.R., Audemard, F., Bowman, S.A., Eisner, P.N., Perez-Cruz, C., 1991. The stratigraphic signatures of tectonics, eustasy and sedimentology—an overview. In: Einsele, G., Ricken, W., Seilacher, A. (Eds.), Cycles and Events in Stratigraphy. Springer-Verlag, pp. 617–659.

Van Wagoner, J.C., Posamentier, H.W., Mitchum, R.M., Vail, P.R., Sarg, J.F., Loutit, T.S., Hardenbol, J., 1988. An overview of sequence stratigraphy and key definitions. In: Wilgus, C.K., Hastings, B.S., Kendall, C.G.S.C., Posamentier, H.W., Ross, C.A., Van Wagoner, J.C. (Eds.), In: Sea Level Changes—An Integrated Approach, vol. 42. SEPM Special Publication, pp. 39–45.

Van Wagoner, J.C., Mitchum Jr., R.M., Campion, K.M., Rahmanian, V.D., 1990. Siliciclastic Sequence Stratigraphy in Well Logs, Core, and Outcrops: Concepts for High-Resolution Correlation of Time and Facies. American Association of Petroleum Geologists Methods in Exploration Series 7, 55 pp.

Wheeler, H.E., 1958. Time stratigraphy. Am. Assoc. Pet. Geol. Bull., v. 42, p. 1047–1063.

Wheeler, H.E., 1959. Unconformity bounded units in stratigraphy. Am. Assoc. Pet. Geol. Bull., v. 43, p. 1975–1977.

Wheeler, H.E., 1964. Baselevel, lithosphere surface, and time-stratigraphy. Geol. Soc. Am. Bull., v. 75, p. 599–610.

Wright, V.P. and Marriott, S.B., 1993. The sequence stratigraphy of fluvial depositional systems: the role of floodplain sediment storage. Sediment. Geol., v. 86, p. 203–210.

Zecchin, M., Catuneanu, O., 2013. High-resolution sequence stratigraphy of clastic shelves I: units and bounding surfaces. Mar. Pet. Geol. 39, 1–25.

Zecchin, M. and Catuneanu, O., 2015. High-resolution sequence stratigraphy of clastic shelves III: applications to reservoir geology. Mar. Pet. Geol., v. 62, p. 161–175.

Zecchin, M., Catuneanu, O. and Rebesco, M., 2015. High-resolution sequence stratigraphy of clastic shelves IV: high-latitude settings. Mar. Pet. Geol., v. 68, p. 427–437.

Zecchin, M., Caffau, M., Catuneanu, O., Lenaz, D., 2017a. Discrimination between wave-ravinement surfaces and bedset boundaries in Pliocene shallow-marine deposits, Crotone Basin, southern Italy: an integrated sedimentological, micropaleontological and mineralogical approach. Sedimentology (in press).

Zecchin, M., Catuneanu, O., and Caffau, M., 2017b. High-resolution sequence stratigraphy of clastic shelves V: criteria to discriminate between stratigraphic sequences and sedimentological cycles. Mar. Pet. Geol., v. 85, p. 259–271.

FURTHER READING

Hunt, D. and Tucker, M.E., 1995. Stranded parasequences and the forced regressive wedge systems tract: deposition during base-level fall—reply. Sediment. Geol., v. 95, p. 147–160.

CHAPTER TWO

Sequence Stratigraphy and Geologic Time

Andrew D. Miall[1]
University of Toronto, Toronto, ON, Canada
[1]Corresponding author: e-mail address: miall@es.utoronto.ca

Contents

Abstract

The maturation of sequence stratigraphy and the development of a refined geological timescale have provided the basis for much more detailed stratigraphic reconstructions than were possible as little as a decade ago. Earlier studies showing that measured sedimentation rates vary inversely in proportion to the timescale over which they are measured may now be placed in a context that relates rates and timescales to actual geological processes. As little as 10% of elapsed time may be represented by a vertical succession at any one location. Sediment breaks account for the remainder, and constitute a much more significant component of stratigraphic processes than is generally assumed. Research methods that assume stratigraphic continuity must take this fact into account.

1. THE EMERGENCE OF MODERN STRATIGRAPHIC METHODS

Two significant developments that reached maturity during the last decade have fundamentally changed the science of stratigraphy:

1. Sequence stratigraphy: The principles of the sequence-stratigraphic approach were introduced by Sloss (1963) and Vail et al. (1977), but their application was characterized by considerable controversy regarding

Stratigraphy & Timescales, Volume 2
ISSN 2468-5178
http://dx.doi.org/10.1016/bs.sats.2017.06.001

terminology and driving mechanisms. Solutions to terminology and classification problems were offered by Catuneanu (2006), and Miall (2010) summarized the debate regarding driving mechanisms and proposed an updated sequence classification. Bhattacharya (2011) discussed practical issues regarding the selection of the key surfaces for correlation and definition of realistic sequence architectures.

2. The geologic timescale: Development of a modern, integrated timescale was facilitated by the introduction of the principle of the Global Stratigraphic Section and Point (GSSP), and has culminated in the creation of an online database and updated timescale incorporating all available means of age determination (www.stratigraphy.org; Gradstein et al., 2004).

Modern stratigraphic approaches involve an integration of these two methodological themes (Miall, 2016), and the result has been the development of an ever more detailed database of local to regional stratigraphic records characterized by sequence interpretations and detailed age determinations. This has facilitated highly detailed examinations of the sedimentary record with a view to clarifying the rates and nature of geological processes, and mass balance calculations relating to the source-to-sink problem.

One of the outcomes of these developments has been a renewed attention to an important discovery by Sadler (1981, 1999), who demonstrated that in the stratigraphic record, measured sedimentation rates vary in inverse proportion to the time span over which they are measured. This relationship holds over all timescales, spanning 14 orders of magnitude, from seconds to tens of millions of years. In a later paper, Sadler and Jerolmack (2015) supplemented the data compilation on vertical aggradation rates (Sadler, 1981) with a similar compilation on lateral accumulation rates (the forward growth of ripples, bars, delta lobes, continental margin clinoforms, etc.). They demonstrated that, in cross-sectional area perpendicular to strike, growth rates of fluvial, coastal, and shelf clastics are fairly constant at around $1\,m^2/a$, at all timescales, as measured over time spans from months to hundreds of millions of years. This continuity of sediment flux is, of course, not evident from the geologic record but, as we discuss here, nondeposition or erosion in one location is expected to be contemporaneous with deposition elsewhere.

The explanation for these relationships is that the stratigraphic record is replete with depositional gaps representing elapsed time on all scales. The new stratigraphy, characterized by detailed sequence and sedimentological relationships and accurate age information, is confirming Sadler's discovery and is providing the basis for in-depth analyses of the range of processes that lead to the accumulation, removal, and ultimate preservation of the depositional record. Miall (2015) suggested that these rates and processes could be grouped into 12 broad classes, which he categorized as a *Sedimentation Rate Scale* (*SRS*) (Table 1; Fig. 1).

The stratigraphic record may be described as the product of a geological "preservation machine" comprising multiple geological processes operating simultaneously at different rates and timescales, as a series of "nested loops" (Fig. 2) (c.f., the "Stratigraphy Machine" of Smith, 1994). The postglacial record, the study of which has formed the basis for much of modern sedimentology, only contains complete records of geological processes up to SRS 7.

This clarification of processes and timescales has led to a process-based classification of unconformities (Table 2).

A focus on the nature and timescale of sedimentary breaks, from regional angular unconformities down to the practically microscopic scale of the boundaries between mud and silt beds formed by wave, tidal, and storm activity, is becoming increasingly detailed in current sedimentological studies, and is providing essential insights into the rates and nature of sequence-generating processes. The construction of Wheeler diagrams (Wheeler, 1958, 1959) to explore and illustrate time-stratigraphic relationships is becoming an essential component of stratigraphic research.

The current classification of stratigraphic sequences (Table 3) is now two decades old, and the on-going spate of detailed stratigraphic work promises to provide important insights into the nature of this classification, with refinements to be expected about the rates and magnitudes of allogenic controls.

It has been suggested that the distribution of sedimentation rates and unconformity durations is fractal in character (Middleton et al., 1995; Plotnick, 1986), which would imply a self-similarity in geological processes over the full range of timescales. However, as argued by Miall (2015) the resemblance may be superficial. The processes that control the accumulation of the sedimentary record, including the climatic and tectonic processes of local to global scope that govern the addition and removal of accommodation, and affect the nature of the sediment supply, comprise a range of genetically unrelated mechanisms. However, Plotnick's (1986) fractal model for generating a distribution of sedimentary gaps creates a pattern of stratigraphic variability that mimics in many ways the character of actual stratigraphic

Table 1 Hierarchy of Depositional Units, Recurrence Interval, Rates of Sedimentation, and Preservation

SRS	Timescale (years)	Inst. Sed. Rate (m/ka)	Dating Method	Sedimentary Process	Preservational Accident	True Cyclicity
1	10^{-6} (30 s)	10^{6}		Burst-sweep cycle	Small-scale autogenic shifts in distribution of grain carpet due to aggradation, turbulent bursts	None
2	10^{-5}–10^{-4} (5–50 min)	10^{5}		Ripple migration	Small-scale autogenic shifts	Repetition by climbing
3	10^{-3} (8 h)	10^{5}		Dune migration, foreset bundles	Small-scale autogenic shifts	Repetition by climbing
4	10^{-2}–10^{-1} (3–36 days)	10^{4}	On the ground observation	Diurnal variability to normal meteorological floods (dynamic events)	(1) Accommodation: lateral space on bars. Preservation: meander cutoff and abandonment by shift in channel during high-discharge event (2) Crevasse splays abandoned by autogenic shifting	Bar accretion
5	10^{0}–10^{1}	10^{2}–10^{3}	Repeated maps, air photos	Seasonal to 10-year flood: macroform growth increment	As for SRS 4, above	Bar accretion
6	10^{2}–10^{3}	10^{2}–10^{3}	Short-lived radionuclides, old maps	Long-term (100-year) flood: macroform, point bar, splay	As for SRS 4, above	Bar accretion

7	10^3–10^4	10^0–10^1	^{14}C	Long-term geomorphic process: channel, delta lobe, coal seam	Avulsion, delta-lobe switching, valley fill	Channel/delta aggradation
8	10^4–10^5	10^{-1}	Magnetostratigraphy, O isotope scale	Channel belt, delta, orbital cycle	(1) Channel belt avulsion (2) Valley fill (3) Intraplate stress changes adjusting paleoslope	Channel/delta aggradation; orbital forcing
9	10^5–10^6	10^{-2}–10^{-1}	Magnetostratigraphy, O isotope scale	Depositional system, alluvial fan, major delta complex, orbital cycle	As for SRS 8, above	Orbital forcing
10	10^6	10^{-1}–10^0	Radiometric methods	Rapid subsidence of convergent margin basins, syntectonic clastic progradation, growth strata	Accommodation generation at geologically rapid rates (e.g., pull-apart basins, growth faults) Coastal clinoform progradation	Crude cyclicity related to tectonic episodicity
11	10^6–10^7	10^{-2}–10^{-1}	Radiometric methods	Basin-fill complex, tectonic cyclothem (e.g., "clastic wedge")	Tectonic changes in location and rate of accommodation Coastal clinoform progradation	Crude cyclicity related to tectonic episodicity
12	10^6–10^7	10^{-3}–10^{-2}	Radiometric methods	Very low-accommodation cratons	Coastal clinoform progradation	As above

From Miall, A.D., 2015, Updating uniformitarianism: stratigraphy as just a set of "frozen accidents". In: Smith, D.G., Bailey, R.J., Burgess, P., Fraser, A. (Eds.), Strata and Time. Geological Society, London, Special Publication 404, pp. 11–36.

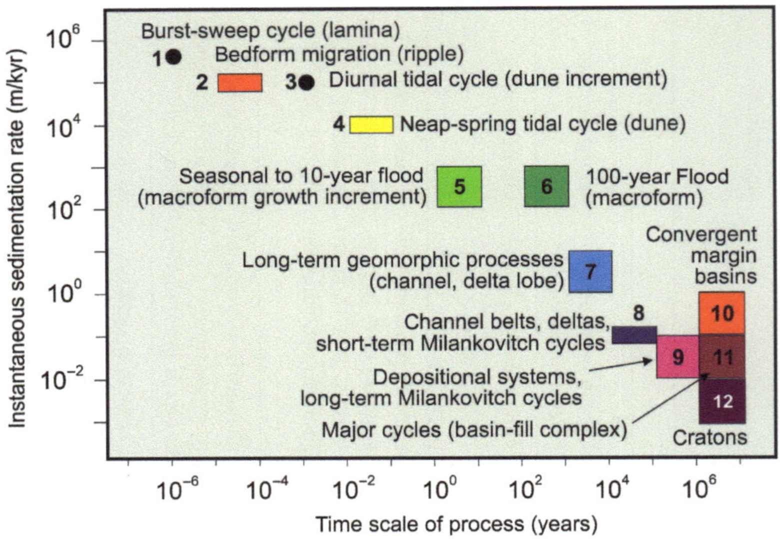

Fig. 1 Rates and durations of sedimentary processes. Numerals refer to the Sedimentation Rate Scale (see also Table 1).

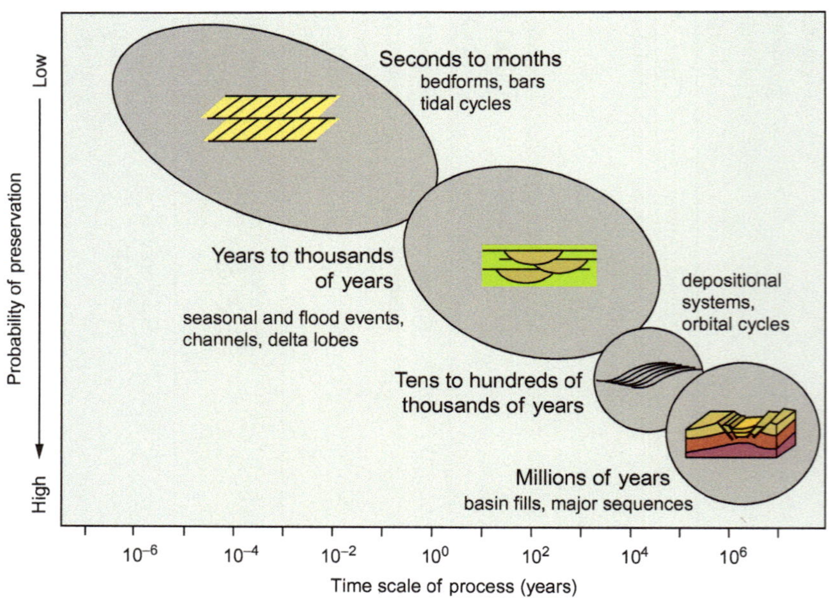

Fig. 2 The grouping of generational and preservational processes in terms of timescales and probability of preservation.

Table 2 A Classification of Unconformities

SRS	Timescale (years)	Inst. Sed. Rate (m/ka)	Process	Description of Break	Field Characteristics of Sedimentary Break and/or of Beds Above and Below
1–4	10^{-6}–10^{-1}	10^{4}–10^{6}	Bedform migration; diurnal to normal meteorological changes in runoff; tidal cycles	Local channel scours	Nesting of channels, macroforms and bedforms within a structure of minor bounding surfaces (ranks 1–5 of Miall, 1996)
5–7	10^{0}–10^{3}	10^{0}–10^{3}	Subtle tectonism, including in-plane stress Autogenic seasonal to long-term geomorphic processes Rare extreme weather events	Migration and switching of depositional systems Minor erosion, mature paleosols Marked facies change, minor regional erosion	Minor cut-and-fill erosion, early cementation, "unconformity paleosols" in nonmarine settings Superimposition of delta and shelf-margin clinoform lobes separated by transgressive ravinement surfaces, rare preservation of falling-stage incised distributary channels, incised valleys Facies blanket, regional marker horizon
7–9	10^{4}–10^{5}	10^{-2}–10^{0}	High-frequency tectonism Regional response to flexural loading/unloading Far-field intraplate stress changes Orbital forcing	Syndepositional unconformities Basin-wide low-angle unconformities Tilting and warping of sequences and sequence sets Continental (potentially global)—scale, nonangular break	Strong but very localized angularity, coarse clastic wedges ("growth strata") Low- to very low-angle clinoform sets Evidence of fluvial or marine erosion, transgressive lag deposits at breaks Widespread shifts in paleocurrent patterns, shoreline trends Cyclothemic facies changes, potentially deep erosion of unconformity surface, coastal and shelf-margin clinoform onlap-offlap cycles

Continued

Table 2 A Classification of Unconformities—cont'd

SRS	Timescale (years)	Inst. Sed. Rate (m/ka)	Process	Description of Break	Field Characteristics of Sedimentary Break and/or of Beds Above and Below
9–12	10^6–10^7	10^{-3}–10^{-1}	Orogenic tectonism	Regional angular unconformity	May be associated with deep erosional relief, clastic wedges
			Dynamic unconformities associated with basin formation	Onlap and offlap caused by basin subsidence	Onlap of extensional margins during flexural subsidence
			Dynamic topography	Subcontinental unconformity	Onlap/offlap during motion of foreland-basin forebulge
			Global eustasy	Global unconformity	Low-angularity (units above and below have similar dip). Commonly little field evidence of major time break
			Long-term environmental change	Regional disconformities	Similar to above
					Eolian supersurfaces, drowning unconformities (carbonates)

SRS = Sedimentation Rate Scale (*from Miall, A.D., 2015, Updating uniformitarianism: stratigraphy as just a set of "frozen accidents". In: Smith, D.G., Bailey, R.J., Burgess, P., Fraser, A. (Eds.), Strata and Time. Geological Society, London, Special Publication 404, pp. 11–36*).

Table 3 Stratigraphic Cycles and Their Causes

Sequence Type	Duration (m.y.)	Other Terminology
A. Global supercontinent cycle	200–400	First-order cycle (Vail et al., 1977)
B. Cycles generated by continental-scale mantle thermal processes (dynamic topography), and by plate kinematics, including: 1. Eustatic cycles induced by volume changes in global mid-oceanic spreading centers 2. Regional cycles of uplift and subsidence induced by dynamic topography 3. Regional cycles of basement movement induced by extensional downwarp and crustal loading	10–100	Second-order cycle (Vail et al., 1977), supercycle (Vail et al., 1977), sequence (Sloss, 1963)
C. Regional to local cycles of basement movement caused by regional plate kinematics, including changes in intraplate stress regime, crustal loading	0.01–10	Third- to fifth-order cycles (Vail et al., 1977). Third-order cycles also termed: megacyclothem (Heckel, 1986), mesothem (Ramsbottom, 1979)
D. Global cycles generated by orbital forcing, including glacioeustasy, productivity cycles, etc.	0.01–2	Fourth- and fifth-order cycles (Vail et al., 1977), Milankovitch cycles, cyclothem (Wanless and Weller, 1932), major, minor cycles (Heckel, 1986)

From Miall, A.D., 2010. The Geology of Stratigraphic Sequences, second ed. Springer-Verlag, Berlin, 522 pp.

sections, and it may be useful to consider the distribution of sedimentation rates and gap lengths as "fractal like." The resulting "Devil's Staircase" (Fig. 3) contains gaps of varying length, in accordance with the fractal distribution model. Sequence boundaries are the more significant gaps characterized by the longest duration and greatest extent.

2. THE CASE OF THE SEVIER CLASTIC WEDGE OF UTAH–COLORADO

The Cenomanian–Maastrichtian succession of Utah–Colorado (Fig. 4) is a classic example of a syntectonic clastic wedge, developed in

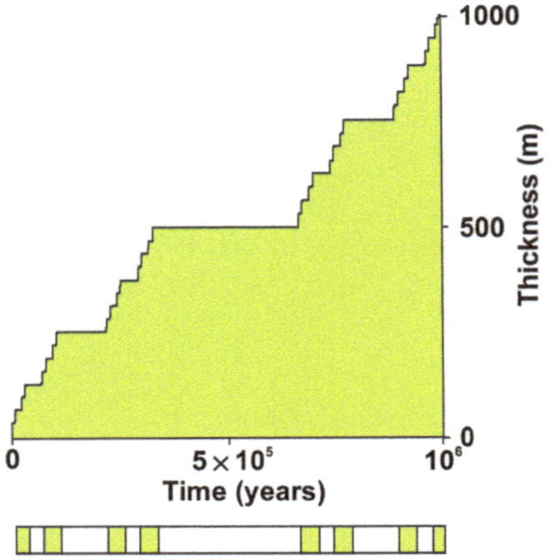

Fig. 3 Stratigraphic thickness accumulation viewed as a Cantor function—what has been termed a Devil's Staircase—constructed with gap length $G = ½$. The corresponding Cantor bar is shown at a lower resolution below the graph. *Adapted from Plotnick, R.E., 1986. A fractal model for the distribution of stratigraphic hiatuses. J. Geol. 94, 885–890, Fig. 5.*

response to the Sevier orogeny. The cross section illustrated here is about 3750 m in thickness, deposited over a time span of 34 m.y., at an average accumulation rate of 0.11 m/ka. This corresponds to SRS 11 of Miall (2015), the rate typical of long-term (10^6–10^7-year) geological accumulation. Aschoff and Steel (2011), who studied the upper part of this succession (Blackhawk to Williams Fork Formations), calculated sedimentation rates for selected intervals in order to explore the variation in response of the foreland basin to tectonism. Their measured rates varied from 0.047 to 0.14 m/ka, calculated over stratigraphic time spans of between 2.1 and 6.5 Ma.

This succession exhibits at least three scales of cyclicity at timescales $\geq 10^3$ years. The first and largest scale of cyclicity is that of the clastic wedge itself, and represents the progradational response of depositional systems to a rising sediment source area and east-verging thrusting and crustal shortening.

The *second scale of cyclicity* is that suggested by the alternating eastward- and westward-directed migration of the coastal facies belts, highlighted in Fig. 5 by the red "shazam" line superimposed on the cross section, and drawn to illustrate in very general terms the variable geographical extent and duration of the cycles. This interpretation shows 17 cycles, averaging 2 m.y. in duration.

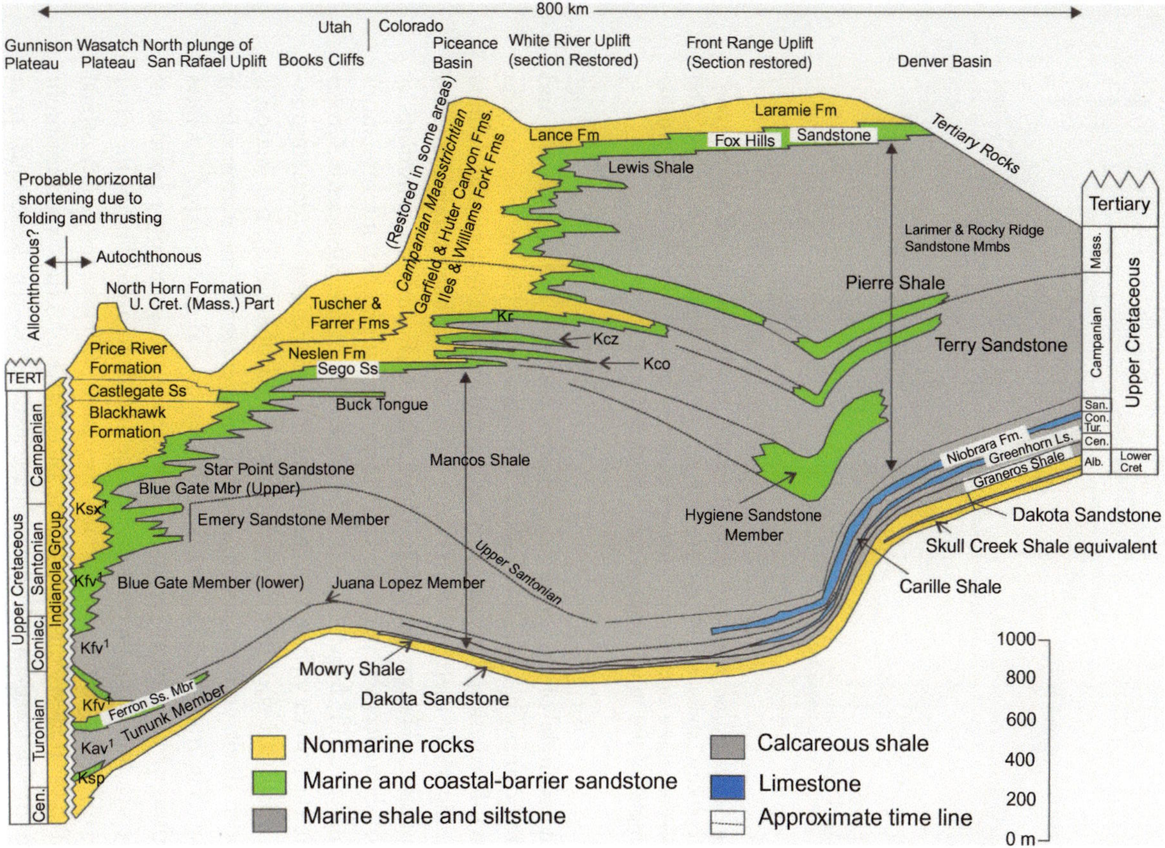

Fig. 4 The Sevier clastic wedge, Utah–Colorado. *Adapted from Molenaar, C.M., Rice, D.D., 1988. Cretaceous rocks of the Western Interior Basin. In: Sloss, L.L. (Ed.), Sedimentary Cover—North American Craton: U.S., The Geology of North America, vol. D-2. Geological Society of America, Boulder, CO, pp. 77–82.*

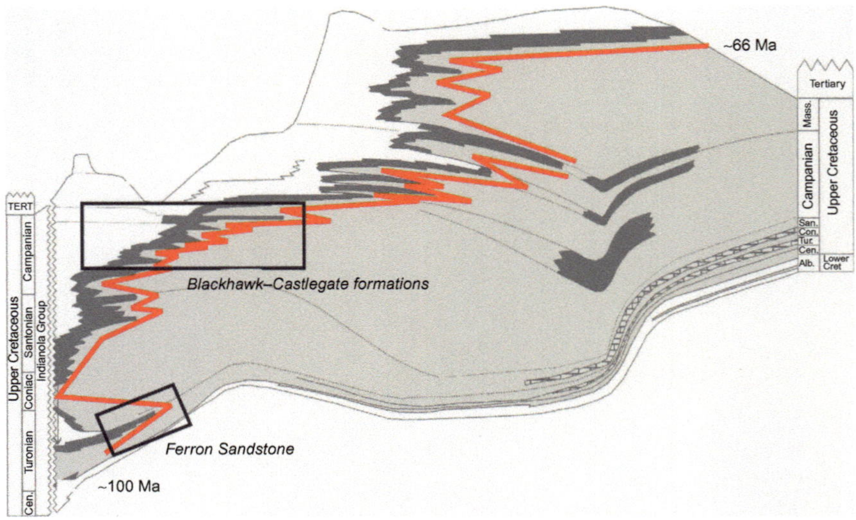

Fig. 5 The Sevier clastic wedge of Utah–Colorado, highlighting in *red* the "second" scale of cyclicity, showing 34 cycles with an average frequency of 2 m.y. The Blackhawk–Castlegate and Ferron intervals outlined by the *two rectangles* provide good examples of the "third" scale of cyclicity, as discussed in the text.

As an example of this scale of cyclicity, consider the Blackhawk and Castlegate Formations, which constitute the Mesaverde Group (Fig. 6). The Blackhawk Formation consists of six sequences (originally formally defined as lithostratigraphic members by Young, 1955) and has been subdivided into 23 submembers (Hampson, 2010). The Castlegate Formation constitutes at least two sequences, depending on different stratigraphic interpretations (Bhattacharya, 2011; Miall, 2014a). The time span of the Mesaverde Group is estimated at 4.86 m.y., based on ammonite chronostratigraphy (data summarized by Miall, 2014a). Near Price, Utah (central Book Cliffs), the section is up to 700 m thick, and was accumulated at an average sedimentation rate of 0.14 m/ka. This rate (10^{-1} m/ka) is at the upper end (more rapid) of rates characteristic of long-term geological subsidence measured over periods of 10^5–10^7 years, including thermal subsidence and flexural loading of the crust (SRS 10–11), and is comparable to rates associated with low-frequency orbital cycles (SRS 9). If all six members of the Blackhawk plus the two constituent sequences of the Castlegate Formation are assumed to represent equal time spans (for which there is no evidence), this equates to an average duration of 607 ka per unit, a value that corresponds to no known geological frequency.

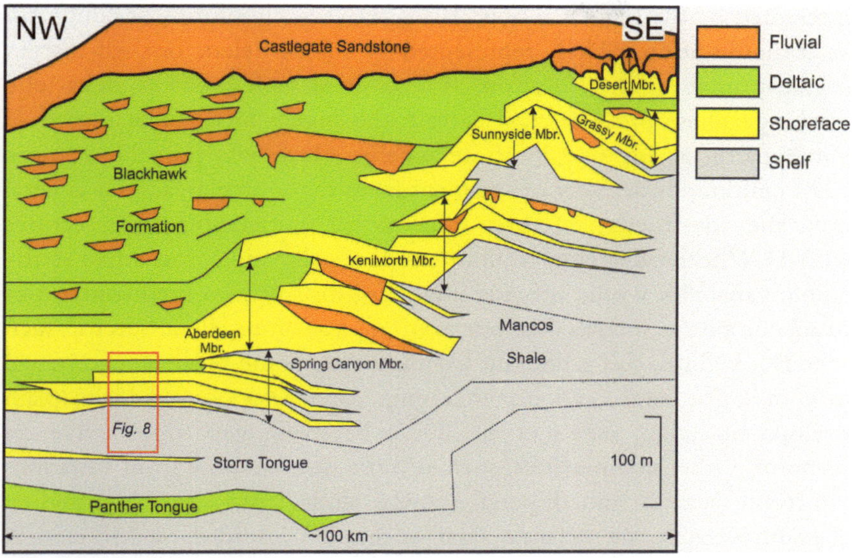

Fig. 6 The Blackhawk and Castlegate Formations of central Utah, showing the subdivision of the two formations into 23 submembers. This stratigraphic assemblage is enlarged from the upper of the *two rectangles* in Fig. 5.

Subdividing the Blackhawk Formation into 23 submembers (the individual shoreface tongues in Fig. 6) yields an average duration for the submembers of 125 ka. At this time there is no method by which the ages and time spans of the constituent units of the Mesaverde Group may be individually dated with greater precision.

A significant proportion of the elapsed time at any given location may be represented by intraformational hiatuses, member/submember boundaries, and other surfaces of erosion or nondeposition. Indeed, as suggested by Miall (2014a) such breaks may encompass as much as 90% of the elapsed time represented by the section. It seems very likely that at least the member and submember boundaries represent significant breaks in sedimentation, with durations in the high 10^4-year range. This is consistent with the pattern that emerged from a forward modeling exercise by Howell et al. (1999). As a preliminary model, the six member boundaries are set here arbitrarily so that the preserved sedimentary record represents a total of approximately half of the elapsed time, with an average sedimentation rate for the preserved deposits of 0.29 m/ka. Sediment bypass may explain much of the time "missing" through any given vertical section. Also, the elapsed time represented by the sequence boundaries between clinoforms or clinoform sets at one location is likely represented by coastal-deltaic or shelf-margin progradation at another location, as a result of such events as delta-lobe switching. These processes can explain the highly incomplete record of sedimentation at any one location, whereas there was probably a more or less continuous delivery of sediment from the source area. As a general point, the construction of three-dimensional stratigraphic cross sections and 3-D Wheeler diagrams to illustrate down-dip and across-strike stratigraphic variability would provide important insights into the evolution of coastal complexes such as this, but even in areas of excellent outcrop such as the Book Cliffs, this is difficult to impossible. Some of the experimental work of Chris Paola and his colleagues to develop coastal delta plain development (e.g., Sheets et al., 2002) has been very informative on this point. Other complexities, such as those that can arise from variations in tectonic activity and dispersal systems along strike, are illustrated in an example from the Atlantic shelf-margin of Trinidad by Moscardelli et al. (2012).

This sedimentation rate of the Blackhawk members is comparable to that of deposits measured at the SRS 8 timescale of 10^4–10^5 years, including the late Cenozoic glacioeustatic cycles of the Wanganui Basin in New Zealand and the Triassic Newark cycles of the eastern United States (details for these and other comparative rates are provided in Miall, 2015).

A speculative chronostratigraphic chart for the Mesaverde Group that includes the suggestion of significant gaps at the sequence boundaries is presented here as Fig. 7, based on the member age distribution suggested by Hampson (2010, Fig. 7).

In the updip coastal plain region represented by the Blackhawk Formation, most of the elapsed time is thought to be completely unrepresented. Some of this unit consists of fluvial deposits formed above marine base level (Hampson et al., 2013) and here, accommodation may be explained by the buffer concept of Holbrook et al. (2006). These deposits are shown schematically in Fig. 7. The pattern shown here in Fig. 7 compares to that of Krystinik and DeJarnett (1995, Fig. 3), in which it was suggested that the proximal region to the west of the basin was a region of uplift and erosion. Units of condensed (slow) sedimentation may be expected to develop offshore at times of regional transgression, and there is also evidence that sediment gravity flows occurred on distal clinoform slopes during the falling stage of some of the sequence cycles (Hampson, 2010, Fig. 7; not shown in Fig. 7). The time span represented by the member (sequence) boundaries

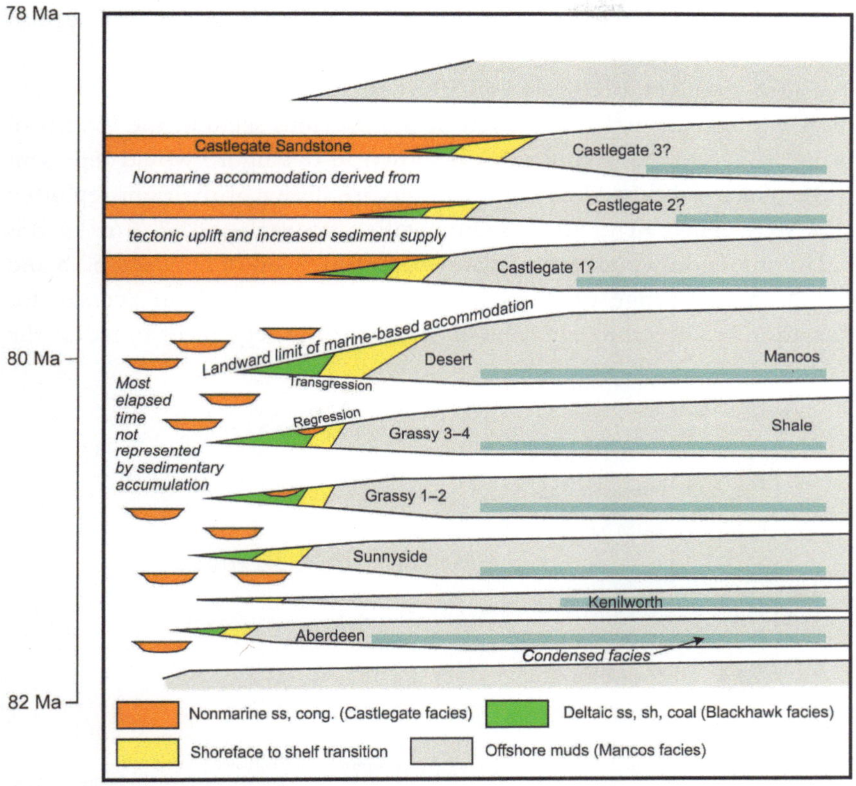

Fig. 7 A speculative chronostratigraphic chart for the Blackhawk–Castlegate Formations. *From Miall, A.D., 2014. The emptiness of the stratigraphic record: a preliminary evaluation of missing time in the Mesaverde Group, Book Cliffs, Utah. J. Sediment. Res. 84, 457–469.*

increases landward, and the undifferentiated proximal, deltaic Blackhawk Formation represents, in total, much less than one half of the available elapsed time, with substantial erosional breaks embedded in this coastal succession.

The third scale of cyclicity in the Utah–Colorado clastic wedge is that of the individual submembers of the Mesaverde Group, exemplified here by the shoreface tongues of the Blackhawk Formation (Fig. 6). The long-term geomorphic evolution of depositional systems, such as major channels, delta lobes, and coastal barrier systems over time periods of 10^3–10^4 years (the timescale suggested by the chronostratigraphic analysis, above), corresponds to SRS 7, with sedimentation rates in the range of 10^0–10^1 m/ka. The oldest member of the Blackhawk Formation, the Springdale Member, as documented in Spring Canyon (the section outlined by the rectangle in Fig. 6) is shown in detail in Fig. 8.

The section is plotted against an SRS 8 timescale. At the SRS 8 sedimentation rate of 0.29 m/ka employed in the previous section, the 79.2 m of Spring Canyon section (units 2–15) shown in this figure would represent 273 ka of elapsed time. The same 15 units are shown at the right replotted using SRS 7 rates. At an SRS 7 sedimentation rate range of 1.5–6 m/ka, this 79.2 m of section would accumulate over a time span of between 52.8 and 13.2 ka. A mid-range of 27 ka used here (arbitrarily) as an average for illustration is one tenth the time assumed for the sequences timed at the

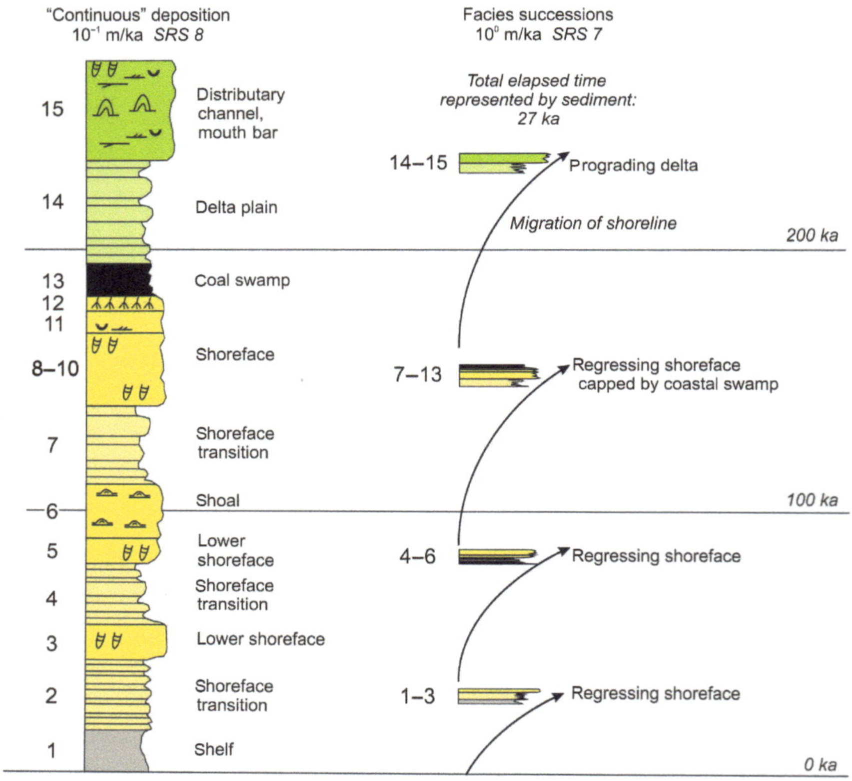

Fig. 8 The Spring Canyon Member dismembered. At *left*, the section is plotted to correspond to an *SRS 8* timescale, suggesting an approximate 273 ka time span for the accumulation of the succession. At *right*, the same section is evaluated in terms of an *SRS 7* timescale, for which sedimentation rates are an order of magnitude more rapid. The section is subdivided into a set of progradational coastal plain and shoreface successions.

longer-term SRS 8 rate in the previous section (and comparable to the 7% of elapsed time represented by sediment that was calculated by de Natris, 2012). How is preservation and nonpreservation distributed through the estimated 273 ka represented by the section?

As suggested in Fig. 8, and following all previous interpretations, the section may be interpreted in terms of four progradational successions or parasequences. Using the same lithofacies unit numbers as in the original section, these are displayed at the right in this figure. Three regressive shoreface successions, the last capped by a coal swamp (unit 13 is the Hiawatha coal), were followed by a progradational delta.

At the SRS 7 scale, the most significant time-related events represented in the Spring Canyon outcrops are the ravinement surfaces that cap the shoreface successions, and include the surfaces that cap each of the Blackhawk constituent members. Ravinement occurs during a rise in relative sea level. Nummedal and Swift (1987), in their original definition of this process, described various examples of ravinement developed during postglacial sea-level rise, a process that during the Holocene, continued for thousands of years. We cannot know the duration of this process during the Cretaceous, but we may speculate. The driving process for episodicity in stratigraphic accommodation seems likely to include flexural subsidence and changes in intraplate stress (Aschoff and Steel, 2011). These are processes that operate over rates and timescales in the SRS 8–9 range, that is, time periods of 10^4–10^6 years. It is conceivable, therefore, and consistent with the interpretations of Howell and Flint (2003) that basinward clinoform progradation and the ravinement and flooding process that preceded each of the progradational shoreface successions could have taken many tens of thousands of years, a significant proportion of the time available for each of these cycles.

The Ferron Sandstone, of Turonian age, near the base of the clastic wedge (Fig. 5), provides additional examples of high-frequency cyclicity. This unit consists of several separate fluvial-deltaic complexes that prograded eastward into the Western Interior Seaway from the Sevier orogen. Famubode and Bhattacharya (2016) summarized a detailed field study of one of these complexes, the Notom delta of south-central Utah. It consists of 1 sequence set, 5 sequences, 18 parasequence sets, and 43 parasequences. Radioisotopic dating of bentonites indicates an age range of 620,000 years, comparable in age range to the individual members of the Blackhawk Formation. Each sequence is estimated to represent a duration in the order of 10^5 years.

Famubode and Bhattacharya (2016) focused on a detailed analysis of the uppermost of the Ferron sequences, a nonmarine succession that detailed mapping demonstrated may in fact be broken down into three sequences, each about 10 m thick, and representing durations of around 30 ka. This indicates an average sedimentation rate of 0.33 m/ka, and corresponds to SRS 8, similar to the members of the Blackhawk Formation. In detail, the sequences are composed of cyclic packages termed facies aggradational cycles (FACs) (Atchley et al., 2004; Cleveland et al., 2007; Prochnow et al., 2006). These are decimeter- to meter-scale units that are bounded either by pedogenically modified surfaces of rooting and burrowing or by surfaces representing erosional truncation. They stack into FAC sets that record periods of gradual channel and floodplain aggradation, indicating increased stability and eventually abandonment of the channel belt. The density of rooting typically increases upward within a FAC set, and sets are typically capped by a widespread paleosoil or a coal, both facies indicating episodes during which clastic sediment supply is limited as a result of sediment bypass or uplift. FACs are estimated to each represent 10^1–10^3 years, accumulated at sedimentation rates estimated from modern analogs at 4–30 m/ka. Pedogenic set boundaries estimated to represent periods of 10^2–10^3 years. FAC sets show estimated sedimentation rates of 10^1–10^2 m/ka, and have durations of 10^3–10^4 years, which puts them in the SRS 6–7 range.

Fig. 9 is a cross section and Wheeler diagram, illustrating the depositional history of the three nonmarine Ferron sequences. FAC sets are identified by the numbers on the right. The first two sequence boundaries (S1C and S1B) are represented by valley incision during relative sea-level fall and valley filling during subsequent rise. The basal boundary of Sequence 1A is represented by a period of nondeposition marked by extensive pedogenic development. The contact between the Ferron Notom and the Blue Gate Shale is a regional unconformity that is likely tectonic. The sequence-scale cyclicity is compared by Famubode and Bhattacharya (2016) to the timescale of the higher-frequency Milankovitch frequencies.

There is an even higher order of repetition, cyclicity, and episodicity in clastic successions, generated by the seasonal, daily, and moment-to-moment variability of traction currents. These are categorized by SRS 1–5 in Table 1. All give rise to discrete depositional units and sedimentary gaps at the appropriate timescales.

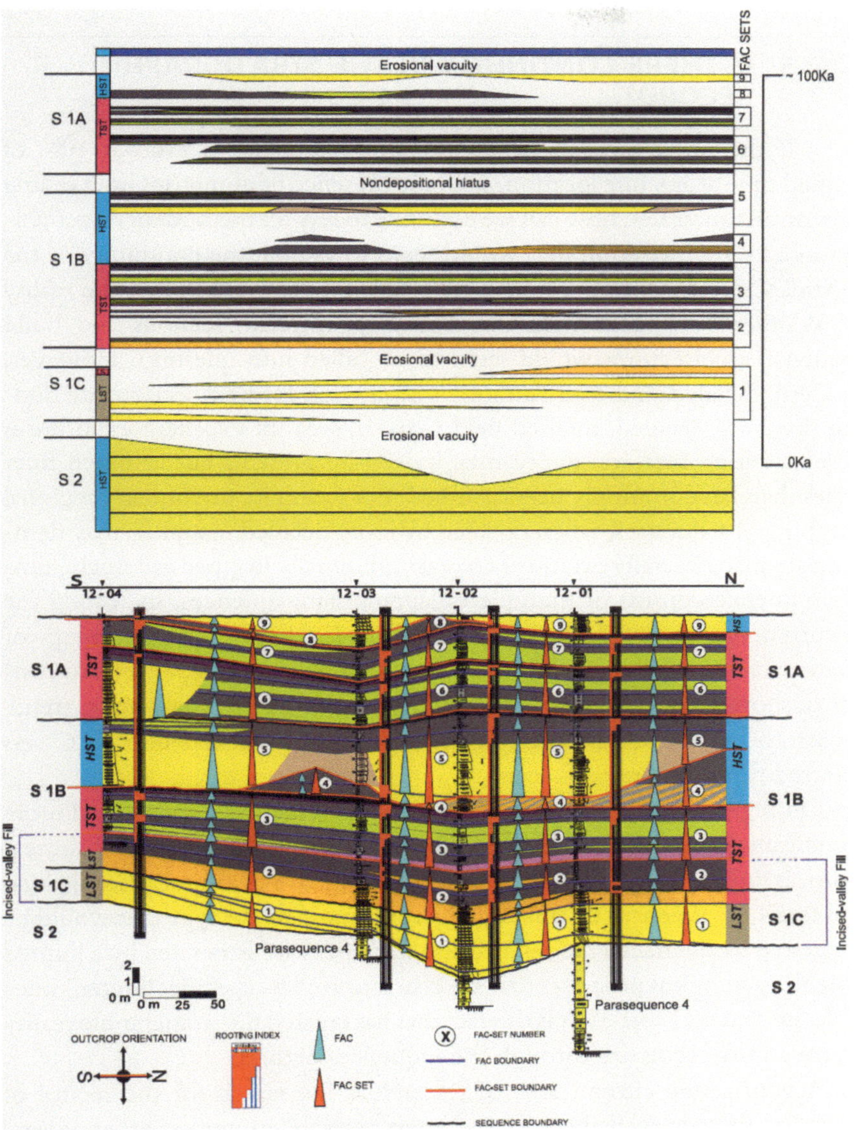

Fig. 9 Stratigraphic cross section (*below*) and Wheeler diagram (*above*) of the nonmarine sequences comprising the top of the Ferron Sandstone Notom delta of south-central Utah. *From Famubode, O.A., Bhattacharya, J.P., 2016. Sequence stratigraphic analysis of the youngest nonmarine sequence in the Cretaceous Ferron Notom Delta, south central Utah, U.S.A. J. Sediment. Res. 86, 168–198.*

3. IS THERE CONTINUITY IN THE STRATIGRAPHIC RECORD?

If the preserved sedimentary record represents as little as 10% of elapsed time at any one location, with the presence of numerous intervening gaps on all timescales, how can we continue to rely on methods of documentation and interpretation that would seem to assume the continuity of the record. Concepts such as fining- and coarsening-upward cycles, the reality of Walther's law, and even the ability to correlate sections and build sequence architectures would seem to be called into question. However, modern, detailed studies, including those cited in the next section, are finding that very detailed, focused field work in areas of excellent exposure is demonstrating that continuity may indeed be present, but at much finer scales than has commonly been assumed. For example, preliminary regional mapping of sequences, when updated by more detailed mapping, may demonstrate that a laterally extensive coastal unit initially mapped as a single, simple sequence consists of multiple offlapping parasequences, indicating the occurrence of repeated high-frequency base-level cycles. Thick fining- or coarsening-upward successions may, on closer examination, be seen to consist of shorter successions that each contains only part of the vertical trend, separated by internal hiatuses, such as the FACs stacked into FAC sets referred to in the previous section.

On regional scales, sequence boundaries represent the longer and more significant of the sedimentary breaks that characterize all stratigraphic successions, as suggested by the Devil's Staircase of Fig. 3. The timescales and physical scales of sequences do not define a continuum of the type that would be suggested by the fractal model, because sequences are generated by a limited range of geological processes that are characterized by specific, limited timescales and rates (Table 3). It is this fact that has created the stratigraphic reality that we now recognize through the sequence method.

A very active current area of research is the search for the record of orbitally forced cyclicity and its interpretation in terms of an astrochronological timescale (Hilgen et al., 2015). The study of cyclostratigraphy requires stratigraphic sections that clearly exhibit facies cyclicity, stable sedimentation rates, and lack sedimentary breaks and out-of-place facies, such as a turbidite bed in the middle of a section of rhythmic pelagic deposits. If these conditions are met, the extraction of the time signal requires the conversion of the "depth domain" of the section to a "time domain" using age calibration points (Strasser et al., 2006, p. 82). This type of analysis should always be treated with considerable caution and, like the other applications described here, requires meticulous measurement and evaluation of the preserved record (see Miall, 2016, Sect. 8.12 for a further elaboration of this point).

4. SOURCE-TO-SINK STUDIES AND THE FULCRUM METHOD

The success of source-to-sink studies of clastic systems, including the fulcrum method of comparing sediment source volumes with the volume of basin accumulations (Holbrook and Wanas, 2014), depends on the ability to reconstruct accurate paleogeographies and to account for sediment delivery paths and timetables. Only the most accurate and detailed stratigraphic synthesis will suffice for such endeavors. Studies now being published (e.g., Bhattacharya et al., 2016; Lin and Bhattacharya, 2017) are based on meticulous field documentation of sequence architecture, down to meter-scale tracing of parasequence boundaries across the nonmarine to marine transition in coastal deposits, in order to link, in as refined a manner as possible, the delivery paths along trunk river systems to the dispersal paths out through distributary or estuarine tidal channels. Calculations of sediment throughput are made based on channel dimensions, grain-size characteristics, and the physics of flow, but questions will typically remain about the significance of the channels identified as the sediment delivery paths. Are these the most important channels? Were there others? How many were operating at the same time? Detailed stratigraphic reconstructions may or may not be able to suggest answers to these questions, which remain as significant unknowns, as indicated by the accumulated potential errors documented by the error "tornado" diagram.

Timing is also a significant part of the reconstruction. What do stratigraphic correlations reveal regarding the sequencing and relative timing of episodes of sediment transfer, sediment bypass, nondeposition, and erosion? As summarized in this paper, modern stratigraphic methods are revealing that significant stretches of elapsed time may be entirely unrepresented by preserved sediment at any given location, whereas the existence of a well-developed coastal paleogeography, reconstructed by, for example, the linking of accumulated sediment to its source area by detailed provenance studies, typically suggests a continuous delivery of sediment to a basin for millions of years. So where does the sediment go at times represented by the sediment gaps? Clearly it is diverted by such processes as lateral delta-lobe switching or long-shore drift. Fine-grained detritus may be widely dispersed into adjacent marine basins. The reconstructions of the stratigraphy required by source-to-sink research must include the exploration of this question.

5. SEQUENCE MODELS

Current sequence models are listed in Table 3. It is anticipated that modifications and refinements in these models will emerge as the modern stratigraphic methods summarized here are increasingly applied to new problems. Two areas of research are touched on here:

Lofer cycles, in the Dachstein Limestone of the Dolomites area in northern Italy, have long been attributed to orbital forcing (Fischer, 1964). However, there has been a continuing controversy regarding the legitimacy of this interpretation, deriving in particular from the age determinations for the unit in which they occur and consequent questions about the validity of interpretations of orbital frequency. A recent examination of this problem by Schwarzacher (2005) focused on the acquisition of a sizeable database of lengthy cycle records and the application of rigorous time series analysis to cycle thicknesses, which appears to lend support to the orbital hypothesis. Hilgen et al. (2015) provided a lengthy discussion of the issues associated with the establishment of an astrochronological timescale. Among the questions to be answered is how orbital frequencies may have varied through geological time. The field context of cyclic successions is critical (facies, bed contacts, possible hiatuses), and it is not enough to build a cyclostratigraphic model solely on a statistical demonstration of cyclicity. This and other points are discussed in more detail in Miall (2016, Chap. 8).

Miall (2014b, Chap. 6) reexamined the accommodation-based models for fluvial sequences that were introduced by Wright and Marriott (1993) and Shanley and McCabe (1994) and argued that modern, accurate data on accumulation rates and avulsion rates in fluvial systems were not consistent with the models. The pattern of dispersed channel bodies that supposedly characterizes "high-accommodation" phases of a base-level cycle (transgressive to highstand systems tracts) requires accommodation rates at least an order of magnitude greater than may be derived from data relating to subsidence histories. As an alternative, it is suggested that variations in channel density (the net-to-gross parameter of reservoir geology) record proximal to distal shifts in facies belts in response to climatic or tectonic controls.

6. IMPLICATIONS FOR FUTURE RESEARCH

Many questions about sequence architecture, on which refined stratigraphic models depend, require exceptional levels of field documentation, including three-dimensional architectural detail down to a submeter scale and unprecedented levels of chronostratigraphic accuracy and precision. These requirements may only be available in areas of excellent exposure, or basins at an advanced stage of subsurface exploration characterized by very dense networks of well records. The construction of three-dimensional Wheeler diagrams should become an essential component of such research. The interpretive effort involved may provide important insights into the nature of depositional processes through time. Modern methods of outcrop mapping, including the use of LiDAR, and drone fly-bys, have become available just in time to meet the needs of the new 21st-century stratigraphy.

REFERENCES

Aschoff, J. L., and Steel, R. J., 2011, Anomalous clastic wedge development during the Sevier-Laramide transition, North American Cordilleran foreland basin, USA: Geol. Soc. Am. Bull., v. 123, p. 1822–1835.

Atchley, S. C., Nordt, L. C., and Dworkin, S. L., 2004, Eustatic control on alluvial sequence stratigraphy: a possible example from the Cretaceous-tertiary transition of the Tornillo Basin, Big Bend National Park, West Texas, U.S.A., J. Sediment. Res., v. 74, p. 391–404.

Bhattacharya, J. P., 2011, Practical problems in the application of the sequence stratigraphic method and key surfaces: integrating observations from ancient fluvial-deltaic wedges with Quaternary and modeling studies: Sedimentology, v. 58, p. 120–169.

Bhattacharya, J. P., Copeland, P., Lawton, T. F., Holbrook, J., 2016, Estimation of source area, river paleo-discharge, paleoslope, and sedimentation budgets of linked deep-time depositional systems and implications for hydrocarbon potential. Earth Sci. Rev., v. 153, p. 77–110.

Catuneanu, O., 2006. Principles of Sequence Stratigraphy. Elsevier, Amsterdam. 375 pp.

Cleveland, D. M., Atchley, S. C., and Nordt, L. C., 2007, Continental sequence stratigraphy of the Upper Triassic (Norian–Rhaetian) Chinle strata, Northern New Mexico, U.S.A.: allocyclic and autocyclic origins of paleosol-bearing alluvial successions: J. Sediment. Res., v. 77, p. 909–924.

de Natris, M.F., 2012. Facies- and time-analysis of the upper part of the Brent Group (mid-upper Jurassic) in the Greater Oseberg area, northern North Sea. M.Sc. ThesisUniversity of Bergen90 pp.

Famubode, O. A., and Bhattacharya, J. P., 2016, Sequence stratigraphic analysis of the youngest nonmarine sequence in the Cretaceous Ferron Notom Delta, south central Utah, U.S.A. J. Sediment. Res., v. 86, p. 168–198.

Fischer, A.G., 1964. The Lofer cyclothems of the Alpine Triassic. In: Merriam, D.F. (Ed.), Symposium on Cyclic Sedimentation. Geological Survey of Kansas, pp. 107–149. Bulletin 169.

Gradstein, F.M., Ogg, J.G., Smith, A.G., 2004. A Geologic Time Scale. Cambridge University Press, Cambridge, 610 pp.
Hampson, G. J., 2010, Sediment dispersal and quantitative stratigraphic architecture across and ancient shelf: Sedimentology, v. 57, p. 96–141.
Hampson, G. J., Jewell, T. O., Irfan, N., Gani, M. R. and Bracken, B., 2013, Modest changes in fluvial style with varying accommodation in regressive alluvial-to-coastal-plain wedge: upper Cretaceous Blackhawk Formation, Wasatch Plateau, central Utah, U.S.A.: J. Sediment. Res., v. 83, p. 145–169.
Heckel, P.H., 1986. Sea-level curve for Pennsylvanian eustatic marine transgressive-regressive depositional cycles along midcontinent outcrop belt. 14, Geology, North America, pp. 330–334.
Hilgen, F.J., Hinnov, L.A., Aziz, H.A., Abels, H.A., Batenburg, S., Bosmans, J.H.C., de Boer, B., Hüsings, S.K., Kuiper, K.F., Lourens, L.J., 2015. Stratigraphic continuity and fragmentary sedimentation: the success of cyclostratigraphy as part of integrated stratigraphy. In: Smith, D.G., Bailey, R.J., Burgess, P., Fraser, A. (Eds.), Strataand Time, Geological Society, London, pp. 157–197. Special Publication 404.
Holbrook, J., and Wanas, H., 2014, A fulcrum approach to assessing source-to-sink mass balance using channel paleohydrologic parameters derivable from common fluvial data sets with an example from the Cretaceous of Egypt: J. Sediment. Res., v. 84, p. 349–372.
Holbrook, J., Scott, R. W., and Oboh-Ikuenobe, F. E., 2006, Base-level buffers and buttresses: a model for upstream versus downstream control on fluvial geometry and architecture within sequences: J. Sediment. Res., v. 76, p. 162–174.
Howell, J.A., Flint, S.S., 2003. Siliciclastics case study: the Book Cliffs. In: Coe, A. (Ed.), The Sedimentary Record of Sea-level Change. Open University and Cambridge University Press, Cambridge, UK, pp. 135–208.
Howell, J., Hodgetts, D., Burgess, P., Flint, S., 1999. In: The Blackhawk Formation of eastern Utah: application of a high-quality outcrop dataset to calibrate modeling studies. American Association of Petroleum Geologists, International Conference and Exhibition, Birmingham, Search and Discovery Article #90923.
Krystinik, L.F., DeJarnett, B.B., 1995. Lateral variability of sequence stratigraphic framework in the Campanian and Lower Maastrichtian of the Western Interior Seaway. In: Van Wagoner, J.C., Bertram, G.T. (Eds.), SequenceStratigraphy of Foreland Basins, American Association of Petroleum Geologists, pp. 11–25. Memoir 64.
Lin, W., Bhattacharya, J.P., 2017. Estimation of source-to-sink mass balance by a fulcrum approach using channel paleohydrologic parameters of the Cretaceous Dunvegan Formation, Canada. J. Sediment. Res. 87, 97–116.
Miall, A.D., 1996. The Geology of Fluvial Deposits: Sedimentary Facies, Basin Analysis and Petroleum Geology. Springer-Verlag, Inc., Berlin, p. 582.
Miall, A.D., 2010. The Geology of Stratigraphic Sequences, Second ed. Springer-Verlag, Berlin. 522 pp.
Miall, A. D., 2014a, The emptiness of the stratigraphic record: a preliminary evaluation of missing time in the Mesaverde Group, Book Cliffs, Utah: J. Sediment. Res., v. 84, p. 457–469.
Miall, A.D., 2014b. Fluvial Depositional Systems. Springer-Verlag, Berlin. 316 pp.
Miall, A.D., 2015. Updating uniformitarianism: stratigraphy as just a set of "frozen accidents". In: Smith, D.G., Bailey, R.J., Burgess, P., Fraser, A. (Eds.), Strata and Time. Geological Society, London, pp. 11–36. Special Publication 404.
Miall, A.D., 2016. Stratigraphy: A Modern Synthesis. Springer-Verlag, Berlin. 454 pp.
Middleton, G.V., Plotnick, R.E., Rubin, D.M., 1995. Nonlinear Dynamics and Fractals; New Numerical Techniques for Sedimentary Data Sets. Society for Sedimentary Geology, Tulsa, OK. Short Course No. 36. 174 pp.

Moscardelli, L., Wood, L. J., and Dunlap, D. B., 2012, Shelf-edge deltas along structurally complex margins: a case study from eastern offshore Trinidad: Am. Assoc. Pet. Geol. Bull., v. 96, p. 1483–1522.

Nummedal, D., Swift, D.J.P., 1987. Transgressive stratigraphy at sequence-bounding unconformities: some principles derived from Holocene and Cretaceous examples. In: Nummedal, D., Pilkey, O.H., Howard, J.D. (Eds.), Sea-Level Fluctuation and Coastal Evolution. Society of Economic Paleontologists and Mineralogists, pp. 241–260. Special Publication 41.

Plotnick, R. E., 1986, A fractal model for the distribution of stratigraphic hiatuses:J. Geol., v. 94, p. 885–890.

Prochnow, S. J., Atchley, S. C., Boucher, T. E., Nordt, L. C., and Hudec, M. R., 2006, The influence of salt withdrawal subsidence on palaeosol maturity and cyclic fluvial deposition in the Upper Cretaceous Chinle Formation, Castel Valley, Utah: Sedimentology, v. 53, p. 1319–1345.

Ramsbottom, W.H.C., 1979. Rates of transgression and regression in the Carboniferous of NW Europe. J. Geol. Soc. Lond. 136, 147–153.

Sadler, P. M., 1981, Sedimentation rates and the completeness of stratigraphic sections:J. Geol., v. 89, p. 569–584.

Sadler, P. M., 1999, The influence of hiatuses on sediment accumulation rates: GeoResearch Forum, v. 5, p. 15–40.

Sadler, P.M., Jerolmack, D., 2015. Scaling laws for aggradation, denudation and progradation rates: the case for time-scale invariance at sediment sources and sinks. In: Smith, D.G., Bailey, R.J., Burgess, P., Fraser, A. (Eds.), Strata and Time. Geological Society, London, pp. 69–88. Special Publication 404.

Schwarzacher, W., 2005, The stratification and cyclicity of the Dachstein Limestone in Lofer, Leogang and Steinernes Meer (Northern Calcareous Alps, Austria): Sediment. Geol., v. 181, p. 93–106.

Shanley, K. W., and McCabe, P. J., 1994, Perspectives on the sequence stratigraphy of continental strata: Am. Assoc. Pet. Geol. Bull., v. 78, p. 544–568.

Sheets, B. A., Hickson, T. A., and Paola, C., 2002, Assembling the stratigraphic record: depositional patterns and time-scale in an experimental alluvial basin: Basin Res., v. 14, p. 287–301.

Sloss, L. L., 1963, Sequences in the cratonic interior of North America: Geol. Soc. Am. Bull., v. 74, p. 93–113.

Smith, D.G., 1994. Cyclicity or chaos? In: de Boer, P.L., Smith, D.G. (Eds.), Orbital Forcing and Cyclic Sequences. International Association of Sedimentologists, pp. 531–544. Special Publication 19.

Strasser, A., Hilgen, F. J., and Heckel, P. H., 2006, Cyclostratigraphy—Concepts, definitions, and applications: Newsl. Stratigr., v. 42, p. 75–114.

Vail, P.R., Mitchum Jr., R.M., Todd, R.G., Widmier, J.M., Thompson III, S., Sangree, J.B., Bubb, J.N., Hatlelid, W.G., 1977. Seismic stratigraphy and global changes of sea-level. In: Payton, C.E. (Ed.), Seismic Stratigraphy—Applications to Hydrocarbon Exploration. American Association of Petroleum Geologists, pp. 49–212. Memoir 26.

Wanless, H.R., Weller, J.M., 1932. Correlation and extent of Pennsylvanian cyclothems. Geol. Soc. Am. Bull. 43, 1003–1016.

Wheeler, H. E., 1958, Time-stratigraphy: Am. Assoc. Pet. Geol. Bull., v. 42, p. 1047–1063.

Wheeler, H. E., 1959, Stratigraphic units in time and space: Am. J. Sci., v. 257, p. 692–706.

Wright, V. P., and Marriott, S. B., 1993, The sequence stratigraphy of fluvial depositional systems: the role of floodplain sediment storage: Sediment. Geol., v. 86, p. 203–210.

Young, R. G., 1955, Sedimentary faces and intertonguing in the Upper Cretaceous of the Book Cliffs, Utah-Colorado: Geol. Soc. Am. Bull., v. 66, p. 177–202.

CHAPTER THREE

Two Approaches to Sequence Stratigraphy

Ashton F. Embry*[,1] and Erik P. Johannessen[†]
*Embry Corp, Calgary, AB, Canada
[†]EP Skolithos, Stavanger, Norway
[1]Corresponding author: e-mail address: ashtonfoxembry@gmail.com

Contents

Abstract

Sequence stratigraphy is a stratigraphic discipline in which the defined stratigraphic surfaces represent either breaks in deposition or changes in depositional trend. Two approaches for defining sequence stratigraphic surfaces and units have evolved, with one being inductive and the other deductive.

The empirical, inductive approach defines surfaces based on observable, physical characteristics. Five surfaces, subaerial unconformity (SU), unconformable shoreline ravinement (SR-U), slope onlap surface (SOS), maximum regressive surface (MRS), and maximum flooding surface (MFS), are used to define inductive sequence units and to construct a correlation framework. Two different types of sequences (depositional T-R, genetic stratigraphic) and two systems tracts (transgressive systems tract, regressive systems tract) are employed in the inductive approach.

In contrast, the deductive approach defines sequence stratigraphic surfaces in terms of theoretical events on a base-level curve (e.g., start base-level fall). Three

Stratigraphy & Timescales, Volume 2
ISSN 2468-5178
http://dx.doi.org/10.1016/bs.sats.2017.08.001

deductive surfaces (SU, MRS, and MFS) used for unit definition and correlation are the same as those employed by the inductive approach. Two surfaces, not recognized in the inductive approach, are also used to delineate units and for correlation. Notably, both are chronostratigraphic surfaces which display no diagnostic physical characteristics. These surfaces include the basal surface of forced regression (BSFR), the depositional surface at the start of base-level fall, and the correlative conformity (CC), the depositional surface at the start of base-level rise. A sequence is defined as a succession of strata deposited during a full base-level cycle and associated change in accommodation or sediment supply. Recognition of the five surfaces has allowed the definition of four types of sequences (depositional sequence type 1 and 2, T-R, genetic stratigraphic), with each type being divided into four systems tracts (lowstand, transgressive, highstand, and falling stage).

The decision as to what approach to use—inductive or deductive—depends on the robustness of the available data and whether the BSFR and CC can be recognized with reasonable objectivity and consistency over the study area. Caution must be exercised to avoid the pitfall of trying to force fit data into the deductive approach by interpreting an inappropriate surface (e.g., facies change) as either the BSFR or CC. In many situations, especially when seismic data are not available, it is not possible to recognize and correlate the BSFR and/or the CC with objectivity. In these cases, the inductive approach is required for unit delineation and correlation.

1. INTRODUCTION

Sequence stratigraphy has become a popular methodology for correlating sedimentary strata and constructing a quasi-chronostratigraphic framework for a basin. Such a framework, in conjunction with the sedimentology of the strata, is fundamental for interpreting paleogeography and depositional history in terms of changes of base level, shoreline movement, and/or sediment supply.

As significant as sequence stratigraphy has become over the past 30 years, controversies regarding sequence stratigraphic terminology and methodology exist (Burgess and Prince, 2015; Catuneanu et al., 2009, 2011; Embry, 1995, 2002, 2009; Embry and Johannessen, 1993; Madof et al., 2016). Such controversies seem in part to be rooted in stratigraphic classification problems addressed by Hedberg (1958) who stressed the need for precise definitions of stratigraphic terms so as "to allow ready communication and clear understanding." He elaborated, "Confusion results from ill-defined terms which are used with a certain meaning by one man and with a different meaning by another." Hedberg (1958) also highlighted the need to avoid "the mixing of the objective and the subjective in our terminology" and lamented "the millions of hours wasted in stratigraphic controversy" merely as a result of such a mixture.

We have used sequence stratigraphy in our basin analysis studies since 1977, when the modern era of sequence stratigraphy began with the watershed publication of the Exxon definitions and methodology (Payton, 1977). By 1990, it was apparent to us that two different approaches to sequence stratigraphic methodology and classification were in use and that the existence of two different approaches might well be contributing to methodological and communication problems associated with sequence stratigraphy. Miall and Miall (2004) drew attention to the two approaches which they termed inductive and deductive. Regarding sequence stratigraphy, Miall (2004) commented that, "two distinct intellectual approaches resulted in the development of two conflicting and competing paradigms which are currently vying for the attention of practicing earth scientists."

The inductive approach to sequence stratigraphy requires a methodology and classification system similar to the other empirical and material-based, stratigraphic disciplines (Embry, 2002, 2010; Embry and Johannessen, 1993). The deductive approach, conceptual and somewhat model-driven (Miall, 2004; Miall and Miall, 2004), has dominated sequence stratigraphic methodology and terminology for the past 30 years (Catuneanu, 2006; Catuneanu et al., 2009, 2011; Helland-Hansen and Gjelberg, 1994; Hunt and Tucker, 1992; Jervey, 1988; Posamentier and Allen, 1999; Van Wagoner et al., 1990). The deductive approach to sequence stratigraphy differs from other stratigraphic disciplines as expressed by Helland-Hansen (2009): "Sequence stratigraphy is and will necessarily remain interpretive in its philosophical essence, far more so than these other methodological branches of stratigraphy."

As demonstrated below, the inductive approach to sequence stratigraphy, which is empirical and material-based, is readily applied using a wide variety of data sets. This contrasts with the deductive approach, which generally relies on very specialized data in order to be applied in a scientific and reliable manner. It is critical that those engaged in sequence stratigraphic investigations are aware of the existence of the two distinctly different approaches to the discipline, and of the problems that might arise if one does not apply that approach most appropriate to their study. This contribution describes both sequence stratigraphic methodologies emphasizing the strengths and weaknesses of each so as to allow the practitioner to make an informed choice as to which approach is best suited to his or her work.

2. INDUCTIVE APPROACH TO SEQUENCE STRATIGRAPHY

2.1 Sequence Stratigraphy as an Inductive Stratigraphic Discipline

Wikipedia defines stratigraphy as "a branch of geology which studies rock layers (strata) and layering (stratification)." We would add that such layers must obey Steno's Law of Superposition (younger strata overlie older strata). Stratigraphic analysis includes the description and interpretation of physical, biological, and chemical properties of strata and the recognition of a variety of stratigraphic surfaces and units on the basis of vertical changes of the aforementioned properties (Hedberg, 1959).

Inductive stratigraphy seeks to define and delineate surfaces on the basis of observable, physical features. Each inductive, stratigraphic discipline concentrates on vertical variations of a specific property of strata. Inductively defined stratigraphic surfaces serve as surfaces of correlation and define the boundaries of stratigraphic units of the discipline. Established, inductive, stratigraphic disciplines and their characteristic, observable properties include lithostratigraphy (changes in lithology), biostratigraphy (changes in fossil content), magnetostratigraphy (changes in magnetic properties such as polarity), and chemostratigraphy (changes in chemical properties such as isotope ratios).

Two types of changes in sedimentation leave a record in the form of sequence stratigraphic surfaces—breaks in deposition (unconformities, diastems) and changes in depositional trend, including fining-upward or coarsening-upward trends. Given the above, we define sequence stratigraphy as the recognition and correlation of stratigraphic surfaces which represent depositional breaks and changes in sedimentation trend in the rock record. Such changes are recognized by sedimentological criteria and geometric relationships.

Finally, it is useful, but not essential, to have a solid theoretical foundation linking the generation of the various surfaces of a given stratigraphic discipline to phenomena that occur on our planet. For example, biostratigraphic surfaces represent changes in fossil content that are due mainly to a combination of evolution and shifting environments of deposition. It is noteworthy that biostratigraphy flourished long before the theory of evolution was developed. Similarly, most sequence stratigraphic surfaces were recognized in the rock record and used for correlation before a model was developed to explain their existence. It is now generally accepted that surfaces included in sequence stratigraphy are generated by the interaction of changing rates of sedimentation and base-level movement (Barrell, 1917; Burgess and Prince, 2015; Catuneanu, 2006; Embry, 2002; Jervey, 1988, 1993; Wheeler, 1958, 1959, 1964).

2.2 Inductive Sequence Stratigraphic Surfaces

It is essential that inductive sequence stratigraphic surfaces are material-based and guided by objectivity and reproducibility. It is also important that such surfaces (1) can be recognized in diverse stratigraphic successions, from undisturbed basin fill to tectonically disrupted strata with only fragmentary continuity, and (2) can be potentially delineated in outcrop sections, mechanical well logs supported by analysis of chip samples and discontinuous core, seismic data, and any combination of these data types.

Seven inductive surfaces of sequence stratigraphy have been described in the literature and are most commonly referred to as: (1) subaerial unconformity (SU), (2) unconformable shoreline ravinement (SR-U), (3) diastemic shoreline ravinement (SR-D), (4) maximum regressive surface (MRS), (5) maximum flooding surface (MFS), (6) slope onlap surface (SOS), and (7) regressive surface of marine erosion (RSME). Each of these surfaces is characterized by a combination of observable attributes that allow it to be distinguished from other surfaces. Reliable recognition of inductive surfaces requires adequate and reliable sedimentological information from outcrop and/or the subsurface as well as geometric relationships from outcrop exposures, correlated cross sections, and/or seismic data. Finally, it must be emphasized that although the surfaces tend to have names which have a genetic overtone (e.g., MRS), their definitions and criteria of recognition in the inductive approach are devoid of any genetic relationships, that is, they are objective.

Embry (2009, 2010) discussed the inductive surfaces, including their various sedimentological and stratigraphic attributes (Fig. 1), their inferred relationships with base-level changes (Figs. 2–4), their relationships with each other (Figs. 3 and 4), their significance as indicators of time (Fig. 1), and their value to regional correlation. Five of the seven surfaces (SU, SR-U, SOS, MRS, MFS) are, by virtue of their low diachroneity, especially useful for establishing a chronostratigraphic relationships and as bounding surfaces of sequence stratigraphic units. The highly diachronous SR-D and RSME surfaces are useful for facies analysis within an established sequence stratigraphic framework (Embry, 2010).

Surface	Contact	Facies Below	Facies Above	Stratal Terminations	Relationship to Time
Subaerial Unconformity	Scoured to weathered	Highly variable	Nonmarine	Major, regional truncation below and onlap above	Approximate time barrier
Regressive Surface of Marine Erosion	Scoured	Coarsening upward offshore marine shelf	Coarsening and Shallowing upward shoreface	Minor, local truncation below and downlap above	Highly diachronous
Unconformable Shoreline Ravinement	Scoured	Variable. Most commonly C.U. marine shelf	Fining and deepening upward shallow marine shelf	Major, regional truncation below and onlap above	Time barrier
Diastemic Shoreline Ravinement	Scoured	Nonmarine	Fining and deepening upward marine shelf	Minor, local truncation below and onlap above	Highly diachronous
Maximum Regressive Surface	Conformable to scoured	Marine—Coarsening and usually shallowing upward Nonmarine—increasing sandstone/shale ratio	Marine—Fining and often deepening upward Nonmarine—decreasing sandstone/shale ratio	Usually conformable. Rare truncation	Low diachroneity
Maximum Flooding Surface	Conformable to scoured	Marine— fining and often deepening upward Nonmarine—decreasing sandstone/shale ratio	Marine—coarsening and often shallowing upward Nonmarine—increasing sandstone/shale ratio	Truncation or conformable below and downlap above	Low diachroneity
Slope Onlap Surface	Scoured	Marine Slope Deposits. Often coarsening - upward	Marine Slope to Basin Deposits. Often coarsening - upward	Truncation to conformable below and onlap above	Time barrier

Fig. 1 A summary of the characteristics of the surfaces of sequence stratigraphy recognized by the inductive approach. *Modified from Figure 5 of Embry, A.F., 2010. Correlating siliciclastic successions with sequence stratigraphy. In: Ratcliffe, K., Zaitlin, B. (Eds.), Application of Modern Stratigraphic Techniques: Theory and Case Histories, SEPM Special Publication 94. pp. 35–53.*

2.3 Inductive Sequence Stratigraphic Units

Sequence stratigraphy relies upon three types of units—sequence, systems tract, and parasequence. However, the inductive approach to sequence stratigraphy utilizes only sequences and systems tracts with the parasequence being classified as a lithostratigraphic unit on the basis of its accepted definition (see discussion below). Inductive sequences and

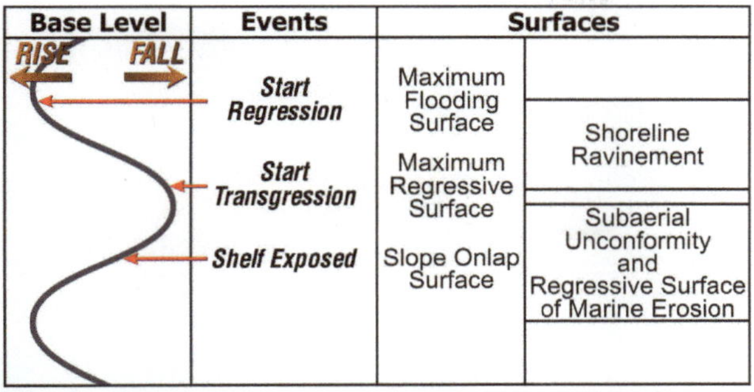

Fig. 2 Base-level change model for the generation of the inductive surfaces of sequence stratigraphy. Each surface is generated during a specific time interval of a base-level cycle as a consequence of the interaction of rates of change of accommodation space and sedimentation. *Modified from Figure 1 of Embry, A., 2002. Transgressive-regressive (T-R) sequence stratigraphy. In: Armentrout, J., Rosen, N. (Eds.), Sequence stratigraphic models for exploration and production. Gulf Coast SEPM Conference Proceedings, Houston, pp. 151–172.*

systems tracts are primarily defined by the inductive, sequence stratigraphic surfaces which bound them.

2.3.1 Sequence

The primary unit of sequence stratigraphy is the sequence, initially defined in an empirical manner by Sloss et al. (1949) as a stratigraphic unit bound by large-scale, regional unconformities. Wheeler (1958) retained this overall definition but extended the term sequence to units bounded by smaller-scale unconformities. Although a particular type of bounding unconformity was not specified by either Sloss et al. (1949) or Wheeler (1958), applications of this concept in the 1950s and 1960s used either SUs or unconformable shoreline ravinements as sequence bounding surfaces (e.g., Sloss, 1963; Wheeler, 1958). These types of unconformities are mainly confined to the flanks of a basin, limiting their correlation over much of the central portions of a basin and therefore their practical application to basin analysis.

In 1977, Exxon researchers demonstrated in a series of articles (Payton, 1977) that the seismic reflectors encompassing basin flank unconformities could be traced basinward into submarine unconformities and conformable surfaces. Thus, the definition of a sequence was modified to include a unit "bounded by unconformities or their correlative conformities" (Mitchum et al., 1977). Accordingly, a sequence boundary was recognized as a combination of surfaces rather than just an unconformity and, importantly, it permitted the correlation of sequence boundaries and their contained

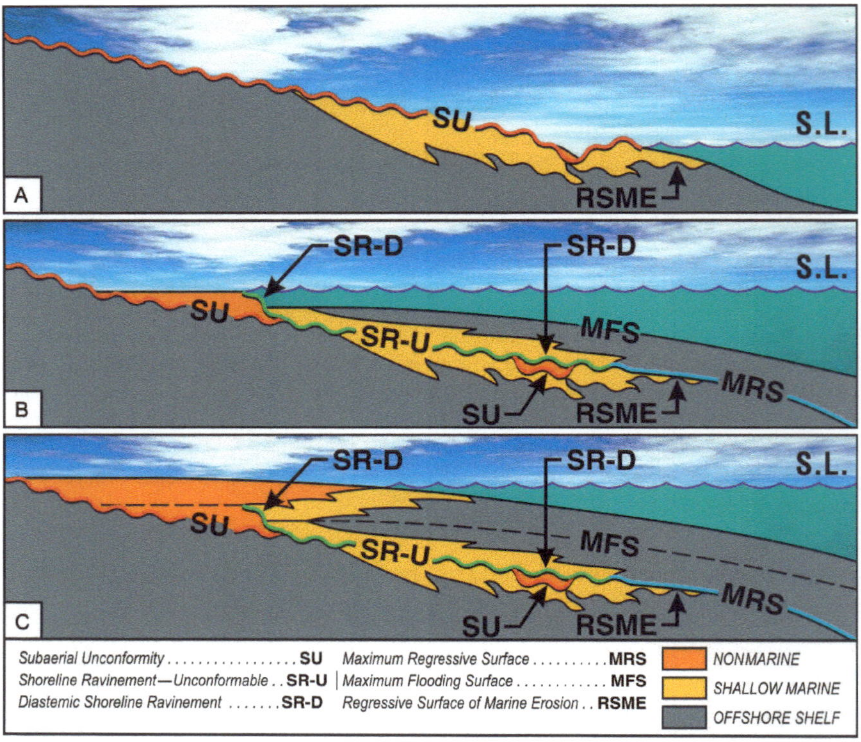

Fig. 3 Schematic evolution of the inductive sequence stratigraphic surfaces in a ramp setting. (A) By the end of base-level fall, the subaerial unconformity (SU) reaches its maximum extent and a regressive surface of marine erosion (RSME), a highly diachronous surface at the base of shoreface deposits, has migrated basinward. (B) An unconformable shoreline ravinement (SR-U) begins to migrate landward as base level starts to rise, eroding portions of the SU. Finer sediment is deposited at any given marine locality and a maximum regressive surface (MRS) is generated in deeper water. A maximum flooding surface (MFS) develops near the time of maximum transgression. (C) In the late phase of base-level rise a progradational, coarsening-upward succession begins to accumulate on the shelf. *Modified from Figure 4, Embry, A.F., 2010. Correlating siliciclastic successions with sequence stratigraphy. In: Ratcliffe, K., Zaitlin, B. (Eds.), Application of Modern Stratigraphic Techniques: Theory and Case Histories, SEPM Special Publication 94. pp. 35–53.*

sequences across an entire basin. This greatly expanded the utility of sequences for basin analysis.

Two types of sequences were defined in the late 1980s. Exxon scientists (Van Wagoner et al., 1988) defined a depositional sequence as a unit bound by SUs or their correlative conformities. Contemporaneously, Galloway (1989) proposed a sequence bounded by MFSs which he termed a genetic

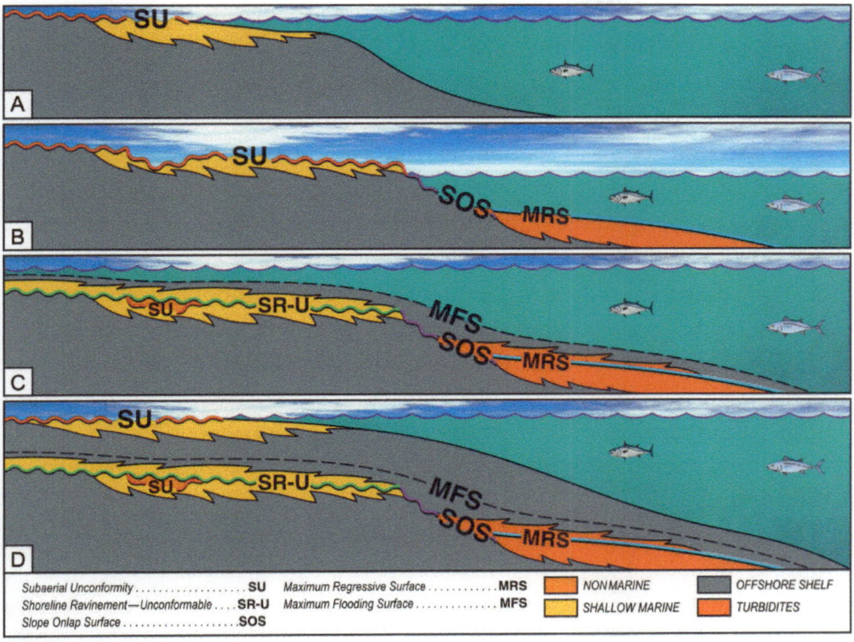

Fig. 4 Schematic evolution of inductive sequence stratigraphic surfaces associated with a shelf/slope/basin setting. (A) Falling base level is accompanied by the basinward migration of the subaerial unconformity (SU). (B) The late stage of falling base level, when much of the shelf is exposed, is marked by the generation of a slope onlap surface (SOS) and deposition of turbidites in the basin. The initiation of rising base level is defined by a maximum regressive surface (MRS) within the basinal turbidite deposits. (C) During transgression, an unconformable shoreline ravinement (SR-U) removes most of the SU. Finer-grained sediment accumulates in the basin as base level rises. The stratigraphic position of the finest sediment defines the maximum flooding surface (MFS). (D) The reversal of rising base level results in the accumulation of coarser-grained sediment only to be eroded as the SU progrades basinward. *Modified from Figure 7.4, Embry, A.F., 2009. Practical Sequence Stratigraphy. Canadian Society of Petroleum Geologists, Calgary, 76 pp.*

stratigraphic sequence. The genetic stratigraphic sequence is considered to be compatible with the general definition of a sequence because a portion of the MFS may be unconformable due mainly to sediment starvation and/or minor scouring. However, the MFS is clearly much different from the depositional sequence boundary of Van Wagoner et al. (1988). Finally, a T-R sequence was defined (Embry, 1993; Embry and Johannessen, 1993) as a sequence bounded by an SU and/or an SR-U on the basin margins and an MRS farther basinward. Because a T-R sequence fits the definition of the earlier defined, depositional sequence, we apply the name depositional T-R sequence to emphasize this equivalence. Notably, this is the only type of depositional sequence so far defined in the literature bounded solely by inductive surfaces.

Given various types of sequences have been defined, a suitable, generic definition of a sequence is required. Accordingly, Embry (2009) defined a sequence as "a stratigraphic unit bounded by a specific type of unconformity or its correlative surfaces." A correlative surface is a sequence stratigraphic surface that passes laterally into a sequence-bounding unconformity, thereby forming a continuous sequence boundary. Consistent with this definition, a specific type of sequence can be defined and named on the basis of different types of bounding unconformities and/or their correlative surfaces. For example, the defining unconformity for a depositional sequence is an SU and that for a genetic stratigraphic sequence is the unconformable portion of the MFS.

The inductive, sequence stratigraphic approach requires that all bounding surfaces of any type of sequence must be material-based surfaces. The fact that the MFS is a relatively easily recognized, inductive surface makes the genetic stratigraphic sequence an acceptable inductive sequence. Indeed, there are many published examples of the use of genetic stratigraphic sequences in basin analysis (e.g., Combellas-Bigott and Galloway, 2006; Partington et al., 1993; Xue and Galloway, 1993).

The defining SU of a depositional T-R sequence is material-based as are its correlative surfaces. Fig. 5 illustrates inductive depositional T-R sequences and their bounding surfaces in both ramp (Fig. 5A) and shelf–slope–basin (Fig. 5B) settings. The relationships of the surfaces to each other are based on our stratigraphic studies over the past 40 years (Embry, 1993, 2010, 2011; Embry and Johannessen, 1993; Hadler-Jacobsen et al., 2005; Johannessen and Embry, 1989; Johannessen and Steel, 2005; Johannessen et al., 1995, 2011; Mjos et al., 1998) and on a published (Embry, 2009, 2010), base-level/sedimentation interaction model for surface generation (Figs. 3 and 4). A depositional T-R sequence in a ramp setting is bounded by an SU and an SR-U on the flanks of a basin. The boundary passes laterally basinward into an MRS which joins with the seaward termination of the shoreline ravinement. A depositional T-R sequence boundary in a shelf–slope-basin setting (Fig. 5B) is bounded by a combination of SU and unconformable shoreline ravinement (SU/SR-U) over part, or all, of the shelf. Where the SU/SR-U does not come close to the edge of the shelf (Fig. 5B, lower boundary), the boundary is most often extended from the

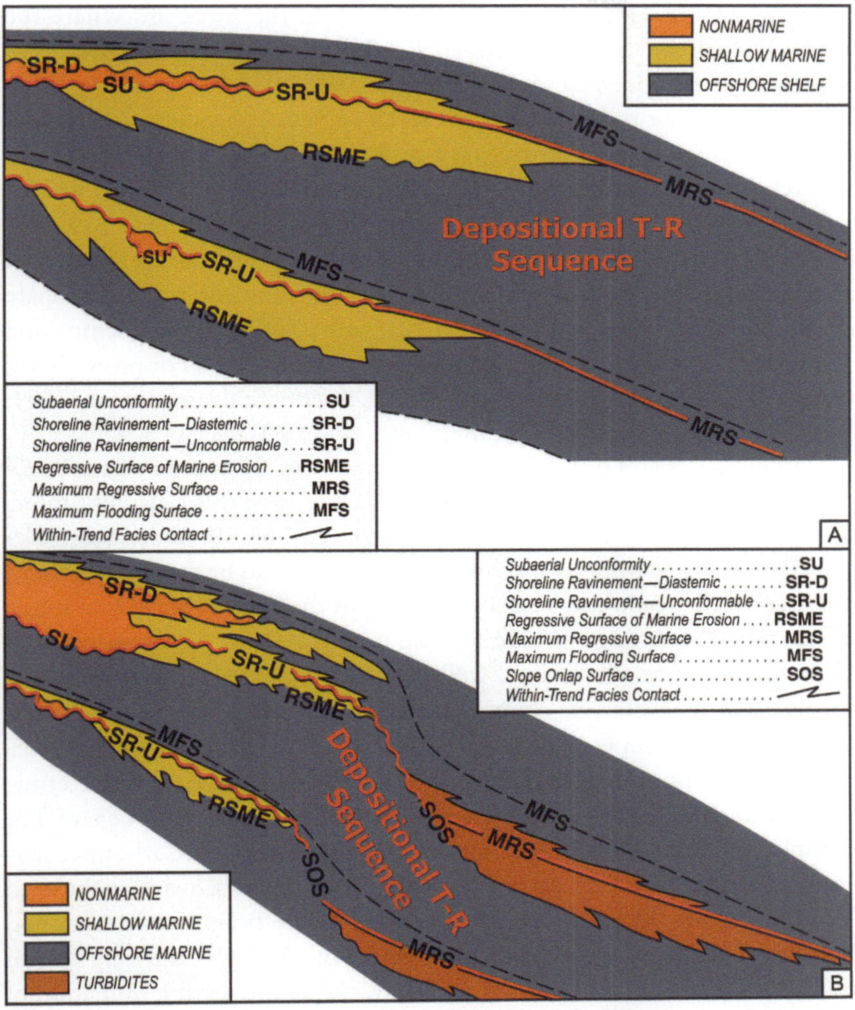

Fig. 5 Depositional T-R sequences. (A) Boundaries of a depositional T-R sequence in a ramp setting are shown in *red*. The unconformable shoreline ravinement (SR-U) truncates the basinward portion of the subaerial unconformity (SU) and is a correlative surface of the SU. The SR-U passes basinward into the maximum regressive surface (MRS). Thus, a continuous depositional T-R sequence boundary in this setting consists of an SU, SR-U, and MRS. (B) Boundaries of a depositional T-R sequence in a shelf/slope/basin setting are shown in *red*. The correlative surfaces of the SU include SR-U, the slope onlap surface (SOS), and the MRS. *Modified from Figure 2, Embry, A.F., 2010. Correlating siliciclastic successions with sequence stratigraphy. In: Ratcliffe, K., Zaitlin, B. (Eds.), Application of Modern Stratigraphic Techniques: Theory and Case Histories, SEPM Special Publication 94. pp. 35–53.*

outer shelf, slope, and basin along the MRS. In the situations where the SU/SR-U can be traced to the shelf edge, the depositional sequence boundary can be traced basinward along an SOS, an unconformable, often channeled, surface onlapped by the MRS in deeper regions of the basin (Fig. 5B). Given that the correlative surfaces of an SU are an SR-U and an MRS (Fig. 5), a unit bounded on top and bottom by an SR-U and/or an MRS without a presence of an SU would also be classified as a depositional T-R sequence. It is noteworthy that many published examples of material-based, depositional T-R sequences in siliciclastic and carbonate successions can be found in the literature (e.g., Beauchamp and Henderson, 1994; Brigaud et al., 2014; Embry, 1993; Hampson, 2016; Lash and Engelder, 2011; Sansom, 2010; Sonnenfeld and Cross, 1993; Wendte and Uyeno, 2005; Zimmermann et al., 2015).

Catuneanu et al. (2009, 2011) have noted that, from a theoretical perspective (e.g., Jervey, 1988), the shoreline ravinement (SR) may not intersect and erode part or all of the SU. Consequently, an SR-U would not be generated and only a highly diachronous, SR-D would be present. In such a case, the MRS would not be continuous with the SU (by way of an intervening SR-U) forming a through-going boundary (Fig. 6). Our review of the literature has revealed a few, well-documented cases of this relationship and all occur in siliciclastic strata characterized by very high rates of deposition (e.g., Pellegrini et al., 2017). In such cases, the inductive, depositional T-R sequence boundary cannot be extended beyond the basinward termination of the SU (Fig. 6). However, such an occurrence seems to be rare. For example, all sequence boundaries in the 13-km-thick, Carboniferous–Cretaceous succession of the Sverdrup Basin of Arctic Canada include an SR-U as part of the unconformable portion of the boundary (Embry and Beauchamp, 2008). However, in the absence of an SR-U, the only option for extending the sequence boundary basinward of the distal termination of the SU is to use the deductive approach as is discussed in a subsequent section.

2.3.2 Systems Tracts

A sequence can be subdivided into component units on the basis of sequence stratigraphic surfaces within a sequence, thereby enhancing the resolution of the sequence stratigraphic framework. The component unit of the sequence is the systems tract (Posamentier and Vail, 1988; Van Wagoner et al., 1988), a unit originally defined by Brown and Fisher (1977), as "a linkage of contemporaneous depositional systems." This definition is not adequate because

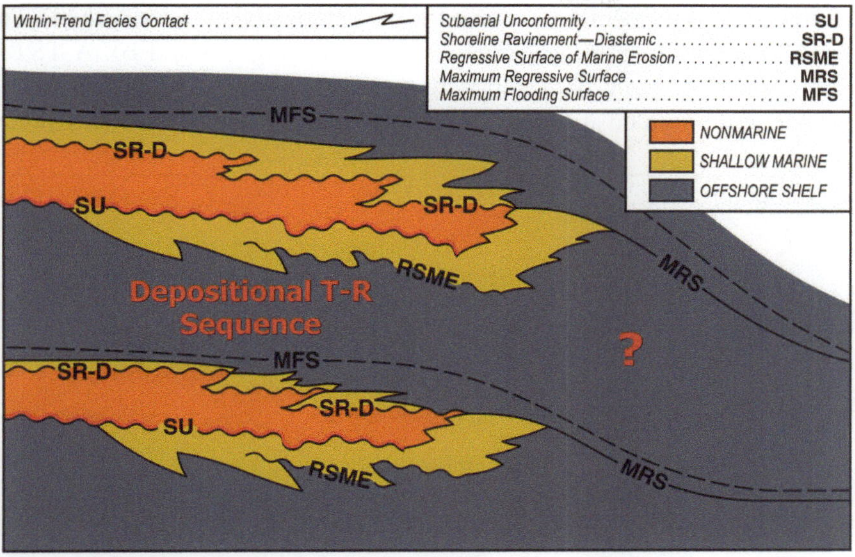

Fig. 6 Depositional T-R sequence with sequence boundaries highlighted in *red*. An unconformable shoreline ravinement (SR-U) has not formed in this model (only a diastemic shoreline ravinement (SR-D)). Sequence boundaries consist only of subaerial unconformities (SU) restricted to the basin flanks. There are no inductive, correlative surfaces, which would allow the sequence boundaries to be extended into the basin. As discussed in the text, such stratigraphic relationships are very rare but have been documented in a few cases. *Modified from Figure 9.7, Embry, A.F., 2009. Practical Sequence Stratigraphy. Canadian Society of Petroleum Geologists, Calgary, 76 pp.*

it does not identify the bounding surfaces of the systems tract. A simpler and more straightforward definition, and one which matches the application of systems tracts, is a component unit of a sequence which is bound by sequence stratigraphic surfaces.

In the inductive approach, the depositional T-R and genetic stratigraphic sequences can be subdivided into systems tracts using inductive, sequence stratigraphic surfaces. A system tract is defined by its bounding stratigraphic surfaces rather than by the strata that comprise it. Still, the characteristics of the strata (e.g., grain size trends) aid in the recognition of various bounding surfaces and, thus, indirectly contribute to the delineation of a given systems tract.

The only inductive, low diachroneity, sequence stratigraphic surface within depositional T-R sequences of ramp and shelf–slope–basin settings is the MFS (Fig. 5A and B). On this basis, Embry (1993) and Embry and Johannessen (1993) proposed that a depositional T-R sequence be subdivided into two systems tracts, a lower transgressive systems tract (TST) and an overlying regressive systems tract (RST) (Fig. 7A and B). A TST, adhering to the definition of Van Wagoner et al. (1988), is a sequence stratigraphic unit bounded at its base by an MRS and its correlative surfaces and

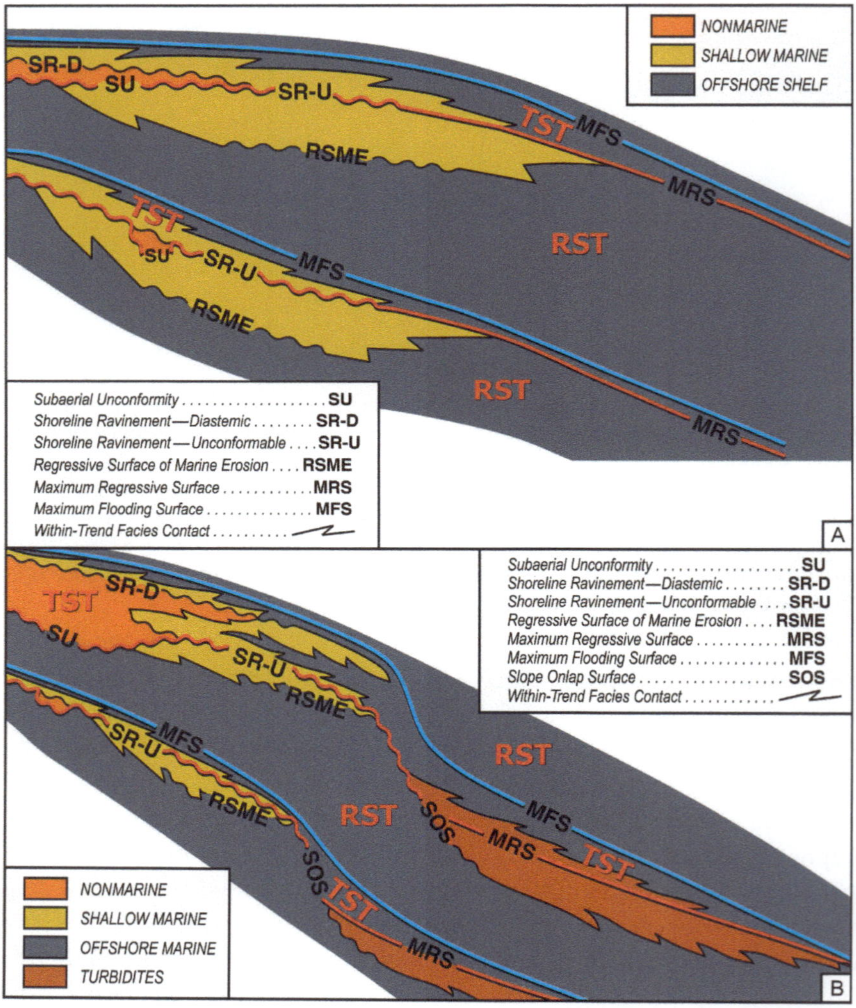

Fig. 7 Inductive systems tracts. (A) The boundaries of a depositional T-R sequence (SU, SR-U, MRS) in a ramp setting are shown in *red*. The occurrence of the maximum flooding surface (MFS) allows the sequence to be subdivided into two systems tracts—a transgressive systems tract (TST) and a regressive systems tract (RST). (B) The boundaries of depositional T-R sequence (SU, SR-U, SOS, MRS) in a shelf/slope/basin setting are shown in *red*. The internal, MFS allows a sequence to be subdivided into a TST and an RST.

an MFS above. The RST is bounded by an MFS below and by an MRS and its correlative surfaces above.

The genetic stratigraphic sequence has MFSs as its bounding surfaces as discussed previously. Like the depositional T-R sequence, it can also be subdivided into a TST and an RST by using the internal, composite boundary of an SU/SR-U/MRS/SOS as the shared boundary for both systems tracts.

The RST, as defined above, includes both the lowstand systems tract (LST) and highstand systems tract (HST) of Van Wagoner et al. (1988). The boundary used by Van Wagoner et al. (1988) to separate our RST into lowstand (LST) and highstand (HST) systems tracts is a highly diachronous, facies change surface in the basin area (e.g., base prograding turbidites). Such a surface is not a valid systems tract boundary, and thus, recognition of the HST and LST by the inductive approach is not possible. As will be subsequently discussed, the HST and LST are part of the deductive approach to sequence stratigraphy.

2.3.3 Parasequence

A parasequence was originally defined by Van Wagoner et al. (1988) as "a relatively conformable succession of beds or bedsets bound by marine-flooding surfaces." A marine-flooding surface or flooding surface (FS) was defined by Van Wagoner et al. (1988) as "a surface separating younger from older strata across which there is an abrupt increase in water depth." Given that all stratigraphic surfaces separate younger from older strata, this leaves "an abrupt increase in water depth," a highly interpretive criterion, as the only means of recognition. Van Wagoner et al. (1990) provided much more information and insight into what they meant by the terms flooding surface and parasequence. It is clear from their diagrams that Van Wagoner et al. (1988, 1990) envisioned a flooding surface as a contact between a marine sandstone and an overlying deeper water, marine shale/siltstone (Fig. 8). Such a contact is a diachronous lithofacies boundary which, in most cases, lies between an MRS below and an MFS above (Fig. 9). Thus, the parasequence as defined by Van Wagoner et al. (1988, 1990) is a lithostratigraphic rather than a sequence stratigraphic unit. This is not meant to diminish the utility of the parasequence to stratigraphic analysis; rather, it is simply not a unit of sequence stratigraphy.

2.4 Sequence Hierarchy in the Inductive Approach

As discussed in detail by Embry (1993, 1995, 2009), it is necessary to assign recognized sequence boundaries and other associated sequence surfaces of a

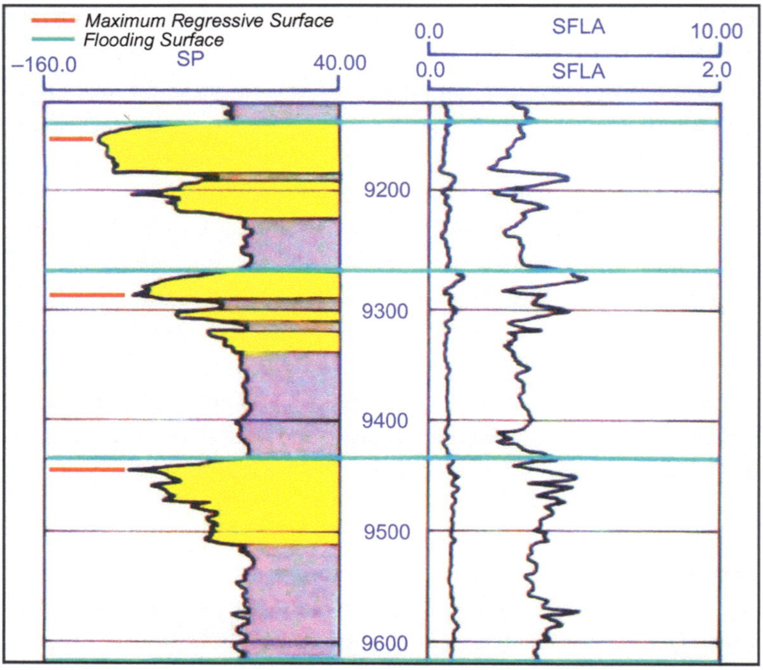

Fig. 8 A portion of Figure 7 of Van Wagoner et al. (1990) showing the parasequence boundaries (FSs in *blue*) as well as MRSs in *red*. Flooding surfaces represent a facies change from sandstone below to shale/siltstone above. *Modified from a portion of Figure 7, Van Wagoner, J.C., Mitchum, R.M., Campion, K.M., Rahmanian, V.D., 1990. Siliciclastic sequence stratigraphy in well logs, cores and outcrops. AAPG Methods in Exploration, No. 7. AAPG, Tulsa, 55 pp.*

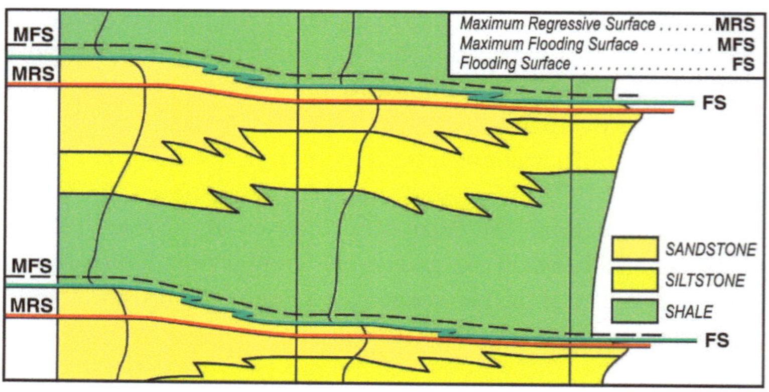

Fig. 9 A schematic cross-section illustrating the correlation of three transgressive–regressive successions using maximum regressive surfaces, maximum flooding surfaces, and flooding surfaces as defined by Van Wagoner et al. (1988, 1990). The diachronous, flooding surfaces (FS in *blue*), which are lithostratigraphic surfaces, bound each parasequence, making them lithostratigraphic units. The depositional T-R sequence is bounded by MRSs, whereas the MFSs bound genetic stratigraphic sequences.

basin to a hierarchy to allow the delineation of various orders of sequences. In the absence of such a hierarchy, any two sequence boundaries, regardless of their magnitude (e.g., any two MRSs in the case of a depositional T-R sequence), could, in theory, be used to form the boundaries of a sequence. This would result in a very large number of sequences and the only way to escape such a chaotic, unmanageable situation is to establish a hierarchy of surfaces.

The hierarchy of boundaries within a given basin is established based on observable criteria and an assessment of the relative magnitudes of the recognized sequence boundaries. Such criteria reflect the relative magnitudes of the base-level changes that generated the boundaries. The criteria include:

1. the degree to which the boundary reflects a changing tectonic regime;
2. the degree of change of the depositional regime and sediment composition across the boundary;
3. the amount of section missing below the unconformity at as many localities as possible, especially those localities proximal to the basin edge;
4. the estimated amount of deepening associated with an MFS that overlies an erosional sequence boundary;
5. the distance that the SU and associated shoreline facies extend into the basin.

The largest magnitude sequence boundaries within a basin are referred to as first order and some basinal successions contain as many as six orders of sequence boundaries (Fig. 10). The reader is referred to Embry (1993, 1995, 2009, 2011) for in-depth discussion of the application of an inductive hierarchy and examples of its use. Catuneanu (2006) and Catuneanu et al. (2011) also favor the use of the inductive approach to determining a sequence hierarchy and recommend the use of similar observable features of the boundaries as discussed above.

3. DEDUCTIVE APPROACH TO SEQUENCE STRATIGRAPHY

3.1 Sequence Stratigraphy as a Deductive Stratigraphic Discipline

A second sequence stratigraphic methodology and classification entails a deductive, conceptual approach. This approach includes surfaces and units documented by the inductive approach, as well as several surfaces and units unique to the deductive approach. The deductive approach is based primarily on a mathematical, stratigraphic model which was developed by Mac Jervey of Exxon in 1979 (Jervey, 1988) to provide a theoretical basis for

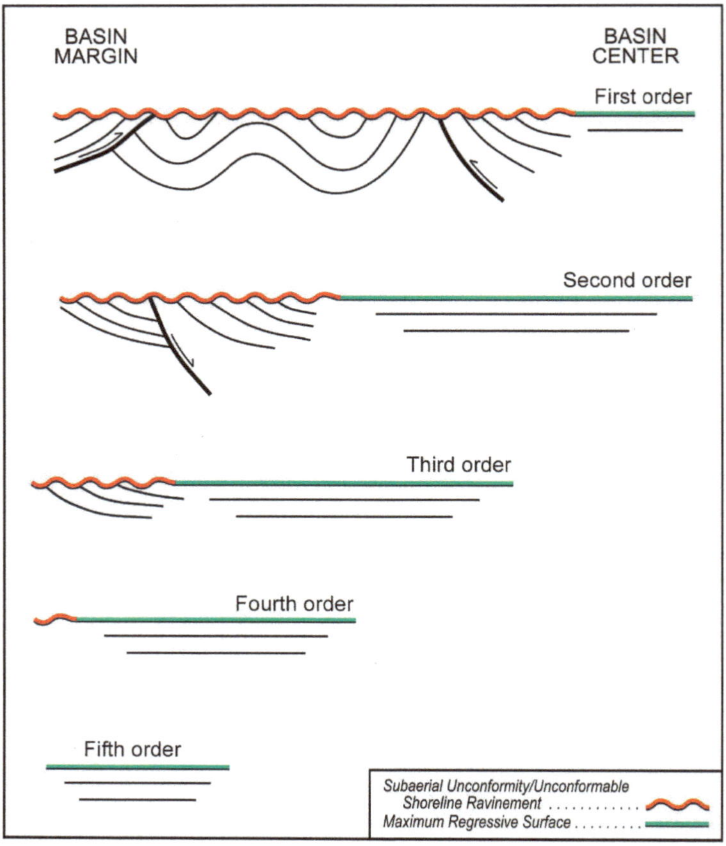

Fig. 10 Schematic depiction of five orders of sequence boundaries of the inductive sequence hierarchy. Each order is determined by observable criteria including degree of tectonic regime change and penetration of the unconformity into the basin. The criteria needed to build such a hierarchy reflect the amount of base-level change that resulted in the formation of the sequence stratigraphic surfaces. *Modified from Figure 7, Embry, A.F., 1995. Sequence boundaries and sequence hierarchies: problems and proposals. In: Steel, R.J., Felt, F.L., Johannessen, E.P., Mathieu, C. (Eds.), Sequence Stratigraphy on the Northwest European Margin. NPF Special Publication 5, pp. 1–11.*

seismic-based, sequence stratigraphic concepts presented by Exxon scientists in 1977 (Payton, 1977). The model used sinusoidal sea-level rise and fall, hinged subsidence that increased basinward, and a constant sediment supply as a priori input parameters. Notably, it predicted the occurrence of stacked, basin flank unconformities and downlap surfaces in central regimes of the basin, the principal surfaces interpreted from seismic sections. As noted earlier, this model provides a reasonable explanation for the inductive surfaces of sequence stratigraphy which had been empirically recognized long before the model was formulated. Exxon researchers (Posamentier and Vail, 1988; Posamentier et al., 1988; Van Wagoner et al., 1988) adopted the Jervey Model to provide the theoretical underpinning of sequence stratigraphy, including the terminology of sequence stratigraphy and sequence stratigraphic models (Posamentier et al., 1988). The reader is referred to Nystuen (1998), Embry (2002, 2009), and Catuneanu (2006) for comprehensive descriptions and discussions of the Exxon sequence stratigraphic methods and terminology.

3.2 Deductive Sequence Stratigraphic Surfaces

The deductive sequence stratigraphic approach recognizes seven surfaces, including SU, SR, RSME, MRS, MFS, basal surface of forced regression (BSFR), and correlative conformity (CC) (Catuneanu, 2006). The first five surfaces are recognized in the inductive approach, although they are defined somewhat differently by the inductive approach. As expressed by Catuneanu (2006), sequence stratigraphic surfaces in the deductive approach "are defined relative to the four main events of the base-level cycle" (Fig. 11). Consistent with the deductive methodology, "the maximum regressive surface is defined relative to the transgression-regressive curve, marking the change from shoreline regression to subsequent transgression" (Catuneanu, 2006, p. 135). Despite the more theoretical

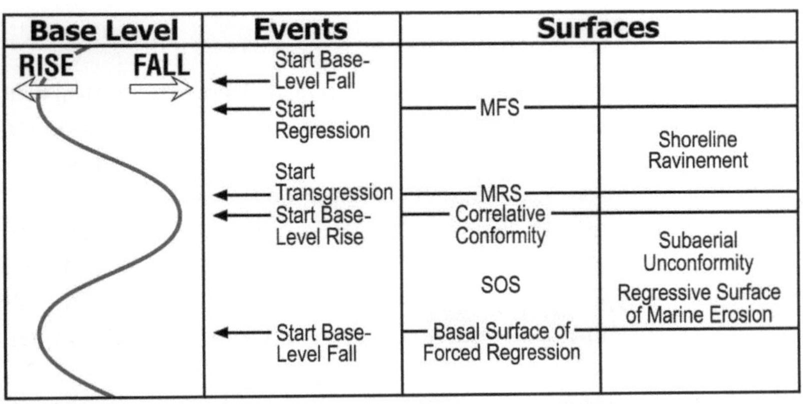

Fig. 11 The seven sequence surfaces of the deductive approach relative to events on base-level curve. As discussed in the text, the basal surface of forced regression (seafloor at start base-level fall) and the correlative conformity (seafloor at start base-level rise) are chronostratigraphic surfaces that display no diagnostic physical attributes.

approach to defining sequence surfaces inherent to the deductive approach, surfaces common to both approaches (SU, SR, MRS, MFS, RSME) are delineated based on objective, physical criteria, as they are in the inductive approach (Catuneanu, 2006; Embry, 2002, 2009, 2010). It is noteworthy that these surfaces are reliably and consistently delineated in studies using either sequence stratigraphic approach.

The two deductive surfaces not recognized in the inductive approach are the BSFR and the CC and both are integral to the deductive approach. They were originally defined by Hunt and Tucker (1992) who had recognized a few, critical inconsistencies in Posamentier et al.'s (1988) application of the Jervey Model. Specifically, Hunt and Tucker (1992) demonstrated that the stratigraphic position of the sequence boundary advocated by Posamentier et al. (1988) placed strata deposited during falling base level below the sequence boundary on the basin flanks and above it in more basinward localities. To rectify the inconsistency, Hunt and Tucker (1992) introduced the BSFR and CC. The BSFR equates to the depositional surface (time surface) at the start of base-level fall, whereas the CC represents the depositional surface at the start of base-level rise (Fig. 11). Recognition of these two chronostratigraphic surfaces is discussed in a subsequent section.

3.3 Deductive Sequence Stratigraphic Units

The deductive and inductive approaches to sequence stratigraphy share common units—sequence and system tract.

3.3.1 Sequence

The deductive approach defines a sequence in terms of the theoretical model of the generation of sequence stratigraphic surfaces as "a succession of strata deposited during a full cycle of change in accommodation or sediment supply" (Catuneanu et al., 2009, 2011). The model-based definition anticipates four main types of sequences: (1) depositional sequence type 1 bounded by a small portion of the SU and the BSFR (Fig. 12A), (2) depositional sequence type 2 bounded by the entire SU and the CC (Fig. 12B), (3) genetic stratigraphic sequence bound by MFSs, and (4) T-R sequence bounded by MRSs.

The genetic stratigraphic sequence is the same in both approaches. The T-R sequence is similar to the depositional T-R sequence of the inductive approach except that it does not include an SR-U as part of the boundary and the MRS onlaps the SU (compare Figs. 6 and 12).

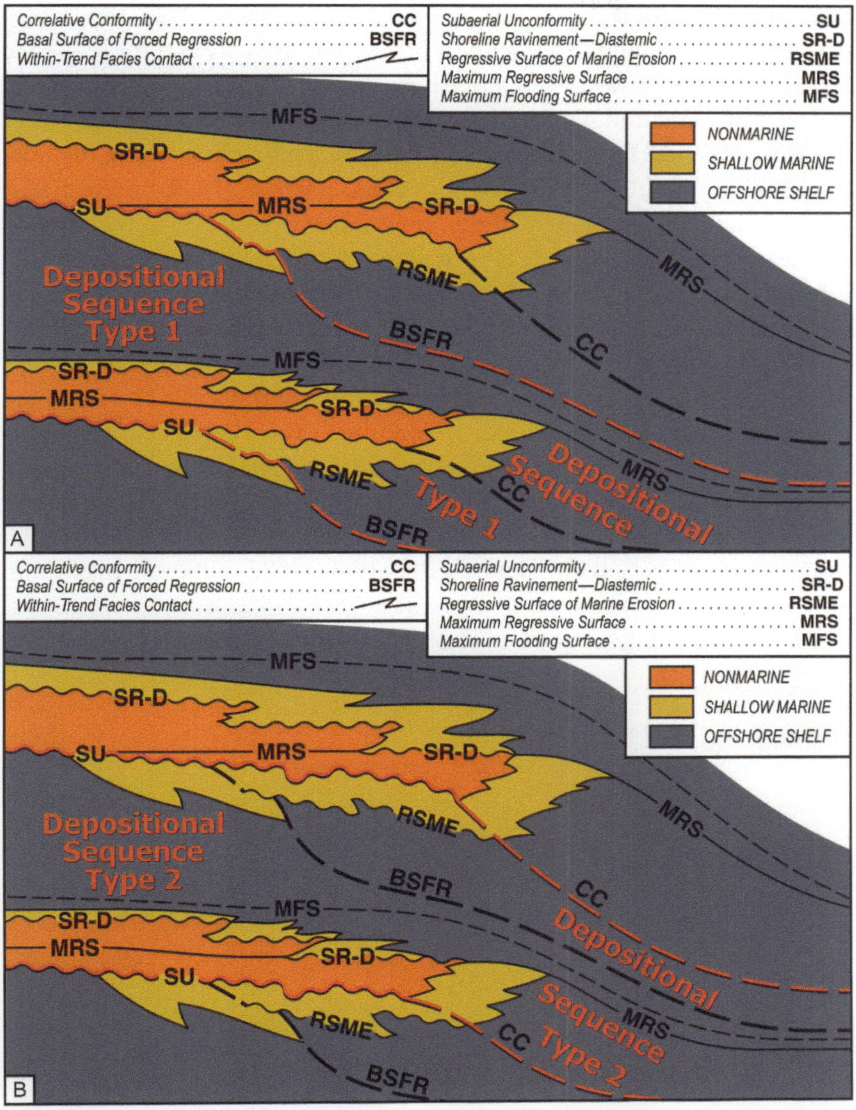

Fig. 12 Deductive depositional sequences. (A) Boundaries of a deductive depositional sequence type 1 in a ramp setting displayed in *red*. The basal surface of forced regression (BSFR) is used as the primary correlative surface of the SU. (B) Boundaries of a deductive depositional sequence type 2 in a ramp setting displayed in *red*. The correlative conformity (CC) is used as the primary correlative surface of the SU.

3.3.2 Systems Tracts

Any of the four deductive sequence types contain as many as three sequence surfaces. The depositional sequence type 2, equivalent to the depositional sequence of Hunt and Tucker (1992), is bounded by the SU on the basin margin and the CC farther basinward (Fig. 13). This sequence type includes an MRS, an MFS, and a BSFR. Hunt and Tucker (1992) defined four systems tracts on the basis of these three internal surfaces. Subsequently, Catuneanu et al. (2009, 2011) attributed the four systems tracts to the deductive approach. Thus, a depositional sequence type 2 comprises, from base to top, the LST, TST, HST, and falling-stage systems tract (FSST) (Fig. 13).

The LST is bounded by the SU/CC below and the MRS above. In the context of the deductive approach, deposits of this system tract accumulated between the onset of base-level rise (CC) and the start of transgression (MRS). The TST is bounded by the MRS below and the MFS (time of maximum transgression) above (Van Wagoner et al., 1988). Strata that comprise the HST were deposited between the time of maximum

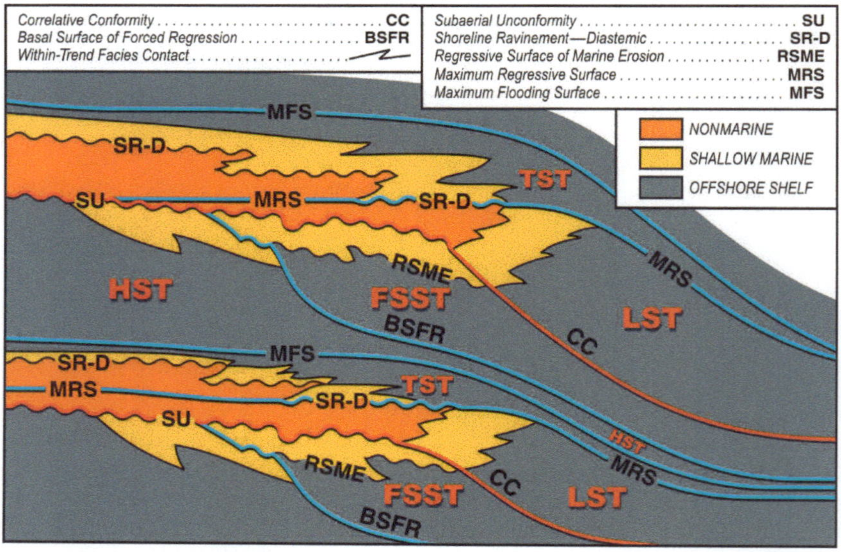

Fig. 13 Boundaries of a deductive depositional sequence type 2 in a ramp setting (SU, CC) are shown in *red*. The sequence contains three sequence stratigraphic surfaces—maximum regressive surface (MRS), maximum flooding surface (MFS), and basal surface of forced regression (BSFR). The sequence comprises four systems tracts: lowstand (LST), transgressive (TST), highstand (HST), and falling-stage (FSST).

transgression (MFS) and the start of base-level fall, defined by the BSFR. The uppermost systems tract, the FSST, is bounded at its base by the BSFR and its top by the CC, both model-derived, chronostratigraphic surfaces. Accordingly, the FSST encompasses all strata deposited during falling base level.

The above discussion demonstrates that the deductive approach to sequence stratigraphy subdivides a given type of sequence into four systems tracts, whereas the inductive approach recognizes only two systems tracts. The TST is the same for both approaches, whereas the RST of the inductive approach encompasses the LST, HST, and FSST of the deductive approach (Fig. 14). However, recognition of the three systems tracts of the regressive portion of a sequence relies upon the scientifically rigorous delineation of the BSFR and CC, chronostratigraphic surfaces. This critical point is addressed below.

Neal et al. (2016) have offered a modest variation of the deductive classification scheme as it relates to the depositional sequence type 2 of Catuneanu et al. (2011). Specifically, although they adopt SU and CC surfaces as sequence boundaries, they do not recognize the BSFR as a valid surface of sequence stratigraphy. Consequently, Neal et al. (2016) divide depositional sequences into three systems tracts rather than four. They

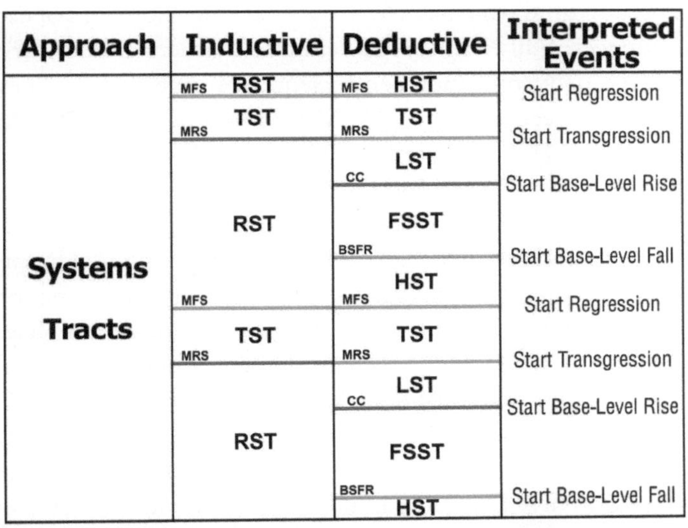

Fig. 14 Comparison of the systems tracts of an inductive, depositional T-R sequence and a deductive, depositional sequence type 2. The sequence boundaries are shown in *red* and systems tract boundaries in *blue*.

combine the HST and FSST, separated by the BSFR (Catuneanu et al., 2011), into a single unit, the HST.

3.3.3 *Parasequence*

The deductive approach adopts Van Wagoner et al.'s (1988) definition of a parasequence. As discussed previously, the parasequence is a lithostratigraphic unit rather than a sequence stratigraphic one.

3.4 Sequence Hierarchy in the Deductive Approach

The deductive approach to determine the hierarchy of sequence boundaries has been described by Vail et al. (1977), Mitchum and Van Wagoner (1991), Vail et al. (1991), and Posamentier and Allen (1999). This model-driven methodology assumes that sequence stratigraphic surfaces are generated by eustasy-driven, sinusoidal, base-level changes and that the amplitude of such eustatic cycles increase as frequency diminishes. Sequence boundaries associated with very large amplitude base-level changes (tectono-eustasy) are assigned to either a first- or second-order category. High-order boundaries (fourth, fifth, and sixth order) are related to climate-driven, high-frequency, eustatic changes in the 20,000–400,000 year band.

In the deductive approach, a sequence is assigned to a given order based on the amount of time which lapsed between the development of each of its bounding surfaces. Vail et al. (1991) defined six orders of boundaries consistent with the deductive approach to sequence stratigraphy. These were first order: 50 MA; second order: 3–50 MA; third order: 0.5–0.3 MA; fourth order: 0.08–0.5 MA; fifth order: 0.03–0.08 MA; and sixth order: 0.01–0.03 MA.

The deductive approach to establishing a hierarchy of sequences is basically unworkable and prone to circular reasoning. Any desired frequency of boundary occurrence can be determined simply by selecting only the boundaries that fit the desired result (Embry, 1995). As emphasized by Embry (2009), "If one wants to determine the frequency of 2nd order sequence boundaries, one must be able to empirically recognize 2nd order boundaries in the first place. Boundary frequency is a conclusion that can be only be reached once the different orders of boundaries are defined with reasonable objectivity. Duration is not an observable characteristic of a sequence." Catuneanu et al. (2011) also pointed out major flaws in the deductive approach to sequence hierarchy.

3.5 Recognizing Sequence Stratigraphic Surfaces in the Deductive Approach

The inductive approach employs five surfaces, two sequence types, and two systems tracts, whereas the deductive approach relies upon seven surfaces, four sequence types, and four systems tracts (Fig. 14). The deductive approach does not recognize the SOS, a useful empirical surface of the inductive approach (Embry, 2009). In contrast, the inductive approach does not utilize the two conceptual chronostratigraphic surfaces of the deductive approach—the BSFR and the CC.

The additional surfaces and units of the deductive approach potentially offer enhanced resolution, assuming that these surfaces can be recognized with a high level of certainty. The BSFR and the CC have been defined as chronostratigraphic surfaces tied to the base-level curve of the Jervey Model (Fig. 11). The chronostratigraphic nature of the BSFR was underscored by Catuneanu (2006, p. 123) who declared that the surface "approximates the paleo-seafloor at the onset of base level fall at the shoreline." The CC of Hunt and Tucker (1992) was envisioned to be a chronostratigraphic surface described by Catuneanu (2006) as the "end-of-fall paleo-seafloor."

Recognition and delineation of chronostratigraphic surfaces is a highly interpretive exercise because such conceptual surfaces are abstract and have few, if any, physical attributes. The BSFR displays no obvious sedimentological variation or grain size change that would permit its ready recognition in outcrop or well logs. Indeed, Posamentier and Allen (1999, p. 90) pointed out that, as the BSFR "exists only as a chronohorizon, precise identification … can be limited." Plint and Nummedal (2000, p. 5) wrote that the BSFR is "difficult or impossible to recognize in outcrops or well logs." Catuneanu (2006, p. 129) echoed this view declaring that "the basal surface of forced regression … has no physical expression in a conformable succession of shallow water deposits."

Similar problems attend recognition of the CC. Catuneanu (2006) noted that, "The main problem relates to the difficulty of recognizing it in most outcrop sections, core or wireline logs." The CC he continued "develops within a conformable prograding package (coarsening-upward trends below and above); lacking any lithofacies and grading contrasts." Catuneanu's (2006) points were underscored by Plint and Nummedal (2000, p. 5) who succinctly stated that "From a practical point of view, this marine surface will be difficult to impossible to identify."

Geologists using the deductive approach to sequence stratigraphy are faced with the prospect of recognizing these chronostratigraphic horizons. Indeed, failure to objectively and reliably delineate these surfaces negates the value of the deductive approach. Notably, erroneous interpretations can result if these surfaces are identified on a speculative basis or if inappropriate proxy surfaces (e.g., a facies contact for the BSFR) are used.

Catuneanu et al. (2009, 2011), referring to the Jervey Model, suggested that the BSFR and CC mark changes in parasequence stacking pattern. Specifically, the BSFR defines a change from a normal regressive stacking pattern to a forced regressive pattern. The CC marks the opposite trend, from a forced regressive stacking pattern to a normal regressive one. This approach, however, suffers from a lack of reasonably objective criteria that can be used to delineate these stacking pattern changes in outcrop or on well logs.

Catuneanu (2006) interpreted the BSFR to be the clinoform (paleo-seafloor) at the start of offlap along a given transect perpendicular to the shoreline. Unfortunately, it is difficult at best to recognize the first clinoform associated with offlap, especially based on outcrop studies. Seismic data seem to offer the best hope for identifying a BSFR in this manner and, in theory, it is approximated by (or contained within) the reflector that intersects the SU at the start of a basinward trajectory. However, such an intersection point can rarely, if ever, be determined with objectivity as it is lost to erosion during falling base level (Catuneanu, 2006).

The viability of the CC as a meaningful chronostratigraphic surface depends on its ease of identification in the rock record. Unfortunately, the CC is not readily delineated on a regional basis by a change in stacking pattern from forced regression to normal regression, principally because physical attributes of such a change remain unknown in the rock record. As with the BSFR, seismic data hold the best potential for approximating the stratigraphic position of a CC, the surface being contained within a reflector that merges with that reflector defining the marginal unconformity (SU/SR-U). Notably, such a reflector would also most likely encompass the MRS, which generally forms simultaneously with, or shortly after, the CC (Figs. 1 and 6).

Catuneanu (2006, Fig. 6.2) and Catuneanu et al. (2011, Fig. 7) referenced a seismic section on which they have identified both a BSFR and a CC (Fig. 15A). However, we interpret their BSFR as an MFS, given the apparent downlap and condensation onto the reflector (Fig. 15B). The surface labeled as the CC on Fig. 15A can be interpreted as an MRS, given the overlying transgressive strata and initiation of a new sequence (Fig. 15B).

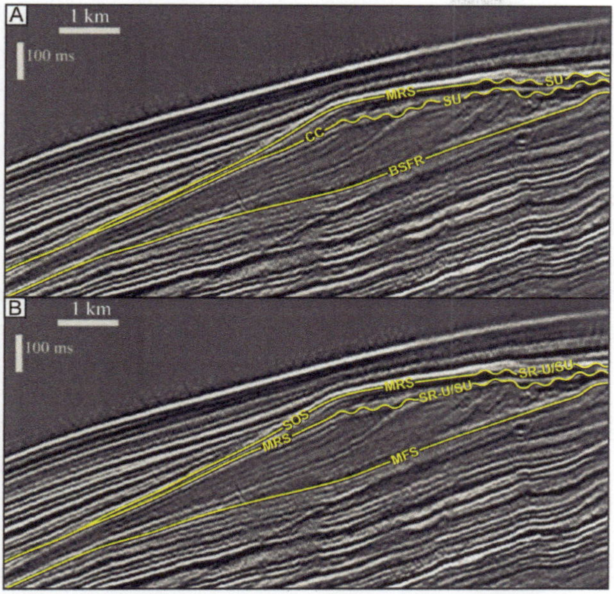

Fig. 15 Deductive interpretation vs inductive interpretation of a seismic section. (A) Deductive sequence stratigraphic interpretation of a seismic section of Pleistocene-Recent sediments in the Gulf of Mexico by Posamentier (2003) and Catuneanu (2006). (B) Inductive interpretation of the same seismic section. Refer to text for discussion. *BSFR*, basal surface of forced regression; *CC*, correlative conformity; *MFS*, maximum flooding surface; *MRS*, maximum regressive surface; *SOS*, slope onlap surface, *SU*, subaerial unconformity; *SR-U*, unconformable shoreline ravinement.

The reflector interpreted to be an MRS by Catuneanu (2006) (Fig. 15A) appears to merge with the SU/SR-U (clear downcutting) and an SOS (clear onlap) (Fig. 15B). The lesson to be learned from this example is that seismic data can be equivocal when used to identify sequence surfaces. The identification of the BSFR and CC requires unequivocal, physical criteria that can be applied in a consistent manner on a regional basis using a variety of databases. Such criteria have yet to be described and documented and, until they are, the widespread application of the deductive approach of sequence stratigraphy will not be reasonable or reliable.

4. COMPARISON OF THE TWO APPROACHES

It is essential for practitioners to be cognizant that two different approaches to sequence stratigraphic methodology exist and that a choice must be made as to which one will be used in any given situation. The inductive approach involves the delineation of five surfaces (SU, SR-U, SOS, MRS, MFS), all of which can be empirically recognized on the basis of physical characteristics and geometric relationships. These surfaces are used for correlation and as bounding surfaces of depositional T-R and genetic stratigraphic sequences, each of which can be subdivided into TST and RST. The great advantage of this approach is that all of these surfaces can be reliably and consistently identified using sedimentological and stratigraphic data from outcrop, well log, and/or seismic sections in undeformed to highly deformed sedimentary successions. Further, the inductive approach is more readily applied in regional work, because only objectively defined surfaces are correlated. The main disadvantage of this approach is that it does not provide the level of stratigraphic resolution and consequent interpretive insight as provided by the deductive approach. An additional, but minor, drawback is that in rare cases, a regional depositional T-R sequence boundary cannot be delineated because the SU does not adjoin the SR-U.

The deductive approach may provide more stratigraphic resolution and hence improved understanding of the depositional history of a sedimentary succession by virtue of its recognition of the BSFR and CC. As noted above, these chronostratigraphic surfaces potentially enhance the time stratigraphic framework of basin fills. The surfaces recognized by the deductive approach, including surfaces embodied in the inductive approach, allow four different types of sequences to be recognized, each of these divided into four systems tracts linked to the base-level and shoreline movement history. Thus, the fact that the deductive approach is potentially more comprehensive than the inductive approach would seem to make the former the preferred approach, all things being equal.

The major weakness of the deductive approach is the great difficulty in delineating the BSFR and CC in a reliable, consistent, and scientifically acceptable manner. Currently, there are no published examples of these surfaces being objectively recognized with rock-based data in a regional context. This stands in stark contrast to the regionally consistent recognition of the inductive surfaces of sequence stratigraphy described in numerous publications (e.g., Brigaud et al., 2014; Hampson, 2016; Ichaso et al., 2016; Lash and Engelder, 2011; Sansom, 2010; Zimmermann et al., 2015). It is noteworthy that every publication we examined purporting to delineate a BSFR and CC using rock-based data was, in fact, describing inductive, material-based stratigraphic surfaces (facies contacts, inductive sequence surfaces). Such misidentification of undoubtedly diachronous surfaces (e.g., lithofacies contacts) as chronostratigraphic surfaces leads to erroneous interpretations of sedimentology and depositional history.

In summary, the choice of whether to use the inductive approach or the deductive approach comes down to the question, "Are my data sufficient to reliably delineate and correlate BSFRs and CCs throughout my study area?" If so, then use of the deductive approach would be best so as to gain maximum resolution and greater insight into basin history. If not, then the practitioner would want to use the inductive approach to avoid nonactualistic and misleading interpretations of facies relationships and depositional history based on the misidentification of BSFRs and CCs.

5. SUMMARY

There are currently two distinctly different approaches to sequence stratigraphic classification and methodology. The inductive approach defines sequence surfaces on the basis of observable characteristics. The deductive approach, on the other hand, defines sequence surfaces that reflect the interaction of base-level change and sedimentation. It is important for any practitioner to be aware of these two different approaches to sequence stratigraphy so that the level of interpretation is justified by the data at hand and to escape the problem of force-fitting data to an inappropriate classification system.

The inductive approach recognizes seven surfaces of sequence stratigraphy in terms of observable characteristics. Five of these surfaces, SU, SR-U, SOS, MRS, and MFS, are used for correlation and as bounding surfaces of various sequence stratigraphic units.

A sequence is defined as a unit bounded by an unconformity or its correlative surfaces. Two types of inductive sequences are recognized: the depositional T-R sequence which is bounded by SUs and/or correlative surfaces (SR-U, MRS, SOS), and the genetic stratigraphic sequence which is bounded by MFSs. Each sequence type can be divided into two systems tracts: a TST and an RST. A parasequence, as originally defined and as currently employed in stratigraphic analysis, is a lithostratigraphic unit rather than a sequence stratigraphic one.

Sequence stratigraphic surfaces of the deductive approach include five surfaces recognized in the inductive approach (SU, SR, MRS, MFS, RSME). However, the deductive approach recognizes two additional chronostratigraphic surfaces: the BSFR which marks on the onset of falling base level, and the CC which represents the seafloor at the start of rising base level. Unlike the surfaces common to both sequence stratigraphic approaches, the BSFR and CC display no diagnostic physical characteristics. Rather, it is suggested that these surfaces can be identified based on interpreted changes in parasequence stacking patterns.

The deductive sequence stratigraphic approach recognizes four types of sequences, each one comprised of four systems tracts: lowstand, transgressive, highstand, and falling stage (LST, TST, HST, and FSST). Thus, the deductive approach is potentially superior to the inductive approach in that it yields greater resolution and interpretive insight. However, the deductive approach suffers from a general inability to reliably recognize the BSFR and the CC. The practitioner must choose between the inductive approach and its readily recognized surfaces, but lower resolution, and the nebulous surfaces but higher resolution of the deductive approach. Ultimately the choice usually comes down to answering the question, "Are my data sufficient to reliably identify and correlate the BSFR and CC throughout my study area?"

ACKNOWLEDGMENTS

First, we would like to acknowledge the fundamental contributions of Hollis Hedberg, the father of modern stratigraphic practice. From the late 1930s to the late 1970s he established the guiding principles for stratigraphic classification and nomenclature. His concepts and advice remain very relevant today. We would like to thank Michael Montenari for inviting our contribution and shepherding it through the publication process. The critical reviewers, especially Gary Lash, provided many helpful suggestions for improvement. We have discussed sequence stratigraphic concepts and practice with many colleagues over the past 40 years and we are grateful for all the insights and guidance they provided. Finally, we would like to thank our former employers, the Geological Survey of Canada (A.F.E.), and Statoil (E.P.J.) for fostering our stratigraphic studies.

REFERENCES

Barrell, J., 1917. Rhythms and the measurements of geologic time. Bull. Geol. Soc. Am. 28, 745–904.

Beauchamp, B., Henderson, C., 1994. The Lower Permian Raanes, Great Bear Cape and Trappers Cove formations. Bull. Can. Petrol. Geol. 42, 562–597.

Brigaud, B., Vincent, B., Carpentier, C., Robin, C., Guillocheau, F., Yven, B., Huret, E., 2014. Growth and demise of the Jurassic carbonate platform in the intracratonic Paris Basin (France): interplay of climate change, eustasy and tectonics. Mar. Pet. Geol. 53, 3–29.

Brown, L., Fisher, W., 1977. Seismic-stratigraphic interpretation of depositional systems: examples from the Brazilian rift and pull-apart basins. In: Payton, C. (Ed.), Seismic Stratigraphy: Application to Hydrocarbon Exploration, pp. 213–248. AAPG Memoir 26.

Burgess, P., Prince, G., 2015. Non-unique stratal geometries: implications for sequence stratigraphic interpretations. Basin Res. 27, 351–365.

Catuneanu, O., 2006. Principles of Sequence Stratigraphy. Elsevier, New York. 386 pp.

Catuneanu, O., Abreu, V., Bhattacharya, J.P., Blum, M.D., Dalrymple, R.W., Eriksson, P.G., Fielding, C.R., Fisher, W.L., Galloway, W.E., Gibling, M.R., Giles, K.A., Holbrook, J.M., Jordan, R., Kendall, C.G.S.C., Macurda, B., Martinsen, O.J., Miall, A.D., Neal, J.E., Nummedal, D., Pomar, L., Posamentier, H.W., Pratt, B.R., Sarg, J.F., Shanley, K.W., Steel, R.J., Strasser, A., Tucker, M.E., Winker, C., 2009. Towards the standardization of sequence stratigraphy. Earth Sci. Rev. 92, 1–33.

Catuneanu, O., Galloway, W.E., Kendall, C.G.S.C., Miall, A.D., Posamentier, H.W., Strasser, A., Tucker, M.E., 2011. Sequence stratigraphy: methodology and nomenclature. Newsl. Stratigr. 44, 173–245.

Combellas-Bigott, R., Galloway, W., 2006. Depositional and structural evolution of the middle Miocene depositional episode, east-central Gulf of Mexico. AAPG Bull. 90, 335–362.

Embry, A.F., 1993. Transgressive-regressive (T-R) sequence analysis of the Jurassic succession of the Sverdrup Basin, Canadian Arctic Archipelago. Can. J. Earth Sci. 30, 301–320.

Embry, A.F., 1995. Sequence boundaries and sequence hierarchies: problems and proposals. In: Steel, R.J., Felt, F.L., Johannessen, E.P., Mathieu, C. (Eds.), Sequence Stratigraphy on the Northwest European Margin, pp. 1–11. NPF Special Publication 5.

Embry, A., 2002. Transgressive-regressive (T-R) sequence stratigraphy. In: Armentrout, J., Rosen, N. (Eds.), Sequence Stratigraphic Models for Exploration and Production. Gulf Coast SEPM Conference Proceedings, Houston, pp. 151–172.

Embry, A.F., 2009. Practical Sequence Stratigraphy. Canadian Society of Petroleum Geologists, Calgary. 76 pp.

Embry, A.F., 2010. Correlating siliciclastic successions with sequence stratigraphy. In: Ratcliffe, K., Zaitlin, B. (Eds.), Application of Modern Stratigraphic Techniques: Theory and Case Histories, pp. 35–53. SEPM Special Publication 94.

Embry, A.F., 2011. Petroleum prospectivity of the Triassic-Jurassic succession of Sverdrup Basin, Canadian Arctic Archipelago. In: Spencer, A.M., Embry, A.F., Gautier, D.L., Stoupakova, A.V., Sørensen, K. (Eds.), Arctic Petroleum Geology. Geological Society, London, pp. 545–558. Memoirs 5.

Embry, A., Beauchamp, B., 2008. Sverdrup Basin. In: Miall, A. (Ed.), Sedimentary Basins of the World. In: The Sedimentary Basins of the United States and Canada, vol. 5. Elsevier, Amsterdam, pp. 451–472.

Embry, A.F., Johannessen, E.P., 1993. T-R sequence stratigraphy, facies analysis and reservoir distribution in the uppermost Triassic-Lower Jurassic succession, western Sverdrup Basin, Arctic Canada. In: Vorren, T., Bergsager, E., Dahl-Stamnes, O.A., Holter, E., Johansen, B., Lie, E., Lund, T.B. (Eds.), Arctic Geology and Petroleum Potential, pp. 121–146. NPF Special Publication 2.

Galloway, W., 1989. Genetic stratigraphic sequences in basin analysis I: architecture and genesis of flooding surface bounded depositional units. AAPG Bull. 73, 125–142.

Hadler-Jacobsen, F., Johannessen, E.P., Ashton, N., Henriksen, S., Johnson, S., Kristensen, J., 2005. Submarine fan morphology and lithology distribution: a predictable function of sediment delivery, gross shelf to basin relief, slope gradient and basin topography. In: Dore, A., Vining, B. (Eds.), Petroleum Geology: Northwest Europe and Global Perspectives. Geological Society, London, pp. 1121–1145.

Hampson, G.J., 2016. Towards a sequence stratigraphic solution set for autogenic processes and allogenic controls: Upper Cretaceous strata, Book Cliffs, Utah, USA. J. Geol. Soc. London 173, 817–836.

Hedberg, H.D., 1958. Stratigraphic classification and terminology. AAPG Bull. 42, 1881–1896.

Hedberg, H.D., 1959. Towards a harmony in stratigraphic classification. Am. J. Sci. 257, 674–683.

Helland-Hansen, W., 2009. Towards the standardization of sequence stratigraphy: discussion. Earth Sci. Rev. 94, 95–97.

Helland-Hansen, W., Gjelberg, J., 1994. Conceptual basis and variability in sequence stratigraphy: a different perspective. Sediment. Geol. 92, 1–52.

Hunt, D., Tucker, M., 1992. Stranded parasequences and the forced regressive wedge systems tract: deposition during base level fall. Sediment. Geol. 81, 1–9.

Ichaso, A.A., Dalrymple, R.W., Martinius, A.W., 2016. Basin analysis and sequence stratigraphy of the syn-rift Tilje Formation (Early Jurassic), Halten Terrace giant oil and gas fields, offshore mid-Norway. AAPG Bull. 100, 1329–1375.

Jervey, M., 1988. Quantitative geological modeling of siliciclastic rock sequences and their seismic expression. In: Wilgus, C., Hastings, B.S., Kendall, C.G., Posamentier, H.W., Ross, C.A., Van Wagoner, J.C. (Eds.), Sea Level Changes: An Integrated Approach, pp. 47–69. SEPM Special Publication 42.

Jervey, M., 1993. Siliciclastic sequence development in foreland basins, with examples from the Western Canada Foreland Basin. In: Macqueen, R.W., Leckie, D.A. (Eds.), Foreland Basins and Fold Belts, pp. 47–80. Am. Assoc. Pet. Geol. Mem. 55.

Johannessen, E.P., Embry, A.F., 1989. Sequence correlation: Upper Triassic to Lower Jurassic succession, Canadian and Norwegian Arctic. In: Collinson, J.D. (Ed.), Correlation in Hydrocarbon Exploration. Norwegian Petroleum Society/Graham and Trotman, London, pp. 155–170.

Johannessen, E.P., Steel, R.J., 2005. Shelf-margin clinoforms and prediction of deep water sands. Basin Res. 17, 521–550.

Johannessen, E.P., Mjos, R., Renshaw, D., Dalland, A., Jacobsen, T., 1995. Northern limit of the Brent delta at the Tampen Spur—a sequence stratigraphic approach for sandstone prediction. In: Steel, R., Felt, V., Johannessen, E., Mathieu, C. (Eds.), Sequence Stratigraphy of the Northwest European Margin, pp. 213–256. Norwegian Petroleum Society Special Publication 5.

Johannessen, E.P., Henningsen, T., Bakke, N.E., Johansen, T.A., Ruud, B.O., Riste, P., Elvebakk, H., Jochmann, M., Elvebakk, G., Woldengen, M.S., 2011. Palaeogene clinoform succession on Svalbard expressed in outcrops, seismic data, logs and cores. First Break 29, 35–44.

Lash, G.G., Engelder, T., 2011. Thickness trends and sequence stratigraphy of the Middle Devonian Marcellus Formation, Appalachian Basin: implications for Acadian foreland basin evolution. AAPG Bull. 95, 61–103.

Madof, A., Harris, A., Connell, S., 2016. Nearshore along-strike variability: is the concept of the systems tracts unhinged? Geology 44, 319–322.

Miall, A.D., 2004. Empiricism and model building in stratigraphy: the historical roots of present-day practices. Stratigraphy 1, 3–25.

Miall, A.D., Miall, C.E., 2004. Empiricism and model-building in stratigraphy: around the hermeneutic circle in the pursuit of stratigraphic correlation. Stratigraphy 1, 27–46.

Mitchum, R., Van Wagoner, J., 1991. High frequency sequences and their stacking patterns: sequence stratigraphic evidence for high frequency eustatic cycles. Sediment. Geol. 70, 131–160.

Mitchum, R., Vail, P., Thompson, S., 1977. Seismic stratigraphy and global changes in sea level, part 2: the depositional sequence as the basic unit for stratigraphic analysis. In: Payton, C. (Ed.), Seismic Stratigraphy: Application to Hydrocarbon Exploration, pp. 53–62. AAPG Memoir 26.

Mjos, R., Hadler-Jacobsen, F., Johannessen, E., 1998. The distal sandstone pinchout of the Mesa Verde Group, San Juan basin and its relevance for sandstone prediction of the Brent Group, northern North Sea. In: Gradstein, F., Sandvik, K., Milton, N. (Eds.), Predictive High Resolution Sequence Stratigraphy, pp. 263–297. Norwegian Petroleum Society Special Publication 8.

Neal, J.E., Abreu, V., Bohacs, K.M., Feldman, H.R., Pederson, K.H., 2016. Accommodation succession (δA/δS) sequence stratigraphy: observational method, utility and insights into sequence boundary formation. J. Geol. Soc. London 173, 803–816.

Nystuen, J.P., 1998. History and development of sequence stratigraphy. In: Gradstein, F., Sandvik, K., Milton, N. (Eds.), Predictive High Resolution Sequence Stratigraphy, pp. 31–116. Norwegian Petroleum Society Special Publication 8.

Partington, M., Mitchener, B., Milton, N., Fraser, A., 1993. Genetic sequence stratigraphy for the North Sea Late Jurassic and Early Cretaceous: distribution and prediction of Kimmeridgian–Late Ryazanian reservoirs in the North Sea and adjacent areas. In: Parker, J. (Ed.), Petroleum Geology of Northwest Europe, pp. 327–370. 4.

Payton, C. (Ed.), 1977. Seismic Stratigraphy: Applications to Hydrocarbon Exploration. AAPG, Tulsa. AAPG Memoir 26. 516 pp.

Pellegrini, C., Maselli, V., Gamberi, F., Asioli, A., Bohacs, K., Drexler, T., Trincardi, F., 2017. How to make a 350-m-thick lowstand systems tract in 17,000 years: the Late Pleistocene Po River (Italy) lowstand wedge. Geology 45, 327–330.

Plint, A., Nummedal, D., 2000. The falling stage systems tract: recognition and importance in sequence stratigraphic analysis. In: Hunt, D., Gawthorpe, R. (Eds.), Sedimentary Responses to Forced Regressions, pp. 1–17. Geological Society of London, Special Publication 172.

Posamentier, H.W., 2003. A linked shelf-edge delta and slope-channel turbidite system: 3D seismic case study from the eastern Gulf of Mexico. In: Roberts, H.H., Rosen, N.C., Fillon, R.H., Anderson, J.B. (Eds.), Shelf-Margin Deltas and Linked Downslope Petroleum Systems: Global Significance and Future Exploration Strategy. Proceedings, 23rd GCSSEPM Foundation Bob. F. Perkins Research Conference, pp. 115–134.

Posamentier, H., Allen, G., 1999. Siliciclastic Sequence Stratigraphy—Concepts and Applications. SEPM Concepts in Sedimentology and Paleontology 7, Tulsa, SEPM. 210 pp.

Posamentier, H., Vail, P., 1988. Eustatic controls on clastic deposition II—sequence and systems tract models. In: Wilgus, C., Hastings, B.S., Kendall, C.G., Posamentier, H.W., Ross, C.A., Van Wagoner, J.C. (Eds.), Sea Level Changes: An Integrated Approach, pp. 125–154. SEPM Special Publication 42.

Posamentier, H., Jervey, M., Vail, P., 1988. Eustatic controls on clastic deposition I—conceptual framework. In: Wilgus, C., Hastings, B.S., Kendall, C.G., Posamentier, H.W., Ross, C.A., Van Wagoner, J.C. (Eds.), Sea Level Changes: An Integrated Approach, pp. 109–124. SEPM Special Publication 42.

Sansom, P.J., 2010. Reappraisal of the sequence stratigraphy of the Humber Group of the UK Central Graben. In: Vining, B., Pickering, S. (Eds.), Petroleum Geology. Mature Basins to New Frontiers: Proceedings of the 7th Petroleum Geology Conference. Geological Society, London, pp. 177–211.

Sloss, L., 1963. Sequences in the cratonic interior of North America. GSA Bull. 74, 93–113.

Sloss, L., Krumbein, W., Dapples, E., 1949. Integrated facies analysis. In: Longwell, C. (Ed.), Sedimentary Facies in Geologic History. Geological Society America, pp. 91–124. Memoir 39.

Sonnenfeld, M., Cross, T., 1993. Volumetric portioning and facies differentiation within the Permian Upper San Andreas Formation of last Chance canyon, Guadaloupe Mountains, New Mexico. In: Loucks, R., Sarg, F. (Eds.), Carbonate Sequence Stratigraphy, pp. 435–474. AAPG Memoir 57.

Vail, P., Mitchum, R., Thompson, S., 1977. Seismic stratigraphy and global changes in sea level. In: Payton, C. (Ed.), Seismic Stratigraphy: Applications to Hydrocarbon Exploration, pp. 49–212. AAPG Memoir 26.

Vail, P., Audemard, F., Bowman, S., Eisner, P., Perez-Cruz, C., 1991. The stratigraphic signatures of tectonics, eustasy and sedimentology—an overview. In: Einsele, G., Ricken, W., Seilacher, A. (Eds.), Cycles and Events in Stratigraphy. Springer-Verlag, New York, pp. 617–659.

Van Wagoner, J.C., Posamentier, H.W., Mitchum, R.M., Vail, P.R., Sarg, J.F., Loutit, T.S., Hardenbol, J., 1988. An overview of the fundamentals of sequence stratigraphy and key definitions. In: Wilgus, C., Hastings, B.S., Kendall, C.G., Posamentier, H.W., Ross, C.A., Van Wagoner, J.C. (Eds.), Sea Level Changes: An Integrated Approach, pp. 39–46. SEPM Special Publication 42.

Van Wagoner, J.C., Mitchum, R.M., Campion, K.M., Rahmanian, V.D., 1990. Siliciclastic sequence stratigraphy in well logs, cores and outcrops. AAPG Methods in Exploration, No. 7, AAPG, Tulsa. 55 pp.

Wendte, J., Uyeno, T., 2005. Sequence stratigraphy and evolution of Middle to Upper Devonian Beaverhill Lake strata south-central Alberta. Bull. Can. Petrol. Geol. 53, 250–354.

Wheeler, H.E., 1958. Time stratigraphy. AAPG Bull. 42, 1208–1218.

Wheeler, H.E., 1959. Stratigraphic units in time and space. Am. J. Sci. 257, 692–706.

Wheeler, H.E., 1964. Base level, lithosphere surface and time stratigraphy. Geol. Soc. Am. Bull. 75, 599–610.

Xue, L., Galloway, W., 1993. Sequence stratigraphic and depositional framework of the Paleocene Lower Wilcox strata, northwestern Gulf of Mexico. GCAGS Trans. 43, 453–462.

Zimmermann, J., Franz, M., Heunisch, C., Luppold, F., Monnig, E., Wolfgramm, M., 2015. Sequence stratigraphic framework of the Lower and Middle Jurassic in the North German Basin: epicontinental sequences controlled by Boreal cycles. Palaeogeogr. Palaeoclimatol. Palaeoecol. 440, 395–416.

FURTHER READING

Nummedal, D., Riley, G., Templet, P., 1993. High resolution sequence architecture: a chronostratigraphic model based on equilibrium profile studies. In: Posamentier, H., Summerhayes, C., Haq, B., Allen, G. (Eds.), Sequence Stratigraphy and Facies Association, pp. 55–68. International Association of Sedimentologists, Special Publication 18.

CHAPTER FOUR

Conceptual Models for Sequence Stratigraphy of Continental Rift Successions

Michael Holz[1], Daniel Bono Vilas-Boas, Edric Brasileiro Troccoli, Vinicius Carneiro Santana, Paulo Augusto Vidigal-Souza
Universidade Federal da Bahia, Salvador, Bahia, Brazil
[1]Corresponding author: e-mail address: michael@cpgg.ufba.br

Contents

Abstract

Due to the basins asymmetry and the subsidence controlled by one major fault zone, the accommodation space within a rift basin is variable, and the popular passive margin-based sequence stratigraphic scheme, where depositional systems tracts develop following a predictable order during a complete cycle of base-level change, is not applicable. For a practical and useful stratigraphic analysis of rift basins, conceptual and methodological adaptations are required. This chapter discusses some main aspects of sequence stratigraphy of continental rift basins and the conceptual adaptation needed, and proposes an integrative sequence stratigraphic model, with two different scenarios (gradual and punctuated subsidence), defining three

Stratigraphy & Timescales, Volume 2
ISSN 2468-5178
http://dx.doi.org/10.1016/bs.sats.2017.07.002

tectonostratigraphic phases: (1) the rift initiation phase, which is characterized by isolated and restricted faults that create incipient half-grabens, infilled with fluvial and deltaic facies, (2) the rift development phase, during which the initial rift faults tend to link and to form a larger and deeper depositional area, recording lacustrine facies with an overall retrogradational trend, and (3) rift termination phase, when the accommodation decreases, the sedimentation regime is progradational, and the rift basin is filled with fluvial–deltaic and eolian facies. The chapter closes illustrating those models with examples from selected Brazilian rift basins.

1. INTRODUCTION

The establishment of conceptual and predictive models is one of the main concerns of scientific analysis in whatever field of expertise. Concerning specifically sedimentary basin analysis, the triumphal march of sequence stratigraphy as an analytic tool during the last decades has two main reasons for its success: the extreme applicability and its predictive character. This applies to hydrocarbon exploration—the main and most popular and well-known application of sequence stratigraphy—but also to other mineral resources associated with sedimentary basins, since the tool may be used to predict the occurrence of metalliferous mineralizations in sedimentary basins (e.g., Ruffell et al., 1998) in general and the rift basins in particular. Predictive rift stratigraphy, in particular, is not only important for hydrocarbon exploration, because fine-grained shales of the main rift phase may be hosting copper mineralization and other hydrothermal mineral deposits, including gold-enriched fluvial deposits (e.g., McIntyre et al., 2005; Pirajno, 1992; Whymark and Frimmel, 2017).

Although rift basins have been intensively studied, there are still some important discrepancies among researchers, as, for instance, concerning the rift initiation: some authors defend that the onset of the rift is marked by rapid subsidence and huge accommodation, while others defend a slow subsidence regime and consequent low accommodation rates for this phase. The infill of a rift basin is also a matter of debate since the early 1990s, when the period of proposition of comprehensive models for the rift initiation and the stratigraphic signature of the rift basin under the viewpoint of sequence stratigraphy began. This chapter intends to contribute to this debate, discussing some main aspects of sequence stratigraphy of continental rift basin, and proposing an integrative sequence stratigraphic model, looking toward a useful and predictive analytic tool. The chapter closes illustrating those models with examples from selected Brazilian rift basins.

2. WHY PASSIVE MARGIN SEQUENCE STRATIGRAPHY DOES NOT WORK IN RIFT BASINS?

Crustal stretching prior to continental breakup, forming elongate crustal depressions bounded by normal faults cutting the pre-rift succession and the crystalline basement, causes rift basins. The crustal stretching, which has rates up to 20 mm/year (Friedmann and Burbank, 1995), produces asymmetric half-grabens where subsidence is controlled by one major boundary fault zone, which may have an extension of tens of kilometers and a dip up to 70°. These structures have alternating dips; hence, tilting of adjacent half-grabens frequently goes to opposite directions, separated by a transfer fault. Therefore, a rift basin is actually composed of a series of depressions forming a system with a very complex geometry (Fig. 1).

A continental rift system develops when one branch of a three-armed rift system fails to open and does not reach the passive margin stage, developing a system of linked half-grabens infilled with continental systems, such as

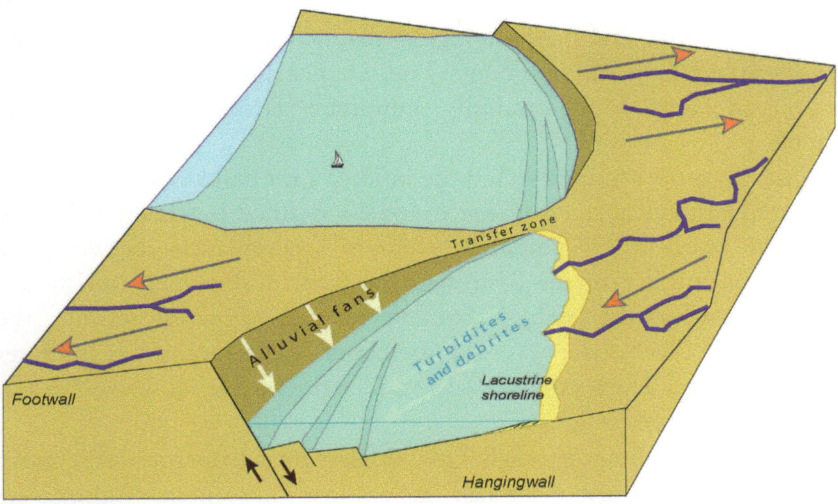

Fig. 1 Morphology and sediment supply of a typical half-graben system of a continental rift basin, showing two tectonic blocks separated by a high-angle normal fault. The block located above the fault plane is the hangingwall, the counterpart is the footwall. Note the inverted polarity of the half-grabens and the control of the rift relief on the direction of drainage. The *arrows* indicate the general direction of sediment yielded by different depositional processes.

alluvial fans, eolian, fluvial, deltaic, lacustrine debrites, and turbidites. Modern examples are East African rift system (mainly the Tanganyika and Malawi rifts) and Baikal Lake in Russia. Due to the asymmetry of half-grabens and subsidence controlled by one major fault zone, the accommodation space within a rift basin is variable, and zones with high accommodation develop close to zones with low accommodation or even erosion. This implies in a synchronous subsidence movement and uplift, therefore, synchronous increase and decrease in accommodation; hence, the popular passive margin-based sequence stratigraphic scheme, where the lowstand, transgressive, highstand, and forced regressive systems tracts develop following a predictable order during a complete cycle of base-level change, is not applicable.

Therefore, the traditional analysis of stacking patterns and the development of the aforementioned geometric systems tracts as proxies for base-level variations are not applicable in a rift basin, because regressive and transgressive systems tracts may form relatively synchronous and very close geographically. This is a pitfall for stratigraphic analysis of rift basins: within the same half-graben, one well may display a clear retrogradational pattern, while in a well a few tens of kilometers apart a progradational pattern may be seen. For a practical and useful stratigraphic analysis of rift basins, conceptual and methodological adaptations are required in order to understand the basins sedimentary infill, especially concerning the prediction of specific economically important facies and depositional systems.

One key aspect to understand the infill of a rift basin is the creation of accommodation space, a concept originally defined by Jervey (1988) as the space available for potential sediment accumulation, between the depositional surface of the basin and the base level, which is a surface defined regarding the geologic context: for passive margin settings, the ultimate base level is the sea level; for continental settings it is the river equilibrium profile; in continental rifts it is the lake level; in eolian settings it is the phreatic level.

The tectonic rifting regime has two fundamental properties that control the depositional history of the half-graben: first, movements (faulting) are fast, geologically almost instantaneous. Normal fault have vertical slip rates between 0.1 and 2 m kyr^{-1} (e.g., DePaolo and Anderson, 2000); second, subsidence is increasingly intense. A crustal stretching of ~30 km, for example, causes ~5.2 km subsidence (cf. Bosence, 1998), with tectonic movements in the order of 2–5 m (vertical) per thousand years, and with footwall uplift reaching expressive 10%–15% of the total rift displacement. Adding the sediment load, subsidence is increased, generating sediment successions with thicknesses up to 16 km.

As hangingwall subsidence is accompanied by synchronous footwall uplift, the creation of accommodation space is variable across the rift graben system, and, more important, creation and destruction of accommodation space is synchronous. Another factor to be considered is that due to the typical asymmetric geometry of the basin, zones with very different accommodation rates are emplaced during the increasing rift subsidence. In the beginning, supply (remembering: it is held constant in whatever sequence stratigraphic model!) outpaces the rate of accommodation due to the typical slow rates of subsidence at the rift onset (Fig. 2A). When subsidence increases, zones with different supply/accommodation ratios will emplace (Fig. 2B), and when subsidence reaches the maximum rates (the epoch of maximum rifting), accommodation will outpace sediment supply in almost the entire rift basin (Fig. 2C). Therefore, the sedimentation regime is progradational during the time of rift onset, becomes more retrogradational with increasing rifting (when progradation is limited to the flexural border of the system), and becomes almost entire retrogradational during the maximum rifting epoch.

Thus, the analysis of parasequence stacking pattern and mapping of different geometric tracts as indicators of the base level of variation is a hard task to be applied in a rift basin. This rises a problem for the stratigraphic analysis in a given half-graben: one borehole may be showing a clear retrogradational pattern, indicating transgression, while in a borehole located some tens of kilometers away the pattern may be undoubtedly progradational, indicating regression. In addition, one must consider that subsidence and creation of accommodation at the hangingwall is synchronous with uplift and erosion of the footwall; hence, zones of huge accommodation form synchronous with erosion surfaces relatively close to each other. Therefore, an unconformity (a sequences limit) in the footwall region may correspond to a transgressive surface in the hangingwall region of a given rift system.

These differences in comparison to a passive margin basin, "cradle" of the classic sequence stratigraphy model from 1988, have been recognized since the mid-1990s, when sequence stratigraphy began to be applied in different geological scenarios.

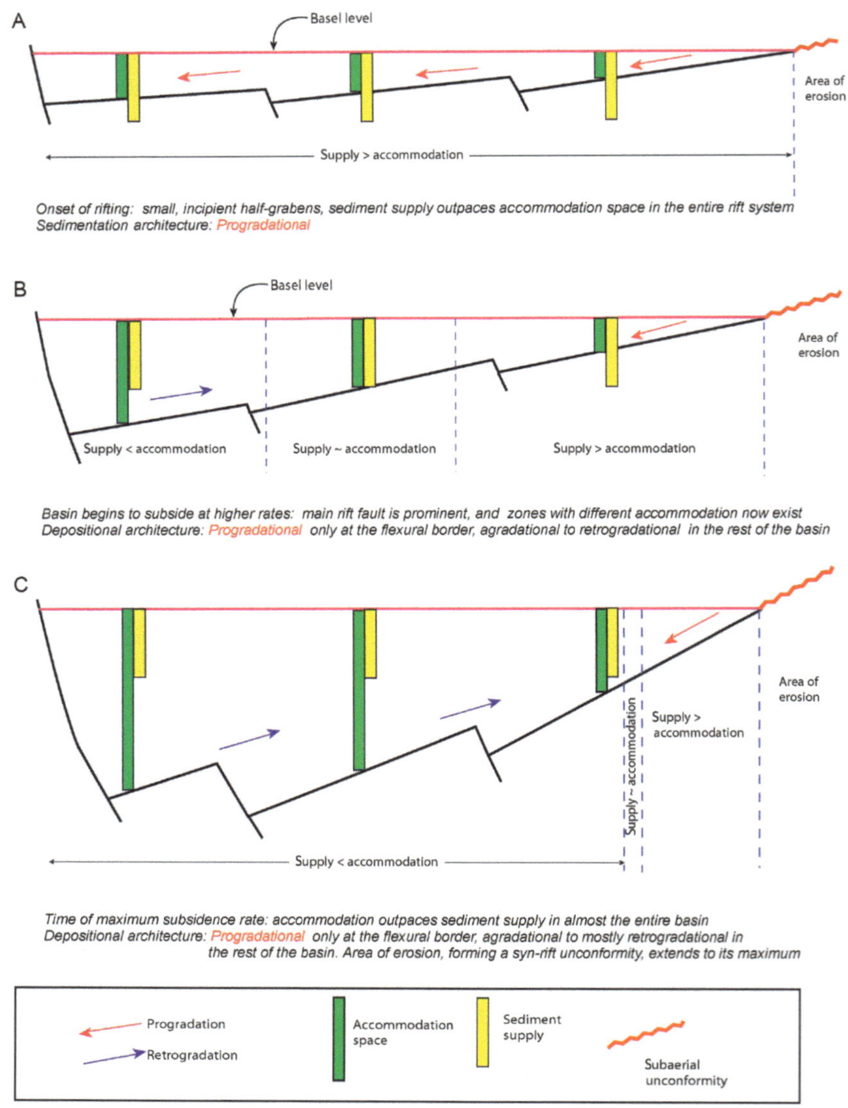

Fig. 2 Accommodation space in a rift basin: the subsidence of the hangingwall is accompanied by uplift the footwall; hence, the synchronous movements of subsidence and uplift of the half-grabens cause synchronous creation and destruction of accommodation space. *Green bars* represent the accommodation rate, variable due to increasing rift subsidence, while the *yellow bars* represent the sediment supply, held constant in sequence stratigraphic models. (A) Onset of rifting with low subsidence rate, hence low accommodation rate and high (and constant) sedimentation rate. (B) Increased subsidence leads to higher accommodation. (C) Maximum subsidence during the time of intense rifting. Note the changing depositional architecture during the increase of the rifting process.

3. CLASSIC TECTONOSTRATIGRAPHIC MODELS FOR RIFT BASINS—CONSENSUS AND CONTROVERSIES

What's the fuss about rift basins? How can there be controversies? Doesn't "everybody" know what a rift basin is?

These kind of questions may arise in the mind of the reader when depicturing a chapter on "controversies" and "consensus" concerning an apparently well-investigated and "solved" topic such as rift basins, which are important hosts for hydrocarbons and therefore intensively studied since many decades. The fact is that the answers for the questions above would be: "There is no consensus about how and when a rift basin begins neither how and when it ends, and there is also not much consensus on the mechanisms which control the sedimentation during the rifting process." This section intents to give a brief overview on these controversies.

First of all, the limits of a rift succession must be addressed. What kind of events delimit the rift basin's infill? What kind of stratigraphic surfaces are implied?

Bosence (1998) proposed the existence of two unconformities associated with a rift basin, the *syn*-rift unconformity and post-rift unconformity (PRU), delimiting three successions: a pre-rift, a *syn*-rift, and a post-rift (Fig. 3). The pre-rift strata are below the syn-rift unconformity and have

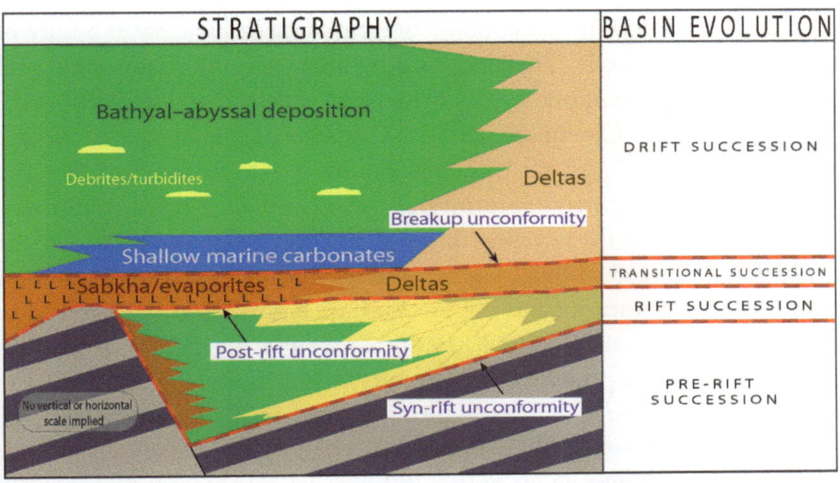

Fig. 3 Schematic overview of the tectonic evolution and stratigraphy of a rifted continental margin, redrawn with modifications from several authors (e.g., Bosence, 1998; Cainelli and Mohriak, 1999; Nottvedt et al., 1995). Not to scale.

an inclined geometric pattern with respect to the *syn*-rift succession. The syn-rift strata overlie the syn-rift unconformity and show a divergent and inclined geometric pattern. Finally, post-rift layers are superimposed on the PRU and feature a horizontal and parallel geometric pattern.

The *syn*-rift unconformity, also known as the "rift onset unconformity" (e.g., Falvey, 1974) marks the beginning of the rift phase. It is characterized by an erosive event that may be local or encompassing the entire basin, and is directly defined as an onlap surface of syn-rift strata characterized by tilting and rotation of structural blocks.

The PRU marks the end of the rift phase and is overlain by post-rift strata, which are initially fluvial–deltaic and shallow marine and evolving to a more oceanic, deepwater environment; therefore, most authors divide the post-rift succession into a "transitional" and a "drift" or "open ocean" phase (e.g., Cainelli and Mohriak, 1999; Nottvedt et al., 1995). The post-rift-unconformity may be equivalent to the "breakup unconformity"—in the case the rift basin evolves into a passive margin stage. The "breakup unconformity" is a basin-wide angular unconformity generated at the time of breakup and formation of the early oceanic crust and the beginning of post-rift sedimentation under conditions of thermal subsidence (Bosence, 1998), across which *syn*-rift deposits and the remaining rift topography are covered by flat-lying post-rift deposits (Falvey, 1974). Development of such a "separation unconformity" on passive margins during their final rifting stage preceding crustal separation is a function of thermal uplift exceeding extension-induced isostatic subsidence, resulting in subaerial exposure of large parts of the rift zone. These regional erosional unconformities are found in many passive margins but are not evident on others (e.g., Swiecicki et al., 1998; Ziegler and Cloetingh, 2004). The eventual equivalence of the "breakup" and the "post-rift" unconformities is due to the fact that the breakup unconformity may overprint and cut below the *syn*-rift unconformity and has the potential for beveling much of the previously deposited syn-rift strata (Nottvedt et al., 1995). Some authors make a distinction between the post-rift unconformity and the breakup unconformity, as depictured in Fig. 3 (e.g., Cainelli and Mohriak, 1999; Nottvedt et al., 1995), while users do not recognize the post-rift unconformity (e.g., Franke, 2012).

However, lower-ranked unconformities may form within the syn-rift succession (e.g., Prosser, 1993; Rosendahl et al., 1986) reflecting periods with reduced or nil subsidence of the rift basin, capable of generating significant geological gaps. These unconformities are dealt with in the conceptual models discussed ahead in this chapter.

The rifting is a progressive and diachronous process, i.e., it does not occur at the same geological time in every area of the rifting zone. The opening of the Atlantic margin between South America and Africa, for example, started in the south and spread to the north, through a system of propagating rifts. Bueno (2004) proposes a "propagating rift system" for the South Atlantic, consisting of three phases associated with implementation of oceanic crust:

(a) a rift phase at 131 Ma, extending from the Falkland Islands to the area of Paraná–Etendeka Province,

(b) a second phase at 126 Ma, ranging from north of the Pelotas Basin to south of the Santos Basin,

(c) a third phase at 115 Ma, extending from the northern basin of Santos to the Alagoas Basin.

Therefore, during the continental rifting process, while a segment is already in the sea-floor spreading phase, others are still in the rift phase. The term "propagation unconformity" has been proposed for that kind of unconformity, which is of interbasinal extension and formed by tectonic and erosional events that reflect the evolutionary diachronism of the propagating rift (Bueno, 2004).

An ongoing discussion among experts concerns the onset of the rifting. What is the diagnostic stratigraphic feature? What is the stratigraphic record of the onset of crustal stretching?

For some authors, the rifting process is marked by stages of rapid mechanical subsidence typically followed by longer periods of tectonic quiescence, when sediment supply gradually fills the available accommodation, showing an overall progradational trend. In this kind of model, flooding surfaces are the rift sequence boundaries, related to rapid tectonic subsidence and the consequent "instantaneous" generation of accommodation. Therefore, a typical rift sequence would develop initially a transgressive systems tract, and during the later rift phase, when accommodation rate decreases due to slower subsidence, would develop a highstand systems tract. This is the idea of Martins-Neto and Catuneanu (2010), applying this model to selected Brazilian rift basins. The problem with this approach is that the initial rift phase has the following characteristics:

1. It is characterized by low subsidence rates, as shown by subsidence curves of several rift basins worldwide (e.g., Izart et al., 2003; Moretti and Colletta, 1987; Strecker et al., 1988; Watts et al., 1993; see Fig. 4), hence low accommodation relative to sediment input so that the initial rift basin is easily filled and does not record extensive lacustrine facies. van Wees et al. (1996) studied the subsidence of the Dnieper–Donets basin formed during the Devonian rifting of the Eastern Europe craton,

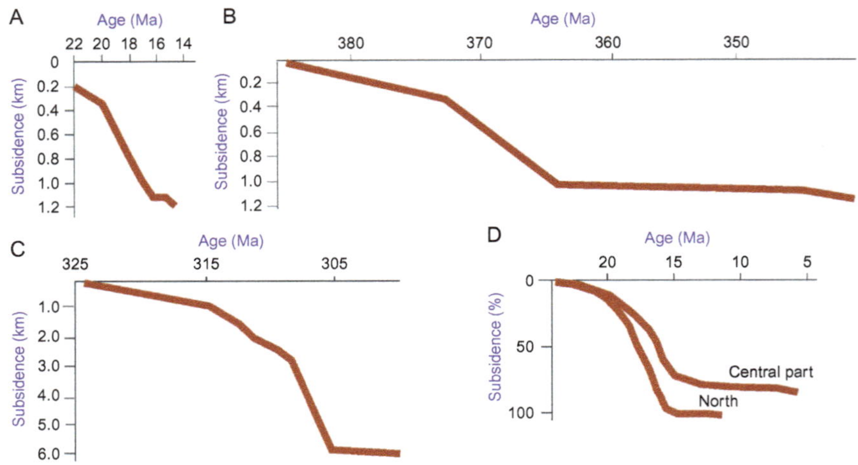

Fig. 4 Subsidence curves obtained by backstripping method in different rift basins, redrawn with the same timescale magnitude. (A) Backstripped tectonic subsidence of a well offshore Andalucia (Spain). (B) Subsidence curve of the late Devonian and early Carboniferous of the Dnieper–Donets Basin. (C) Subsidence curve of the Asturias Basin (total subsidence without corrections). (D) Curves of different Suez rift zones. Note the initial phase of low subsidence with relative short duration compared to the complete rifting process, while the main rifting phase, characterized by huge subsidence rates, represents more than 50% of the total time of rifting. *Panel (A): redrawn from* Watts, A.B., Platt, J., Buhl, P., 1993, Tectonic evolution of the Alboran Sea Basin. Basin Res. 5, 153–177. *Panel (B): redrawn from van* Wees, J.D., Stephenson, R.A., Stovba, S.M., Shymanovskyi, V.A., 1996. Tectonic variation in the Dniepr-Donets Basin from automated modelling of backstripped subsidence curves. Tectonophysics 268, 257–280. *Panel (C): redrawn after* Izart, A., Stephenson, R., Vai, G.B., Vachard, D., Nindrfe, Y.L., Vaslet, D., Sues, P., Fauvel, P.-J., Kossovaya, O., Chen, Z., MAslo, A., Stovba, S., 2003. Sequence stratigraphy and correlation of late Carboniferous and Permian in the CIS, Europe, Tethyan area, North Africa, Arabia, China, Gondwanaland and the USA. Palaeogeogr. Palaeoclimatol. Palaeoecol. 196 (1–2), 59–84. *Panel (D): redrawn from* Moretti, I., Colletta, B., 1987. Spatial and temporal evolution of the Suez rift subsidence. J. Geodyn. 7, 151–168.

between the Ukraine range and the Voronezh massive. The rift clearly had an initial phase of lower subsidence rate and then almost tripled the rate from the end of the Frasnian (Fig. 4B). This initial slow subsidence phase is also supported by modeling. Gupta et al. (1998), for instance, found that, even with constant extension rates, slow fault displacement rates prevail during an initial phase of distributed extension, followed by an increase in displacement rates; hence, rift basin stratigraphy commonly records an early stage of slow subsidence followed by an abrupt increase in subsidence rate. Therefore, the core conclusion would be: the onset of the rift succession is marked by a slow subsidence regime and consequent low accommodation rates of this initial tectonic phase, resulting in fluvial–deltaic successions with a progradational patterns. The rift initiation is not characterized by rapid subsidence and huge accommodation, filled with deep lacustrine facies, but by river and delta deposits with a progradational pattern, as demonstrated by ancient (e.g., Melchor, 2007; Prosser, 1993) and modern examples (e.g., Rohais et al., 2007).

2. The initial rift system seems to be formed by locally restricted faults, resulting in small, shallow, and disconnected basins rather than a basin huge and deep enough in order to record extensive lacustrine facies, as discussed later in this chapter (e.g., Gawthorpe and Leeder, 2000) and supported elsewhere (e.g., Kinabo et al., 2008; Morley et al., 2007). These initial depressions tend to connect during the rifting process, forming large half-grabens where deep lacustrine facies may develop.

It is noteworthy that for some authors (e.g., Kuchle et al., 2011; Morley, 2002) the rift onset is marked by the generation of a large and wide depression, resembling a syneclise. For instance, Scherer et al. (2014) have studied the continental rift system of the Brazilian Araripe Basin and corroborated the proposition of Morley (2002) that the first stage of rifting is the development of a wide, syneclise-like basin. A modern example of this model is the region in southern Tanzania known as the Usangu plains.

The problem here is that this kind of broad topographic depression does not always occur, as Morley clearly points out in his article of 2002 and of 2007, when he states that other areas of African rift system do not show the wide depression of the "Usangu Flats." Corti (2009), studying the rift of Ethiopia, also does not recognize the syneclise stage, depicting a rift system that starts with volcanism and uplift of the resulting volcanic plateau, followed by emplacement of small and laterally restricted faults that will connect and expand, as depictured by the tectonic model of Gawthorpe and Leeder (2000). Other African rifts studied in detail (e.g., Abdunaser and McCaffrey, 2014) also do not show the broad initial depression in the evolution of the rift succession. Arche and López-Gómez (1996) studied the intracratonic rift of the Iberian Basin, which developed from Lower Permian to Jurassic. Again, there is no recognizable feature of an "initial syneclises," but "very small basins along the initial breakup lineaments" quoting the authors. The rifts segments were connected during the Early Triassic forming a continuous structure, with transfer faults accommodating differences in stretching rates of that rift system. Similar context is reported for growth of the normal faults in the axial part of the Ethiopian Rift (Acocella et al., 2003).

Young et al. (2003) demonstrated that the Thal Fault, a rift border fault of the Suez Rift, has evolved by the progressive linkage of at least four fault segments during ca. 6 Myr of the initial rifting phase of that system. The model of small faults generating isolated half-grabens has also been corroborated by numerical modeling techniques. Starostenko et al. (1996), for instance, used a numerical model to predict that graben growth is divided into two stages: one preceding brittle failure stage, involving faulting, and a main phase with intense faulting. The initial stage is similar to what Gawthorpe and Leeder (2000) call the "fault initiation stage" characterizing rift onset, although Starostenko et al. (1996) label this phase as "pre-rift."

For the South Atlantic opening and the formation of the Brazilian rift basins, Cainelli and Mohriak (1999) describe as a "pre-rift megasequence" the sedimentary succession formed in a widespread syneclises prior to the main rift phase, although recognizing that the development of regional sags is related to an early stretching which preceded the main rift phase. The same is recognized for the West African South Atlantic margin (e.g., Beglinger et al., 2012).

From the above discussion, one can draw the conclusion that the interpretation of a wide depression (syneclise) as characteristic of the onset of a continental rift is a controversial proposal, defended by some authors, but not supported by others. Anyway, considering the tectonostratigraphic signature to be disclosed and interpreted when carrying out a basin analysis, "rift" implies in faulting, in tilting of blocks, and in formation of grabens, so when there is no direct evidence of faults being generated and controlling the drainage hence the sedimentation, why call this stage "rift onset"?

Another ongoing discussion among experts involves the end of the rift and the transition to the drift phase. For some authors (e.g., Nottvedt et al., 1995; Prosser, 1993) the rifting ends with the creation of large and deep half-grabens, with extensive lacustrine sedimentation including turbidites and sandy debrites. When this phase is over and the stretching-related subsidence is declining until reaching a complete stop, the reduced available accommodation space is filled by continental facies (eolian and fluvial–deltaic), a depositional phase these authors considered to be "post-rift." Hence, in this line of thought, the rift phase ends when the tectonic activity related to crustal stretching ends, even if the continental breakup has not occurred yet. Other authors (e.g., Bosence, 1998) argue that the sedimentation filling the final accommodation space of the grabens generated by the rift tectonics is "*syn*-rift" sedimentation.

So the distinction between syn-rift and early post-rift strata is not simple nor consensus. As stated by Nottvedt et al. (1995), theoretically, it is easy to define the boundary between syn-rift and post-rift strata in a half-graben, because it occurs after the end of faulting and a change in the tilting of fault blocks. However, basins that have suffered some degree of sediment starvation are usually characterized by remnant topography at this stage, and the early post-rift strata will therefore display wedge-shaped geometries. The divergence of *syn*-rift strata against the active footwall, in contrast with the parallel buildup and onlap of post-rift strata against this footwall after the end of movement, may be used as criteria for delimiting the two kinds of strata. Another important criterion would be the tilting of strata due to fault activation: while rotation of strata is away from the graben axis during the main rift stage, during the post-rift stage strata rotation is directed toward the basin axis.

The post-rift stage may be subdivided into an early post-rift sub-stage of sediment infilling of the remnant basin topography inherited from the active stretching stage and a late post-rift sub-stage of draping of the entire graben system (Nottvedt et al., 1995; Prosser, 1993).

Cainelli and Mohriak (1999) proposed a geodynamic model specifically for sequential evolution of the opening of the South Atlantic, recognizing five main phases with different patterns of tectonics and sedimentation. Phase I is marked by the beginning of extensional processes, marked by small asthenospheric uplift and regionally distributed thinning of the continental crust and upper mantle, with incipient faults in the upper crust controlling local depocenters associated with widespread, thin sedimentary sequences labeled the pre-rift megasequence. The next phases, forming the "syn-rift megasequence," are the formation half-grabens and deep lacustrine environments (Phase II) and increase in the lithospheric extension that is marked by large faults, which rotate the rift blocks, and the sedimentary layers previously deposited (Phase III). Phase IV begins with erosion of rift blocks by a regional unconformity that levels the topography, separating continental from transitional to marine environments of deposition. This surface is the breakup unconformity, overlain by fluvial–deltaic and sabkha-originated deposits, which form the "transitional megasequence." Phase V is marked by an increase in the bathymetry, shutting down the carbonate factory by the end of the Albian and resulting in the deepening of the environment of deposition, with abyssal–bathyal sedimentation.

For some authors (e.g., Karner et al., 2003), evaporites may be part of a late-stage *syn*-rift sediment package and topped by the breakup unconformity, so the definition and characterization of the "transitional sequence" is the most tricky one.

Although the existence of a transitional phase recording the change from *syn*-rift to post-rift conditions is reported for several divergent margins, difficulty may arise in recognizing "breakup unconformities" on continental margins in the presence of multi-phased rifting and breakup, as pointed out by Soares et al. (2012). Similar observation has been made by Kyrkjebø et al. (2004) for the Jurassic–Cretaceous syn-rift to post-rift transition in the central part of the northern North Sea basin, where the breakup unconformity is regionally known as the "base Cretaceous" or "late Cimmerian" unconformity; it is in fact a complex of unconformities. Hence, the boundary of the "transitional sequence" is often diffused, and therefore Soares et al. (2012) proposed the term "lithospheric breakup surface" to replace the term "breakup unconformity." It is acknowledged that the stratigraphic surface created by this lithospheric breakup event is not developed as an unconformity in all rift-related basins on the proximal, outer proximal, and distal margins, and they propose the concept of "breakup sequence" which comprises the complete seismic and stratigraphic expression of lithospheric breakup.

Concerning sequence stratigraphy, Prosser (1993) was the first to demonstrate that it is better to simply identify and describe the depositional systems in terms of facies and stacking pattern, and to interpret them in terms of local changes in the basic level (in this case, the lake level), without trying to recognize depositional systems tracts and link them to specific positions of the base-level variation curve, as it is done in sequence stratigraphy of a passive margin basin. The author based her model of stratigraphic evolution upon data from the North Sea rift basins, a geologic context in principle different from the Brazilian rift basins, but the model is effectively valid because of the basic premise—the sedimentation regime is controlled by the interaction of supply rate vs space accommodation rate. In this conception, genetically linked depositional systems would form in response to variation of subsidence of the rift zone, the so-called *tectonic systems tract*, i.e., a set of depositional systems linked to specific tectonic phase. For the active phase of rifting, three main stages of development are recognized by Prosser (1993):

1. *Rift initiation systems tract (RIST)*: at the beginning of rifting, faults are developing, but the typical rift relief does not exist yet. The space creation rate is still relatively low, and the resulting space is easily filled by

sediment yield, resulting in a progradational to aggradational stratal geometry. Sedimentation is dominantly fluvial and deltaic.

2. *Rift climax systems tract*: considered the main phase of the rift, with the establishment of systems of large and increasingly deeper half-grabens and development of significant fault scarps. Subsidence is growing and intense; the rift lakes form and become deeper. Subsidence outpaces sedimentation due to intense movement of faults and huge space creation. With the increasing rate of creation of space, sedimentation is initially aggradational but becomes increasingly retrogradational at the hangingwall, while in the footwall, close to the border fault, aggradation of alluvial fans occurs. The end of this tectonic tract is marked by the maximum expansion of the lake, in a similar situation to the maximum transgressive surface in the general model of sequence stratigraphy.
3. *Post-rift systems tract*: marks the end of sedimentation of the active phase of the rift, when tectonic movement gradually ceases and space creation rate is greatly reduced, resulting in progradational stratal geometry. Alluvial fans cease its formation due to denudation of the relief. After the phase of active rifting, a phase of tectonic quietude sets in, where the movement of faults decreases and ceases completely, ending the creation of space linked to the rift phase. Only the thermal subsidence is responsible for creating space in this phase. With tectonic stability and the consequent stability of the relief, drainage network can now be established very effectively and begins to erode the relief that was left of the rift phase, which results in increasingly sandy successions, filling up the remaining accommodation space.

Nottvedt et al. (1995) proposed a very similar tectonostratigraphic scheme, involving three distinct stages of basin development that may be or not separated by unconformities. The first stage labeled the proto-rift stage describes the rift onset. The tectonic evolution of this initial stage depends on the regional thermomechanical properties of the lithosphere (Cloetingh et al., 1997), sometimes characterized by deposition in a wide, slowly subsiding basin with only minor fault activity, as in some segments of the East African Rift System, or thermally induced upward of rift domes as it occurs at the Red Sea rift. The *syn*-rift stage is characterized by development of half-grabens, formed within the hanging walls of major rift-bounding or intra-rift basin faults. The post-rift stage is marked by decreasing subsidence and may be subdivided into an early post-rift sub-stage of sediment infilling of the remnant basin topography inherited from the active stretching stage and a late post-rift sub-stage of draping of the entire graben system, similar to the model depictured by Prosser (1993).

An interesting tectonic model was given by Gawthorpe and Leeder (2000) developing ideas raised by other authors (e.g., Crossley, 1984; Watson et al., 1987). The key point is that observations of fault geometry in rift zones have shown that the fault zone is formed by segments, usually of the order of 20–25 km long, with the highest displacement normally located in the center of the fault zone. Studies indicate a direct relationship between displacement (*D*) and length (*L*) of the fault segment. Initially, this relationship is more erratic, producing faults with a spread *D*/*U* ratio, but to the extent that those faults interact and join, the *D*/*U* is more correlative, i.e., higher *D* corresponds to higher *L* values. This led Gawthorpe and Leeder (2000) to propose three phases in the development of a rift basin. A large number of small faults, with no significant creation of topographic relief, characterize the initial stage or "fault initiation stage." The second stage is the "fault interaction and linkage stage," where the deformation in the fault arrangement starts to become more localized. Thus, some faults end up inactive, while others grow and join, forming a larger fault zone. The final stage is a continuous fault zone (through-going fault zone stage) where the deformation is localized along some large areas, forming large half-grabens with significant depocenters, while other faults die completely. Similar observations on the tectonic evolution of a rift basin were also made by other authors (e.g., Kinabo et al., 2008; Morley et al., 2007; Wang and Zhou, 2014; Withjack et al., 2002; Zhang, 2002), although the terminology differs. Zhang (2002), for instance, describes a rift succession with three stages of sequence development: (i) the "early rift stage megasequence" (where areas with faults are restricted and accommodation space is not large, hence distributed in smaller areal extent), (ii) the "intense rift stage megasequence" (with a huger creation of accommodation space and larger distribution area), and (iii) the "stable rift stage megasequence" (when the whole basin subsides slowly and accommodation space decreases, the topography of the basin flattens and the difference in thickness of strata in the steep side and slop side of the basin is not obvious).

Summarizing the discussion in this section:

1. For the rift onset there are two main interpretations—one considering small and isolated half-grabens as the rift initiation phase (e.g., Gawthorpe and Leeder, 2000; Watson et al., 1987), while the other interprets the syneclises prior to the active rifting phase already as the beginning of the rift (e.g., Morley, 2002).

2. Concerning the upper limit of the rift succession, there are two currents of thought: one argues that, with the termination of crustal stretching, the final filling of the rift basin is not rift, but "post-rift," disregarding the fact that it still takes a long time until the formation of breakup unconformity and the effective continental split-up. The other line of thought defends that the final infilling of the rift system, even though it is only under thermal subsidy and with little block tilting and stretching activity, represents the "final stage of rifting" or "rift termination."
3. Any basic tectonostratigraphic model of a continental rift system has to be threefold—the initial rift system is formed by locally restricted faults resulting in small, relatively shallow and separated basins, rather than a large basin deep enough for extensive lacustrine facies to be deposited. Due to the small amount of topographic relief during the initial phase, there is also no alluvial fan deposition. Toward the phase of rift maximum, the initial rift faults tend to link and form a larger and deeper depositional area, recording lacustrine facies with an overall retrogradational trend. Hence, locally restricted fluvial sandstone at the base of the stratigraphic succession records the onset of the rift phase, while the basal lacustrine mudstones found in other parts of the rift basin register an already more advanced rift phase. When the rift termination approaches, subsidence will slow down and accommodation rate decreases, the sedimentation regime is mainly progradational, and the rift basin is filled with fluvial, deltaic, and eolian facies.

Fig. 5 is an attempt to give a correlative overview on this tectonostratigraphic "controversies" and "consensus" among rift basin specialists, comparing the tectonostratigraphic evolution of a rifted margin with the several theoretical schemes discussed herein, including the conceptual sequence stratigraphic model presented ahead in this chapter.

3.1 The Role of Climate Variations

As discussed above, tectonics has an essential role as control mechanism of the rift basin infill. But what about climate? Considering that rift basins normally evolve to deep and widespread lacustrine systems, and considering that lakes are very sensitive to climate changes, what about climatic influence on the stratigraphy of a rift basin? That is what this section intends to discuss.

Earth history teaches us that paleoclimate changed profoundly during geologic time (e.g., Frakes, 1979). Since cyclical variations in the Earth's climate occur at different timescales, involving magnitudes from 10^1 to 10^6 years, it is important to get a grip on the climate cycle hierarchy in order

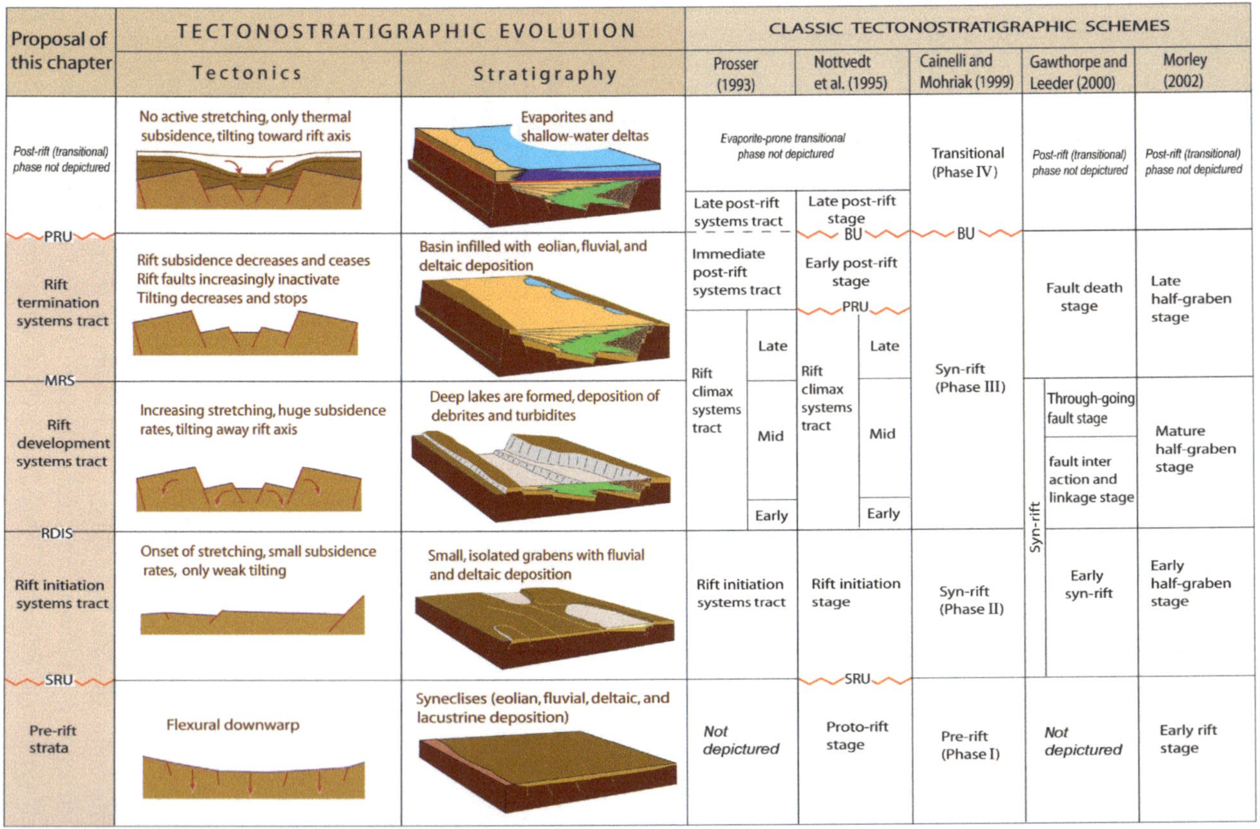

Fig. 5 Attempt toward a correlative overview chart on tectonostratigraphic "controversies" and "consensus" among rift basin specialists, comparing the tectonostratigraphic evolution of a rifted margin with the several theoretical schemes discussed herein. The first column shows the conceptual sequence stratigraphic model presented ahead in this chapter. *BU*, breakup unconformity; *PRU*, post-rift unconformity; *SRU*, syn-rift unconformity.

to discuss its influence on the stratigraphic record. Cycles at each scale are caused by a variety of physical mechanisms. For the large-scale global climatic cycles, Fischer (1982) introduced the nowadays widely used terms "greenhouse" and "icehouse" to characterize the two major climatic states of the Earth during the Phanerozoic. Kidder and Worsley (2010) proposed that there are actually three basic states for Earth climate: Icehouse, Greenhouse (subdivided into cool and warm states), and Hothouse. These stages have a 10^6 magnitude since they have a duration of tens of million years and could be labeled first-order climatic cycles. Therefore, the variations within the first-order stages (e.g., the glacial and interglacial stages during an Icehouse cycle of the earth) would be second-order cycles, with an duration of hundreds of thousands years (i.e., a 10^5 magnitude). Within these second-order cycles, minor climatic variations could be labeled third-order cycles, with a magnitude of 10^4 years and durations of tens of thousands of years. By this logic and only for the sake of organization of the different climatic signatures into a hierarchical scheme, the fourth-order cycles would have a duration of a few thousand years. Smaller cycles can also be considered and will have a fifth to sixth order of magnitude, such as ~1000 years "Century Cycle" and the Schwabe or El Nino cycles.

Table 1 shows a summary of this hierarchic scheme. Many of the climatic cycles are not quite well understood, neither have a defined duration, but the attempt of a hierarchy we show is important to understand the possible role of climate variation on the stratigraphic signature.

Climate cycles are important to be considered because lakes are depositional systems very sensitive to climatic changes, and lake levels vary according to multi-hierarchical control mechanisms, and during time, lake systems can be hydrologically "open" or "closed." An open lake is a system with a constant water flow, forming in areas where precipitation outpaces evaporation. Open lakes typically have stable water levels since hydric input is always matched by outflow to rivers. On the other hand, closed lake are systems where no water flows out, and hence salinity is significant. Closed lakes typically form in areas where evaporation is greater than precipitation. Water level in closed lakes is characteristically variable and very climatically driven. In time, a lake may change from open to close. Lake Malawi, for instance, is currently an open but it is known that during very dry periods such as the last glacial maximum, outflow from Lake Malawi (the Shire River) has ceased entirely. In fact, all major East African rift lakes (Malawi, Tanganika, Vitoria, Albert, etc.) have undergone open and closed phases many times during their history (e.g., Talbot and Allen, 1996).

Table 1 Tentative Scheme of Hierarchy for Earth's Climatic Cycles

Climatic Cycle	Characterization Description	Examples	Duration	Mean Control
First order	Icehouse, Greenhouse, and Hothouse Earth	Triassic Hothouse, Carboniferous icehouse (e.g., Frakes, 1979; Kidder and Worsley, 2010)	Tens to hundreds of millions years	Tectonics (supercontinent configuration, volcanism)
Second order	Huge glacial/interglacial epochs during an icehouse stage, or cool/warm periods during a greenhouse epoch	Triassic humid phase (Anysian humid pulse, Carnian Pluvial Event) (e.g., Haas et al., 2012; Szulc, 1999)	Hundreds of thousands to a few millions years	
Third order	Eccentricity cycle	Glacial and interglacial cycles of the last 800,000 years (e.g., Berger et al., 2016)	Few hundred thousand years	Earth's orbital (Milankovitch) cycles
Fourth order	Precession cycle Obliquity cycle		Few ten thousand years	
Fifth order	Century-scale cycle	Medieval Warm Period (AD 900–1300) and the Little Ice Age (AD 1450–1900) (Bond et al., 2001; Wanner et al., 2008)	200–1500 years	Changes in solar activity and ocean circulation patterns
Sixth order	Schwabe Cycle, El Niño, Brückner Cycle	Maunder Solar Minimum (e.g., Eddy, 1976; Usoskin et al., 2007)	5–100 years	
Seventh order	Seasonal cycle	Summer/winter	1 year	Earth's translation

The water level of the East African rift lake Malawi fluctuated as much as 250–400 m in the course of its history, while Lake Victoria had a water level 18 m higher 40,000 years ago (Zaarur et al., 2013). For Lake Malawi, important low stands probably lasting several tens of thousands of years have been reported by Delvaux (1995). The water level of the lakes Nakuru, Elmenteita, and Naivasha of the Central Kenya Rift also varied significantly during the last million years, as depictured by Bergner et al. (2009): during most of the time span, dry conditions reigned, leaving almost no significant lake record. During the main three wet periods—Mid Pleistocene (0.98–0.97 Ma), Late Pleistocene (0.14–0.073 Ma), and Early Holocene (0.015–0.009 Ma)—the lakes had a depth up to 180 m. For the Newark Basin Smoot and Olsen (1994) report cycles of rise and fall of lake level in response to climatic fluctuations at intervals of 20,000 years and larger patterns of 100,000- and 400,000-year intervals, recorded by up to 10-m-thick successions of lacustrine mudstone and shale.

Such large oscillations of the lake level produce two contrasting environmental scenarios. During periods of lake highstands, deltaic deposition occurs at the lake shoreline and sediment gravity flows (debrites, turbidites) in the depocenter, while during lowstands, coarse sediment accumulates in fluvial channels on the basin margins and in delta systems prograding into lakes that are considerably reduced in size. In these shrinked lakes, under extreme lowstand conditions, rift systems could develop saline alkaline water chemistry and even evaporite deposits (Cohen et al., 1997; Scholz et al., 1990). The lowstand deltas and associated deposits are most likely preserved in the record, covered by the sediments of the next highstand (Fig. 6).

The vertical lake-level changes of hundreds of meters may occur over time spans of hundreds of thousands of years, producing what can be termed "climatic sequences," corresponding to the third-order climatic cycle as discussed before (Table 1). The cyclicity within these third-order climatic sequences would be of fourth-order magnitude.

The first to describe cyclic lacustrine sequences was Van Houten (1962), studying successions of laminated shale grading to massive mudstone in the Newark Basin. Olsen (1986) quantified this cyclicity using Fourier analysis and linked the sedimentary cyclicity to Milankovitch periodicities of the Earth's orbit, concluding that cycles of 3–7 m thickness, which he termed "Van Houten cycles," are comparable to 20,000-year precession cycles (i.e., fourth-order climatic sequences), and that these are grouped into compound cycles equivalent to 100,000- and 400,000-year eccentricity cycles (i.e., third-order climatic sequences in the terms we use herein).

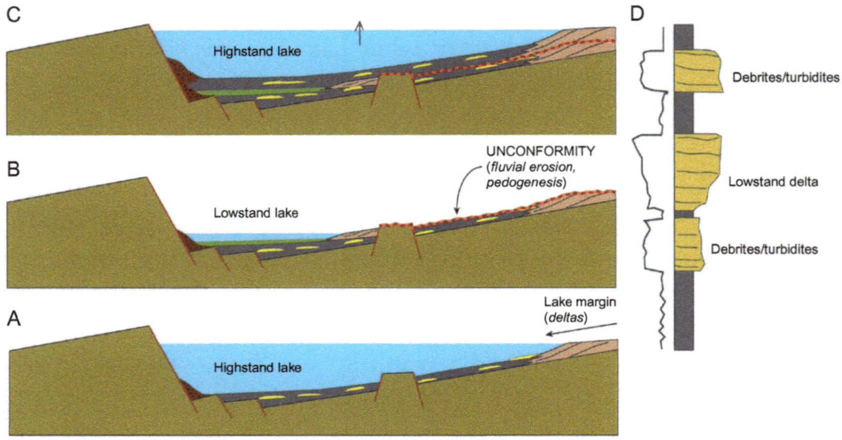

Fig. 6 Climate-controlled highstands and lowstand in rift lakes and their main depositional systems. (A) During periods of lake highstands, deltaic deposition occurs at the shoreline and gravity-driven deposits (debrites, turbidites) occur in the depocenter. (B) The lake shrinked due to climate change. The lowstand period is characterized by erosion, ravines, sediment bypass, incised fluvial channels, and lowstand delta systems prograding into a relatively shallow lake, where carbonates and even evaporite deposits may form. (C) During the subsequent highstand period, the lake will again expand and become deep, and the lowstand deposits are covered and preserved. Therefore, in a given profile as indicated by the well symbol, a lithologic log (D) may show deepwater deposits covered by the lowstand deltas. No vertical or horizontal scale implied.

The cause of such huge oscillations is the change in precipitation. However, hard data on precipitation rates in the geologic past are hard to find. For Holocene lake-level highstands, for instance, lake-balance modeling developed by Duehnforth et al. (2006) suggests a long-term increase in precipitation up to 45% as the main cause for lake-level oscillation. Hence, this kind of control may also be important for ancient rift successions. Data on global temperature (e.g., Veizer et al., 2000) show important modifications during the Cretaceous, the main period of rifting of the South Atlantic and the consequent formation and infill of basins, which will be used as examples ahead in this chapter.

Analyzing the global temperature variation during the time span encompassed by the sedimentary succession of the Recôncavo Basin, for instance, it is noticeable that important temperature variations have occurred during the Early Berriasian to Late Aptian (145–125 Ma) infill of that basin (Fig. 7).

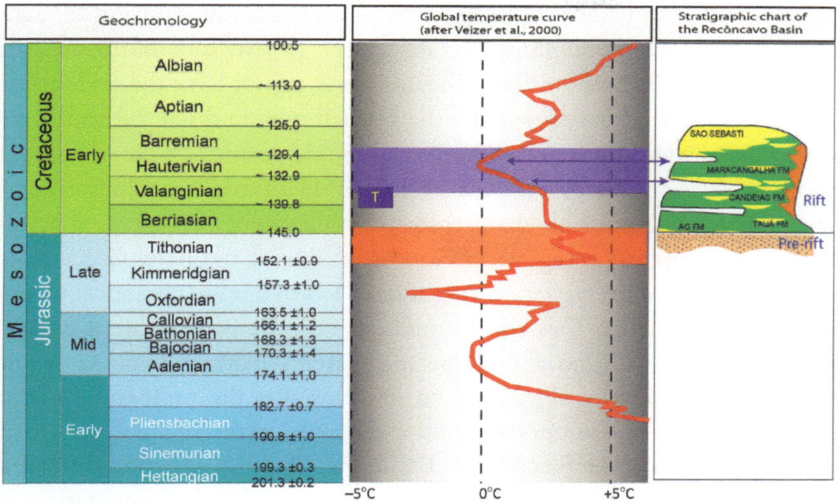

Fig. 7 Global temperature variation during the time span encompassed by the sedimentary succession of the Recôncavo Basin (northeastern Brazil). The *red* and *blue strips* enhance the warmest and coldest epochs of the time span recorded by the basin's sediments. *T*=stratigraphic position of the tillite described by Alley and Frakes (2003) in Australia. Note that during the pre-rift phase hot climate prevailed, and that the important temperature drop during the Berriasian–Valanginian–Hauterivian seems to correlate with important unconformities in the rift succession.

While the Oxfordian–Kimmeridgian (Early Late Jurassic) was a time of cool greenhouse conditions (Valdes and Sellwood, 1992), the Tithonian (Late Jurassic) experienced a climate sift toward more warm and humid conditions (Moore et al., 1992), which persisted until the beginning of the Cetaceous. The Berriasian (Early Cretaceous) was a relatively warm time, but toward the Valanginian the temperature dropped and in the Hauterivian cool greenhouse conditions ruled. Alley and Frakes (2003) described a tillite in southern Australia of Berriasian to Valanginian age based upon palynological and paleobotanical dating. After this "cold snap," increasingly warm conditions persisted, with a temperature increase of ~10°C from the Hauterivian toward the Albian (Veizer et al., 2000; Wang et al., 2012).

Comparing the described paleoclimatic evolution with the sedimentary record of the Recôncavo basin, two aspects seem noticeable. One is that the pre-rift strata seem to have been deposited during quite warm conditions, while the rift phase of that basin has developed under increasingly cooler temperatures. The other aspect is the gradual temperature drop which peaks during the Hauterivian. Biostratigraphic data of the basin, crosschecked with borehole and seismic data, indicate unconformities formed during the Late Valanginian–Hauterivian (Silva et al., 2007). These unconformities are restricted to the flexural border of the basin and do not have the widespread character as in other basins, such as the Camamu basin discussed ahead, for instance. However, the approximate stratigraphic position of these marginal unconformities correlates with the Valangian–Hauterivian cooling event It is speculative but admissible that a colder climate might have affected significantly the precipitation regime of Gondwana during the Early Cretaceous, and that the loss of precipitation might have led to a huge climatically controlled lake lowstands. This climate influence on the generation of unconformities in the basin has already be speculated by others (e.g., Silva and Picarelli, 1990; Silva et al., 2007), so it seems that the temperature drop may have enhanced the unconformities recorded at the flexural margin of the rift succession of the Recôncavo Basin (Fig. 7). That does not imply that climate rather than tectonics is the main driving force of generation of unconformities, but that in specific cases the climate overprint may enhance the expressions of unconformities, as it seems to be the case for the Recôncavo Basin.

As stated by Bergner et al. (2009), interpretation of lake records from tectonically active extensional regions requires a careful assessment of tectonic and other factors, such as volcanic forcing on sedimentation and lake history, even on short timescales of 10^4–10^3 years, and that changes of lake level must be interpreted with care and do not automatically reflect the effects of climatic change. Kreuser (1995) studying the rift of the Karoo basin states that controlling factors in the development of lacustrine systems—climate vs tectonics—cannot be estimated reliably for this basin, but comments on depositional pattern which may represent sporadic subsidence rather than cyclic climatic change. In contrast, Olsen (1986) linked depositional mechanisms for laminated lacustrine sequences in the Newark Basin to climatic changes induced by orbital forcing, as mentioned before. Schlische and Olsen (1990) later presented an opposing model citing several features that favor a dominant tectonic control for that basin. These include: (1) stratal onlap; (2) transition from initial fluvial to deep lacustrine conditions, followed by a progressive decrease in water depth during lake evolution; and (3) marked asymmetry linked to increased extension and depositional rates in rift development. Hamblim (1992) interpreted Lower Carboniferous lacustrine rift sediments with multiple shallowing-upward sequences, in Nova Scotia, Canada, to have formed mainly under tectonic rather than climatic control. Kreuser (1995) studied the lacustrine phases in the

East African rifts and confirmed a strong tectonic influence on lake development. Samanta et al. (2016) have studied the effect of climate on sequence development in the Cambay rift basin, western India, during the Early Paleogene super greenhouse globe. Detailed sequence stratigraphic analysis has indicated that climate-induced high siliciclastic supply possibly enhanced the autocyclic reorganization and hindered the development of key sequence stratigraphic surfaces across the basin during climate extremes of the Paleocene.

From the above discussion, one can conclude that for rift basins there is no consensus about how a depositional sequence develops under strong climatic control, considering that the rate of accommodation in rift basins is very high due to the intensively of the rift tectonic. It seems that on a short-term scale (i.e., third order or less, considering the climate cycles hierarchy cf. Table 1) the climatic overprint is clearly present, i.e., that climatically controlled lake-level fluctuations may be expressive enough to imprint a stratigraphic signature to the rift basin's sedimentary record. However, the stratigraphic signature of a complete rift succession is primarily driven by tectonics and the consequent accommodation rate, i.e., on a long-term perspective, tectonics is the main control on the stratigraphic signature of the rift succession, as discussed in Section 4.

4. CONCEPTUAL SEQUENCE STRATIGRAPHIC MODELS FOR CONTINENTAL RIFT SUCCESSIONS

Sequence stratigraphy is a tool for basin analysis, where the main goal is not to reach the "absolute truth" about the stratigraphic record (because that is beyond possibility), but to understand what happened during the basins infill. A meaningful sequence stratigraphic model must rely on the identification of the sequence stratigraphic surfaces that can be traced across the entire basin and hence support the quest for understanding. This section discusses and exemplifies the adaptations regarding the classical sequence stratigraphic model, and discusses the tectonic and stratigraphic aspects related to rift basins. Initially, it is necessary to clarify the concept of "tectonic systems tract" as the conceptual model equivalent to the geometric systems tracts of the classical sequence stratigraphic model. The necessity derives from the fact that the sequence stratigraphic terminology uses the term "systems tract" in two different hierarchical levels, but not always the texts (such as articles, etc.) make a proper differentiation. Hence, the origin of the term "systems tract" will be addressed here.

Broadening the concept of "depositional system" of Fisher and McGowen (1969), the concept of "depositional systems tract" was introduced by Brown and Fisher (1977), defined as a set of contemporary depositional systems. The sedimentary unit is known as "parasequence," as defined by Van Wagoner et al. (1988); it is essentially the same as "depositional systems tract." That is because "parasequence" is also defined as a set of genetic and temporally linked depositional systems, with the only difference that its limits are marine flooding surfaces, a nonexistent element in the concept of "depositional systems tract."

In a higher hierarchical level, there is the concept of the "geometric systems tract," which is the fundamental building block of a depositional sequence, i.e., it is the first subdivision within a depositional sequence, usually based on the standard geometric stacking patterns of the strata (retrogradational, etc.), which in turn can be well detected by the correlation of parasequences (i.e., the set of depositional systems tracts delimited by marine flooding surfaces). This brief discussion is needed to make clear and define that "tectonic systems tract" as a stratigraphic unit that groups all the depositional systems and formed parasequences during a certain tectonic phase of the rift basin is conceptual and hierarchically equivalent to the "geometric systems tract" in the classic model of stratigraphic sequences.

In the basic model of sequence stratigraphy, the four geometric systems tracts develop in a cyclical order, and the recognition of lowstand and highstand normal regressions, forced regression, and transgression. The establishment of the framework of the geometric systems tracts is extremely useful in the analysis of passive margin or intracratonic basins, because they allow mapping base-level variations, thus allowing the understanding of the distribution of facies associations and economically important depositional systems.

Nevertheless, in rift basins, the main control factor of base level is not the sea-level change, and the main aspects influencing accommodation space are:

1. the peculiar tectonics of half-grabens, with conceptually synchronous creation and destruction of accommodation space, as discussed earlier in this chapter and
2. the peculiar subsidence of a rift basin, which begins with low rates, increases to very high rates, and decreases during the final stage of the rift process.

Based on these characteristics, in this chapter we propose a threefold stratigraphic scheme of tectonic systems tracts, inspired by the classic scheme of Prosser (1993) and Nottvedt et al. (1995), integrated to the threefold tectonic model of Gawthorpe and Leeder (2000). We propose an integrative sequence stratigraphic model with three tectonostratigraphic phases:

1. the rift initiation phase, which is characterized by isolated and restricted fault which creates incipient half-grabens, infilled with fluvial and deltaic facies,
2. the rift development phase, during which the initial rift faults tend to link and to form a larger and deeper depositional area, recording lacustrine facies with an overall retrogradational trend, and
3. rift fill-up phase, when the accommodation rate decreases, the sedimentation regime is progradational and the rift basin is filled with deltaic, fluvial, and eolian facies.

   Each of these phases is recorded by a specific tectonic systems tracts, which is defined as a unit gathering all sedimentary deposits formed during a specific tectonic phase of the rift basin, as preconized by Prosser (1993). This tripartite stratigraphic architecture has also been recognized by other authors (e.g., Lambiase, 1990; Schlische and Olsen, 1990; Schlische and Withjack, 1999), showing that numerous continental rift basins show an infill, which begins with fluvial deposits followed by a relatively abrupt deepening-upward lacustrine successions, overlain by a gradual shallowing-upward lacustrine and fluvial succession, independently of the geographic setting or the geologic age.

Concerning the classic papers, three main modifications introduced by this chapter are:

- a change of nomenclature of the tectonic systems tracts is introduced,
- a terminology for the stratigraphic surfaces involved in this model is proposed,
- the sediment phase after the climax of the rift that Prosser (1993) and Nottvedt et al. (1995) consider as "post-rift" is herein still considered as belonging to the rift phase, as detailed ahead in this chapter.

4.1 Tectonic Systems Tracts

The sedimentation developed during each of the three above depictured tectonostratigraphic phases is linked to a so-called tectonic systems tract, a concept first proposed by Prosser (1993) and applied by several authors during the 1990s (e.g., Dinis and Trincão, 1995; Nottvedt et al., 1995; Sopeña and Sánchez-Moya, 1997).

4.1.1 Rift Initiation Systems Tract

The onset of the rift is characterized by the installation of isolated and restricted faults that create incipient half-grabens, which are characterized by a low accommodation rate and infilled by fluvial and deltaic facies under a general progradational sedimentation regime. During this initial phase, significant parts of the future area of the rift basin are not yet deposition zones, and due to the relatively low topographic relief, no alluvial fan deposition yet occurs (Fig. 8A).

4.1.2 Rift Development Systems Tract

This systems tract was called the "rift climax" in the classic paper of Prosser (1993); however, in this research it was chosen to change this nomenclature because the concept of "climax" involves the notion of a fast and very short event, but not the most enduring and intense tectonic phase of the rift. In the matter of fact, it can represent more than 50% of the total time of rifting (Fig. 4).

The sedimentary record of this tectonic tract occurs during the main rift phase, when the initial faults tend to connect, forming a larger and deeper depositional area under high subsidence rates. The previously non-depositional areas are now also subsiding and turned into depositional areas due to the increasing rifting rate. The sedimentary fill during this phase is characterized by the development of shallow to deep lacustrine facies with a general retrogradational trend. Thus, during this phase, the areas that recorded the initial fluvial sedimentation and the previously nondepositional areas now are part of the rift basin and are mostly filled with lacustrine and gravity-driven deposits, such as debrites and turbidites (Fig. 8B).

4.1.3 Rift Termination Systems Tract

This tract was labeled "immediate post-rift systems tract" in the classic paper of Prosser (1993), but herein we interpret the final infill of the basin as belonging to the rift phase, despite the decrease of tilting and of mechanical subsidence induced by crustal stretching. The motive is that this decrease is not an abrupt but rather a gradual phenomenon during this phase, since subsidence will continue due to the lithospheric thermal cooling, and because the unconformity separating the rift from the transitional phase has not been formed yet, neither has the crust really split-up, and the process of tearing apart two continents still goes on.

Hence, herein, all sedimentation prior to the timing of the post-rift unconformity (PRU) (which might be overprinted by the breakup unconformity sensu Falvey, 1974) is gathered in the Rift Termination Systems Tract (RTST). It is formed because when rifting ends, there is a reduction of accommodation rates and the overall sedimentation regime will be progradational. The rift basin is filled initially by deltaic sedimentation and ending with fluvial and eolian sedimentation. Since the rifting decreases until

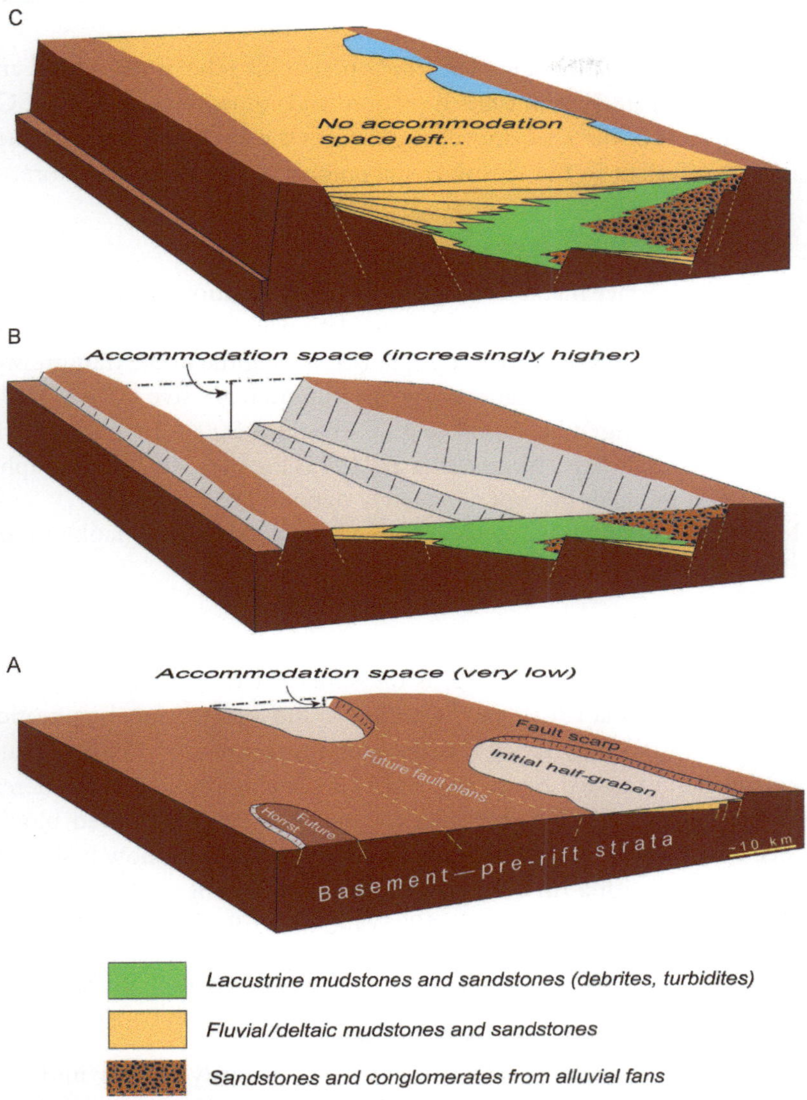

Fig. 8 The conceptual model adopted for the stratigraphic analysis of a continental rift basin depictures three phases, characterized by tectonic systems tracts. (A) Rift initiation systems tract, characterized by isolated and restricted faults that create incipient half-grabens, filled mostly by fluvial and deltaic sedimentation under a progradational regime. During this initial phase, a significant part of the future area of the basin is not yet a depositional zone. (B) Rift development systems tract, where the initial faults tend to connect, forming a larger and deeper depositional area under high subsidence rates. The previously nondepositional areas are now also rapidly subsiding. This phase is characterized by the development of lacustrine facies with a general retrogradational trend. (C) Rift termination systems tract, with decrease in accommodation rates and a mainly progradational sedimentation regime. The basin is filled with fluvial, deltaic, and eolian sediments.

there is no fault activation anymore, the topographic relief is denudated and the deposition of alluvial fan facies is reduced and comes to an end (Fig. 8C). The denudation of the remnant relief generates the PRU (cf. Fig. 3) which might be equivalent to the "breakup unconformity" as discussed before.

4.2 Stratigraphic Surfaces

A stratigraphic surface marks a change of the sedimentation regime. The surface which marks the shift from the progradational/aggradational regime of the lowstand systems tract to the retrogradational regime of the transgressive systems tract, for instance, is labeled the "maximum regressive surface" in the basic sequence stratigraphic model (e.g., Catuneanu, 2006). In the classic rift stratigraphy models (e.g., Nottvedt et al., 1995; Prosser, 1993) stratigraphic surfaces were implied, but not adequately defined nor labeled.

Herein, the stratigraphic surfaces of the proposed rift stratigraphy model are the following.

4.2.1 Syn-Rift Unconformity

The syn-rift unconformity (SRU) is the surface that separates the pre-rift strata from the initial rift strata, involving tilting and erosional truncation of the pre-rift strata, reiterating that the broad syneclises at the base of many typical rift basin formed by graben and half-graben composites are herein interpreted as pre-rift, because of the absence of fault-delimited blocks, absence of active tilting, and sedimentation in broad and shallow basin with no resemblance with a rift basin. The term is equivalent to the "proto-rift unconformity" (cf. Nottvedt et al., 1995) and the "rift onset unconformity" (cf. Withjack et al., 2002).

4.2.2 Rift Development Initiation Surface

That surface separates the locally restricted and relatively shallow initial rift lakes with a dominantly progradational pattern (the rift initiation systems tract—RIST) and the widely distributed increasingly broader and deeper lacustrine facies with a retrogradational pattern (the rift development systems tract—RDST).

4.2.3 Maximum Rifting Surface

Separate the deep lacustrine retrogradational phase of the rift development systems tract (RDST) from the progradational pattern of the rift termination systems tract (RTST), when the lakes are being filled-in due to reduced accommodation rate.

4.2.4 Post-rift Unconformity

That hiatal surface separates the final rift strata of the RTST from post-rift strata and may be equivalent to the "breakup unconformity. Nottvedt et al. (1995) labeled that unconformity as the "syn-rift unconformity," despite the fact that this surface marks not a "syn"-rift phase but rather the end of the rift and the beginning of a new tectonic phase.

4.3 Conceptual Models for Single- and Multi-rift Successions

Some rift successions appear to have undergone a more gradual and uniform subsidence regime due to rapid crustal stretching, with few base-level variations recorded at the flexural border, while other rift successions record several phases of more or less marked subsidence, followed by phases without creation of accommodation space and by generation of subaerial unconformities. Hence, two models are shown here, one dealing with a rather gradual subsidence, where much of the basin has a quite continuous history of subsidence, and one of subsidence punctuated by phases of subsidence stillstands.

4.3.1 The "Gradual Subsidence" Model

A rift basin with a regime of "gradual subsidence" has an initial phase of low subsidence rate, followed by an epoch of increasing subsidence due to intense rifting, and finally reaching a decreasing phase, when the rifting process comes to its end. As shown before, several ancient and modern rift systems are reported to have this kind of subsidence history (see Fig. 4). For that scenario, Holz et al. (2015) had proposed an integrative sequence stratigraphic model where three tectonostratigraphic phases record different tectonic systems tract. Fig. 9A and B shows the hypothetical steady subsidence curve and the implied conceptual stratigraphic framework as discussed above.

4.3.2 The "Punctuated Subsidence" Model

A rift basin with a regime of "punctuated subsidence" is marked by periods where tectonic subsidence ceases. In this tectonic scenario, the subsidence history of the rift basin is more complex, because after the initial phase characterized by the incipient half-grabens and low accommodation rates (forming the RIST), intra-rift unconformities are formed during stillstand periods of the subsidence. After a stillstand, the resuming subsidence increases and creates accommodation space, resulting in larger and deeper depositional area, recording the RDST. When approaching a new period of stillstand, accommodation rate decreases and the sedimentation regime will be progradational again, recording the RTST. The result of that kind

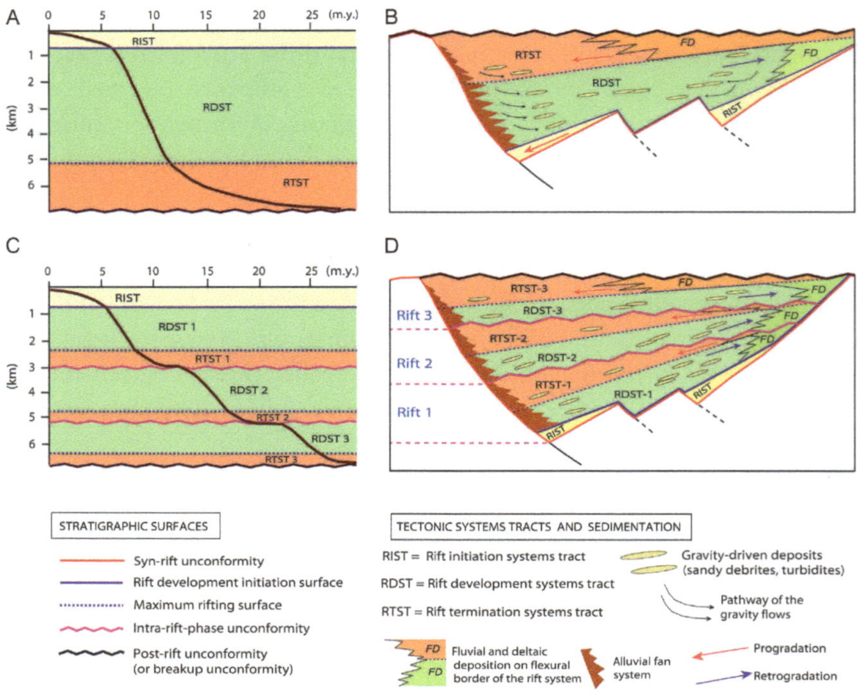

Fig. 9 The conceptual tectonostratigraphic model with a hypothetical subsidence curve and the implied conceptual stratigraphic framework as discussed in the text. (A) and (B) illustrate the "gradual subsidence model," while (C) and (D) show depicture the "punctuated subsidence model."

of punctuated subsidence (Fig. 9C) is a rift succession with several intra-rift unconformities, each one overlying an RTST, eroding most of the progradational (fluvial–deltaic) record of that systems tract (Fig. 9D). The multiple rifting events and the punctuated sedimentary record are the reflex of the fact that the stress built up and released to the stretching crust is not a uniform process (e.g., Nottvedt et al., 1995), so the fault movement and the rotation of blocks are not uniform nor gradual, but episodic. The possibility of thermal uplift of rift zones must be remembered here. As pointed by Ziegler and Cloetingh (2004), models of purely mechanical, "passive" stretching of the lithosphere do not take into account the possibility of thermal upward displacement of the asthenosphere–lithosphere boundary. The authors state that stress-induced extension of the lithosphere, causing upwelling of the asthenosphere, is accompanied by thermal destabilization of the asthenosphere–lithosphere boundary, which contributes to its progressive upward displacement and thus to doming of a rift zone. Moreover, intrusion of melts at intralithospheric levels, such as the crust–mantle boundary, can also cause uplift of a rift zone. Massive thermal doming of rift zones can commence 15–60 Ma after the beginning of crustal extensions (e.g., Ziegler, 1992), with uplift of rift flanks in excess of 2 km.

On seismic data, seismic reflectors that exhibit rotation and divergence, indicative of differential subsidence and block rotation, characterize each rift event. Hence, what is generically called "the rift succession" of a given basin may have several unconformities, so the term "rift successions" (i.e., plural) would be more adequate. Moreover, more important is to enhance the fact that a basin with several intra-rift unconformities may be adjacent to a basin with one single rift succession, as will be shown by the regional examples ahead.

4.4 Hierarchy of Rift Sequences

The first proposals on depositional sequence hierarchy emerged within the framework of seismic stratigraphy (e.g., Payton, 1977) and were based on the duration of the eustatic cycles. Several authors address the issue, considering the major eustatic cycles, with an duration of 50 million years or more, as the driving mechanism for first-order sequences, with subdivisions into smaller cycles: second-order sequences of 3–50 million years and third-order sequences of 1–3 millions of years (e.g., Duval et al., 1992; Einsele et al., 1991; Meyers and Peters, 2011; Vail et al., 1977). There are three problems in this approach of organizing the cycles by their duration. The first is that one needs to have relatively consistent temporal data, i.e., the time resolution of the sedimentary succession must be precise so that the proper hierarchy can be made. This is not always the case, especially in those basins with low-resolution biostratigraphic zoning, or those that do not have any biostratigraphic data, as the Precambrian basins, which lack fossils. The second problem is that the cycles do not always close with the theoretical scheme, that is, not necessarily a third-order cycle falls within the range of 200,000 to 3 million years, as predicted by the time-based hierarchy scheme. The third problem stems from the second: cycles do not always have the duration advocated by cyclical studies like those of the aforementioned authors. Literature shows a very wide range of papers that assign different times to cycles of the same order, including huge overlappings—what for one author is first order, for another it is already second order. Schlager (2004) summed these controversies up by comparing data on the duration of cycles published in articles since the 1970s, showing that the duration of a cycle is very “controlled” by the opinion of the author, which used, in a very loose way, that order of magnitude and appear to be subdivisions for convenience rather than actual organization.

In order to solve this problem, a different conception of the hierarchy of sedimentary cycles emerged, originally presented by Embry (1995). This conception does not consider the time involved in the formation of the sequences, but the magnitude of changes in the base level and the extent of the respective sequence limits. To classify the hierarchy of a sequence boundary, attributes such as the intensity of change of the sedimentary regime above and below a given stratigraphic surface or the degree of change of the tectonic regime are decisive. Embry (1995) originally proposed this hierarchical scheme for his T-R sequences, created a scheme where the first-order units are delimited by subaerial unconformities associated with significant deformations, while the order of lower hierarchy would be the transgressive surfaces of only local expression within the basin. The main problem in this conception of hierarchizing the stratigraphic record is the following: in order to be applied, it is necessary that there is a good preservation of the proximal succession of the basin, because otherwise there is no way to evaluate the extension of sequences limits or the degree of modification across the boundary. However, it is precisely the proximal portion along the edges of the basin that is usually the first to be eroded during the tectonic phases subsequent to the formation of a basin.

One can summarize the controversial issue of hierarchies as follows: both hierarchical schemes have merits and failures. The scheme based on the duration of the cycles of change of the base level is “tied” too temporarily and is inapplicable when there is no good temporal resolution. On the other hand, the hierarchical scheme based on the magnitude of the base-level changes does not work very well when there is little preservation of the proximal part of the basin. In order to solve this dilemma, Catuneanu (2006) argues in favor of abandoning a rigid and formal scheme, and proposes that hierarchy of the boundary surfaces of the sequences should be established on a “case-by-case basis,” i.e., for each basin, hierarchy must be determined based on the relative importance of the areas, taking into account the signature of the surface and, when possible, duration can also be taken into account.

That concept was adopted here and is used to describe the rift stratigraphy as shown in Section 5. The relative importance determines hierarchical greatness. Under this viewpoint, the first hierarchical order is the tectonic phase, i.e., the rift phase. The whole record of this phase is affixed to the label of “first-order sequence.” In Cenozoic Brazilian rift basins, the first order is a 20–30-million-year record. Within the first-order rift succession, major surfaces such as the unconformities predicted in the "punctuated subsidence model" described before divide the sedimentary record into second-order rift sequences, each on composed of tectonic systems tracts (RIST, RDST, and RTST).

5. EXAMPLES OF RIFT BASIN SEQUENCE STRATIGRAPHY

The breakup of southwestern Gondwana occurred during the Cretaceous, when several rift basins were generated along the equatorial and Atlantic margin of South America. Rifting between Africa and Brazil occurred in three phases: Berriasian (~145–140 Ma), Hauterivian (~135–137 Ma), and Late Barremian to Early Aptian (~130–125 Ma), so the rift phases of the Brazilian basin record at least three phases of basin infill, separated by intra-rift unconformities. In order to illustrate the applicability of the tectonic tracts scheme and the implied stratigraphic surfaces, we selected four basins with different geotectonic and stratigraphic contexts. These are, from north to south, the Ceará, Recôncavo, Camamu, and Pelotas basins.

The first rifting event records the three tectonic systems tracts as depictured by the "punctuated subsidence" model, while the other rifting phases record a huge RDST and the RTST, overlain by an intra-rift unconformity or, in case of the last rifting event, by the PRU. However, there are important discrepancies between the basins. While most of the Recôncavo rift is well depictured by the "gradual subsidence" model, the northern Camamu basin records seven rifting events, as discussed ahead. Fig. 10 shows a simplified scheme of the stratigraphy of these basins.

The Ceará Basin was chosen because it is located on the equatorial margin, where the rifting began later in comparison to the opening of the south Atlantic margin. The Recôncavo basin represents an aborted rift, while the Camamu Basin, directly adjacent to the Recôncavo basin, developed the

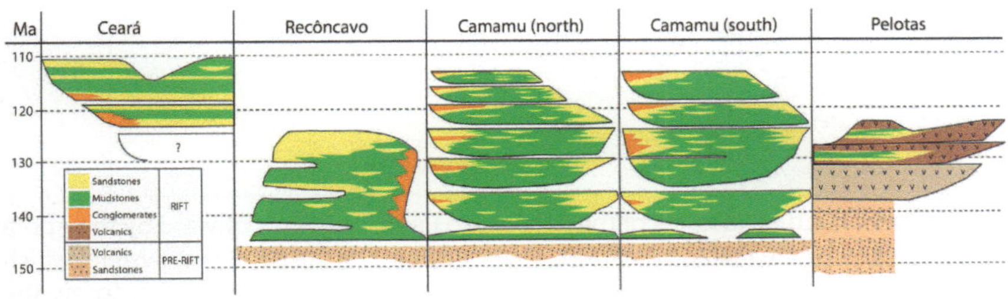

Fig. 10 Stratigraphic overview of the five Brazilian rift basins discussed herein.

breakup unconformity and the passive margin succession. The Pelotas basin, on the southernmost shorezone of the Brazilian coast, underwent a relatively rapid rifting process, accompanied by intense volcanism, so that it has a rather different stratigraphic signature compared to the other Brazilian marginal basins. For each basin, a representative seismic line and a well log will be shown and the sequence stratigraphy of the rift succession will be briefly discussed.

5.1 Recôncavo Basin

The Recôncavo Basin, located in northeastern Brazil, with an area of ~11,500 km^2, is Cretaceous rift basin in northeastern Brazil with a sedimentary record from Late Jurassic to mid-Cretaceous, spanning a period of about 20 Ma. That area forms the southern end of the inland "Recôncavo–Tucano–Jatobá aulacogene," a huge continental rift system which extends from Bahia to Pernambuco states, embracing 1200 km of aborted rift basins. The structured NE–SW elongated half-graben is limited by the structural "Aporá High" toward the adjacent Tucano Basin and by several fault systems (Fig. 11) and records pre-, syn-, and post-rift sediments overlying a

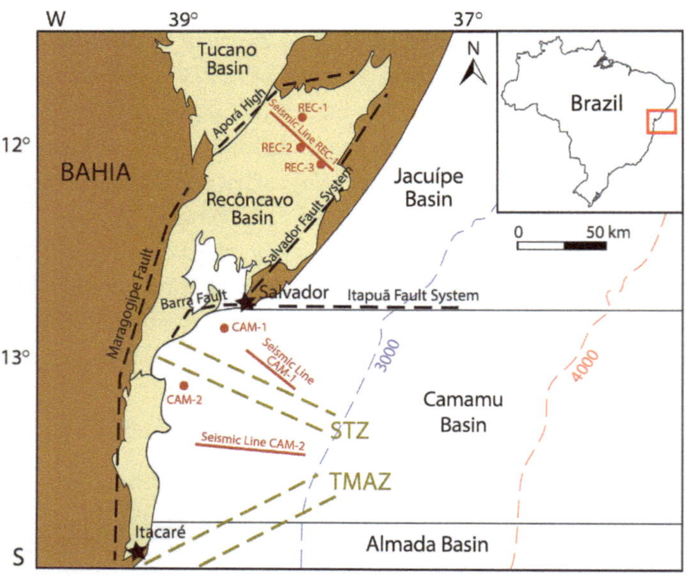

Fig. 11 Basic structural features of the Recôncavo Basin, limited northward by the Aporá High, westward by the Maragogipe fault, and southward by the Barra fault system, which separates Recôncavo from the Camamu Basin. The Salvador Fault System forms the border fault zone of the rift basin. Location of well and seismic data discussed in the text is also shown. *STZ*, Salvador transfer zone; *TMAZ*, Taipus-Mirim Accommodation Zone.

Paleozoic sequence and a cratonic basement (e.g., Magnavita et al., 2012). The Reconcavo Basin has been an exploratory target from the beginning of the last century and is the birthplace of the Brazilian oil industry.

5.1.1 Rift Sequence Stratigraphy of the Recôncavo Basin

Seismic and well data have been analyzed and integrated. Fig. 12 shows a schematic correlation section based on three selected well logs, using the gamma ray pattern to depict the stratigraphic surfaces of the conceptual model. The well log analysis has revealed three main depositional patterns, beginning with a progradational pattern interpreted as the reflex of low accommodation rates due to the typically slow subsidence rate during the initial rift phase, followed by a thick retrogradational succession, evidencing a significant increase in subsidence rates during the main rift phase, and ending with a progradational succession, demonstrating a gradual decrease in basin subsidence rates. This variation of the depositional trends, without intra-rift unconformities, is an example of the proposed "Gradual Subsidence Model," where these three phases of different depositional trends result in a basin stratigraphy with three different Tectonic Systems Tract (Fig. 12).

The RIST is limited at the base by the SRU that separates it from pre-rift strata. The RIST is interpreted as fluvial–deltaic depositional systems prograding over locally restricted lakes formed during the early stages of rift subsidence, as depictured by wells REC-1 and REC-3.

The RDST is limited at the base by the rift development initiation surface (RDIS). This surface marks the change from progradational to retrogradational patterns, with predominance of lacustrine facies and significant gravity-driven facies such as sandy debrites and turbidites, with up to several hundred meters of amalgamated deposits. Well REC-2 is located in deepest region of the half-graben and shows a clear retrogradational pattern after the RDIS, reflecting the increasing extension of the lacustrine system during the RDST. Well REC-3 is located close to the border fault, hence showing the influence of fan-delta systems, which has an aggradational trend because of the approximately vertical stacking of the fan system's strata during the development of the main rift phase.

The RTST is limited at the base by the Maximum Rifting Surface (MRS), over which an increasingly progradational succession develops, formed by deltaic and fluvial facies. The MRS, representing the time of maximum expansion of the lacustrine system, is a regional surface with chronostratigraphic significance and can be used as a datum for stratigraphic

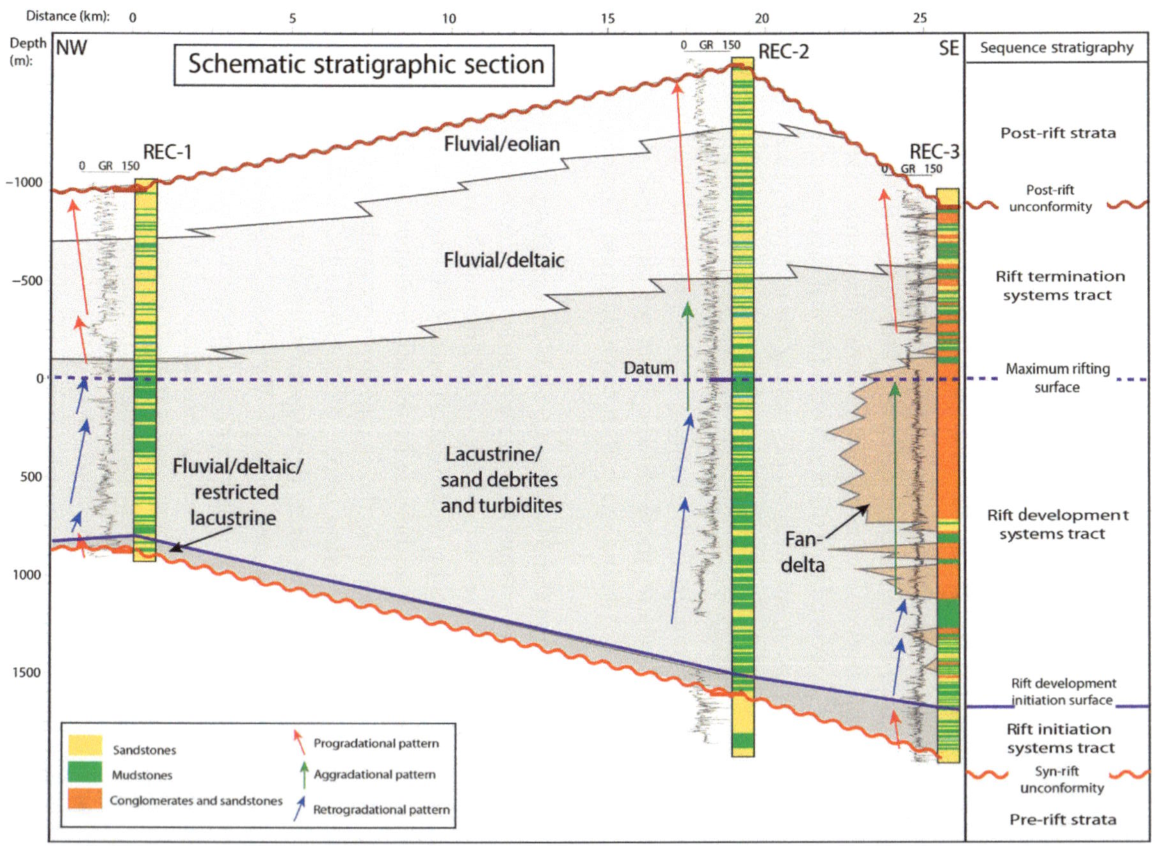

Fig. 12 Schematic stratigraphic section in northeast region of Recôncavo Basin using the maximum rifting surface as datum for correlation. This rift succession, with changes between depositional trends controlled by accommodation space and without intra-rift unconformities, illustrates the "gradual subsidence" model. For location of wells REC-1 to REC-3 see Fig. 11.

correlation. In a well log, the MRS is marked on the change from retrogradational to progradational pattern (i.e., at the highest argilosity of the gamma log, reflecting the lake expansion). While in wells close to flexural margin, such as REC-1, the change from retrogradation to progradation is relatively fast, wells located in the deepest regions of half-graben, like the REC-2, initially display an aggradational pattern, which evolves to a progradational regime. In boreholes close to the border fault, as REC-3, retrogradational stacking pattern is observed, suggesting back-stepping of the fan-delta systems due to decrease in tectonic activity toward the rift termination.

5.1.1.1 Seismic Stratigraphy

Basic seismic interpretation procedure was applied to a 2D seismic data set. Fig. 13 shows a selected regional seismic section and its interpretation,

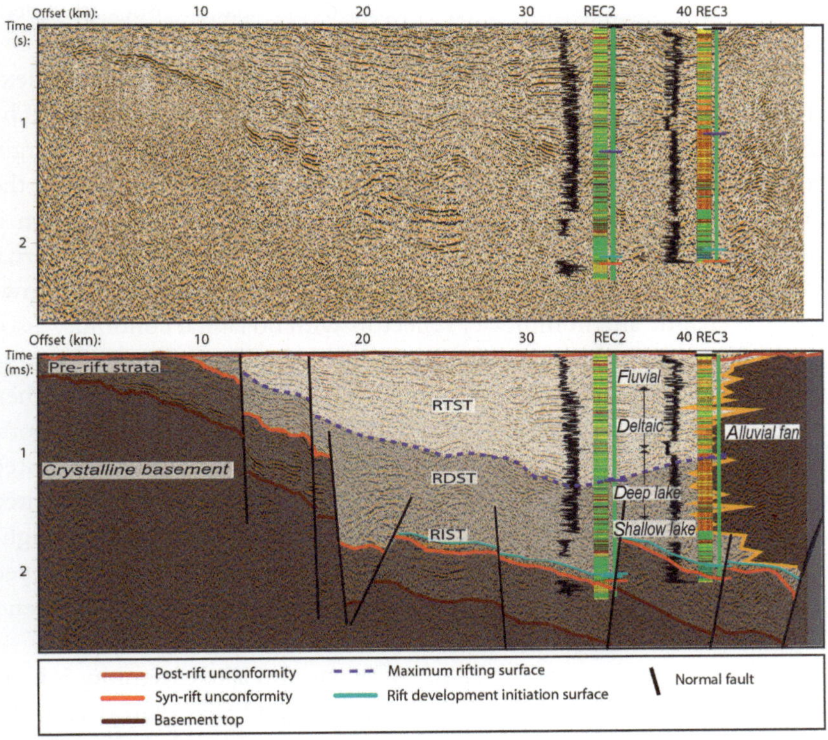

Fig. 13 *Above*: not interpreted dip-directed seismic line REC-1 in northeast region of Recôncavo Basin; *below*: interpretation of the main sequence stratigraphic elements of the "gradual subsidence model." For location of seismic line REC-1 see Fig. 11.

displaying the half-graben system with synthetic and antithetic faults and the border fault.

Fig. 14 summarizes the criteria—reflector terminations and configurations pattern—used for the seismic stratigraphic interpretation and the recognition of the tectonic systems tracts.

The pre-rift seismic facies have subparallel, high- to low-amplitude reflectors with variable continuity, and display several internal reflector terminations (as onlaps, downlaps, and truncations), seemly the result of conjugation of fluvial, lacustrine, and eolian systems (Figs. 13 and 14A and C).

The SRU is characterized by erosive truncations.

Due to its reduced thickness (~150m), the RIST is hardly mapped due to low vertical seismic resolution. Downlaps and toplaps are barely visible and scattered (Figs. 13 and 14A). Low-amplitude and continuity reflectors are characteristic, with typically subparallel internal configuration representing restricted fluvial and lacustrine deposits. The external shape is slightly wedged, and occasional reflectors with progradational characteristics (shingled) occur.

The RDIS is characterized as an onlap surface in regions closest to flexural margin, reflecting the transgression of the lacustrine system. The increasing angles of these onlap terminations toward the flexural margin indicate progressive tilting. Toward the depocenter, reflectors above the RDIS are concordant (Figs. 13 and 14A). A general divergent pattern is detectable in deposits related to RDST, which reflects in a general external wedge shape (Figs. 13 and 14E). Other mapping configurations are: low-amplitude chaotic and hummocky reflectors with no lateral continuity associated with deep lake sedimentation, including mounds representing gravity-driven deposits (slides, slumps, sand turbidites, and debrites when with large thicknesses) and regions deformed due to argilokinesis. Chaotic reflectors, with no continuity and variable amplitude, sometimes presented configurations in mounds, representing deposits associated with the faulted margin. Toward shallow platform regions, parallel, continuous, high-amplitude, and frequency reflectors and external sheet drape can be mapped (Fig. 14B), suggesting alternation between distal lake system sediments and/or carbonate interspersed with the mudstone lake.

The MRS, representing the epoch of maximum lake extension, is marked when there are no more reflectors with divergent characteristics. At the flexural margin it is depictured by downlap terminations, with discrete sigmoids suggesting the shift to a progradational regime. At the faulted margin, delta-fan systems that were predominantly aggradational during the RDST

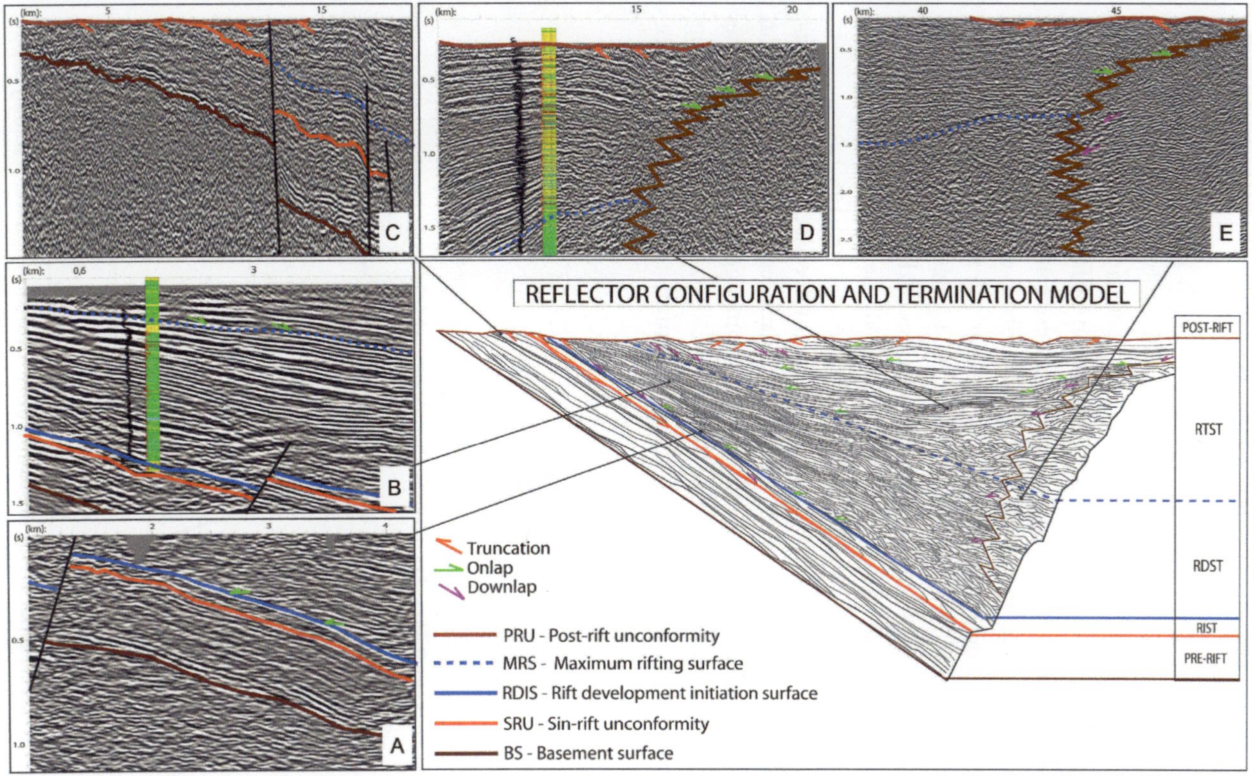

Fig. 14 Seismic sequence stratigraphic model of the Recôncavo basin, showing representative reflector termination and configurations. *Insets* (A–E) show examples taken from several seismic lines of the basin. *Modified from Troccoli, E.B., 2015. Sismoestratigrafia em Bacias Rifte e Análise de Atributos Sísmicos e Filtros na Interpretação de Dados 2D e 3D na Bacia do Recôncavo. UFBA, Salvador, Brazil. M.Sc. dissertation available at http://www.pggeofisica.ufba.br/publicacoes/detalhe/311.*

become increasingly retrogradational after MRS, hence in this region the MRS is marked where a decrease of the influence of the overlying delta-fan facies is noted.

Subparallel reflectors represent the RTST with good continuity and high-to-medium amplitude in most of the half-graben. The presence of downlaps and onlaps above the MRS is constant, representing the progradational infill of the basin. Retrogradation of fan-delta facies can be seen at the border fault. Mounded seismic facies related to gravity-driven deposits are very rare in this tectonic tract. Erosive truncations can be mapped in the topmost zone of the RTST (Figs. 13 and 14D and E).

The top of the RTST is marked by the RTU characterized by intense tilting and uplifted blocks rise so that large portions of the RIST, RDST, and RTST are eroded at the flexural margin; hence, all the stratigraphic surfaces discussed above are truncated by the RTU (Figs. 13 and 14C).

The RDST is marked at the onset of significant deep lacustrine sedimentation, followed by parasequences with retrogradational stacking patterns. At the border fault, fan-delta deposits change from aggradation during the RDST to retrogradation during the RTST (Fig. 15), due to the increasing uplift of the relief and increasing accommodation toward the epoch of

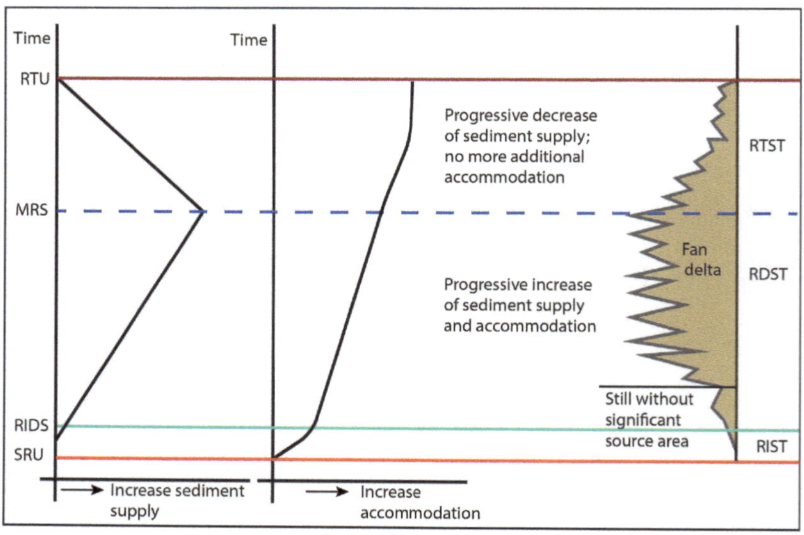

Fig. 15 Conceptual explanation for the stratigraphic evolution and the resulting depositional architecture of the fan-delta deposition at the border fault of a half-graben.

maximum rifting and a decrease of these factors during the rift termination, when the main fault does not grow anymore and the rift relief is progressively eroded; hence, the alluvial fan deposition decreases and disappears.

5.2 Camamu Basin

The Camamu Basin is located southward the Recôncavo Basin, on the shoreline, between the cities of Salvador and Itacaré, extending over an area of around 13,000 km^2 (Fig. 11). Currently, the basin is considered as a new exploratory frontier, despite the presence of important gas fields in production. It is limited northward by the Barra Fault System (the limit with the Recôncavo Basin) and the Itapuã Fault System (separating Camamu from the Jacuípe Basin). The western limit is the Maragogipe Fault, and to the east, the 3000 m bathymetry. The Salvador Transfer Zone (STZ) is located in the northern region of the Camamu Basin, showing transcurrent (flower) structures (Ferreira et al., 2009). This structure splits the rift zone, controlling depocenters and basement highs. The southern limit of the basin is still under discussion. Ferreira et al. (2009) describe the so-called Taipus-Mirim Accommodation Zone (TMAZ), as the limit toward the Almada Basin.

The sedimentary record of the basin encompasses a intracratonic sag phase formed during the Permian, the pre-rift strata similar to Recôncavo basin, a complex rift phase discussed ahead, and a thick shallow to deep marine succession (post-rift and drift).

The basement of the Camamu Basin is mainly compound by gneissic rocks of the São Francisco Craton, with several fold belts and shear zones, especially the TMAZ and the Salvador Transfer Zone (STZ) (see Fig. 11 for location). The STZ is associated with the Salvador Triple Junction, and configures a large part of the rift section in the Northern Camamu Basin with a structural framework characterized by a series of synthetic and antithetic NE–SW-directed high-angle normal. Such faults rotate blocks and tilt the rift strata toward major faults, where most *syn*-rift sequences show a high-angle northwestern dip. An important system of normal faults, labeled the "Northern Camamu Fault System" by Vilas-Boas (2016), crosses the northern portion of Camamu basin parallel to the Barra Fault System, but with greater vertical throws than the Barra Fault System. This structural feature was mapped in dip seismic sections, limiting some syn-rift sequences and presenting structures of syn-sedimentary growth strata and associated alluvial fan deposits. The STZ is the geographical limit between northern and southern areas of the Camamu basin, as originally described by Ferreira et al. (2009). Both areas show several intra-rift unconformities, attributed to footwall uplift in response to extensional unloading of the lithosphere (e.g., Ziegler and Cloetingh, 2004).

Interpretation and mapping of Camamu Basin's basement, sag phase and pre-rift sequences followed the same procedures performed for the Recôncavo Basin interpretation (Section 5.1). In Northern Camamu basin, however, the syn-rift phase is characterized by seven third-order sequences, limited by unconformities interpreted from erosive truncations, especially in the proximal region, and subordinated onlap and downlap terminations. Southern Camamu basin has less rift sequences; therefore, the discussion on the northern and southern parts the basin will be discussed separately. Both, however, are examples for the "punctuated subsidence" sequence stratigraphy model.

5.2.1 Northern Camamu Basin

For this part of the basin, seven rift sequences have been mapped and are briefly described below.

5.2.1.1 Rift Sequence 1

The rift unconformities 1 and 2 limit the first syn-rift sequence, characterized by a predominant progradational pattern, commonly presenting high-amplitude reflectors (HARs), approximately parallel to the pre-rift sequence (Fig. 16). These deposits appear limited in area, increase to southeast, and are formed basically by shallow lacustrine shales and fluvial–deltaic sandstones from the flexural margin. The tectonic systems tracts could not be mapped due to low seismic resolution. For instance, onlap reflectors mark the RDISs; however, the low seismic resolution precludes reliable mapping. It is noteworthy that often the tract does not exceed 200 ms of seismic thickness, resulting in no more than three pairs of reflectors, thus precluding the identification of onlaps other than the base of the sequence.

Analysis of well data, however, leads to the identification of the three tectonic systems tracts depictured by the model, as shown by well CAM-1 (Fig. 17). In this well (see figure for location), the RIST reaches 300 m of thickness. The RDIS-1 is also mappable due to the shift from progradational to retrogradational patterns, indicating the transition from the predominant proximal fluvial–deltaic system of the RIST to a more widespread lacustrine deposition of the RDST. The maximum rifting surface of the sequence Rift 1 (MRS-1) is depictured by the shift from predominantly retrogradational to more progradational depositional patterns (Fig. 17).

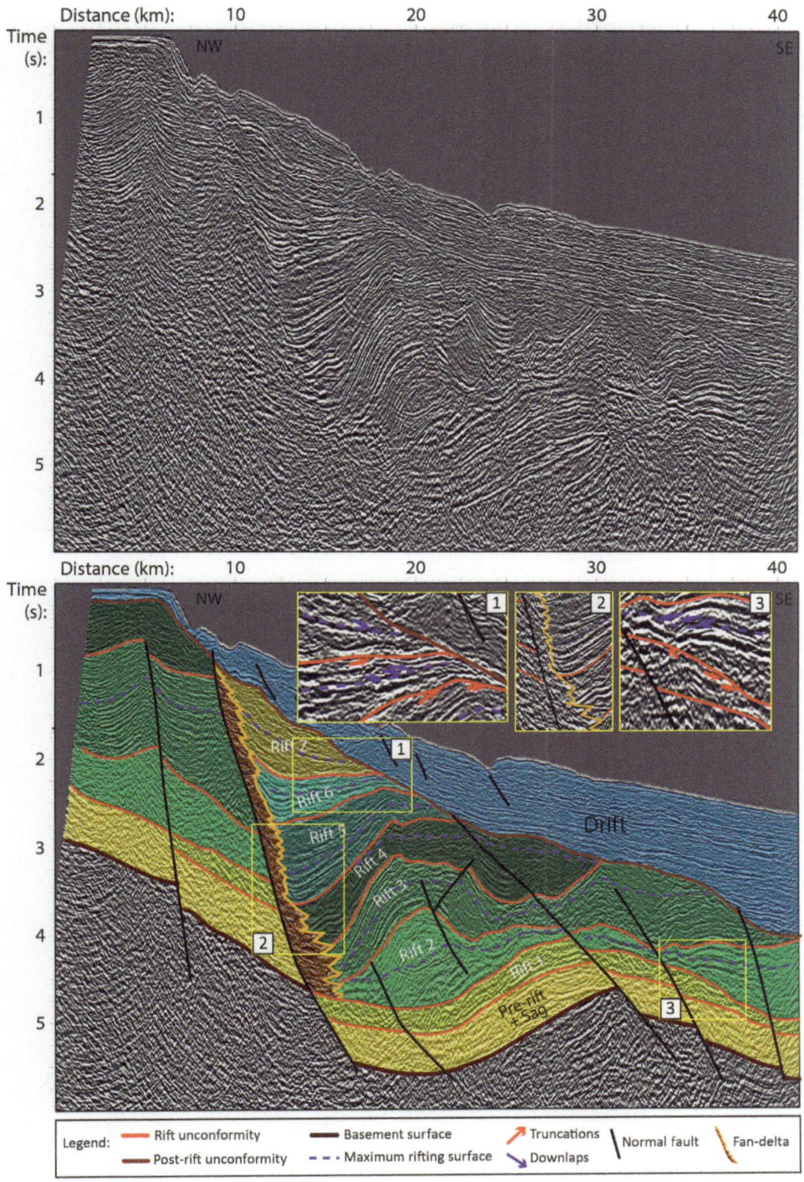

Fig. 16 Noninterpreted and interpreted dip seismic section CAM-1 of the Northern Camamu Basin. Zoom frames: (1) Note the truncations under rift and post-rift unconformities, besides, downlaps above maximum rifting surfaces; (2) note the chaotic seismic facies close to the main fault, related to the fan deposits; (3) observe the different seismic facies between the first two rift sequences. For location of seismic section CAM-1 see Fig. 11.

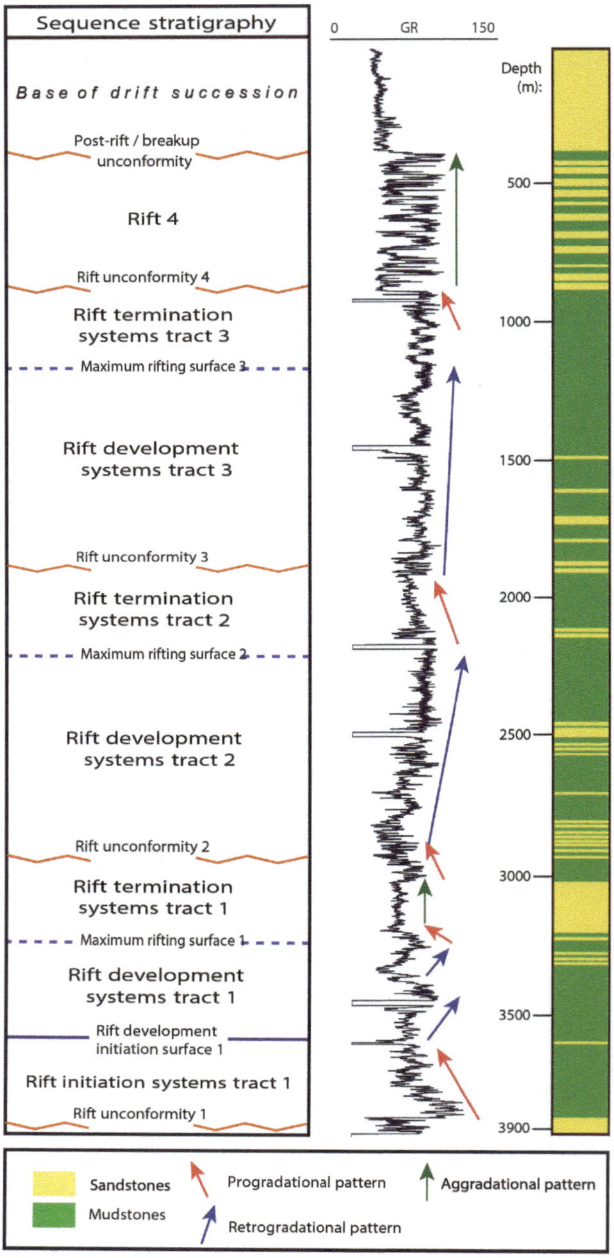

Fig. 17 Stratigraphic interpretation of well CAM-1 (for location, see Fig. 11). Note the rift unconformities and tectonic system tracts associated with stacking patterns as interpreted from the gamma ray log.

5.2.1.2 Rift Sequences 2, 3, and 4

After the deposition of the first rift sequence, the predominant depositional system became deep lacustrine, which results in lower amplitudes and continuity of reflectors relative to the previous sequence. However, the presence of proximal fluvial–deltaic progradations, sandy debrites, and turbidites as depictured by well data results in several high-amplitude and continuity reflectors interfingered to the low-amplitude seismic facies representing the background lacustrine deposition.

Very evident erosive truncations along the entire area lead to the recognition of the rift unconformities 2, 3, and 4, delimiting rift sequences 2, 3, and 4, respectively. These sequences show a thickening trend toward southeast. Rift sequences 2 and 3 have mostly slightly chaotic seismic facies, with lower frequency and continuity reflectors, while Rift Sequence 4 is characterized by an increase in amplitude, continuity, and frequency of reflectors. The RDIS could not be mapped due to low seismic resolution, but some reflectors with downlap terminations preferentially toward southeast were used to mark the MRS of the sequences (Fig. 16).

The well log interpretation was fundamental for the identification of the tectonic systems tract associated to each sequence. The RIST is not present, since the main fault has been installed and the area of the rift basin is structurally defined, as depictured by the "punctuated subsidence" model. The RDST and the RTST, however, can be recognized on well data (Fig. 17). The well log analysis also permitted the identification of relevant sandy debrites and turbidites in such sequences, indicated by an abrupt decrease in the gamma ray log, displaying an aggradational pattern within a general context of expansion of the lacustrine system. Starting in Rift Sequence 2, alluvial fan deposits ware mapped associated to the North Camamu Fault System. These fan deposits, which extend throughout the top of the *syn*-rift succession, were interpreted from a lateral changing seismic facies characterized by a chaotic reflector pattern and minor internal subparallel reflectors (Fig. 16-2).

5.2.1.3 Rift Sequences 5, 6, and 7

After the development of Rift Sequence 4, renewed tectonic pulses dislocated the half-graben system of northern Camamu to an opposite direction in relation to the first four *syn*-rift sequences. In this epoch, the main acting fault configured a half-graben system with a flexural margin dipping toward northwest. This block rotation is reflected in layers thickening next to the main fault, including features of growth strata and maintenance of alluvial fan deposits along the faults, as illustrated in Fig. 16.

Rift sequences 5, 6, and 7 are recorded only in a restricted region in the central part of Northern Camamu Basin, and well data that record these three syn-rift sequences are available. The sequences have a similar seismic facies, with high continuity and frequency reflectors, interpreted as deep lacustrine shales interfingered with fluvial–deltaic sandstones in the proximal portion, besides debrites and turbidites close to the depocenters. The extension of the preserved depositional area of these sequences decreases to the top, expressing a retreat of the lacustrine system of the basin. The end of the syn-rift phase is marked by a series of evident erosive truncations along the entire Camamu Basin, beneath the post-rift unconformity. The interpretation of the post-rift unconformity is supported by the well log analysis, where an abrupt decrease of the gamma ray log can be depictured. In some wells, the post-rift unconformity is overlain by a thin layer of anhydrite or carbonates from the basal drift succession.

5.2.2 Southern Camamu Basin

For the southern Camamu Basin, less rift sequences have been recognized. For this area, Fig. 18 displays a representative interpreted seismic section, showing that seismic mapping of the southern area of the basin leads to the recognition of a syn-rift phase with five third-order depositional sequences. Pre-rift sediments generally have a tabular pattern, without growth strata, and are limited by the basement and the SRU. The basement was mapped using changes in seismic facies and marking the deepest continuous reflector. The SRU shows onlap terminations and significant change in the seismic facies, especially on marginal regions. The five rift sequences are briefly described below.

5.2.2.1 Rift Sequence 1

Rift 1 overlies the SRU and is limited topward by rift unconformity RU2. This unconformity shows noticeable truncations and onlaps in distal regions. The sequence is not present in the entire basin, appearing only locally as isolated, small rift basins and usually associated with large faults (Fig. 18). In marginal regions, fluvial–deltaic sandstones and shallow lacustrine shales are recorded, as shown by well data. Continuity, high-amplitude, and low-frequency reflectors, similar to the pre-rift strata, characterize the respective seismic facies. The main difference between pre-rift strata and rift sequence 1 is the presence of growth strata.

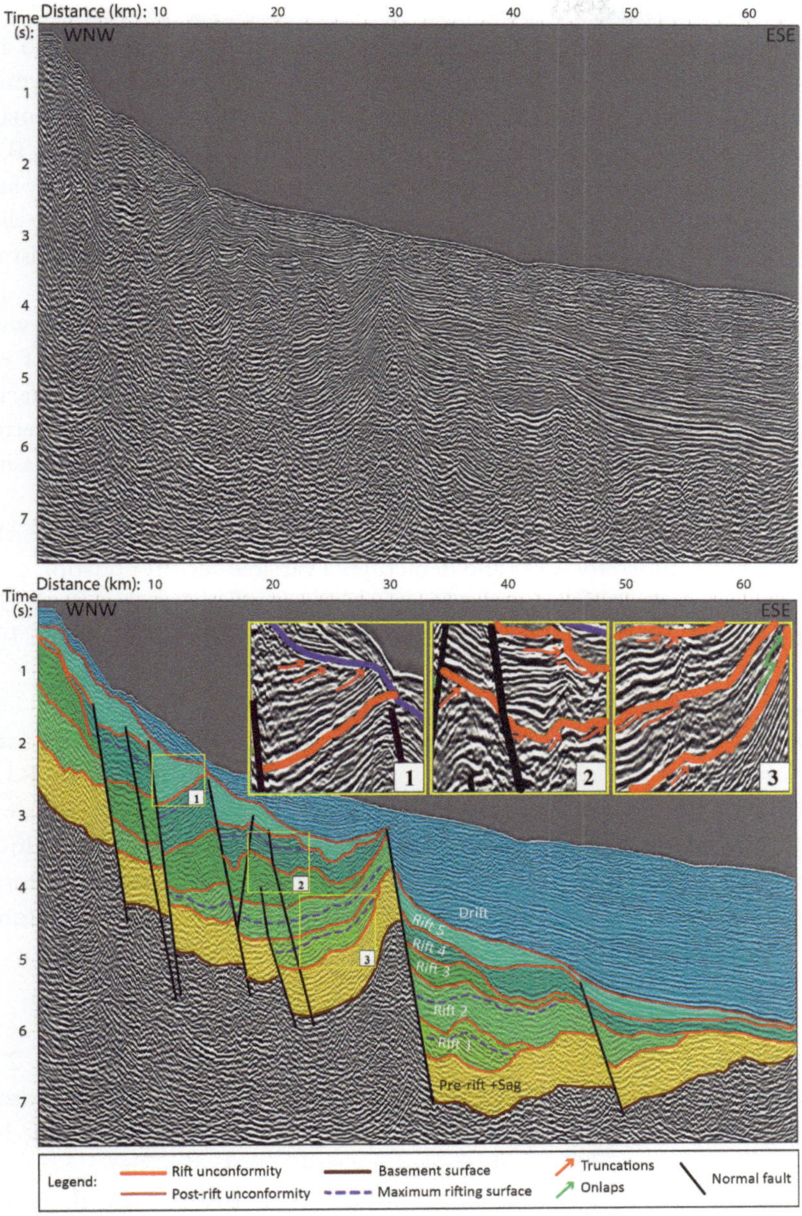

Fig. 18 *Above*: not interpreted dip-directed seismic section in southeast region of Camamu Basin; *below*: interpretation of the main sequence stratigraphic elements of the "punctuated subsidence model." In the zoom frames: (1) notable truncations in post-rift unconformity delimiting Rift sequence 5 and the drift stage. (2) Truncations of rift unconformities RU5 and RU4, delimiting Rift successions 3 and 4. (3) Truncations in U3, U2, and SRU delimiting Rift 2, Rift 1, and pre-rift stage and onlaps in SRU.

5.2.2.2 Rift Sequences 2 and 3

Both sequences are bounded by unconformities (RUs 2, 3, and 4) and are present in almost the entire basin, representing evidences of huge tectonic activity, such as frequent and expressive growth strata (Fig. 18). The main facies are deep lacustrine shales, debrites, and turbidites. Seismic facies in Rift sequence 2 has high amplitude, moderate continuity, and low frequency and is moderately deformed, while Rift sequence 3 is intensely deformed by diapiric structures, which may have originated the observed chaotic seismic pattern.

Well log interpretation is again essential to delimit systems tracts with good reliability. Well CAM-2 (Fig. 19) shows the interpretation of rift sequences 2 and 3. A very abrupt change in the gamma ray and seismic facies separates the pre-rift from Rift sequence 2 at 2180 m depth. The retrogradational gamma ray pattern toward the MRS reflects the increasing accommodation.

Rift 3 presents aggradational and retrogradational pattern above the intra-rift unconformity RU3 and depictures a variable accommodation until the MRS-3, which delimits the RDST 3. Intra-rift unconformity RU4, mapped on seismic data, crosses well CAM-2 eroding the RTST 3 (Fig. 19).

5.2.2.3 Rift 4 and Rift 5 Sequences

Unconformities RU4 and RU5 delimit the Rift 4. High continuity and amplitude reflectors characterize seismic facies. Rift 5 is limited by RU5 and ends with the overlying PRU. The seismic facies of that sequence is characterized by high-frequency, low–amplitude, and, sometimes, low-continuity reflectors. Growth strata are present indicating the *syn*-tectonic deposition. Due to low seismic resolution the delimitation of the tectonic systems tracts is not possible.

5.3 Pelotas Basin

The Pelotas Basin is located at the southernmost Brazilian coast (Fig. 20). Exploration began in 1958, but the basin remains relatively underexplored, with few borehole data and mostly pre-1990 seismic data. The basin has 210,000 km^2 to the 2000 m isobath, and covers 40,000 km^2 offshore, between latitude 28° and 34° South. The northern boundary is formed by the Florianópolis High (with the Santos Basin), which is part of a larger structure, including the São Paulo Crest, defining a zone of fracture and crustal east–west elevation. The morphology developed along this fracture zone during the relatively rapid initial opening of the South Atlantic Ocean formed a barrier against ocean circulation northward so that during the early post-rift phase

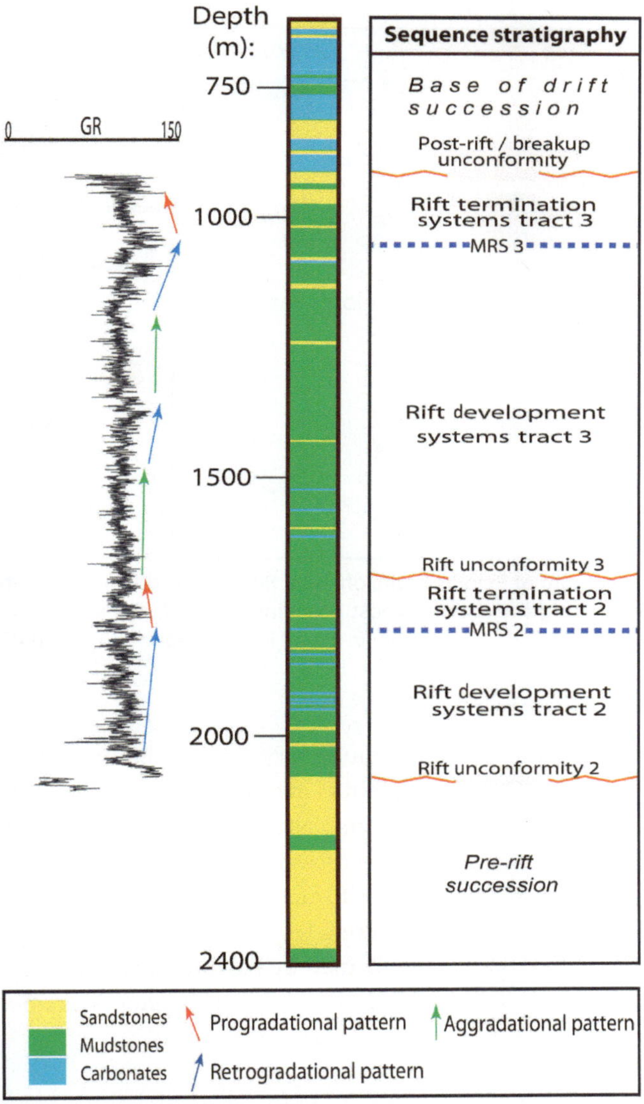

Fig. 19 Well CAM-2 displaying lithological and gamma ray logs. No shifts in stacking patterns as a reflex of changing tectonic systems tracts. The delimitation of the rift unconformities within the muddy sections (e.g., RU3) is based on seismic interpretation.

the Pelotas basin experienced open oceanic condition, while in the adjacent Santos Basin evaporites were formed. The southern limit of the basin is formed by the border with Uruguay (Salado River Basin). Its African counterpart is represented by the Walvis–Namibe Basin (offshore Namibia).

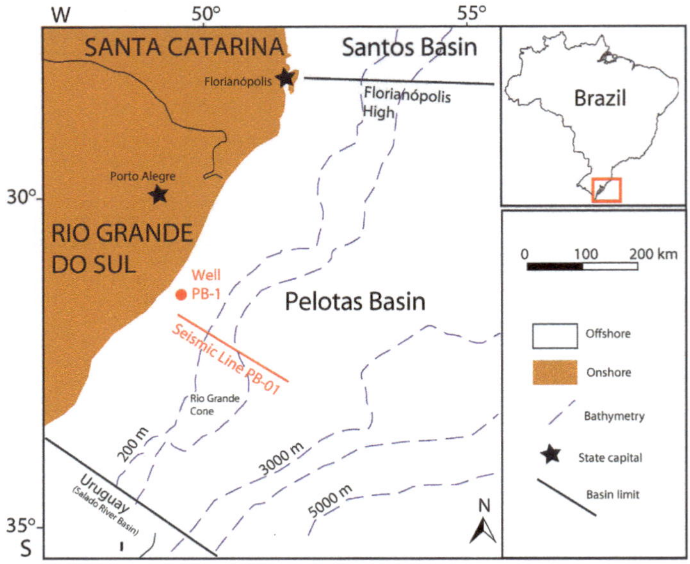

Fig. 20 Location map of Pelotas basin, offshore Rio Grande do Sul and southern Santa Catarina States, along the southernmost shoreline of Brazil. The log and stratigraphic interpretation of Well PB-1 is shown in Fig. 21, while seismic line PB-01 is shown in Fig. 22.

The Pelotas Basin had an evolutionary history differentiated from the other basin on the Brazilian Atlantic margin. The south-to-north rifting was accompanied by intense volcanic activity during the pre- and syn-rift phases, recorded by the Serra Geral (pre-rift) and Imbituba (syn-rift) formations (Fig. 10). The record of the rift succession is restricted to the central-southern region of the basin, where a few half-graben developed.

The infill of these grabens is formed by basaltic rocks (Imbituba Formation) and coarse- to fine-grained siliciclastic (Cassino Formation), interpreted as alluvial fans, fan-deltas, and fluvial deposits.

5.3.1 Rift Sequence Stratigraphy of the Pelotas Basin

Two rifts sequences are recognized and described according to the "punctuated subsidence" model. Rift sequence 1 occurs widespread and seemly has contact with the oceanic crust, while Rift sequence 2 shows a very restricted record due to erosion. Fig. 21 shows a representative seismic line, while Fig. 22 displays the only well that drilled the rift phase of the basin.

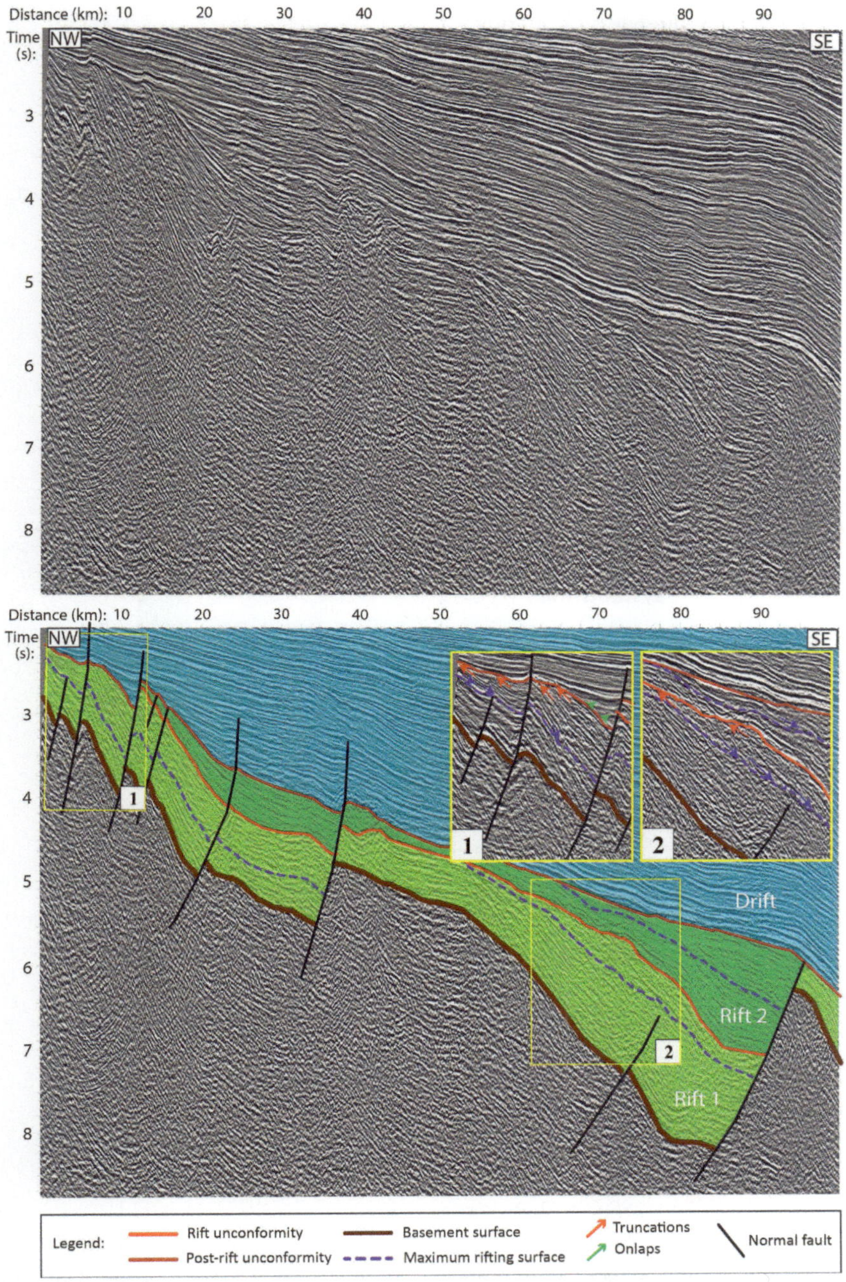

Fig. 21 Seismic line PB-01 of central region of the Pelotas Basin, showing the rift succession and the interpreted sequence stratigraphy. In the zoom frames: (1) downlap terminations marking the first maximum rifting surface, and truncations marking the post-rift unconformity. *Note*: Absence of rift 2 in that part of the basin. (2) Downlap terminations marking the maximum rifting surfaces of the rift sequences, and truncations marking the intrarift unconformity. Note that the post-rift unconformity truncates the second maximum rifting surface.

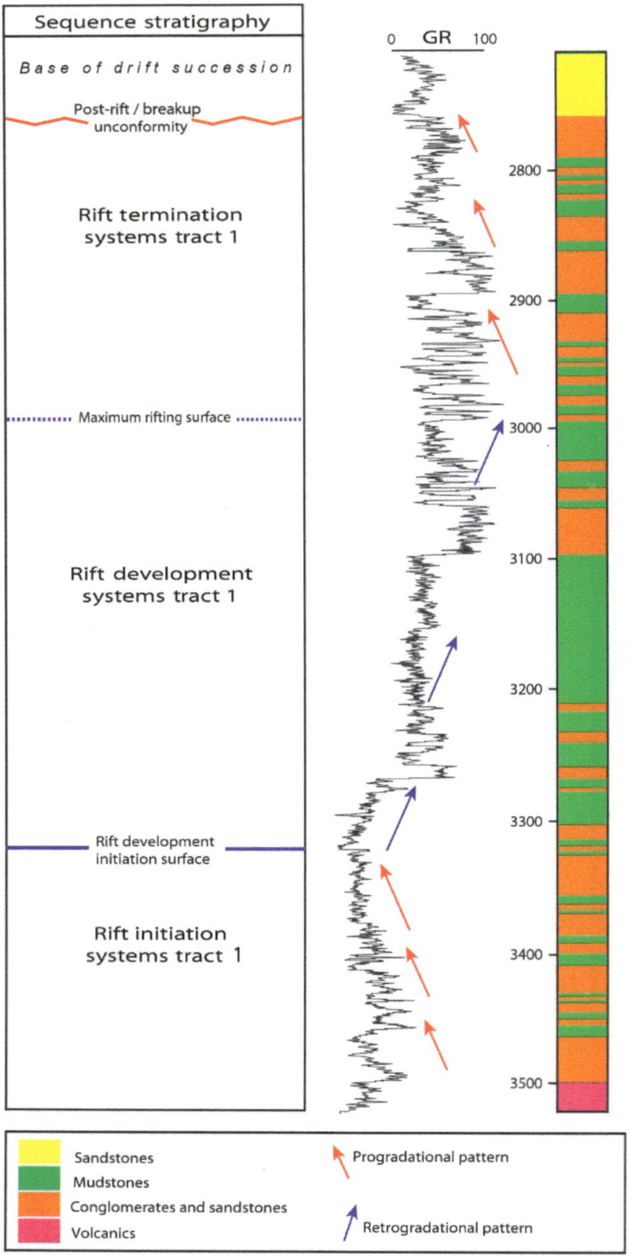

Fig. 22 Representative well log of the Rift 1 succession of the Pelotas basin. Unfortunately this is the only one borehole that actually reached the rift succession, but it serves an as example for the applicability of the stratigraphic model. Note the progradational pattern at the base (recording the rift initiation systems tract), changing to more retrogradational pattern in the middle portion of the well log (recording the rift development systems tract), and changing back to more progradational pattern toward the top of the rift succession, which characterizes the rift termination systems tract.

5.3.1.1 Rift Sequence 1

The sequence has a relatively widespread distribution and shows the typical seismic facies expected for a rift succession (Fig. 21-1). Above the basement low-frequency, high-amplitude, and continuous reflectors associated with basaltic rocks occur, followed by a low-amplitude/continuous seismic facies representing the RDST, delimited by the MRS, which is well marked by downlaps (Fig. 21-1). This surface segments Rift sequence 1 in a third seismic facies with continuous, high frequency, and amplitude interpreted as the record of fluvio-deltaic deposition. The top of this sequence is marked by several truncations due to erosion on the flexural border in the proximal area, where erosion generated the PRU and Rift sequence 2 does not occur.

The only borehole available that reached the rift succession in the Pelotas basin has a very proximal location (Fig. 20), where only rift sequence 1 is preserved; however, it is used herein as an example of application of our stratigraphic model.

The RIST is delimited at the basal portion of the well (Fig. 22) by the progradational pattern of the gamma ray profile, changing to a more aggradational pattern toward the top. The RDST shows the expected retrogradational pattern, reflecting the increase in tectonic activity and the corresponding increase of accommodation space and the consequent expansion of the rift lake. The retrogradational pattern is also evidenced by the increase of mudstones within the conglomerate succession, culminating in more than 100 m thick continuous mudstone succession. The RTST is characterized by an overall progradational gamma ray pattern.

5.3.1.2 Rift Sequence 2

Truncations at the top of Rift 1 marked Rift Unconformity 1, the basal surface of Rift sequence 2. This sequence is restricted to the distal zone of the basin, since the proximal area was eroded forming the post-rift event triggered by the rift tectonics that generate sequence 2.

Since no wells reached this sequence, the limit between the RDST and the RTST was marked solely by downlaps over and seismic facies change across the MRS (Fig. 21-2).

The PRU marks the top of this sequence. The erosion affected most of the RTST and the MRS, and in some areas the drift succession is in direct contact with the RDST.

5.4 Ceará Basin

The Ceará basin is located at the Brazilian equatorial margin (Fig. 23), between the cities of Fortaleza and Parnaiba, with an area of approximately 34,000 km^2. It is geological limited by crystalline basement in the south, the Romanche Fracture Zone in the north, and two structural highs: in the southeast, the limit with the Potiguar basin is marked by the Fortaleza high, and in the northwest the Tutoia high separates the Ceará basin from the Barreirinhas basin. Due to distinct tectonic characteristics and structural features the Ceará basin is divided into four sub-basins: Piauí-Camocim, Acaraú, and Icaraí, with transpressive structures, and Mundaú Sub-basin, situated in west part, with distensive structures. This sub-basin, due to its tectonic characteristic, is dealt with herein.

The sedimentary record has been traditionally divided into three tectonic stages: rift, post-rift, and drift.

The rift phase of the Mundaú Sub-Basin started at ~125 Ma (Early Aptian) and is represented by fluvial–deltaic and lacustrine deposits labeled the Mundaú Formation.

The post-rift stage initiated ~115 Ma, encompassing fluvio-deltaic to lacustrine siliciclastic sediments (Paracuru Formation) as well as lacustrine carbonates (Trairi Member). Very few evaporites are recorded fort this phase (thin halite layers in only two wells). Some authors (e.g., Antunes, 2004; Condé et al., 2007) have observed the presence of growth strata in seismic sections; hence, it is controversial if Paracuru Formation really represents post-rift sedimentation controlled by thermal subsidence.

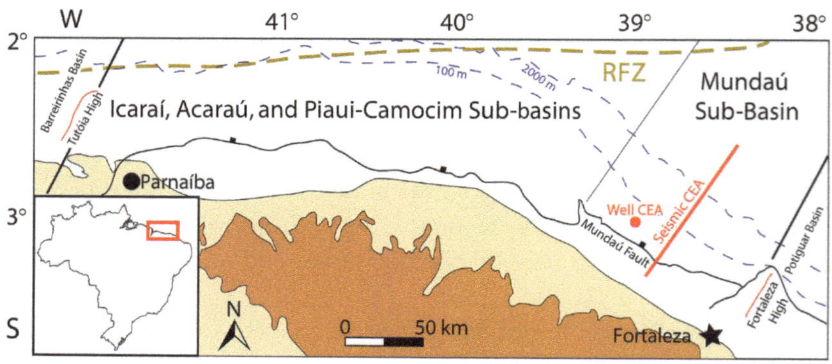

Fig. 23 Location map of Ceará Basin showing its limits and main structural features. The well and seismic section are shown in Figs. 24 and 25, respectively. *RFZ*, Romanche Fracture Zone.

During the Early Albian (~110 Ma) the drift phase started with a transgression, depositing coastal and deepwater sediments punctuated by frequent sea-level variations, resulting in regional unconformities. Magmatic events are reported for the Eocene (Costa et al., 1990).

A series of deep continuous reflectors may indicate the presence of older sequences, such as pre-rift strata, Barremian Rift (similar to the adjacent Potiguar Basin), or a Paleozoic sag (e.g., Beltrami et al., 1994; Condé et al., 2007), although no supporting well data are available. The presence of these older sequences, however, is supported by correlation with the Accra–Keta Basin on the African equatorial margin (Brownfield and Charpentier, 2006).

5.4.1 Rift Sequence Stratigraphy of the Ceará Basin

The seismic and log analysis leads to the recognition of at least two rift sequences in Ceará Basin, and a lower sequence that possibly represents a first rift stage (Fig. 24).

5.4.2 Basement and the "Lower Sequence"

Since none of the wells drilled in this basin reached the basement, it was mapped at the deepest continuous reflectors and crosschecked with gravimetric modeling.

The rift unconformity 1 bounds the "Lower Sequence" at the top, with erosional truncation displayed specially in the distal part of the basin (Fig. 24-inset 1). The "Lower Sequence" is characterized by continuous reflectors with low frequency and low amplitude, eventually with HARs. The sequence displays thinning toward northeast.

As mentioned before this "Lower Sequence" may represent pre-rift strata, a Paleozoic Sag, or a Barremian Rift, possibly the latter due to the features observed in seismic lines such as thickness variation and growth strata to the fault. However, the lack of well data precludes precise identification; hence, tectonic systems tracts have not been mapped.

5.4.3 Rift Sequence 1

This sequence has a seismic facies with low frequency, low amplitude, and high continuity, eventually with internal HARs. This facies seems similar to the "Lower Sequence" facies, but differs due to higher concentration of HARs, possibly linked to the presence of thin magmatic intrusions. Most wells drilled into the sequence show a massive presence of shale, but since neither well crosses the entire sequence, it is not possible to delimit the

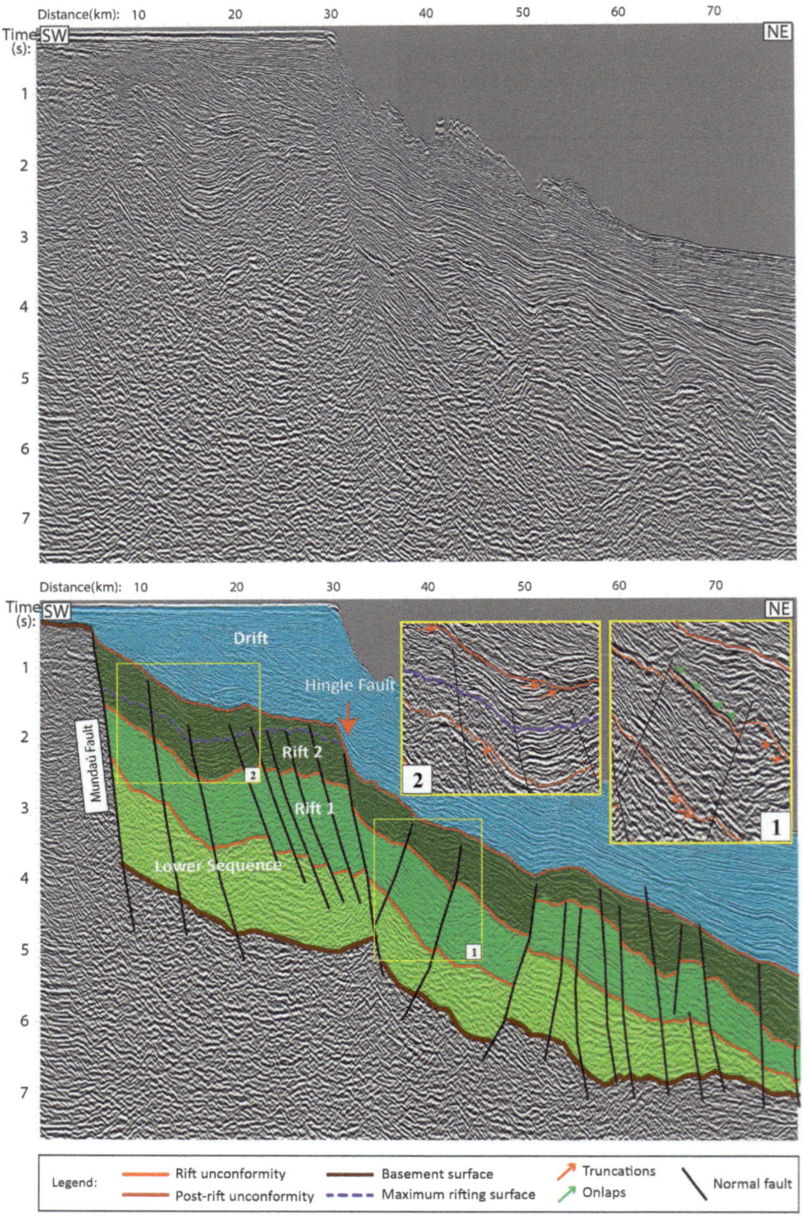

Fig. 24 (A) Not interpreted and (B) interpreted dip-directed seismic section CEA of the Mundaú Sub-Basin showing the Lower Sequence and Rifts 1 and 2. The *yellow rectangles* indicate zooms on specific parts of the seismic section: (1) Truncation on rift unconformities 1 and 2. Onlap in Rift unconformity 2. (2) RDST 2 and RTST 2 seismic facies and truncations on post-rift unconformity.

MRS. However, due to the features described above, Rift sequence 1 probably has a well-developed RDST, with a less expressive RTST. This can be observed in well CEA (see Fig. 25), where the final deposition of Rift sequence 1 is marked by an increase of coarse sedimentation. An erosive surface labeled "Rift Unconformity 2" truncates the top of the Rift 1 succession and separates it from rift sequence 2.

5.4.4 Rift Sequence 2

Deposition of Rift sequence 2 begins with strata onlapping Rift Unconformity 2 (Fig. 24), especially in the distal half-graben (Fig. 24-inset 1). The sequence is divided into an RDST and an RTST, as depictured by well CEA (Fig. 25), which is in a depocenter located approximately 7 km from the Mundaú Fault, where lacustrine shales and gravity-driven deposits predominate.

On seismic sections, the RDST shows predominantly a low-amplitude seismic facies. The retrogradational pattern is noticeable (Fig. 25) and extends up to ~2000 m depth, where decreasing gamma ray values and the changing shape of the well log indicate a shift toward progradational pattern, and where the MRS is marked. The RTST is characterized by progradational and aggradational depositional patterns depictured in well logs, and low-frequency, high-continuity, and high-amplitude reflectors in seismic sections. The presence of high-amplitude seismic facies suggests the characteristic increase in fluvial–deltaic sedimentation during the development of the RTST. Angular truncations mark the PRU eroding the top of Rift Sequence 2 (Fig. 24-inset 2).

6. FINAL REMARKS

There are some important discrepancies among rift basin specialists despite the intensive studies developed ever since the economic importance of this kind of basin has been recognized. Some hydrocarbon-prone rift basins are studied since the 1960s, such as the Viking Graben in the North Sea (Pegrum and Spencer, 1990), and in general huge data sets formed by well logs and seismic lines are available, but problematic aspects concerning rift stratigraphy still remain.

One problematic issue among the specialist concerns the rift initiation: some authors defend that the onset of the rift is marked by rapid subsidence and huge accommodation, while others defend a slow subsidence regime and consequent low accommodation rates for this phase. The infill of a rift

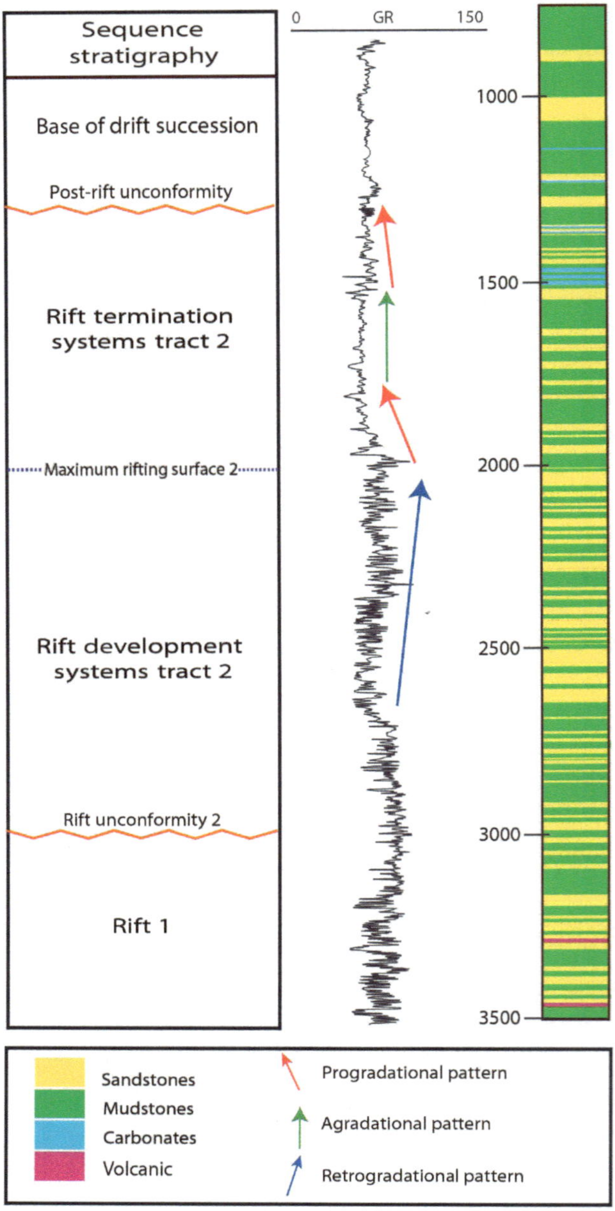

Fig. 25 Well CEA with gamma ray log and sequences and tracts interpretations. The depositional patterns on Rift 2 can be mapped on well log that shows a retrogradation on RDST and progradation on RTST. The sandy strata of the RDST are gravity-driven deposits and the distal facies of alluvial fan, while the sandstones of the RTST are delta and fluvial deposition.

basin is another matter debate since the early 1990s, when the period of proposition of comprehensive models for the rift initiation and the stratigraphic signature of the rift basin under the viewpoint of sequence stratigraphy began (e.g., Nottvedt et al., 1995; Prosser, 1993). The rift termination and the precise limit between the rift succession and the overlying transitional strata originated during thermal subsidence is also no object of consensus. In this chapter, we have discussed these problematics and have proposed a descriptive sequence stratigraphic model with the necessary comprehensiveness and feasibility, which works as a useful tool for description and predictive interpretation of continental rift successions. This model holds three so-called tectonic systems tracts, each gathering the sedimentation occurred under a specific tectonic regime, i.e., rift initiation, rift development, and rift termination. Nomenclature for the stratigraphic surfaces delimiting these systems tracts is also proposed, since descriptive terms are needed for labeling during regional interpretation of both well data and seismic lines. Their surfaces are, from bottom to top of a rift sequence, the *syn*-rift unconformity, the rift development initiation surface, the maximum rifting surface, and the post-rift unconformity. The latter might be overprinted by the breakup unconformity.

The stratigraphy of rift successions may be simple or more complex, depending on subsidence history. Some rift successions appear to have undergone a more gradual and uniform subsidence regime due to rapid crustal stretching, with few base-level variations recorded only at the flexural border, while others record several phases of more or less marked subsidence, followed by phases without creation of accommodation space and by generation of subaerial unconformities. Hence, two models were shown here, one dealing with a rather gradual subsidence, where much of the basin has a quite continuous history of subsidence, and one of subsidence punctuated by phases of subsidence stillstands.

These concepts have been applied to several Brazilian rift basins to illustrate the feasibility and usefulness of the systems tract concept and the proposed gradual and punctuated subsidence models.

Our study has shown that a huge portion of the Recôncavo basin is an example for the "gradual subsidence" model of rift sequences stratigraphy, where lithostratigraphically, the RIST is related to Itaparica and Água Grande formations. This tectonic tract is very thin in the Recôncavo Basin (<200 m), with thickness variations related to minor fault activity, as already mentioned by Ghignone (1979), suggesting an incipient tectonic control for that lithostratigraphic units. Silva (1996) related Água Grande Formation to the rift phase, and Wiederkehr (2010) depictured a gradual contact between the two formations, registering progradation (Água Grande Formation) into a tectonically controlled lake (Itaparica Formation) with the depocenter in the southern zone of the basin, representing the onset of the rifting process.

Concerning the Camamu basin, the results of seismic stratigraphic interpretation as discussed in the previous section lead to two conclusions: first, that the basin is an illustrative example for the "punctuated subsidence" model, since several episodes of tectonic activity, with block tilting and erosion of the rift relief forming intra-rift unconformities; and second, that adjacent and relatively close regions within a given basin can have a significantly different sequence stratigraphy. While in southern Camamu five rift sequences were recognized, closer to the STZ, in northern Camamu, seven rift sequences were mapped, suggesting that stratigraphic differences do not only exist between different basin, even if adjacent (such as Recôncavo and Camamu), but there might be significant differences in the sequence stratigraphic framework within a given basin. Pelotas Basin is an example of rapid rifting, where the opening of the southern zone of the Atlantic ocean and the significant presence of volcanism during the transition of the pre-rift to the syn-rift deposition have resulted in a quite different stratigraphic signature when compared to other rift successions on the Brazilian margin. However, the "punctuated subsidence" model is applicable. For the Ceará Basin, the application of the proposed rift sequence stratigraphy scheme has also proven feasible. Three unconformities encompassing two rift sequences were mapped.

However, it is important to underline that the proposed model is by no means a panaceia for problems in rift sequence stratigraphy. No model is a definitive problem-solver. Based upon classic work by authors as Prosser (1993) and Nottvedt et al. (1995), we intended to resume the conceptual discussion that seems relatively abandoned since the late 1990s, with only sparse publications on the issue (e.g., Martins-Neto and Catuneanu, 2010). This sensation of relative abandonment is enhanced by the fact that never before a nomenclature for stratigraphic surfaces in rift succession has been proposed and integrated to a tectonostratigraphic model as we did herein. So hopefully this publication may trigger a renewing discussion and hence will bring an improvement on sequence stratigraphic models for rift successions.

ACKNOWLEDGMENTS

The authors thank editor Michael Montenari for the invitation to contribute to this volume of "Stratigraphy and Time Scales" and also acknowledge two anonymous reviewers. GEOPARK E&P BRASIL (Rio de Janeiro) is acknowledged for support of research projects on Brazilian rift basins. M.H. acknowledges CNPq (Brazilian Council for research) for study grant (304657/15-8). P.A.V.-S. and D.B.V.-B. thank Brazilian agencies CAPES and FAPESB for M.Sc. study grants. The Brazilian National Petroleum Agency (ANP-Academic program) is acknowledged for supplying additional geological data. Our undergraduate students from GETA/UFBA staff are acknowledged for the assistance during the day-by-day work, and technician Joaquim Lago for keeping our workstations and softwares updated and functional.

REFERENCES

Abdunaser, K.M., McCaffrey, K.J.W., 2014. Rift architecture and evolution, the Sirt Basin, Libya, the influence of basement fabrics and oblique tectonics. Afr. Earth Sci. 100, 203–226.

Acocella, V., Korme, T., Salvini, F., 2003. Formation of normal faults along the axial zone of the Ethiopian Rift. J. Struct. Geol. 25, 503–513.

Alley, N.F., Frakes, L.A., 2003. First known Cretaceous glaciation, Livingston Tillite, South Australia. Aust. J. Earth Sci. 50, 134–150.

Antunes, A.F., 2004. Evolução tectono-estrutural do Campo de Xaréu (Sub-bacia de Mundaú, Bacia do Ceará—NE do Brasil: abordagem multiescala e pluriferramental). Tese (Doutorado em Geodinâmica; Geofisica), Universidade Federal do Rio Grande do Norte, Natal. 384 pp.

Arche, A., López-Gómez, J., 1996. Origin of the Permian Triassic Iberian Basin, central-eastern Spain. Tectonophysics 266, 443–464.

Beglinger, S.E., Doust, H., Cloetingh, S., 2012. Relating petroleum system and play development to basin evolution, West African South Atlantic basins. Mar. Pet. Geol. 30, 1–25.

Beltrami, C.V., Alves, L.E.M., Feijó, F.J., 1994. Bacia do Ceará. B. Geoci. Petrobrás. Rio de Janeiro 8 (1), 117–125.

Berger, A., Crucifix, M., Hodell, D.A., et al., 2016. Interglacials of the last 800,000 years. Rev. Geophys. 54 (1), 162–219.

Bergner, A.G.N., Strecker, M.R., Trauth, M.H., Deino, B., Gasse, F., Blisniuk, P., Duenforth, M., 2009. Tectonic and climatic control on evolution of rift lakes in the Central Kenya Rift, East Africa. Quat. Sci. Rev. 28, 2804–2816.

Bond, G., Kromer, B., Beer, J., Muscheler, R., Evans, M., Showers, W., Hoffmann, S., Lotti-Bond, R., Hajdas, I., Bonani, G., 2001. Persistent solar influence on North Atlantic climate during the Holocene. Science 294, 2130–2136.

Bosence, D.W.J., 1998. Stratigraphic and sedimentological models of rift basins. In: Purser, B.H., Bosence, D.W.J. (Eds.), Sedimentation and Tectonics of Rift Basins, Red Sea—Gulf of Aden. Chapman & Hall, London. 630 pp.

Brown, L.F. & Fisher, W.L. 1977. Seismic-stratigraphic interpretation of deposicional systems, examples from Brazilian rift and pull-apart basins. In, Payton, C.E. (Ed.) 1977. Seismic Stratigraphy—Applications to Hydrocarbon Exploration. American Association of Petroleum Geologists Memoir, Tulsa, OK vol. 26. pp. 213–248.

Brownfield, M.E., Charpentier, R.R., 2006. Geology and total petroleum systems of the Gulf of Guinea Province of west Africa. US Geol. Surv. Bull. 2207-C. 32 pp.

Bueno, G.V., 2004. Diacronismo de eventos no rifte Sul-Atlântico. B. Geoci. Petrobrás. Rio de Janeiro 12 (2), 203–229.

Cainelli, C., Mohriak, W.U., 1999. Some remarks on the evolution of sedimentary basins along the Eastern Brazilian continental margin. Episodes 22, 206–216.

Catuneanu, O., 2006. Principles of Sequence Stratigraphy. Elsevier, Amsterdam. 375 pp.

Cloetingh, S., Balen, R., Voorde, M., et al., 1997. Mechanical aspects of sedimentary basin formation, development of integrated models for lithospheric and surface processes. Geol. Rundsch. 86, 226–240.

Cohen, A.S., Lezzar, K.-E., Tiercelin, J.-J., Soreghan, M., 1997. New palaeogeographic and lake-level reconstructions of Lake Tanganyika, implications for tectonic, climatic and biological evolution in a rift lake. Basin Res. 9 (2), 107–132.

Condé, V.C., Lana, C.C., Pessoa Neto, O.C., Roesner, E.H., Morais Neto, J.M., Dutra, D.C., 2007. Bacia do Ceará. B. Geoci. Petrobrás. Rio de Janeiro 15 (2), 347–355.

Corti, G., 2009. Continental rift evolution: from rift initiation to incipient break-up in the Main Ethiopian Rift, East Africa. Earth-Sci. Rev. 96 (2009), 1–53.

Costa, I.G., Beltrami, C.V., Alves, L.E.M., 1990. Evolução TectonoSedimentar e o habitat do óleo da Bacia do Ceará. B. Geoci. Petrobrás. Rio de Janeiro 4 (1), 65–74.

Crossley, R., 1984. Controls on sedimentation in the Malawi Rift Valley, central Africa. Sediment. Geol. 40, 33–40.

Delvaux D. 1995. Age of Lake Malawi (Nyasa) and Water Level Fluctuations. Royal Museum for Central Africa, Department of Geology and Minerology, Tervuren, Belgium, Annual Report 1993–1994, 99–108 (1995).

DePaolo, C.M., Anderson, J.G., 2000. Estimating the slip rates of normal faults in the Great Basin, USA. Basin Res. 12, 227–240.

Dinis, J.L., Trincão, P., 1995. Recognition and stratigraphical significance of the Aptian unconformity in the Lusitanian Basin, Portugal. Cretac. Res. 16 (2–3), 171–186.

Duehnforth, M., Bergner, A.G.N., Trauth, M.H., 2006. Early Holocene water budget of the Nakuru–Elmenteita basin, Central Kenya Rift. J. Paleolimnol. 36, 281–294.

Duval, B., Cramez, C., Vail, P.R., 1992. Types and hierarchy of stratigraphic *cycles*. In: Mesozoic and Cenozoic Sequence Stratigraphy of European Basins.Basins International Symposium (Dijon, France), Abstract, 44–45.

Eddy, J.A., 1976. The Maunder minimum. Science 192 (4245), 1189–1202.

Einsele, G., Ricken, W., Seilacher, A. (Eds.), 1991. Cycles and Events in Stratigraphy. Springer Verlag, Berlin. 955 p.

Embry, A.F., 1995. Sequence boundaries and sequence hierarchies: Problems and proposals. In: Steel, V.L.F.R.J., Johannesen, E.P., Mathieu, C. (Eds.), Sequence Stratigraphy: Advances and Applications for Exploration and Production in Northwest Europe, pp. 1–11.

Falvey, D.A., 1974. The development of continental margins in plate tectonic theory. J. Aust. Pet. Explor. Assoc. 14, 95–106.

Ferreira, T.S., Caixeta, J.M., Lima, F.D., 2009. Controle do embasamento no rifteamento das bacias de Camamu e Almada. B. Geoci. Petrobrás. Rio de Janeiro 17 (1), 69–88.

Fischer, A.G., 1982. Long-term climatic oscillations recorded in stratigraphy. In: Berger, W. (Ed.), Climate in Earth History. National Research Council, Studies in Geophysics, pp. 97–104.

Fisher, W.L., McGowen, J.H., 1969. Depositional systems in the Wilcox Group of Texas and their relationship to occurrence of oil and gas. AAPG Bull. 53, 30–54.

Franke, D., 2012. Rifting, lithosphere breakup and volcanism: comparison of magma-poor and volcanic rifted margins. Mar. Pet. Geol. 43, 63–87.

Frakes, L.A., 1979. Climates Throughout Geologic Time. Elsevier Scientific, Amsterdam, New York. 304 pp.

Friedmann, S.J., Burbank, D.W., 1995. Rift basins and supradetachment basins: intracontinental extensional end-members. Basin Res. 7 (2), 109–127.

Gawthorpe, R.L., Leeder, M.R., 2000. Tectono-sedimentary evolution of active extensional basins. Basin Res. 12, 195–218.

Ghignone, J.I., 1979. Geologia dos sedimentos fanerozóicos do estado da Bahia. In: INDA, H.A.V. Geologia e Recursos Minerais do Estado da Bahia. Textos Básicos, vol. 1. SME/COM, Salvador, pp. 23–117.

Gupta, S., Cowie, P.A., Dawers, N.A., Underhill, J.R., 1998. A mechanism to explain rift-basin subsidence and stratigraphic patterns through fault-array evolution. Geology 26 (7), 595–598.

Haas, J., Budai, T., Raucsik, B., 2012. Climatic controls on sedimentary environments in the Triassic of the Transdanubian Range (Western Hungary). Palaeogeogr. Palaeoclimatol. Palaeoecol. 353–355, 31–44.

Hamblim, A.P., 1992. Half-graben lacustrine sedimentary rocks of the lower Carboniferous Strathlorne Formation, Horton Group, Cape Breton Island, Nova Scotia, Canada. Sedimentology 39 (2), 263–284.

Holz, M., Moreira, F., Troccoli, E.B., 2015. In: A conceptual sequence stratigraphy model for continental rift successions based on the Reconcavo Basin, Cretaceous, Brazil. STRATI 2015—II International Congress on Stratigraphy, Graz. http://dx.doi.org/10.13140/RG.2.1.3728.9846.

Izart, A., Stephenson, R., Vai, G.B., Vachard, D., Nindrfe, Y.L., Vaslet, D., Sues, P., Fauvel, P.-J., Kossovaya, O., Chen, Z., Maslo, A., Stovba, S., 2003. Sequence stratigraphy and correlation of late Carboniferous and Permian in the CIS, Europe, Tethyan area, North Africa, Arabia, China, Gondwanaland and the USA. Palaeogeogr. Palaeoclimatol. Palaeoecol. 196 (1–2), 59–84.

Jervey, M.T., 1988. Quantitative geological modeling of siliciclastic rock sequences and their seismic expression. In: Wilgus, C.K., Hasting, B.S., Kendall, C.G.S.C., Posamentier, H.W., Ross, C.A., Van Wagoner, J.C. (Eds.), Sea-Level Changes, An Integrated Approach. Society of Economic Paleontologists and Mineralogists, Special Publication No. 42, Tulsa, OK, pp. 47–69.

Karner, N.W., Driscoll, N.W., Barker, d.H.N., 2003. Syn-rift regional subsidence across the West African continental margin, the role of lower plate ductile extension. In: Arthur, T.J., Macgregor, D.S., Cameron, N.R. (Eds.), Petroleum Geology of Africa, New Themes and Developing Technologies. Geological Society, London, Special Publications, 207, pp. 105–129.

Kidder, D.L., Worsley, T.R., 2010. Phanerozoic large igneous provinces (LIPs), HEATT (Haline Euxinic Acidic Thermal Transgression) episodes and mass extinctions. Palaeogeogr. Palaeoclimatol. Palaeoecol. 295 (2010), 162–191.

Kinabo, B.D., Hogan, J.P., Atekwana, E.A., Abdelsalam, M.G., Modisi, M.P., 2008. Fault growth and propagation during incipient continental rifting. Tectonics 27, TC301.

Kreuser, T., 1995. Tectonic and climatic controls of lacustrine sedimentation in pre-rift and rift settings in the Permian-Triassic of East Africa Thomas. J. Paleolimnol. 13, 3–19.

Kuchle, J., Scherer, C.M.S., Born, C.C., Alvarnga, R.S., Adegas, F., 2011. A contribution to regional stratigraphic correlations of the Afro-Brazilian depression—The Dom João Stage (Brotas Group and equivalent units—Late Jurassic) in Northeastern Brazilian sedimentary basins. J. South Am. Earth Sci. 31 (4), 358–371.

Kyrkjebø, R., Gabrielsen, H., Faleide, J.I., 2004. Unconformities related to the Jurassic–Cretaceous synrift–post-rift transition of the northern North Sea. J. Geol. Soc. Lond. 161, 1–17.

Lambiase, J.J. 1990 A model for tectonic control of lacustrine stratigraphic sequences in continental rift basins. In, Katz, B.J. (ed.) Lacustrine Basin Exploration—Case Studies and Modern Analogs. Tulsa, OK American Association of Petroleum Geologists, Memoir, n. vol. 50, 265–276.

Magnavita, L.P., Szatmari, P., Cupertino, J.A., Destro, N., Roberts, D., 2012. The Recôncavo basin. In: Phanerozoic Rift Systems and Sedimentary Basins, pp. 383–419.

Martins-Neto, M.A., Catuneanu, O., 2010. Rift sequence stratigraphy. Mar. Pet. Geol. 27 (1), 247–253.

McIntyre, A., Debenham, S., Gibson, G., Henson, P., Lambeck, A., Southgate, P., 2005. In: Paleogeography, sequence stratigraphy and sedimentology of an evolving rift system, Mt Isa Western Succession. pmd*CRC I1 Project Final Report, March 2005.

Melchor, R.N., 2007. Changing lake dynamics and sequence stratigraphy of synrift lacustrine strata in a half-graben: an example from the Triassic Ischigualasto–Villa Union Basin, Argentina. Sedimentology 54, 1417–1446.

Meyers, S.R., Peters, S.E., 2011. A 56 million year rhythm in North American sedimentation during the Phanerozoic. Earth Planet. Sci. Lett. 303 (3–4), 174–180.
Moore, G.T., Hayashida, D.N., Ross, C.A., Jacobson, S.R., 1992. Paleoclimate of the Kimmeridgian/Tithonian (Late Jurassic) world: I. Results using a general circulation model. Palaeogeogr. Palaeoclimatol. Palaeoecol. 93, 113–150.
Moretti, I., Colletta, B., 1987. Spatial and temporal evolution of the Suez rift subsidence. J. Geodyn. 7, 151–168.
Morley, C.K., 2002. Evolution of large normal faults, evidence from seismicv reflection data. AAPG Bull. 86 (6), 961–978.
Morley, C.K., S., Gabdi, Seusutthiya, K., 2007. Fault superimposition and linkage resulting from stress changes during rifting, examples from 3D seismic data, Phitsanulok Basin, Thailand. J. Struct. Geol. 29 (4), 646–663.
Nottvedt, A., Gabrielsen, R.H., Steel, R.J., 1995. Tectonostratigraphy and sedimentary architecture of rift basins, with reference to the northern North Sea. Mar. Pet. Geol. 12 (8), 881–901.
Olsen, P.E., 1986. A 40-million-year lake record of early mesozoic orbital climatic forcing. Science 234 (4778), 842–848.
Payton, C.P. (Ed.), 1977. In: Seismic Stratigraphy—Applications to Hydrocarbon Exploration, vol. 26. American Association of Petroleum Geologists Memoir. 516 p.
Pegrum, R.M., Spencer, A.M. 1990. In, Brooks, J. (ed.), 1990, Classic Petroleum Provinces, Geological Society Special Publication 50, 441–470.
Pirajno, F., 1992. Hydrothermal Mineral Deposits. Springer-Verlag, Berlin. 709 pp.
Prosser, S., 1993. Rift-related linked depositional systems and their seismic expression. In: Williams, G.D., Dobb, A. (Eds.), Tectonics and seismic sequence stratigraphy Geological Society Special Publication No 71, pp. 35–66.
Rohais, S., Eschard, R., Ford, M., Guillocheau, F., Moretti, I., 2007. Stratigraphic architecture of the Plio-Pleistocene infill of the Corinth Rift: implications for its structural evolution. Tectonophysics 440 (5), 5–28.
Rosendahl, B.R., Reynolds, D.J., Lorber, P.M., Burgess, C.F., Mcgill, J., Scott, D., Lambiase, J.J., And Dersken, S.J., 1986, Structural expressions of rifting, lessons from Lake Tanganyika, Africa, in Frostick, L.E., Renault R.W., Reid, I., and Tiercelin, J.J., eds. 1986. Sedimentation in the African Rifts, Geological Society of London, Special Publication 25, 29–43.
Ruffell, A.H., Moles, N.R., Parnell, J., 1998. Characterisation and prediction of sediment-hosted ore deposits using sequence stratigraphy. Ore Geol. Rev. 12, 207–223.
Samanta, A., Bera, M.K., Sarkar, A., 2016. Climate-modulated sequence development in a tropical rift basin during the Late Palaeocene to Early Eocene super greenhouse Earth. Sedimentology 63 (4), 917–939.
Scherer, C.M.S., Jardim de Sá, E., Córdoba, V.B., Souza, D.C., Aquino, M.M., Cardoso, F.M.C., 2014. Tectono-stratigraphic evolution of the Upper Jurassic Neocomian rift succession, Araripe Basin, Northeast Brazil. J. South Am. Earth Sci. 49, 106–122.
Schlager, W., 2004. Fractal nature of stratigraphic sequences. Geology 32 (3), 185–188.
Schlische, R. W., and Olsen, P. E., 1990, Quantitative filling model for continental extensional basins with applications to early Mesozoic rifts of eastern North America, J. Geol., v. 98, 135–155.
Schlische, R.W., Withjack, M.O., 1999. Rift basin architecture and evolution. In: Olsen, P.E., Kent, D.V. (Eds.), Climatic, Biotic, and Tectonic Pole-to-Pole Coring Transect of Triassic-Jurassic Pangaea, International Continental Drilling Program.
Scholz, C.A., Rosendahl, B.R., Scott, D.L., 1990. Development of coarse-grained fades in lacustrine rift basins, examples from East Africa. Geology 18, 140–144.

Silva, H.T.F., 1996. Caracterização do início da fase sinrifte na Bacia do Recôncavo, estado da Bahia: discussão sobre a formação Água Grande e o reconhecimento da erosão do início do rifteamento. In: Congresso Brasileiro de Geologia, vol. 39. Sociedade Brasileira de Geologia, Salvador, Anais, São Paulo, pp. 325–328.
Silva, H.T.F., Picarelli, A.T., 1990. Variações da linha de costa do lago do Recôncavo e taxa de acumulação durante o Andar Rio da Serra, fase rifte. B. Geoci. Petrobrás. Rio de Janeiro 4 (2), 205–216.
Silva, O.B., Caixeta, J.M., Milhomem, P.S., Kosin, M.D., 2007. Bacia do Recôncavo. B. Geoci. Petrobrás. Rio de Janeiro 15 (2), 423–431.
Smoot, J.P. & Olsen, P.E. 1994. In, Lomando, A. Harris, M. 1994. Lacustrine Depositional Systems. SEPM Core Workshop Notes vol. 19, 201–237.
Soares, D.M., Alves, T.M., Terrinha, P., 2012. The breakup sequence and associated lithospheric breakup surface, their significance in the context of rifted continental margins (West Iberia and Newfoundland margins, North Atlantic). Earth Planet. Sci. Lett. 355–356, 311–326.
Sopeña, A., Sánchez-Moya, Y., 1997. Tectonic systems tract and depositional architecture of the western border of the Triassic Iberian Trough (central Spain). Sediment. Geol. 113 (3–4), 245–267.
Starostenko, V.I., Danilenko, V.A., Vengrovitch, D.B., Poplavskii, K.N., 1996. A fully dynamic model of continental rifling applied to the syn-rift evolution of sedimentary basins. Tectonophysics 268, 211–220.
Strecker, M.S., Berthelot, F., Lyberis, N., Lepichon, X., 1988. Subsidence of the Gulf of Suez, implication for rifting and plate kinematics. Tectonophysics 153, 249–270.
Swiecicki, T., Gibbs, P.B., Farrow, G.E., Coward, M.P., 1998. A tectonostratigraphic framework for the Mid-Norway region. Mar. Pet. Geol. 15, 245–276.
Szulc, J., 1999. Anisian–carnian evolution of the germanic basin and its eustatic, tectonic and climatic controls. In, Bachmann, G.H., Lerche, I. (Eds.), Epicontinental Triassic, Zbl. Geol. Paläont. Teil I, Stuttgart, 1998, pp. 813–852.
Talbot, M.R., Allen, P.A. 1996 – Lakes. In, Reading, H.G. 1996. Sedimentary Environments, Processes, Facies and Stratigraphy. Blackwell Science. P, 83–124.
Usoskin, I.G., Solanki, S.K., Kovaltsov, G.A., 2007. Grand minima and maxima of solar activity, new observational constraints. Astron. Astrophys. 471 (1), 301–309.
Vail, P.R., Todd, R.G., Sangree, J.B., 1977. Seismic stratigraphy and global changes of sea level, part 5: Chronostratigraphic significance of seismic reflections. In: Payton, C.E. (Ed.), In: Seismic Stratigraphy—Application to Hydrocarbon Exploration, vol. 26. American Association of Petroleum Geologists Memoir, pp. 99–116.
Valdes, P.J., Sellwood, B.W., 1992. A palaeoclimate model for the Kimmeridgian. Palaeogeogr. Palaeoclimatol. Palaeoecol. 95 (1–2), 47–72.
Van Houten, F.B., 1962. Cyclic sedimentation and the origin of analcime-rich Upper Triassic Lockatong Formation, west-central New Jersey and adjacent Pennsylvania. Am. J. Sci. 260, 561–576.
Van Wagoner, J.C., Posamentier, H.W., Mitchum, R.M., Vail, P.R., Sarg, J.F., Loutit, T.S., Hardenbol, J., 1988. An overview of sequence stratigraphy and key definitions. In: Wilgus, C.K., Hastings, B.S., Kendall, C.G.St.C., Posamentier, H.W., Ross, C.A., Van Wagoner, J.C. (Eds.), In: Sea-Level Changes: An Integrated Approach, vol. 42. Society of Economic Paleontologists and Mineralogists Special Publication, pp. 39–45.
van Wees, J.D., Stephenson, R.A., Stovba, S.M., Shymanovskyi, V.A., 1996. Tectonic variation in the Dnieper-Donets Basin from automated modelling of backstripped subsidence curves. Tectonophysics 268, 257–280.
Veizer, J., Godderis, Y., François, L.M., 2000. Evidence for decoupling of atmospheric CO_2 and global climate during the Phanerozoic Eon. Nature 408, 698–701.

Vilas-Boas, D.B.R., 2016. Correlação sismoestratigráfica entre as bacias do Recôncavo e de Camamu. Universidade Federal da Bahia, Salvador, Brazil. Master's Dissertation.
Wang, W., Zhou, M.F., 2014. Provenance and tectonic setting of the Paleo- to Mesoproterozoic Dongchuan Group in the southwestern Yangtze Block, South China: implication for the breakup of the supercontinent Columbia. Tectonophysics 610, 110–127.
Wang, C., Feng, Z., Zhang, L., Huang, Y., Cao, K., Wang, P., Zhao, B., 2012. Cretaceous paleogeography and paleoclimate and the setting of SKI borehole sites in Songliao Basin, northeast China. Palaeogeogr. Palaeoclimatol. Palaeoecol. 385, 17–30.
Wanner, H., Beer, J., Butikofer, J., Crowley, T.J., Cubasch, U., Fluckiger, J., Goosse, H., Grosjean, M., Joos, F., Kaplan, J.O., Kuttel, M., Muller, S.A., Prentice, C., Solomina, O., Stocker, T.F., Tarasov, P., Wagner, M., Widmann, M., 2008. Mid- to Late Holocene climate change, an overview. Quat. Sci. Rev. 27, 1791–1828.
Watson, M.P., Hayward, A.B., Parkinson, D.N., Zhang, Z.M., 1987. Plate tectonic history, basin development, and petroleum source rock deposition, onshore China. Mar. Pet. Geol. 4, 205–225.
Watts, A.B., Platt, J., Buhl, P., 1993. Tectonic evolution of the Alboran Sea Basin. Basin Res. 5, 153–177.
Withjack, M.O., Schlische, R.W., Olsen, P.E., 2002. Rift-basin structure and its influence on sedimentary systems. In: Renaut, R.W., Ashley, G.M. (Eds.), Sedimentation in Continental Rifts, vol. 73. SEPM Special Publications, pp. 57–81.
Whymark, W.E., Frimmel, H.E., 2017. Regional gold-enrichment of conglomerates in Paleoproterozoic supergroups formed during the 2.45 Ga rifting of Kenorland. Ore Geol. Rev. doi.org/10.1016/j.oregeorev.2017.04.003
Wiederkehr, F. 2010. Análise tectono-estratigráfica das formações Itaparica e Água Grande, Bacia do Recôncavo, Bahia). UFRGS, PPGEO. 98p. M.Sc. dissertation available at http://www.lume.ufrgs.br/handle/10183/28182.
Young, M.J., Gawthorpe, R.L., Sharp, I.R., 2003. Normal fault growth and early syn-rift sedimentology and sequence stratigraphy, Thal Fault, Suez Rift, Egypt. Basin Res. 15, 479–502.
Zaarur, S., Affek, H.P., Tryon, C., Peppe, D.J., Faith, J., 2013. In: Causes of Late Pleistocene water level change in Lake Victoria, Equatorial East Africa, derived from clumped isotopes of land snails and fresh water mollusks. American Geophysical Union, Fall Meeting 2013, abstract #PP21D-03.
Zhang, S., 2002. In: The genetic mechanism of the sequence stratigraphy of the Rift Lacustrine Basin in Jiyang Depression, East China. Offshore Technology Conference, Houston, 2002. Paper OTC 14071.
Ziegler, P.A., 1992. Plate tectonics, plate moving mechanisms and rifting. Tectonophysics 215, 9–34.
Ziegler, P.A., Cloetingh, S., 2004. Dynamic processes controlling evolution of rifted basins. Earth Sci. Rev. 64, 1–50.

CHAPTER FIVE

Ordovician Sequence Stratigraphy of the Siberian and Russian Platforms

Andrei Dronov[1]
Geological Institute, Russian Academy of Sciences, Moscow, Russia
[1]Corresponding author: e-mail address: dronov@ginras.ru

Contents

Stratigraphy & Timescales, Volume 2
ISSN 2468-5178
http://dx.doi.org/10.1016/bs.sats.2017.07.005

Abstract

The comparative analysis of the Ordovician successions of the Siberian and North American platforms demonstrates a striking similarity in the long-term lithological development and the sea-level curve interpretation. On both platforms, the Ordovician succession commences with tropical stromatolite-bearing carbonates, which abruptly change to siliciclastic deposits and the succession terminates with cool-water carbonates. Numerous K-bentonite beds in the Upper Ordovician of the North American and Siberian platforms emphasize this similarity even further. The sea-level curve for the Ordovician of the North American Platform assumes a prominent sea-level drop at the base of the Middle Ordovician and a long-term lowstand during the Dapingian and Darriwilian. The sea-level curve for the Ordovician of the Siberian Platform looks roughly the same. On the other hand, sea-level curves for the Ordovician of the Gondwanan platforms (North Africa, Yangtze Platform, Tarim, South America, Avalonia, and Baltica) seem to share different patterns. The Middle Ordovician represents there rather a highstand interval in these reconstructions. As a result, instead of one global sea-level curve for the Ordovician, it would be probably more reasonable to suggest two semi-global curves for two major tectonic regions: one of which includes the Siberian and the North American platforms and the other combines Baltica and the Gondwanan platforms. This subdivision probably reflects the position of the main Ordovician lithospheric plates.

1. INTRODUCTION

Sea-level changes are the main mechanism controlling facies dynamics, stratal geometry, and temporal variations in fossil assemblages in marine sedimentary basins. Besides having a direct influence on the water depth sea-level changes could serve as a trigger for various oceanographic, climatic, sedimentary, chemical, and biotic events. Patterns of large-scale evolutionary radiation and mass extinction may be related in complex ways to large-scale fluctuations of the sea level. In recent years a number of new sea-level reconstructions for specific Ordovician basins around the world (Munnecke et al., 2010) as well as the refinement of the global chronology of the Ordovician depositional sequences (Haq and Schutter, 2008) have been suggested.

On the territory of Russia, there are two separate paleocontinents, which are represented by the East European or Russian Platform and the Siberian Platform (Fig. 1).

These two paleocontinents were located far away from each other throughout the Early Paleozoic and this provides an opportunity to test ideas

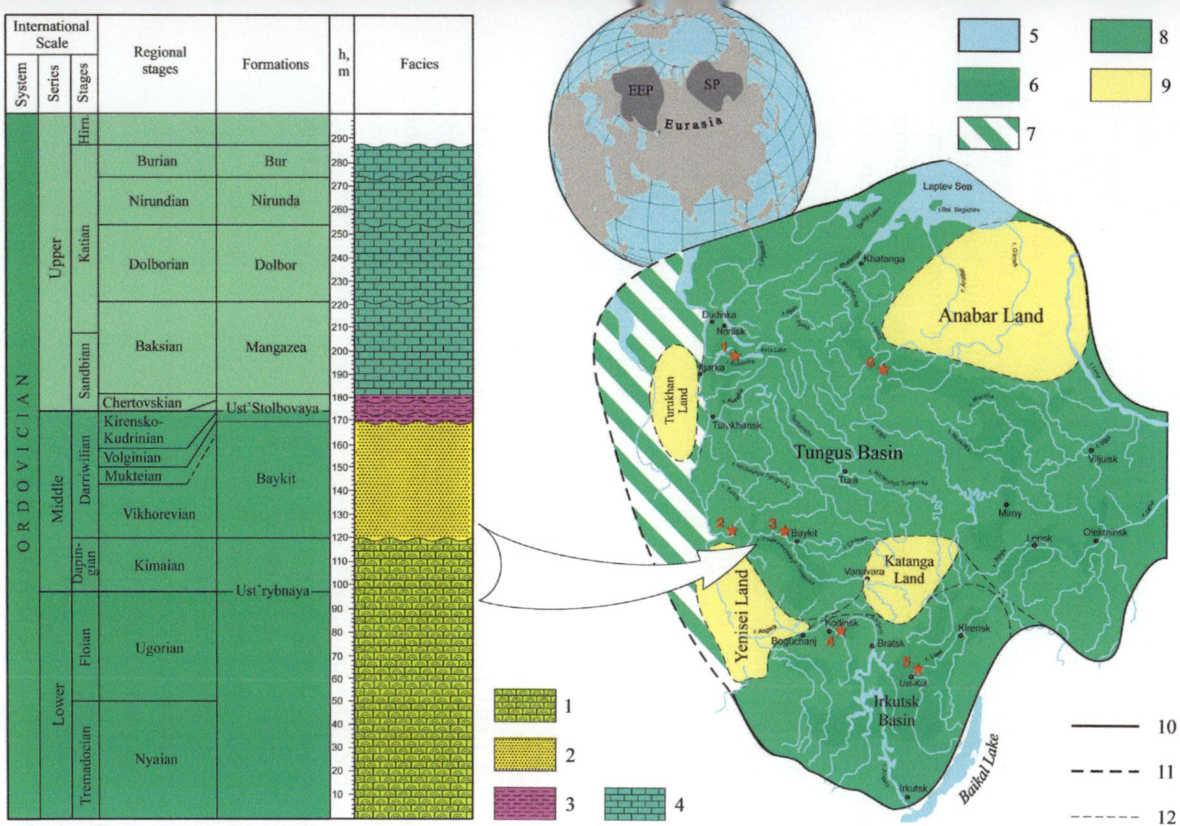

Fig. 1 Distribution of the Ordovician rocks on the Siberian Platform. Legend: 1. warm-water carbonates; 2. quartz sandstone; 3. variegated (*green* and *red*) shale with scattered phosphate pebbles; 4. cool-water carbonates; 5. areas covered by waters of seas, lakes, and estuaries; 6. Ordovician outcrop areas; 7. supposed continuation of the Siberian Craton beneath the Mesozoic cover of the West Siberian basin; 8. Ordovician deposits in the subsurface areas; 9. land areas without Ordovician deposits; 10. boundary of the Siberian Craton; 11. provisional boundaries of the Siberian Craton; and 12. boundaries of the land areas and sedimentary basins. *EEP*, East European (Russian) Platform; *SP*, Siberian Platform. *Red stars* mark the position of the main studied Ordovician sections: (1) Kulumbe River section; (2) Podkamennaya Tunguska and Stolbovaya river sections; (3) Bolshaya Nirunda River section; (4) Angara River section; (5) Lena River section; and (6) Moyero River section.

about synchronous eustatic sea-level changes on both of them during the Ordovician. In order to solve this task abundant research and field work on the Ordovician rocks of the Siberian Platform have been conducted since 2006 (Dronov, 2013; Dronov et al., 2009, 2011a; Kanygin et al., 2010a,b). One of the surprising results of this work was the discovery of widespread cool-water carbonates in the Upper Ordovician of the Tungus basin, which seems to be located at an equatorial latitude at that time (Cocks and Torsvik, 2007). The other result was the elucidation of the fact that the long-term sedimentological changes in the Ordovician of the Tungus basin on the Siberian Platform remind us of the contemporaneous evolution of sediments in North American (Laurentia) Platform rather than on the Russian Platform (Baltica) or any other Gondwanan Platform. The discovery of abundant K-bentonite layers in the Upper Ordovician of the Tungus basin only emphasizes this striking similarity (Dronov et al., 2011a).

In order to construct a sequence stratigraphic chart for the Ordovician of the Siberian Platform the key sections of the Tungus and Irkutsk basins have been studied. In the Tungus basin, the Kulumbe River section in the vicinity of the Turukhan Land, the Moyero River section in the vicinity of the Anabar Land, the main outcrops in the Podkamennaya Tunguska River valley and its tributaries in the vicinity of the Yenisei Land, and finally the Katanga Land have been investigated. In the Irkutsk basin the main natural outcrops and quarries near the town of Bratsk, as well as in the Angara River valley, and in the Lena River valley between the towns of Ust'kut and Kirensk have been studied (Fig. 1). During the field work, particular attention was paid to the recognition, interpretation, and correlation of the sequence boundaries. An identification of the different systems tracts was not always possible. The amplitudes of regressions and transgressions were estimated based on the depth of erosion of the beds underlying unconformities and on the distribution of relatively deep-water facies.

Precise biostratigraphic correlation of the sequence boundaries between Baltica and Siberia also remains problematic due to the absence of common species in faunal assemblages, but the number of sequences in broader stratigraphic intervals is comparable. This makes it possible to compare a general shape of sea-level curves for the Ordovician of these paleocontinents as well as their long-term lithological evolution and geological history. The aim of this chapter is to present an updated summary of recent results for the Ordovician of the Siberian Platform, to provide a comparative analysis with the Ordovician of the Russian Platform, and to place both into a global context.

2. GEOLOGICAL SETTING AND STRATIGRAPHY

The Siberian paleocontinent, which is represented today in Eurasia by the Siberian Platform, included several separate Early Paleozoic sedimentary basins. The largest ones were the Irkutsk and Tungus epicontinental basins, which are also known as the Irkutsk Amphitheater and the Tungus Sineclyse, respectively (Kanygin et al., 2007, 2010b) (Fig. 1). The Irkutsk basin occupies a restricted area at the very south of the Siberian Platform (in present-day orientation). The Tremadocian part of the Lower Ordovician succession of the basin is represented by warm-water tropical-type carbonates with abundant stromatolites (Ust'kut Formation). In this respect, it is similar to the contemporaneous parts of the Siberian Ordovician from the Tungus basin (Fig. 2).

The Floian part of the Lower Ordovician and the Middle Ordovician of the Irkutsk basin are represented mainly by shallow-water quartz sands and gray- and red-colored siltstones. Unfortunately, this part of the succession is only poorly documented in outcrops and contains almost no stratigraphically significant fossils. This makes correlations even with the neighboring Tungus basin rather uncertain. The Upper Ordovician of the Irkutsk basin is represented by restricted basin (lagoon) deposits characterized by fine-grained predominantly red-colored terrigenous sediments sometimes containing pseudomorphic halite crystals. The absence of any kind of fauna points to hypersaline conditions and precludes detailed biostratigraphic correlations. Therefore, the Irkutsk basin will not be considered any further within the present study.

The Tungus basin represents an extensive intracratonic sedimentary sag basin which occupies the central part of the Siberian Craton (Kanygin et al., 2007; Markov, 1970). The Ordovician of the Tungus basin is subdivided into 12 regional stages based predominantly on trilobites, brachiopods, and ostracods. The uppermost (Hirnantian) part of the Ordovician succession on the Siberian Platform is eroded in exposed sections (Kanygin et al., 2007). However, this does not exclude the possibility that it exists in the central parts of the Tungus basin and may be reached in the future by boreholes. Siberian regional stages allow the correlation of Ordovician sections in the Tungus basin with all other sections of the Siberian Platform, but the endemic character of the Siberian fauna makes precise correlations with the global chronostratigraphic chart difficult. A provisional correlation of

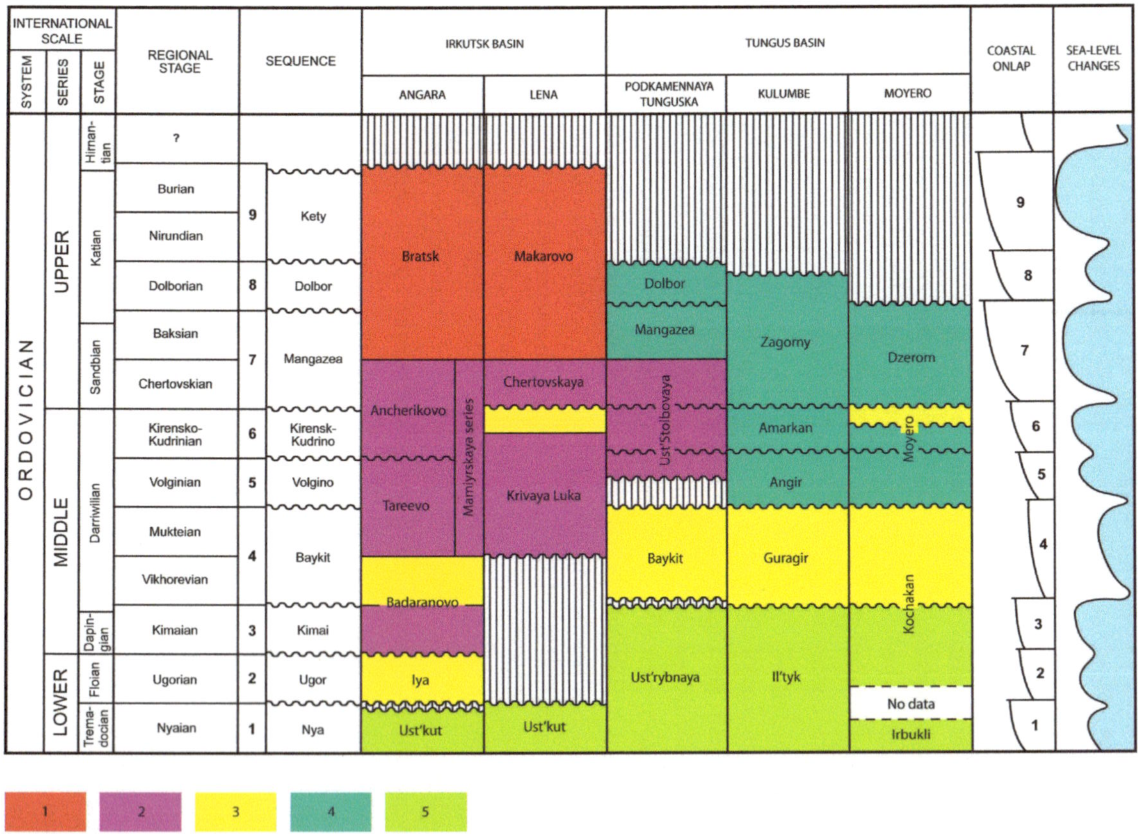

Fig. 2 Stratigraphic chart showing selected correlations of the studied Ordovician successions of the Irkutsk and Tungus basins. It also shows depositional sequences, sea-level fluctuations, and long-term lithological changes. Legend: 1. red siltstones; 2. variegated siltstones; 3. quartz sandstones; 4. cool-water carbonates; and 5. warm-water carbonates.

the Siberian regional stages and the global Ordovician stages was published recently (Bergström et al., 2009).

Currently, for intercontinental biostratigraphic correlation of the Ordovician of the Siberian Platform, only four reliable levels exist so far: (1) the base of the Nyaian Regional Stage that correlates to the base of the Ordovician according to recent conodont investigations (Tolmacheva and Abaimova, 2009); (2) the base of the Volginian Regional Stage that correlates to the base of the *Hustedograptus teretiusculus* graptolite zone; (3) the base of the Chertovskian Regional Stage that correlates to the base of the *Nemagraptus gracilis* graptolite zone; and (4) the base of the Silurian. These four stratigraphic levels create three major stratigraphic intervals that can be correlated globally.

The lower boundary of the Ordovician is conventionally placed within the lower part of the *C. linstroemi* Zone in the following succession of species of *Cordylodus* lineage: *C. andresi*, *C. proavus*, *C. lindstroemi*, and *C. angulatus*. All the representatives of this succession have been found in Siberia. *C. lindstroemi* has recently been documented from sediments of the upper part of the Nyaian Regional Stage in the Kulumbe River section on the northwestern margin of the Tungus basin (Tolmacheva and Abaimova, 2009). Accordingly, the base of the Ordovician on the Siberian Platform can be defined at a level no lower than the base of the Nyaian Regional Stage (Kanygin et al., 2010b).

The base of the Volginian Regional Stage is the most distinct correlatable level within the Middle Ordovician of the Siberian Platform. It corresponds to the base of the Volginian depositional sequence represented on the western margin of the Tungus basin by a prominent transgressive surface (Dronov et al., 2009; Kanygin et al., 2010a). The surface marks the onset of a transgression that caused significant biotic changes across the Tungus basin. This level is marked by the appearance of numerous new taxa of brachiopods, which have proven to be of particular value for wider regional correlations, as well as new taxa of trilobites, crinoids, bryozoans, and conodonts. The Volginian and overlying Kirensko-Kudrinian associations of brachiopods and ostracods exhibit close affinities to those of the Lachug Regional Stage of the Verkhoyano-Chukotka region in the Russian Northeast. Correspondence of this stratigraphic interval with the *H. teretiusculus* graptolite zone of the upper Darriwilian is based on the occurrence of *Hustedograptus* aff. *teretiusculus* (Hisinger) in the shale deposits of the Lachug Regional Stage at the outcrops along the Kolyma River and its tributaries (Kanygin et al., 2010a; Oradovskaya, 1988).

The base of the Upper Ordovician corresponds with the base of the Chertovskian Regional Stage and the base of the Mangazea depositional sequence (Dronov et al., 2009; Kanygin et al., 2010a). It coincides with the most distinctive transgression in the Ordovician of Siberia. The lower boundary of the Chertovskian Stage is well defined by the appearance of numerous new species of brachiopods and ostracods as well as trilobites and conodonts. In the Verkhoyano-Chukotka region (Kolyma River and its tributaries) this fauna is associated with graptolites of the *N. gracilis* Zone. This suggests that the base of the Chertovskian Regional Stage is close to the Middle/Upper Ordovician boundary.

The precise correlation of the Upper Ordovician Siberian regional stages with the global chronostratigraphic stages is problematic. But recently found zircon crystals from K-bentonite beds within the upper part of the Baksian Regional Stage provides a $^{206}Pb/^{238}U$ age of 450.58 ± 0.27 Ma (Huff et al., 2014). This date confirms the Katian age of the upper part of the Mangazea Formation.

The Ordovician of the Tungus basin consists of two carbonate series of contrasting lithology separated by a unit of quartz sand or quartz sand with siltstone intercalations (Fig. 2). A near complete succession is known from the southwestern margin of the Tungus basin (in present-day orientation) where the Ordovician is exposed in the natural outcrops along the Podkamennaya Tunguska River and its tributaries. This succession is represented by the following lithostratigraphic units (Fig. 2):

1. The Ust'rybnaya Formation (Nyaian, Ugorian, and Kimaian regional stages) is only poorly exposed and characterized by contrasting lithologies. In the western, most shallow-water part of the basin in the proximity of the Yenisei Land it is represented by coarse-grained mainly red-colored quartz sandstones intercalated with layers and lenses of conglomerates (quartz gravelstones) and layers of siltstone and limestone. Further to the east, it becomes less coarse-grained and less siliciclastic with predominantly carbonate sedimentation. Limestones are represented by thick-bedded well-washed oolitic grainstones with abundant stromatolites, flat-pebbled conglomerates, and ripple marks pointing to a shallow-water tropical environment. Dolomitization is widespread. The Ust'rybnaya Formation rests on the Upper Cambrian deposits with a regional unconformity. In the westernmost part of the basin even a low-angle angular unconformity is reported (Markov, 1970). The thickness of the formation varies from several meters in the west to 120 m in the east. On the northwestern margin of the Tungus basin along the

Kulumbe River section this stratigraphic interval is represented by limestones and dolomites of the Il'tyk Formation, about 500 m thick. On the northeastern margin (Moyero River section) this part of the succession is poorly known. It is represented by the Irbukli Formation (Nyaian Regional Stage) and the lower part of the Kochakan Formation (Ugorian and Kimaian stages). Both formations consist of oolitic grainstones and dolomites with stromatolites sometimes intercalated with gypsum layers.

2. The Baykit Formation (Vikhorevian and Mukteian regional stages) constitutes a sedimentary body that continues laterally and traceable in the basin of the Podkamennaya Tunguska River over a distance of more than 600 km. It is represented by a uniform succession of light-gray and yellowish quartz sandstones which are 5–80 m thick. The sandstones are composed of angular, less commonly well-rounded grains of quartz (up to 80% of the total rock weight) with rare grains of feldspar and tiny fragments of muscovite crystals. The sandstones are often massive or coarse-bedded. Several levels display cross-bedded series and sometimes lenses of conglomerate, especially at the base. The fabric of the Baykit Sandstone is poorly known because of its relatively monotonous composition and poor exposure. The Baykit Sandstone roughly corresponds to the Guragir Formation in the proximity of the Turukhan Land and to the upper part of the Kochakan Formation in the proximity of the Anabar Land (see Figs. 1 and 2). Both Guragir Formation and the upper part of the Kochakan Formation are represented by intercalations of pure quartz sandstones and red siltstones, sometimes with abundant pseudomorphic halite crystals. For the upper part of the Kochakan Formation, gypsum layers are also typical. The thickness of the siliciclastic units is 200 m for the Guragir Formation and about 60 m for the upper part of the Kochakan Formation.

3. The Ust'Stolbovaya Formation corresponds to the Volginian, Kirensko-Kudrian, and Chertovskian regional stages. It is represented by a relatively condensed (usually 11–15 m) succession of red and green argillites with lenses of quartz and phosphate conglomerates at the base and thick (1.5–2 m) layers of phosphatic conglomerates at the top. The Ust'Stolbovaya Formation includes two depositional cycles (sequences) corresponding to the Volginian and Kirensko-Kudrian regional stages. Deposits of the Chertovskian Regional Stage are similar to the underlying deposits of the Kirensko-Kudrinian Stage and are included in the Ust'Stolbovaya Formation, but the sequence boundary coincides with the boundary of the Kirensko-Kudrian and Chertovskian stages (Dronov et al., 2009). In the Kulumbe River section, this stratigraphic interval is represented by the Angir Formation (Volginian Regional Stage, 30 m in thickness) and the Amarkan Formation (Kirensko-Kudrinian Regional Stage, 50 m thick) of predominantly gray-colored highly bioturbated bioclastic limestones (wackestone). In the Moyero River section, it corresponds to the Moyero Formation (about 80 m thick) that is composed of predominantly carbonate lithologies.

4. The Mangazea Formation (Baksian Regional Stage) is represented by rhythmic intercalations of greenish-gray siltstones and gray micritic (in the lower part) and bioclastic limestones. The bioclasts are mainly fragments of brachiopod shells, trilobite carapaces, and bryozoans. Ostracodes and crinoids are also numerous. On top of the limestone layers in many cases ripple marks can be observed. Several layers contain glauconite grains. Three K-bentonite layers have been found in the upper part of the Mangazea Formation. They are represented by thin beds (1–2 cm) of soapy light-gray or yellowish plastic clays. The beds are different by consistence and color from the enclosing sediments and usually easily identifiable in the outcrops. They were traced over a distance of more than 60 km along the Podkamennaya Tunguska River valley (Dronov et al., 2011a; Huff et al., 2014). The thickness of the Mangazea Formation varies from 40 m in the west to 70 m in the east. In the northern part of the Tungus basin, this stratigraphic interval is represented by the Zagorny Formation (about 70 m thick) in the Kulumbe River section and the Dzherom Formation (about 100 m thick) in the Moyero River section. Both formations share similar lithological characteristics to the Mangazea Formation. However, bentonite layers are typical only for the southwestern margin of the Tungus basin.
5. The Dolbor Formation roughly corresponds to the Dolbor Regional Stage (Dronov et al., 2009; Kanygin et al., 2010a). The Dolbor Formation is represented by yellowish-gray fine-grained sandstones and siltstones, sometimes with a carbonate cement. In the western section, the base of the formation is a sequence boundary with traces of erosion and a visible change in sedimentation. A series of bioclastic limestones interbedded with greenish-gray siltstones, typical for the Mangazea Formation, is overlain by a series of mainly sandstone and siltstone beds with scattered bioclasts and sporadic bioclastic limestone layers. In the eastern sections, the Dolbor Formation contains more greenish-gray siltstones

with intercalations of thin (0,5 – 2 cm) limestone layers. The appearance of tabulate corals, stromatoporoids, and chert lenses is typical for the Dolbor Formation in the east. Like the underlying Mangazea Formation, it contains numerous K-bentonite beds. The thickness of the formation is up to 75 m. In the northern part of the Tungus basin, this part of the Ordovician succession is completely removed by erosion.

6. The Nirunda Formation corresponds to the Nirunda Regional Stage. It is only partly exposed in natural outcrops along the Bolshaya Nirunda River and is represented mainly by cherry-red and greenish-gray siltstones with greenish-gray thin laminated bioclastic limestone layers at some levels. The macrofauna is dominated by bryozoans and brachiopods and the overall thickness of the formation is about 20 m.
7. The Bur Formation coincides with the Bur Regional Stage and is represented by greenish-gray siltstones and intercalated gray bioclastic limestone beds. An abundance of large (up to 20–30 cm in diameter) tabulate coral colonies is typical for the Bur Formation. Some of the colonies are overturned and obviously affected by storms, while others are still in their natural growth position. Some of the outcrops (Yukteli River section, the tributary of the Bolshaya Nirunda River) contain K-bentonite layers. The top of the Bur Formation is a regional unconformity and it is directly overlain by Silurian deposits. The thickness of the formation is about 10–12 m.

During the Cambrian, Ordovician, and Silurian times the Siberian Platform, which constitutes the core of the Siberian paleocontinent, was located in the low-latitude tropical area (Fig. 3), migrating slowly from the southern hemisphere in the Cambrian and Lower Ordovician to the northern hemisphere in the Upper Ordovician and the Silurian (Cocks and Torsvik, 2007).

3. LONG-TERM LITHOLOGIC CHANGE

During the Ordovician period, the mode of sedimentation in relatively shallow-water settings of the Tungus basin on the Siberian Platform underwent significant changes. The Lower Ordovician and the lower part of the Middle Ordovician are represented by well-washed oolitic grainstones and dolomites with abundant stromatolites, cyanobacterial mats, and occasional gypsum layers. This lithology is typical for the Ust'rybnaya Formation in the southwest of the basin, for the Il'tyk Formation in the northwest, and for the Kochakan Formation in the northeast. In all these regions the lower

Fig. 3 Paleogeographical reconstruction for the Late Ordovician showing the distribution of cool-water carbonates, the position of the supposed Taconic-Yenisei Arc, and possible cool-water currents and upwelling. *Based on Chernes, L., Wheeley, J.R., 2009. Early Palaeozoic cooling events: peri-Gondwana and beyond. In: Basset, M.G. (Ed.), Early Palaeozoic Peri-Gondwana Terranes: New Insights From Tectonics and Biogeography. Geological Society, London, Special Publications 325, 257–278; Cocks, L.R.M., Torsvik, T.H., 2007. Siberia, the wandering northern terrane, and its changing geography through the Paleozoic. Earth Sci. Rev. 82, 29–74; Cocks, L.R.M., Torsvik, T.H., 2011. The Palaeozoic geography of Laurentia and western Laurussia: a stable craton with mobile margins. Earth Sci. Rev. 106, 1–51; Harper, D.A.T., Owen, A.W., Bruton, D.L., 2009. Ordovician life around the Celtic fringes: diversifications, extinctions and migrations of brachiopod and trilobite faunas at middle latitudes. In: Basset, M.G. (Ed.), Early Palaeozoic Peri-Gondwana Terranes: New Insights From Tectonics and Biogeography. Geological Society, London, Special Publications 325, 157–170 with modifications.*

and lowermost Middle Ordovician carbonate series are covered by a thick predominantly siliciclastic unit. It is represented by the Baykit Sandstone in the Podkamennaya Tunguska River valley and its tributaries, by the Guragir Formation in the Kulumbe River section, and by the upper terrigenous part of the Kochakan Formation in the Moyero River valley. The upper part of the Middle Ordovician and the Upper Ordovician are again represented by predominantly carbonate lithologies, but the carbonates are gray bioclastic limestones (wackestone and packstone) without oolites, stromatolites, and cyanobacterial mats. The lower carbonate series was interpreted as so-called warm-water carbonates and the upper carbonate series as cool-water carbonates (Dronov, 2013; Dronov et al., 2009).

3.1 The Warm-Water Carbonates

The carbonate sediments can be differentiated into so-called warm-water, or tropical-type (photozoan), and cool-water, or temperate-type (heterozoan), carbonates based on lithological composition, associated minerals, and a set of primary sedimentary structures (Dronov, 2001; Dronov and Rozhnov, 2007; James, 1997; Lindström, 1984). Tropical-type carbonates are characterized by a wide range of carbonate grain types including skeletal grains, intraclasts, peloids, and ooids. Different types of reefs are abundant in tropical-type carbonates including stromatolites (Fig. 4A and B) and cyanobacterial mats. Carbonate mud occurs in abundance in tropical-type carbonates. Sediment grain sizes vary widely from mud to sands to gravels and flat-pebble conglomerates (Fig. 4C). Traces of karstification are also typical for subaerially exposed tropical carbonates (Fig. 4D).

The warm-water tropical-type carbonates (Ust'kut Formation, Ust'rybnaya Formation, Il'tyk Formation, Kochakan Formation) are represented by light gray, sometimes pink or yellowish usually thick-bedded limestones and dolostones intercalated with red siltstones and sometimes quartz sandstone layers. The dominant lithotype found is well-washed and well-sorted oolitic grainstone (Fig. 4E and F). Stromatolite buildups of different morphologies are very common as well as cyanobacterial mats (Fig. 4A, B, and E). Ripple marks on the top of oolitic grainstone beds indicate a shallow-water environment. Another typical feature is the occurrence of flat-pebble conglomerates. They are products of the destruction and redeposition of carbonate crusts caused by tropical storms in the tidal flats of the carbonate platforms (Fig. 4C). The primary depositional structures, biotic components, style and scale of cyclicity, and facies associations are consistent with a shallow subtidal to intertidal and rare supratidal paleodepositional setting on the inner part of the broad carbonate platform. This style of deposition was inherited from the Cambrian and even earlier times and continued without significant changes until the middle of the Middle

Fig. 4 Warm-water, tropical-type carbonates of the Siberian Platform. (A) Big columnar stromatolites (3 m high) in the Ust'kut Formation. Angara River valley. (B) Growth stratification in stromatolites. Kochakan Formation. Moyero River valley. (C) Flat-pebbles conglomerate in Il'tyk Formation. Kulumbe River valley. (D) Karst breccia from the Ust'kut Formation. Angara River valley. (E) Stromatolite crusts and ooids from the Ust'kut Formation. Lena River valley. (F) Oolitic grainstone from the Ust'rybnaya Formation. Podkamennaya Tunguska River valley.

Ordovician. There is no erosional unconformity or any kind of facies change at the Cambrian/Ordovician boundary (Kanygin et al., 2010b; Tolmacheva and Abaimova, 2009). The lower warm-water carbonate series in the Ordovician of the Tungus basin varies from 120 to 700 m in thickness, being about 440 m on average.

3.2 The Cool-Water Carbonates

Until the very recently the Ordovician limestones of the Tungus basin on the Siberian Platform have traditionally been considered to be tropical shelf limestones (shallow-water warm sea) and interpreted using facies models derived from recent tropical carbonate environments (Kanygin et al., 2010b; Markov, 1970). However, depositional rates, grain types, faunas, erosion surfaces, and mineralogy of the uppermost Middle Ordovician and the Upper Ordovician carbonates are more compatible with a temperate or even cold shelf environment (Dronov, 2013; Dronov et al., 2009; Kanygin et al., 2010a). Cool-water carbonates have been largely neglected during the "golden age" of carbonate sedimentology between the 1950s and 1980s, but now it has become obvious that they played a much more important role in recent seas and in the geological record than was previously expected. Shallow-marine temperate and cool-water carbonates constitute about one-third of modern global shelf sediments (Nelson, 1988). Large areas of cool-water carbonates are known today from the southern Australian margins, the northwestern continental margin of Europe, and the Barents Sea (Henrich et al., 1995; James, 1997). Cool-water carbonates have also been described from Cenozoic, Mesozoic (Pedley and Carannante, 2006), and Paleozoic (Beauchamp and Desrochers, 1997; Brandley and Krause, 1997; Lindström, 1984) successions.

Temperate-type carbonates demonstrate a limited spectrum of skeletal grains, which are dominated by fragments of bryozoans, brachiopods, trilobites, ostracods, and echinoderms. Ooids and peloids are absent or scarce. Reefs, stromatolites, and cyanobacterial mats are extremely rare or absent. In the shallow shelf areas, insignificant amounts of carbonate mud could be accumulated owing to bioerosion below the storm wave base. Sediment textures are mainly sands and gravels. In addition to water temperature, carbonate sediments are also affected by water transparency and the number of available nutrients. Increased amounts of nutrients and terrigenous input can cause a temperate-type appearance even to tropical carbonates (James, 1997).

The cool-water carbonate series (from the Volginian to the Burian stages) in the Upper Ordovician of the Tungus basin is characterized by fine-grained terrigenous sediments with subordinate carbonates represented mostly by temperate-type lithologies. Bioclastic wackestones and packstones dominate. Stromatolites, dolomites, and organic buildups are absent. Ooids and peloids are scarce. At some levels, glauconite grains and films occur. The upper series has a relatively reduced thickness of 90–300 m and is about 170 m on average. The decrease in overall thickness probably reflects reduced rates of sedimentation and, correspondingly, a diminished productivity of the shallow-water "carbonate factory." The shift from tropical-type carbonate sedimentation to temperate-type was preceded by the destruction of the warm-water "carbonate factory" and a large input of siliciclastic material (Dronov, 2013).

The best and most well-exposed examples of cool-water carbonates in the Ordovician of the Siberian Platform belong to the Baksian, Dolborian, and, partly, Burian stages. The Mangazea and Dzherom formations are the most typical examples being dominated by the intercalation of greenish-gray siltstones and gray bioclastic limestones. The limestones are medium to thickly bedded (3–30 cm, average 12 cm) packstones and rudstones (Fig. 5A, D, and E). The beds have sharp bases and are commonly amalgamated. The deposits show abundant sedimentary structures such as scouring basal lags, normal grading, thick parallel and cross-lamination with either high- or low-angle planar, and trough types. The rocks are composed of various bioclasts and, less commonly, intraformational particles. The allochems are fine sand- to pebble-sized with coarse sand-sized material being most abundant. Sorting within individual beds is commonly good to fair. Trilobites, brachiopods, and ostracods are the most common bioclasts with a significant number of bryozoans, pelmatozoans, and mollusks (Fig. 5B and C). Most of the bioclasts are broken and abraded. Some of them show more-or-less intensely micritized rims. Intraclasts comprise rounded fragments of biomicrite. The size of the allochems ranges from fine sand to centimeter-sized bioclasts.

The lithology of the Mangazea and Dzherom formations shows sedimentary structures (coarsening- and fining-upward, parallel and low-angle cross-lamination, as well as hummocky cross-stratification) indicative of storm-induced sedimentation (Aigner, 1985; Kreisa, 1981). Wave ripples developed at the top of some beds (Fig. 5A) suggest deposition by oscillatory or combined currents (Lavoie, 1995). As the most typical and widespread sedimentary structures of the Mangazea and Dzherom formations are indicative of high-energy conditions, they are interpreted as being formed at a high-energy carbonate ramp, comparable to that described from the Late Ordovician successions of Quebec (Canada) by Brookfield (1988), Lavoie (1995), and Lavoie and Asselin (1998). As a whole, biotic components in the Mangazea and Dzherom formations are typical of the *bryomol* association defined from modern temperate-water settings (Nelson et al., 1988). The

Fig. 5 Cool-water, temperate-type carbonates of the Siberian Platform. (A) Ripple marks at the top of bioclastic packstone bed. Mangazea Formation. Bolshaya Nirunda River valley. (B) Bryozoan and fragments of trilobites and brachiopods on the surface of bioclastic packstone bed within the Dzerom Formation. Moyerokan River valley about 3 km upstream from the place where it merges into the Moyero River. (C) Fragments of pelmatozoans, trilobites, and brachiopods on the top surface of a bioclastic packstone bed within the Dzerom Formation. Moyerokan River valley about 3 km upstream from the place where it merges into the Moyero River. (D) Intercalation of siltstones and beds of bioclastic packstones and rudstones (storm beds). Mangazea Formation. Podkamennaya Tunguska River valley. (E) Thick storm bed of bioclastic packstone within the Mangazea Formation. Stolbovaya River valley. (F) Chert nodule and colonies of tabulate corals in the Dolbor Formation. Bolshaya Nirunda River valley.

proposed temperate-water-dominated faunal association is also supported by the almost total lack of typical *chlorozoan* elements (Lees and Buller, 1972). The lithology of the contemporaneous Dzherom Formation in the proximity of the Anabar Land demonstrates no principal differences from the lithology of the Mangazea Formation. The lithology of the Zagorny Formation in the Kulumbe River section is also very similar, but sediments of the Zagorny Formation seem to have been deposited in a more deep-water environment.

The lithology of the Dolbor Formation is represented by planar and wavy limestone beds ranging from 1 to 12 cm in thickness (average 3 cm). The limestones are separated by millimeter- to centimeter-thick interbeds of greenish-gray shales. Internal sedimentary structures include fine parallel lamination, low-angle cross-lamination, and hummocky cross-stratification. The bioclastic material is commonly concentrated in centimeter-sized lenses at the bottom of the limestone beds. The macrofauna is dominated by bryozoans, brachiopods, and trilobites with lesser amounts of pelmatozoans and corals. Chert nodules are abundant at some levels and siliceous sponge spicules are closely associated with the silicified areas (Fig. 5F). The lithology of the Dolbor Formation indicates deposition below fair-weather wave base but above storm wave base (Aigner, 1985; Kreisa, 1981). The preservation of articulated pelmatozoans indicates a rapid burial of the fauna probably induced by sudden storm events. In places with predominantly mud sedimentation the preservation of erect fragile pelmatozoans and bryozoans and an overall lack of mechanical breakdown suggest rapid burial in a low-energy setting.

The Bur Formation, although poorly exposed, is with regard to its overall lithological composition similar to the Mangazea, Dzherom, and Dolbor formations. The thickness of the carbonate beds here is less than in the Mangazea and Dzherom formations (average 5–7 cm) but higher than in the Dolbor Formation. The most typical feature of the Bur Formation is the abundance of big (20–30 cm in diameter) colonies of favosite corals. Some of these colonies are in living position but others are overturned, which clearly demonstrates the effects of storm activity. The Bur Formation probably represents the most shallow-water part of the ramp. Despite the abundance of colonial and solitary corals in the Bur deposits, no reefs, patch reefs, or any other kind of coral buildups have been reported so far from these deposits. The combined evidence for a *bryomol* skeletal community deposited on a high-energy, slowly accreting unrimmed shelf are strong arguments to suggest a temperate- or cold-water setting for the deposition of the Mangazea, Dzherom, Dolbor, and Bur formations.

Fig. 6 Bentonite bed visible in the outcrop of Mangazea Formation. Stolbovaya River valley about 6 km upstream from the place where it merges with the Podkamennaya Tunguska River.

The cool-water Upper Ordovician succession of the southwestern part of Tungus basin also contains K-bentonite layers. These layers are generally represented by thin beds (1–2 cm) of soapy light-gray or yellowish plastic clays. They are different by consistence and color from the enclosing sediments and usually easily identifiable in the outcrops (Fig. 6). To date, 12 K-bentonite layers have been identified in the outcrops of the Mangazea, Dolbor, and Bur formations. The lowermost (K-bentonite-I) bed of the Mangazea Formation was traced over a distance of more than 60 km along the Podkamennaya Tunguska River valley. The other two beds (K-bentonite-II and K-bentonite-III) were traced over at least 40 km (Dronov et al., 2011a,b). The K-bentonite beds from the Upper Ordovician of the Tungus basin in Siberia were derived from the alteration of volcanic ash falls. Their appearance points to the intensive explosive volcanism on or near the western (in present-day orientation) margin of the Siberian Craton in Late Ordovician time. The association of cool-water carbonates in the Upper Ordovician of Siberia with K-bentonite beds is similar to the North American succession (Huff et al., 2010).

It is worth noting that the Upper Ordovician succession of the eastern margin (in present-day orientation) of the Siberian Craton contains no

K-bentonite layers. That means that contrary to some popular paleogeographic interpretations (e.g., www.scotese.com) the subduction zone in the Late Ordovician was located along the western but not the eastern margin of the Siberian paleocontinent.

3.3 Comparison to Baltica and Laurentia

Comparative analysis of the Ordovician successions of the Siberian and Russian Platforms demonstrates an opposite trend in the long-term lithologic changes (Fig. 7). On the Russian platform, the average thickness of the regional stages increases persistently upward throughout the Ordovician succession, reflecting a rise in the average rate of sedimentation. In addition, predominantly terrigenous sediments are successively replaced first by cool-water and then by tropical carbonates in shallow-water environments (Dronov and Rozhnov, 2007) (Fig. 8). These climate-controlled changes in sedimentation reflect a drift of the Baltic paleocontinent during the Ordovician from the near-polar latitudes of the Southern Hemisphere into the Equatorial Zone (Cocks and Torsvik, 2005). On the Siberian Platform, the regional stages of the Lower and lower Middle Ordovician have considerably greater thickness than those of the upper Middle and Upper Ordovician. In addition, upward in succession, the predominantly tropical-type warm-water carbonates are replaced by terrigenous and temperate-type carbonate deposits (Fig. 7).

On the other hand, comparative analysis of the Ordovician successions of the Siberian and North American platforms demonstrates a striking similarity in the long-term lithological development (Fig. 7). On both platforms, the Ordovician succession starts with tropical stromatolite-bearing carbonates, which abruptly change to siliciclastic deposits (Baykit Sandstone in Siberia and Eureka Sandstone in North America) and terminate with cool-water carbonates (Ettensohn, 2010; Herrmann et al., 2004; Kanygin et al., 2010a). Numerous K-bentonite beds in the Upper Ordovician of North American and Siberian platforms stress this similarity (Dronov et al., 2011a,b; Huff et al., 2010, 2014).

In Siberia, the cool-water carbonate deposition started in the Late Darriwilian (Volginian Regional Stage). In North America, an enrichment in phosphates coincides with the onset of cool-water carbonate deposition. In Siberia, phosphate-rich deposits are also associated with the onset of cool-water carbonates and appear in the sections during the Late Darriwilian–earliest Sandbian (Volginian, Kirensko-Kudrinian, and Chertovskian

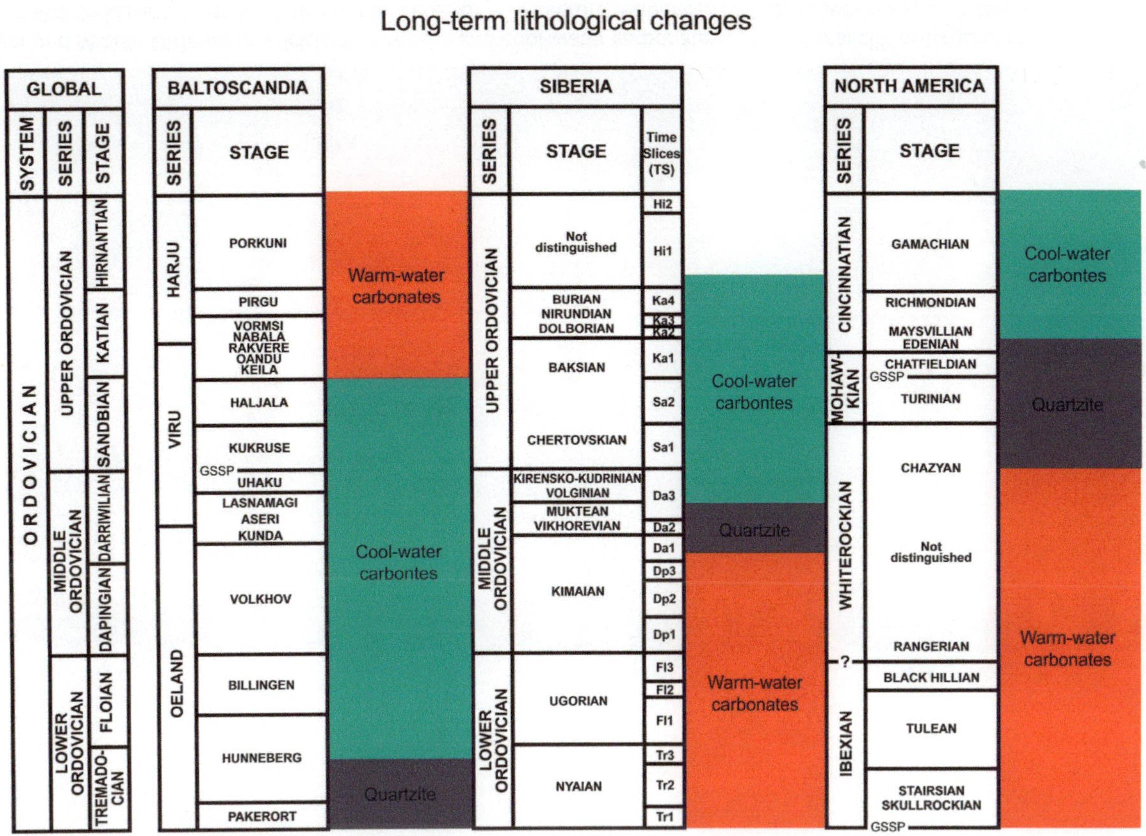

Fig. 7 Long-term lithological changes in the Ordovician of Baltica, Siberia, and Laurentia: a comparison.

Fig. 8 Lower and Middle Ordovician siliciclastic deposits and cool-water carbonates at the coastal cliff outcrop of the Baltic-Ladoga Klint in the cape of Pakri (Pakerort), northwestern Estonia. Legend: 1—Pakerort Sequence; 2—Latorp Sequence; 3—Volkhov Sequence; 4—Kunda Sequence; and 5—Tallinn Sequence. *SB*, sequence boundaries; *TS*, transgressive surface.

regional stages). Moreover, in southeastern Laurentia the onset of cool-water carbonates coincides with the destruction of the tropical carbonate platform (Lexington Platform). In Siberia, the destruction of the carbonate platform and the abrupt change to siliciclastic sedimentation occurred around the base of the Darriwilian when the Baykit Sandstone starts to form (Figs. 1 and 2).

3.4 Discussion

According to paleomagnetic data, the Siberian paleocontinent was located in the tropical zone during the Cambrian, Ordovician, and Silurian (Cocks and Torsvik, 2007). Therefore, the shift from tropical-type to temperate-type carbonates in the Middle and Late Ordovician (Figs. 3 and 7) can only be explained by the upwelling of cold oceanic waters along the margin of the continent. The fact that the underlying Upper Cambrian–Lower Ordovician and overlying Silurian deposits of the Siberian Platform are represented by typical warm-water tropical-type carbonates indicates that in the absence of upwelling the tropical-type carbonates should have formed and did form in these paleolatitudes. The upwelling probably began with destruction of the preexisting extensive warm-water carbonate platform and deposition of the Baykit Sandstone. The upwelling was probably significantly enhanced during an extensive transgression, the start of which is reflected in the transgressive surface at the base of the Volginian Regional Stage (Dronov, 2013; Dronov et al., 2009; Kanygin et al., 2010a).

Sea-level rise itself, however, even though significant, cannot necessarily create an upwelling. On the Siberian Platform, the upwelling seems to have started during the regression marked by the formation of the Baykit Sandstone. The regression seems to have been caused, or at least been reinforced, by the tectonic uplift of the Yenisei Land area at the western margin of the Siberian Platform. The tectonic activity at the platform margin was probably caused by the attachment of some terrane and/or island arc. This tectonic reorganization may have altered the directions of large oceanic currents, which would have caused the upwelling. Volcanic ash beds associated with cool-water carbonates in the Upper Ordovician of the Siberian Platform also point out the tectonic activation of the platform margin. The end of the upwelling at the Ordovician/Silurian boundary was probably related to the eustatic sea-level fall, resulting from the Hirnantian glaciation. The following sea-level rise in the Silurian did not restore the Ordovician system of oceanic currents and, correspondingly, the upwelling. In the absence of upwelling, the warm-water tropical-type carbonate accumulation typical for the low paleolatitudes was reestablished in the shallow-water epicontinental Silurian seas.

The same situation was recorded for the North American Platform. In the Ordovician this platform was also located in the tropical zone, and similarly, the warm-water carbonate sedimentation of the Early and Middle Ordovician was replaced by a cool-water one in the early Late Ordovician (Ettensohn, 2010; Herrmann and Haupt, 2010; Holland and Patzkowsky, 1996). Likewise, the tropical carbonates were accumulated again in the Latest Ordovician–Silurian. This situation on the American platform was explained by upwelling that also resulted from an extended transgression reflected in the M-5 sequence (Ettensohn, 2010; Herrmann and Haupt, 2010; Holland and Patzkowsky, 1996). The transgression took place after a large regression and tectonic rearrangement related to the beginning of the Taconian orogeny, i.e., the initial attachment of the Taconian island arc to the North American continent. In North America, the Taconian orogeny and the subsequent cold-water upwelling started later, in the early Upper Ordovician, while on the Siberian Platform these events occurred slightly earlier in the Middle Ordovician. However, the general succession of the events and their character are very much alike.

4. ORDOVICIAN SEQUENCE STRATIGRAPHY OF THE SIBERIAN PLATFORM

Based mainly on outcrop data, nine depositional sequences bounded by regional unconformities or their correlative conformities have been distinguished in the Ordovician of the Siberian Platform (Dronov et al., 2009; Kanygin et al., 2010a,b). The sequences correspond to sea-level fluctuations of the third order (according to Vail et al., 1977) and have an average duration between 1 and 10 myr. For the purpose of convenience, individual terms derived from the names of regional stages, series, and formations have been assigned to them. From the base upward these are: (1) Nya; (2) Ugor; (3) Kimai; (4) Baykit; (5) Volgino; (6) Kirensk-Kudrino; (7) Mangazea; (8) Dolbor; and (9) Kety sequence (Dronov et al., 2009; Kanygin et al., 2010a) (Fig. 2).

4.1 The Nya Sequence

There are no apparent signs of significant erosion near the Cambrian–Ordovician boundary on the Siberian Platform in contrast to the Russian

Platform where an erosional unconformity is well developed (Fig. 8). In the Kulumbe River valley, the Upper Cambrian–Lower Ordovician interval is represented by a 50–70-m-thick unit of limestones and dolomites with abundant stromatolites, oolitic grainstones, and flat-pebbled conglomerates formed on a tropical shallow-water carbonate platform (Il'tyk Formation). Some transgressive–regressive cycles can be recognized within this unit, but no significant gaps have been recorded. The position of the Cambrian–Ordovician boundary on the Siberian Craton has long been a matter of discussion. Based on the first appearance of dendroid graptolites in the Loparian Regional Stage, the boundary has been drawn at the base of the underlying Mansian Regional Stage where distinct lithological changes and traces of erosion occur (Kanygin et al., 1982, 2006). However, recent data on the distribution of conodonts of the *Cordylodus* lineage indicate a considerably higher position for the boundary, approximately near the base of the Nyaian Regional Stage (Tolmacheva and Abaimova, 2009). The Nyaian Regional Stage has always been interpreted as a separate phase in the evolution of the biota on the Siberian Platform (Kanygin et al., 2006). Most likely it also represents a separate depositional sequence. Detailing the internal structure and development of this sequence, however, requires further investigation. The Nya Sequence correlates approximately with the Pakerort and the Lower Latorp sequences of the Russian Platform.

4.2 The Ugor Sequence

The Ugor depositional sequence corresponds to the Ugorian Regional Stage. Its lower boundary is best exposed in the Irkutsk basin, where it is represented by a regional unconformity separating the Ust'kut and Iya formations. This surface is well exposed at the Angara River valley, 25 km upstream of the town of Kodinsk (Fig. 9E). It is interpreted as a Type I sequence boundary and marks a shift from predominantly carbonate to siliciclastic sedimentation. The shift was probably caused by a rapid sea-level fall, which resulted in subaerial exposure and karstification of a tropical carbonate platform. A subsequent rapid sea-level rise was followed by the deposition of the coarse-grained and cross-bedded quartz sandstones of the Iya Formation.

In the northwestern part of the Tungus basin (the Kulumbe River section) the Ugor Sequence has indistinct boundaries. The Ugor Regional Stage deposits correspond to the middle part of the Il'tyk Formation and are lithologically very similar to the underlying Nyaian Regional Stage as

Fig. 9 Regional unconformities and sequence boundaries in the Ordovician of Siberian Platform. (A) Sequence boundary at the base of the Dolbor depositional sequence in the Tunguska basin (right (northern) bank of the Podkamennaya Tunguska River 5 km upstream of the mouth of the Stolbovaya River). Siliciclastic tempestites of the Dolborian Regional Stage (above) overlie bioclastic calcareous tempestites of the Baksian Regional Stage (below). (B) Sequence boundary at the base of the Mangazea depositional sequence. Shallow-water quartz sands of the Kirensk-Kudrino depositional sequence with *Skolithos* and cross-stratifications overlain by relatively deep-water intercalations of siltstones and bioclastic packstones which represent the Mangazea depositional sequence. Left bank of the Moyero River valley. (C) Hardground penetrated by boring organisms which mark the base the Kirensk-Kudrino depositional sequence. Right bank of the Moyero River valley. (D) Transgressive surface coinciding with the sequence boundary at the base of the Volgino depositional sequence in the northwestern part of the Tungus basin (left (southern) bank of the Kulumbe River). (E) Regional unconformity at the base of the Ugor depositional sequence in the Irkutsk basin (left (southern) bank of the Angara River 25 km upstream of the town of Kodinsk). Sequence boundary separates oolitic grainstones with stromatolites of the Ust'kut Formation (below) and coarse-grained quartz sandstone and gravelstone with cross-stratification of the Iya Formation (above). (F) Erosional surface at the base of the Volgino depositional sequence. Reddish rock is a bioclastic limestone enriched with iron ooids and iron-impregnated bioclasts. Right bank of the Moyero River valley.

well as the overlying Kimaian Regional Stage deposits of the same formation. The Ugorian deposits are represented here by yellowish-gray, locally clay-rich dolomites and marls with rare intercalations of gray oolitic grainstones and marls. The Ugor Sequence correlates approximately with the Upper Latorp Sequence of the Russian Platform.

4.3 The Kimai Sequence

The Kimai depositional sequence coincides with the Kimai Regional Stage. In the Angara River valley, the base of the stage corresponds to the boundary between the Iya and Badaranovo formations. The sequence boundary coincides with a transgressive erosional surface at the base of a 0.6–1.5-m-thick unit of glauconite-enriched sandstone with interbeds of bioclastic, glauconitic limestone and pebbly limestone conglomerate. The underlying Iya Formation and overlying Badaranovo Formation are represented by similar pure quartz sandstones with well-developed unidirectional cross-stratification. The glauconite-rich unit is interpreted as a condensed section representing the upper part of the transgressive systems tracts (Loutit et al., 1988; Schutter, 1996; Van Wagoner et al., 1988). The glauconite-rich deposits were formed during rapid sea-level rise when siliciclastic supply from the adjacent land area ceased and sedimentation rate was very slow. The overlying cross-bedded quartz sandstones of the Badaranovo Formation are considered to be deposits of the subsequent highstand systems tract. In most regions of the Siberian Platform, the Kimaian deposits mark a transgressive episode (Kanygin et al., 2006), which was favorable for a wide distribution of relatively uniform benthic and pelagic assemblages. The Kimai Sequence can be correlated to the Volkhov Sequence of the Russian Platform. It should be noted that both sequences exhibit a Type II sequence boundary (Dronov and Holmer, 1999), i.e., with no considerable erosion of underlying beds and no signs of subaerial exposure.

4.4 The Baykit Sequence

The Baykit depositional sequence corresponds to the Baykit Formation, which includes deposits of the Vikhorevian and Mukteian regional stages. The best outcrops of the formation are located along the Podkamennaya Tunguska River valley, where monotonous light-gray and yellowish colored quartz sandstones, 5–80 m thick, are exposed on both sides of the river. The sandstones are coarsely bedded and frequently massive. At certain levels a well-developed cross-stratification occurs and locally, especially at the base, conglomerates are typical (Markov, 1970).

The Baykit Sandstone constitutes a distinctive sedimentary body extending for over 600 km along the Podkamennaya Tunguska River valley. The lower boundary of the sequence is poorly exposed in the natural outcrops. Near the village of Sulomai in the vicinity of the Yenisei Land, however, the Baykit Sandstone overlies different Lower Ordovician strata and even Cambrian units with an angular unconformity (Markov, 1970). The monotonous composition of the sporadically exposed Baykit Sandstone prevents the identification of systems tracts. The Baykit Sandstone represents an overall regressive stage in the development of the Tunguska basin. It should also be noted that the underlying Kimai depositional sequence is partly eroded from almost the entire Siberian Platform. The Kimaian deposits were completely removed by erosion in the eastern and northeastern parts of the platform (Kanygin et al., 2006). The unconformity at the base of the Baykit depositional sequence probably reflects one of the most significant forced regressions in the Siberian Platform during the Ordovician. The Baykit Sequence roughly correlates with the Kunda Sequence of the Russian Platform.

4.5 The Volgino Sequence

The Volgino depositional sequence coincides with the Volginian Regional Stage, whose deposits constitute a distinct and characteristic transgressive–regressive cycle that is traceable over the entire Siberian Platform. An erosional unconformity at the base of the sequence is well developed in the Podkamennaya Tunguska River valley where it is marked by conglomerates with pebbles of the underlying Baykit Sandstone (Kazarinov et al., 1969; Markov et al., 1971). An erosional surface at the base of the Volgino depositional sequence is very distinct also in the Kulumbe River section on the northwestern margin of the Tungus basin (Fig. 9D) and in the Moyero River section on the northeastern margin of the Tungus basin (Fig. 9F). The lowermost transgressive part of the Volgino Sequence in the Moyero River section is enriched with iron ooids and iron-impregnated bioclasts and pebbles which make the rock look reddish (Fig. 9F) and similar to the iron oolite bed at the base of the Tallinn Sequence in the Ordovician basin of Baltoscandia. The Volginian deposits of the Kulumbe River section belong to the Angir Formation. The upper beds of the Angir Formation show abundant signs of shallowing, including quartz sandstones and multidirectional cross-bedding. In the Moyero River section, the top of the sequence is marked by a hardground surface with traces of borings (Fig. 9C). In the Podkamennaya Tunguska River valley the uppermost Volginian deposits include a thick bed of medium-grained quartz sandstone. The bed may correspond to a highstand systems tract of the Volgino depositional sequence based on its shallow-water depositional environment and stratigraphical position directly below a transgressive surface at the base of the overlying Kirensk-Kudrino depositional sequence. The sequence marks the beginning of one of the most extensive transgressions on the Siberian Platform. The faunal assemblage of the Volginian Stage is easily recognizable and markedly different from the assemblage of underlying strata. The Volginian depositional sequence partly corresponds to the Tallinn Sequence in the Ordovician basin of Baltoscandia.

4.6 The Kirensk-Kudrino Sequence

The Kirensk-Kudrino depositional sequence coincides with the Kirensko-Kudrinian Regional Stage. Like the underlying Volgino Sequence, it forms a full cycle of deposition. In the outcrops along the Podkamennaya Tunguska River, the sequence represents a distinct regressive–transgressive–regressive cycle with three components corresponding to lowstand systems tract, transgressive systems tract, and highstand systems tract. The lowstand systems tract is a uniform unit of greenish-gray and bluish-gray siltstones with rare interbeds of quartz sandstone and/or bioclastic limestone, which directly overlie the coarse-grained quartz sandstones of the uppermost Volgino Sequence. The transgressive systems tract is mainly composed of red siltstones with abundant nautiloid shells. It was formed in an open marine environment and represents a deep-water setting, as is evident from the abundant cephalopod fauna as well as fine-grained sediments and their red color. Phosphatization is also associated with these sediments. The highstand systems tract is represented by greenish-gray and bluish-gray siltstones similar to those of the lowstand systems tract but coarser grained in composition. In the section along the Moyero River section the base of the Kirensk-Kudrino depositional sequence coincides with a prominent hardground surface marked by traces of boring organisms (Fig. 9C). In the section along the Kulumbe River valley the sequence is represented by the Amarkan Formation. Similar to the situation observed in the Podkamennaya Tunguska River sections, its lower boundary is a transgressive surface, along which greenish-gray siltstones of the Amarkan Formation overlie nodular limestones and cross-bedded quartz sandstones of the Angir Formation. The middle and upper parts of the Amarkan Formation are composed mostly of red siltstone. In the Irkutsk basin, the boundary between the Volgino and Kirensk-Kudrino sequences is represented by the regional unconformity in the middle of the Mamyr Series (Kanygin et al., 1984). The lack of natural outcrops and adequate drill core data prevents the identification of any systems tracts. The Volgino and Kirensk-Kudrino sequences are correlative with the Tallinn Sequence of the Russian Platform.

4.7 The Mangazea Sequence

The Mangazea depositional sequence includes the Chertovskian and Baksian regional stages. Its lower boundary coincides with that of the Chertovskian Regional Stage and is represented by a transgressive surface over the entire Siberian Platform. In the Ordovician succession at the Kulumbe and Lena rivers, this level shows clear signs of erosion and a conglomerate, which can be interpreted as a transgressive lag deposit. In the Moyero River section, the base of the Mangazea depositional sequence is represented by a transgressive surface separating shallow-water quartz sandstones of the uppermost Moyero Formation and relatively deep-water intercalation of bioclastic limestones and shale of the Dzerom Formation (Fig. 9B). In the Podkamennaya Tunguska sections, the Mangazea Sequence includes the upper part (corresponding to the Chertovskian Stage) of the Ust'Stolbovaya Formation and the Mangazea Formation. The upper part of the Ust'Stolbovaya Formation is interpreted as a transgressive systems tract and the Mangazea Formation as a highstand systems tract. The transgressive systems tract consists of two members: (1) greenish-gray siltstones interbedded with fine-grained yellowish-gray sandstones and black shales containing carbonate concretions with Chertovskian fauna and (2) red siltstones with scattered rounded phosphate pebbles 0.5–2 cm in diameter and beds of red conglomerates of phosphate pebbles. Both the black shale and the red phosphate conglomerates are interpreted as relatively deep-water deposits. At the Podkamennaya Tunguska region, the highstand systems tract of the Mangazea Sequence is a unit of greenish-gray siltstones alternating with bioclastic limestones. The bioclasts are predominantly fragments of brachiopods and trilobites as well as echinoderms, ostracods, and bryozoans. The limestone interbeds sometimes show ripple marks on their upper bedding planes. At some levels the bioclastic limestones contain glauconite grains. The intercalations of siltstones and bioclastic limestones of the Mangazea Formation are interpreted as cool-water calcareous tempestites.

The Mangazea Sequence correlates with the Kegel and Wesenberg sequences of the Russian Platform.

4.8 The Dolbor Sequence

The Dolbor depositional sequence corresponds to the Dolbor Formation and approximately to the Dolborian Regional Stage. The deposits are yellowish-gray, fine-grained sandstones and siltstones, locally with carbonate cement. In the outcrops along the right (northern) bank of the Podkamennaya Tunguska River, between the Stolbovaya and Listvennichnaya rivers, the base of the sequence is rather sharp and easily recognizable (Fig. 9A). The alternating bioclastic limestones and greenish-gray siltstones of the Mangazea Formation are replaced by predominantly yellowish fine-grained sandstones and siltstones. According to earlier observations based on studies of the outcrops higher upstream along the Podkamennaya Tunguska River, the uppermost beds of the Baksian Regional Stage were eroded and a sequence boundary is indicated by erosional pockets up to 0.4 m deep, filled with fragments of the underlying rocks (Bgatov, 1973). The Dolbor Sequence correlates approximately with the Fjäcka and Lower Jonstorp sequences of the Russian Platform.

4.9 The Kety Sequence

The Kety depositional sequence comprises the Nirundinian and Burian regional stages. They were referred to as substages of the Kety Regional Stage in earlier publications (Sokolov and Tesakov, 1975). The sequence is best exposed in the outcrops along the Bolshaya Nirunda and Nizhnyaya Chunku river valleys, the tributaries of the Podkamennaya Tunguska River. The lower boundary is defined by a sharp change of yellowish-gray and greenish-gray sandstones and siltstones of the Dolbor Formation to the cherry-red mudstones of the Nirunda Formation. By analogy with the cherry-red mudstone facies of the Kirensk-Kudrino Sequence, the Nirundian deposits seem to be related to a transgressive systems tract. The overlying deposits of the Bur Regional Stage are interpreted as the highstand systems tract of the Kety depositional sequence. The Silurian deposits rest unconformably on the deeply eroded Ordovician strata. This does not exclude the presence of younger Ordovician deposits, belonging to the one or two Hirnantian depositional sequences, in the deeper parts of the Tungus basin. The Kety Sequence is interpreted to correlate with the Upper Jonstorp Sequence of the Russian Platform.

5. COMPARISON WITH THE RUSSIAN PLATFORM

The Ordovician successions of the Russian and Siberian platforms show a certain similarity in the number of depositional sequences and the stratigraphic position of their boundaries. However, the Baltic and Siberian paleocontinents belong to different paleobiogeographic provinces and share no common species in the assemblages of trilobites, ostracods, and brachiopods. In the Early Ordovician no common genus between the two paleocontinents has been reported. Some similarity of trilobites and brachiopods at the generic level occurred in the Middle Ordovician. Six genera of trilobites and five genera of brachiopods were common for both continents, but their stratigraphic distribution is quite different. This fact prevents precise biostratigraphic correlation. To date there have been no direct correlations between the Ordovician successions of the Siberian and Russian platforms.

For intercontinental biostratigraphic correlation of the Ordovician of the Siberian Platform, we have so far only four reliable levels: (1) the base of the Nyaian Regional Stage that correlates with the base of the Ordovician based on most recent conodont investigations (Tolmacheva and Abaimova, 2009); (2) the base of the Volginian Regional Stage that correlates with the base of the *Didymograptus bifidus* graptolite zone; (3) the base of the Chertovskian Regional Stage that correlates with the base of the *N. gracilis* graptolite zone; and (4) the base of the Silurian. These four stratigraphic levels constitute three major stratigraphic intervals that can be correlated globally. It is noteworthy that these major stratigraphic intervals include a comparable number of sedimentary sequences in the two different regions. This suggests the synchronous formation of the sequences on both paleocontinents (Fig. 10).

The main turnover events of the Ordovician biota and sedimentation patterns were different on the Russian and Siberian platforms. The global Hirnantian regression is well displayed on both platforms. The turnover event in the mid-Darriwilian is also very distinct. On the Siberian Platform, it is marked by a transgressive surface at the base of the Volgino depositional sequence, which indicates an extended transgression, upwelling, and a significant change in sedimentation. On the Russian Platform, it is marked by a transgressive surface at the base of the Aseri Stage, which coincides with an important change in shelly fossils across the basin. The change allowed the definition of the regional Tallinn Stage with the lower boundary coinciding with the base of the Aseri Stage (Männil, 1966). Also distinct is the base of the Sandbian Stage (the base of the Caradoc Series of Great Britain). This

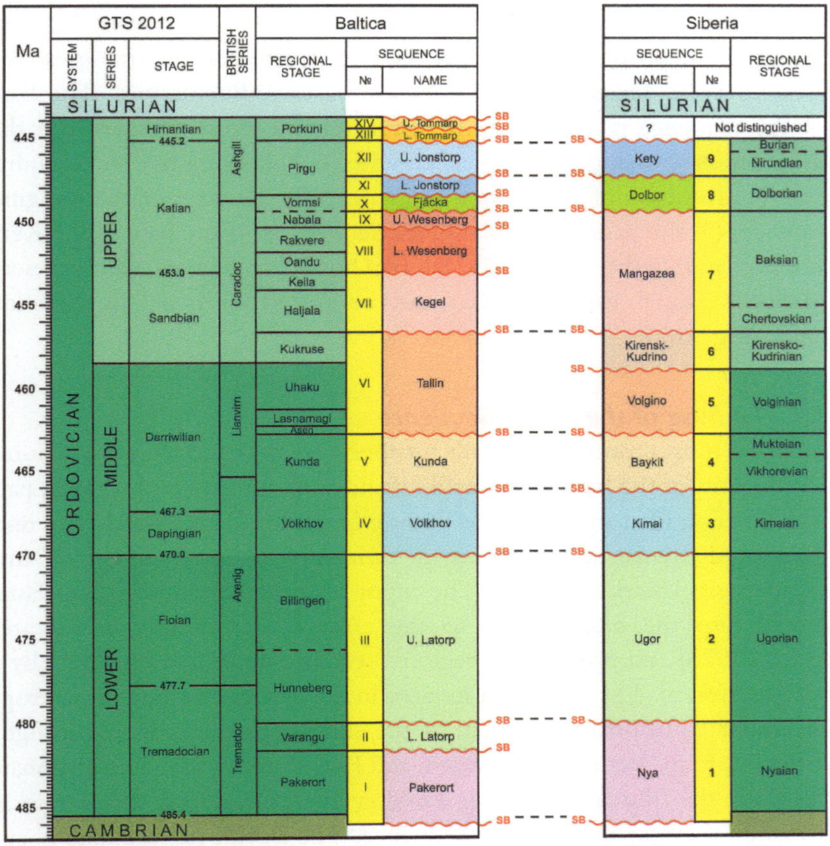

Fig. 10 Comparison between the Ordovician sequence stratigraphic charts of Baltica and Siberia.

level probably marks a global eustatic transgression reflected in the basal beds of the *N. gracilis* Zone (Barnes et al., 1996; Fortey, 1984; Haq and Schutter, 2008).

By contrast, the important level at the base of the Baykit Sequence of the Siberian Platform, which was related to the sedimentation change, destruction of the carbonate platform, and input of enormous siliciclastic masses into the basin, was less pronounced on the Russian Platform. The very distinct unconformity at the Volkhov-Kunda sequence boundary and the important faunal change (Männil, 1966) were not combined with major changes in sedimentation. In general, eustasy seems to have been the main controlling factor for depositional sequence formation on both platforms, but it was combined with regional tectonic effects.

5.1 Detecting Eustasy

Each of the described depositional sequences represents a transgressive–regressive cycle deposited in response to sea-level fluctuations. The base of every sequence is a level of marked facies and biotic changes. But the most prominent changes are associated with the following five stratigraphic levels: (1) the base of the Baykit Sequence; (2) the base of the Volgino Sequence; (3) the base of the Mangazea Sequence; (4) the base of the Kety Sequence; and (5) the Ordovician/Silurian boundary. These levels represent remarkable faunal turnovers that can be traced all over the Siberian Craton (Fig. 10).

5.1.1 The Base of the Baykit Sequence

One of the most pronounced faunal turnovers is associated with the regional unconformity at the base of the Baykit Sandstone. In the southwestern part of the Siberian Platform (near the Yenisei Land) the Baykit Sandstone rests with a slight angular unconformity on various Ordovician and even Cambrian deposits (Markov, 1970). The input of great volumes of siliciclastic material into the basin points to the enlargement of the source area (Bgatov, 1973) and probably to regional tectonic uplift, accompanied by a forced regression. The regression resulted in the final destruction of the tropical carbonate platform that existed on the Siberian paleocontinent throughout the Riphean, Vendian, Cambrian, and Early Ordovician. The carbonate platform in the Irkutsk basin disappeared even earlier, during the regression marked by the base of the Ugor Sequence. The Baykit regression is documented in the northwestern part of the Tungus basin in the vicinity of the Turukhan Land by the quartz sandstones and siltstones of the Guragir Formation. In the eastern part of the basin, near the Anabar Land, it is marked by a considerable hiatus (Kanygin et al., 2007). The underlying Kimaian deposits have been entirely or partly eroded in many areas of the Siberian Platform (Kanygin et al., 2006, 2007). The regression seems to have had a major amplitude (>75 m) according to the semi-quantitative classification of Haq and Schutter (2008). It was one of the largest Ordovician regressions on the Siberian Craton and resulted in significant changes in the development of sedimentation patterns and the biotic evolution.

5.1.2 The Base of the Volgino Sequence

The transgressive surface at the base of the Volgino depositional sequence marks the beginning of an impressive transgressive event that succeeded the lowstand phase displayed by the deposition of the Baykit Sandstone. The transgression is associated with abrupt changes in faunal assemblages and sedimentation. It is therefore not surprising that the base of the Volginian Regional Stage has been suggested as the boundary between the Lower and the Upper Ordovician Series on the Siberian Platform (Sokolov and Tesakov, 1975). These series differ in lithological composition, facies, and thickness as well as in the composition and structural organization of their respective biota (Kanygin et al., 2006). Although the transgression that produced the Volgino Sequence was not the largest in amplitude (probably it was in the range of 25–75 m) according to the classification by Haq and Schutter (2008), the transgressive surface at the base of the sequence represents one of the most significant biotic events in the Ordovician of the Siberian Platform.

5.1.3 The Base of the Mangazea Sequence

The transgressive surface at the base of the Mangazea Sequence is well developed all over the platform. In many places, it is associated with the erosion of underlying deposits and a basal conglomerate. This surface was probably formed during a forced regression and was reworked afterward into a transgressive surface of erosion. The phosphate pebble conglomerate at the base of the Chertovskian Regional Stage has a widespread distribution and is interpreted as a transgressive lag deposits. It is usually overlain by a relatively deep-water black shale or red bed facies. The transgressions of the Volginian and the Chertovskian stages are the most prominent in the Ordovician of the Siberian Platform (Kanygin et al., 2006). But the amplitude of the Mangazea sea-level rise seems to exceed the Volginian one, reaching the amplitude of >75 m. The minimal facies differentiation enabled the uniform distribution of the benthic and pelagic assemblages over the entire Siberian Platform.

5.1.4 The Base of the Kety Sequence

The cherry-red siltstone of the Nirunda Formation seems to represent a facies analog of the relatively deep-water marine red bed facies of the Kirensk-Kudrinian and Chertovskian stages. That implies that the sharp Dolbor/Kety sequence boundary is also a transgressive surface that reflects a rapid sea-level rise. This sequence boundary has not been adequately studied because this stratigraphic interval has a limited number of natural exposures. The best of them are located in the inner part of the Siberian Platform and are not easily accessible.

5.1.5 The Ordovician–Silurian Boundary

The Ordovician–Silurian boundary is marked by a distinct gap generated by a high-amplitude forced regression. The sea-level fall was of eustatic origin, supposedly connected with continental glaciation of Gondwana. The uppermost beds of the Ordovician (Hirnantian) are almost absent on the Siberian Platform. The onset of the Silurian Period was characterized by a transgression and an essential reorganization of marine ecosystems.

The sea-level curve for the Ordovician of the North American Platform assumes a prominent sea-level drop at the base of the Middle Ordovician and a long-term lowstand during all the Dapingian and Darriwilian (80–100 m lower than in the Lower and Upper Ordovician) (Ross and Ross, 1992, 1995; Vail et al., 1977). The sea-level curve for the Ordovician of the Siberian Platform looks roughly the same (Dronov et al., 2009; Kanygin et al., 2010a).

5.2 Baltica: Contradictory Interpretations

The bathymetry of the Ordovician basin of Baltoscandia is considered in numerous papers and has been summarized in regional sea-level curves (Dronov and Holmer, 2002; Dronov et al., 2011b; Nestor and Einasto, 1997; Nielsen, 2004) which differ in several stratigraphic levels. In the following, a short review of the sequence stratigraphy and sea-level changes for the relatively shallow-water Estonian part of the basin is presented.

5.3 Ordovician Basin of Baltoscandia

The Baltoscandian Paleobasin is one of the largest and best-studied Lower Paleozoic sedimentary basins in the world. It represents a typical epicontinental interior sag basin, which covers a vast area (over 1,000,000 km^2) and is characterized by a relatively low rate of subsidence. Despite a large number of small breaks in shallow shelf areas, the Ordovician succession is almost complete from a biostratigraphic point of view, with no essential parts missing. A general facies and bathymetric zonation of the basin was described by Männil (1966), Jaanusson (1982), and Harris et al. (2004) (Fig. 11).

Deep-water off-shore facies are confined to the Scanian Confacies belts whose lithology is dominated by black shales (Fig. 12A and B). The Central Baltoscandian Confacies belt occupies an intermediate position between the Scanian and North Estonian ones. The dominant facies types are marine red beds (Fig. 12C and D) with thin black shale units and gray limestone–marl intercalations at some levels. The sediments seem to be deposited in a hemipelagic environment below the storm wave base. The shallow-water

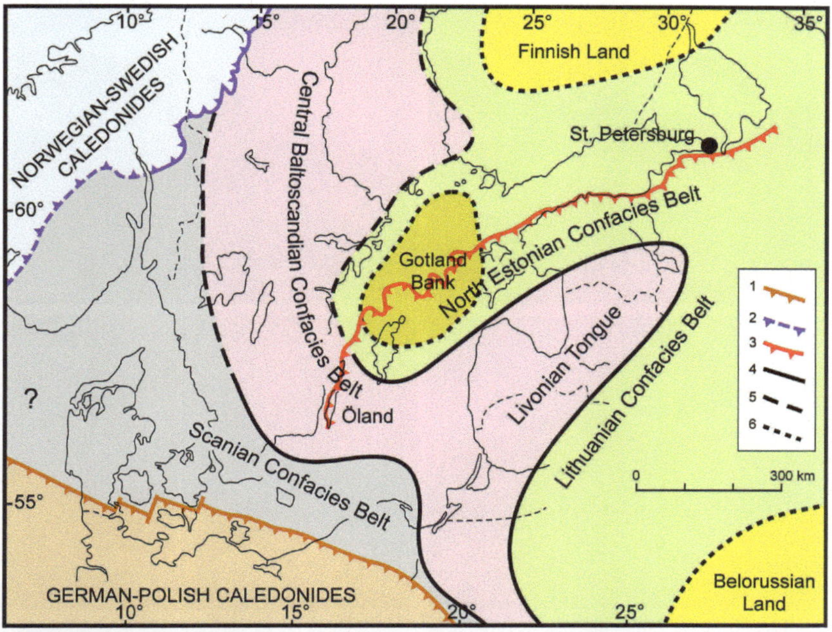

Fig. 11 Facies zonation of the Ordovician basin of Baltoscandia. 1. Front of German-Polish Caledonides. 2. Front of Norwegian-Swedish Caledonides. 3. Baltic-Ladoga Klint line. 4. Boundaries of the main confacies belts. 5. Provisional boundaries of the confacies belts. 6. Provisional shorelines of Finish and Belorussian landmasses. *Modified from Männil, R.M., 1966. Istorija razvitija baltijskogo basseina v ordovike [The Ordovician History of the Baltic Basin]. Valgus, Tallinn (in Russian); Jaanusson, V., 1976. Faunal dynamics in the Middle Ordovician (Viruan) of Balto-Scandia. In: Basset M.G. (Ed.), The Ordovician System: proceedings of a Palaeontological Association symposium, Birmingham, September 1974. University of Wales Press and National Museum of Wales. Cardiff, pp. 301–326; Jaanusson, V., 1995. Confacies differentiation and Upper Middle Ordovician correlation in the Baltoscandian basin. Proc. Estonian Acad. Sci. Geol. 44 (2), 73–86; and Nielsen, A.T., 2004. Ordovician sea-level changes: a Baltoscandian perspective. In: Webby, B.D., Paris, F., Droser, M.L., Persival, G. (Eds.), The Great Ordovician Biodiversification Event. Columbia University Press, New York, pp. 84–93.*

deposits with traces of storm-, wave-, and tidal activity occur in the North Estonian Confacies belt (Figs. 8 and 12E and F). In this environment, the development of regional unconformities at sequence boundaries and shifts of facies at transgressive surfaces are better expressed. This makes a shallow-water setting more favorable for detecting major sea-level fluctuations which principally should affect the whole basin. Latitudinal migration of the Baltic continent is reflected in the succession of shallow-water facies from siliciclastic sands in the Tremadoc through cool-water carbonates in

Fig. 12 Typical lithofacies of the Ordovician basin of Baltoscandia. Scanian Confacies Belt: (A) black graptolitic shale and intercalations of black shale and gray limestones (mudstones). Oslo fjord, Norway. (B) Closer view on intercalation of black shale and gray limestones. Oslo fjord, Norway. Central Scandinavian Confacies Belt: (C) marine red bed facies of the Volkhov and Kunda regional stages (Lanna and Holen formations); Hällekis quarry, Västergötland, Sweden. (D) Closer view on marine red bed facies in Hällekis quarry. Volkhov Regional Stage, Lanna Formation. Västergötland, Sweden. North Estonian Confacies Belt: (E) white or light-gray micritic limestone. Nabala Regional Stage, Saunja Formation. K-3 borehole (Kuressaare) on the Saaremaa island in western Estonia. (F) Intercalations of bluish-gray clayey limestones and marls. Pirgu Regional Stage, Halliku Formation. K-3 borehole (Kuressaare) on the Saaremaa island in western Estonia.

the Floian–Sandbian (Fig. 8) to warm-water tropical carbonates in the Katian–Hirnantian. The shift of depocenters from relatively deep-water to shallow-water settings, clearly visible on the profile across the basins margin (Fig. 13), reflects a transition from cool-water to warm-water carbonate factories (Schlager, 2007).

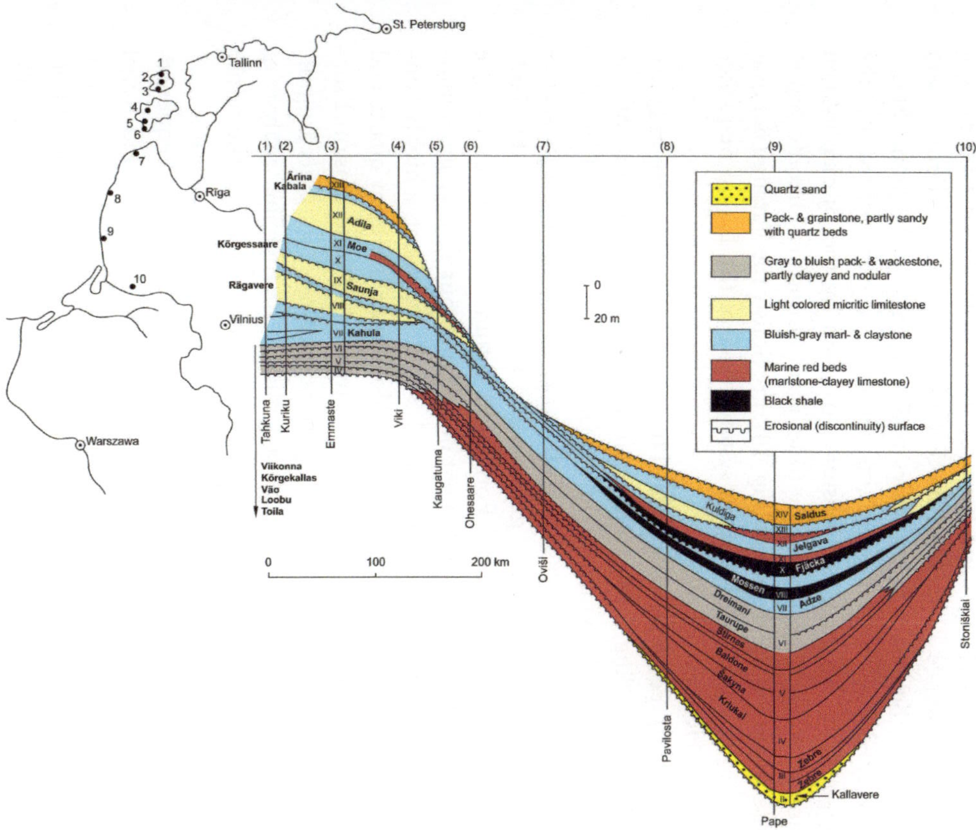

Fig. 13 Profile across the Livonian Tongue displaying facies of shallow-water North Estonian and deep-water Central Baltoscandian Confacies belts. *Vertical lines* mark boreholes and *Roman numbers* mark the sequences.

5.4 Ordovician Sequence Stratigraphy of Baltoscandia

On the basis of outcrop and drill core data, the Ordovician succession of Baltoscandia was subdivided into 14 depositional sequences (Dronov et al., 2011b), modifying the earlier models by Dronov and Holmer (1999) and Harris et al. (2004). The sequences represent third-order cycles of relative sea-level changes and have durations from 0.9 to 12 my (Figs. 14 and 15). Sequence I equals to the Pakerort Sequence of Dronov and Holmer (1999). Sequence II coincides with the Varangu Regional Stage. In relatively deep-water settings this sequence is represented by the "*Ceratopyge* shale" (glauconite sandstone) overlain by the Bjørkåsholmen Formation (former "*Ceratopyge* limestone") and bound by unconformities at the base and top of the sequence (Fig. 14). Previously the Varangu deposits have been interpreted as a lowstand systems tract of the Latorp Sequence (Dronov and Holmer, 1999).

Sequence III corresponds to the Hunneberg and Billingen regional stages. The bounding surfaces are an unconformity on top of the Bjørkåsholmen Formation and an erosional surface at the base of the Volkhov Regional Stage. In the eastern Baltic area sequence III comprises sandstones of the Leetse Formation grading into the limestones of the Toila Formation, and the upper Zebre Formation within the Livonian Tongue (Fig. 14). The sequence has the longest duration (about 12 my) and a minimum average thickness rarely exceeding 1.5 m. It represents a highly condensed stratigraphic interval of diverse lithology, varying from quartz and glauconite sands to clays and bioclastic limestones. Gaps and erosional surfaces are common. Only fragments of previous sedimentary units are preserved and the small thickness makes a sequence stratigraphic analysis of this interval extremely difficult. It could not be excluded that in the future two or three separate depositional sequences will be identified in it.

Sequences II and III make up the former Latorp Sequence. Sequences IV, V, VI, and VII match with the Volkhov, Kunda, Tallinn, and Kegel sequences of Dronov and Holmer (1999), respectively (Fig. 14). Sequence VIII comprises the Oandu and Rakvere regional stages. The Hirmuse Formation and lower parts of the Mossen and Variku formations can be interpreted as a transgressive systems tract, while the shallow-water light-colored micritic limestones of the Rägavere Formation, together with the upper parts of the Mossen and Variku formations, seem to represent a highstand systems tract. The basal unconformity of this sequence is one of the best-developed regional unconformities in the entire Ordovician succession of Baltoscandia (Lashkov and Paškevičius, 1989).

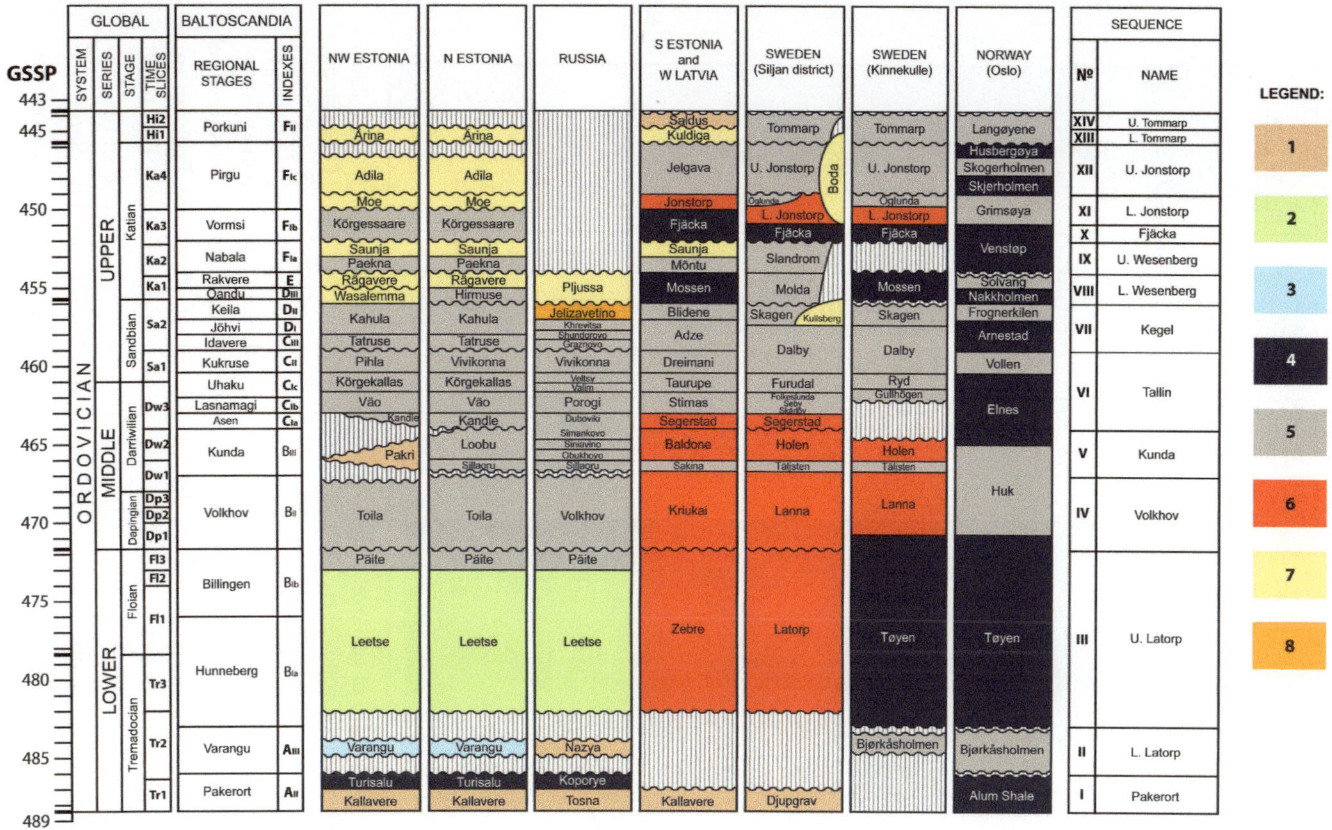

Fig. 14 Ordovician stratigraphy of the Baltoscandian basin. Legend: 1—quartz sandstone; 2—glauconite sandstone; 3—bluish-gray siltstones; 4—black shale; 5—bluish-gray limestones and intercalation of limestones and siltstones; 6—marine red beds (limestones and siltstones); 7—white and light-gray micritic limestones; and 8—yellow dolomites.

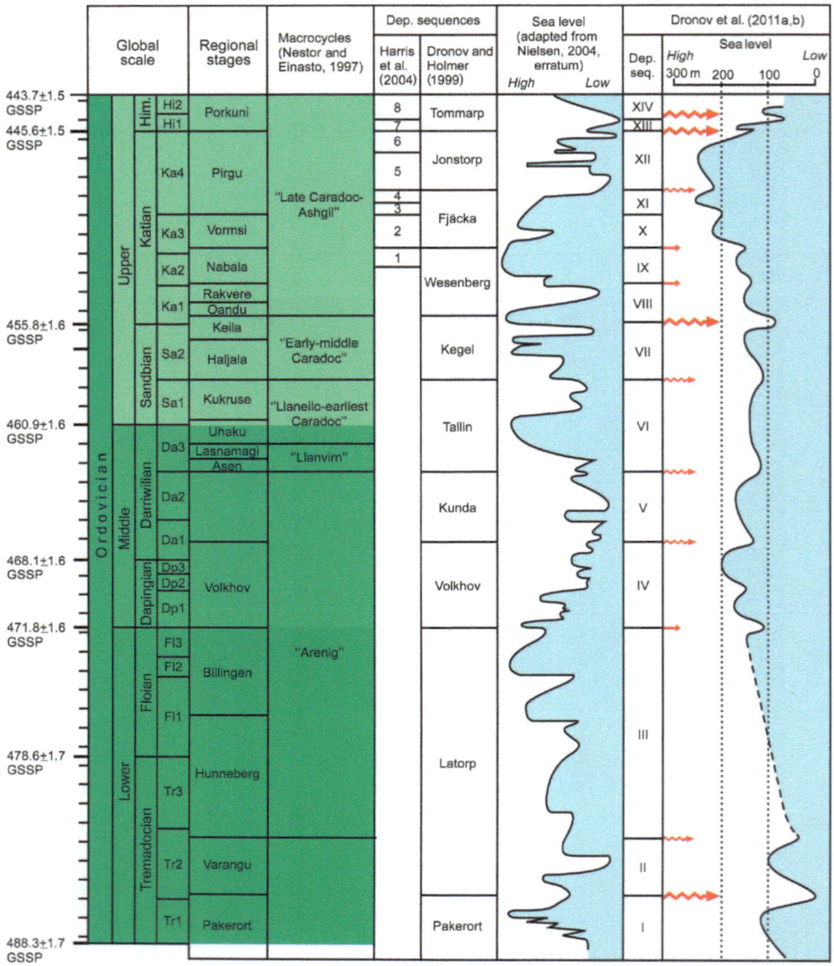

Fig. 15 Comparison of sea-level curves reconstructed for deep-water (Nielsen, 2004) and shallow-water (Dronov et al., 2011a,b) parts of the Ordovician basin of Baltoscandia.

Sequence IX comprises the Nabala Regional Stage. Argillaceous limestones of the Paekna and Mõntu formations seem to represent transgressive systems tract deposits. Similar to the limestones of Rägavere Formation, micritic limestones of the Saunja Formation are interpreted as a highstand systems tract deposit. Sequences VIII and IX have been included in the former Wesenberg Sequence of Dronov and Holmer (1999). Sequence X comprises the Vormsi Regional Stage. It includes the Fjäcka Formation, a black shale, which seems to represent a condensed section in the transgressive systems tract and probably the lower part of the highstand systems tract in its upper calcareous transition interval. In more shallow-water settings the sequence is represented by the Tudulinna and Kõrgessaare formations (Figs. 13 and 14). The lower sequence boundary possesses characteristics of a transgressive surface. The sequence coincides with the sequence 2 of Harris et al. (2004) (Fig. 15).

Sequence XI comprises the lower part of the Pirgu Regional Stage (the Moe Formation in a shallow-water setting and the Jonstorp Formation in a deeper-water environment). In the shallow-water setting two cycles were identified within the sequence (comprising sequences 3 and 4 by Harris et al., 2004), but they cannot be distinguished in deep-water red limestones and marls of the Jonstorp Formation. Sequences X and XI were assigned to the Fjäcka Sequence by Dronov and Holmer (1999). Sequence XII corresponds to the sequences 5 and 6 of Harris et al. (2004) and the Jonstorp Sequence of Dronov and Holmer (1999). Sequence XIII comprises the lower part of the Porkuni Regional Stage. This sequence includes the Ärina Formation in a shallow-water setting and the Kuldiga Formation in a relatively deep-water setting, corresponding to sequence 7 of Harris et al. (2004) (Figs. 14 and 15). The basal unconformity is represented by an erosional surface with deep erosional valleys cutting the underlying deposits (Lashkov and Paškevičius, 1989). Sequence XIV corresponds to the upper part of the Porkuni Regional Stage. It is represented by the Saldus Formation which is mainly distributed in the central parts of the Livonian Tongue and corresponds to sequence 8 of Harris et al. (2004). The basal unconformity of the sequence represents an erosional surface that cuts different underlying deposits. Dronov and Holmer (1999) have attributed sequences XIII and XIV to the Tommarp Sequence.

5.5 Sea-Level Curve for the Ordovician of Baltoscandia

The reconstructed sea-level curve for the shallow-water part of the basin (North Estonian Confacies belt) (Fig. 15) is based on the following assumptions: (1) The main regional unconformities reflect substantial sea-level drops and forced regressions. More widespread unconformities and deeper erosion of the underlying beds indicate a higher magnitude of sea-level fall. (2) The main transgressions are recognized according to the widening of the deep-water facies belts. The absolute magnitude of individual sea-level events is difficult to estimate, but the relative magnitude of each transgression and regression can be derived from the assumptions above.

The approach of Haq and Schutter (2008) who classified each event semi-quantitatively as minor (<30 m), medium (30–75 m), or major (>75 m) has been followed.

Analysis of the regional unconformities and associated gaps shows that the greatest unconformities and deepest erosion occur at the bases of the sequences II, VIII, XIII, and XIV. The Hirnantian sea-level drops seem to be comparable to that induced by the Quaternary glaciations, which require a sea-level fall up to 100–125 m. Erosion at the base of sequence XIII was deeper than at the base of sequence XIV, so the relative magnitude of the sea-level drop at the base of sequence XIII was higher. A magnitude of 120 and 110 m for the bases of sequences XIII and XIV, respectively, can be assumed.

A comparable regional unconformity but without such a deep down-cut was reported from the base of sequence VIII. A major sea-level drop can be assumed for this level. Deep erosion associated with the most distal shift of facies was also reported from the base of sequence II. The shoreline migrated over a distance of more than 700 km which means a sea-level drop of a major magnitude. Less pronounced unconformities were recognized at the bases of sequences III, V, VI, VII, and XII. All of them were connected with regressions of medium magnitude. Among these regressions, the highest relative magnitude was attributed to the base of sequence III, considering the almost complete erosion of the Varangu deposits in the shallow-water settings. The regression at the base of sequence V resulted in traces of sufficient erosion. Sea-level falls of a minor magnitude are reflected at the bases of sequences IV, IX, X, and XI, more remarkable being the regression at the base of sequence IV. The analysis of the spatial distribution of the basinal deep-water facies enables the classification of transgressive events into those of a major magnitude (sequences I, II, III, IV, VIII, X, and XII), medium magnitude (sequences V, VII, XI, XII, and XIV), and minor magnitude (sequences VI and IX).

The most remarkable sea-level rise occurs at the base of sequence III. At that time deep-water red bed facies for the first time occupied wide areas in the Ordovician basin of Baltoscandia. Mass immigration of a new fauna into the basin supports this interpretation. High-magnitude transgressions are indicated by the wide distribution of black shales in shallow-water environments during sequence I and the reestablishment of marine conditions after a high-magnitude regression at the base of sequence II. The magnitude of the Volkhovian transgression (sequence IV) was less than the magnitude of the Lower Ordovician transgressions. The regression at the basal Volkhov sequence boundary was minor and the following sea-level rise enhanced the previous transgression, which led to a deepening of the basin.

The depth of the basin and the lateral extent of the deep-water red bed facies reached its maximum in Volkhov time. Major transgressions are also recorded for sequences XIII, VII, and X. The two latter events are marked by the invasion of black shale facies (lower Mossen and Fjäcka formations, respectively). The highest sea level is marked by the appearance of marine red bed facies in the middle part of the Livonian basin (Jonstorp Formation). The Lower Jonstorp transgression inherited the Fjäcka sea-level rise and the depth of the basin reached its maximum at that time. Moderate individual transgression events occurred in sequences V, VII, XI, XII, and XIV. Transgressions of a minor magnitude are associated with sequences VI and IX. The lack of accommodation space in a shallow-water setting can be recognized at these levels.

5.6 Shallow-Water and Deep-Water Models

A comparison of the sea-level curves published in the last decades (Dronov and Holmer, 2002; Dronov et al., 2011b; Nestor and Einasto, 1997; Nielsen, 2004) displays some disagreements between the curves reconstructed for the shallow-water part of the basin and those based on relatively deep-water sections. The main contradictions are in the Dapingian–Lower Darriwilian and in the Upper Katian intervals. For those intervals (the Volkhov, Kunda, and Pirgu regional stages) the shallow-water model proposes the highest sea-level stands, whereas the deep-water model assumes the lowest sea-level stands, termed the "Late Arenig–Early Llanvirn Lowstand Interval" and "Ashgill Lowstand Interval" (Nielsen, 2004), respectively. This contradiction reflects opposite opinions in the interpretation of limestone units within the Scanian Confacies belt (Fig. 14).

The invasion of carbonate facies into the black shale realm is interpreted as a shallowing event in the deep-water model, assuming that limestones represent more shallow-water facies than the black shales (Nielsen, 2004). On the other hand, the same episodes in shallow-water areas are characterized by the expansion of the relatively deep-water marine red bed facies into the shallow-water realm, suggesting deepening events. The invasion of limestone facies into the deep-water black shale environment could be explained through the mechanism of "highstand shedding" (Schlager, 2007). According to this view carbonates were transported from a shallow-water environment into a deep-water setting only at the time of maximum carbonate production in the shallow-water environment, i.e., during sea-level highstand. In this case, the limestones in a deep-water Scanian Confacies belt mark maximum flooding events rather than regressive ones. The sea-level curve for the Ordovician of Baltoscandia elaborated in the framework of the deep-water model is thought to be in a good agreement with the sea-level curve suggested for the North American Craton by Ross and Ross (1995). This coincidence was regarded as an evidence for the eustatic nature of the displayed sea-level fluctuations. Both curves demonstrate sufficient sea-level lowstand during almost the entire Middle Ordovician. In the Ordovician basin of Baltoscandia, however, this lowstand could be regarded as an artifact. In Laurentia the Middle Ordovician lowstand was apparently deduced from the significant gap comprising this stratigraphic interval in the Ordovician succession of the New York State. This gap is maximal in the pericratonic basin of the New York State and becomes less pronounced toward the middle part of the continent. The Ordovician succession is almost complete in the Great Basin (Nevada and western Utah). Consequently, this pattern could be regarded as an evidence for tectonic uplift and erosion in the New York State basin during the Middle Ordovician, probably caused by the Taconic orogeny.

5.7 Discussion

In the last few decades, numerous sea-level curves for the Ordovician from different paleogeographic domains all over the world have been developed and refined. For Laurentia, a detailed sea-level curve was produced by Ross and Ross (1992, 1995); for Baltica by Nestor and Einasto (1997), Dronov and Holmer (2002), Nielsen (2004), Dronov et al. (2011a,b); for Avalonia by Woodcock (1990); for Western Gondwana by Heredia and Beresi (1995); for Northern Gondwana by Videt et al. (2010); for the Yangtze Platform by Su (2007); for the Tarim Platform by Yu et al. (2001); and for Siberia by Dronov et al. (2009) and Kanygin et al. (2010a,b). Because of the great disparity in quantity and detail of the data available for the different regions the resulting sea-level curves differ greatly in their resolution. In some regions, like North America and Baltoscandia, the number of described depositional sequences and the degree of knowledge about their architecture are very high. In other regions, like Siberia, they have not yet been studied in such detail.

There are also differences in resolution with regard to the precise worldwide chronostratigraphic correlation of established depositional sequences.

The new timescale for the Ordovician System (Bergström et al., 2009) helps a lot in this respect and provides an international standard for intercontinental correlation. However, due to the endemism of fauna and a lack of U/Pb dating for Siberia and some other regions, there are still uncertainties in the precise correlation of depositional sequences and their boundaries. For the moment being, the very eustatic nature of a given depositional sequence and the related sea-level fluctuation are far from being established beyond doubt.

Nevertheless, the material that has already been collected from different Ordovician sedimentary basins all over the world allows us to analyze and compare the general shape of sea-level curves from the majority of large Ordovician stable cratons. A preliminary result of this comparison is that there are two principal groups of regional Ordovician sea-level curves: (1) the curves with sea-level highstands in the Early and Late Ordovician separated by a prominent sea-level lowstand in the Middle Ordovician (Dapingian and/or Darriwilian), and (2) the curves with a prominent sea-level highstand in the Middle Ordovician (Late Dapingian or Early Darriwilian).

A typical example for the first group is an Ordovician sea-level curve for Laurentia (Fig. 16). The same characteristics are clearly visible on the early interpretations by Vail et al. (1977) as well as on the more advanced reconstructions of Ross and Ross (1992, 1995). This regional sea-level curve, probably because of its relatively high resolution, is also traditionally regarded as a basis for a global eustatic sea-level curve for the entire Ordovician (Cooper and Sadler, 2012; Haq and Schutter, 2008; Nielsen, 2004; Vail et al., 1977). The sea-level curve for the Ordovician of the Siberian Platform, though not so detailed, seems to follow these patterns closely (Dronov et al., 2009; Kanygin et al., 2010a,b). The second group consists of the Ordovician sea-level curves for Avalonia (Woodcock, 1990), Western Gondwana (Heredia and Beresi, 1995), the Yangtze Platform (Su, 2007) (Fig. 16), as well as for the Tarim Platform (Yu et al., 2001) and Northern Gondwana (Videt et al., 2010).

The special case is represented by Baltica which has two different sea-level curve interpretation for the deep-water (Nielsen, 2004) and shallow-water settings (Dronov and Holmer, 2002; Dronov et al., 2011a, b) (Figs. 15 and 16). Based on the long-term lithological changes and the paleogeographical and plate tectonic development it seems more logical to place Baltica in the second group of platforms that rifted and drifted from Gondwana in the Early Paleozoic. Following this subdivision, Laurentia and Siberia which share not only similar sea-level curve patterns but also long-

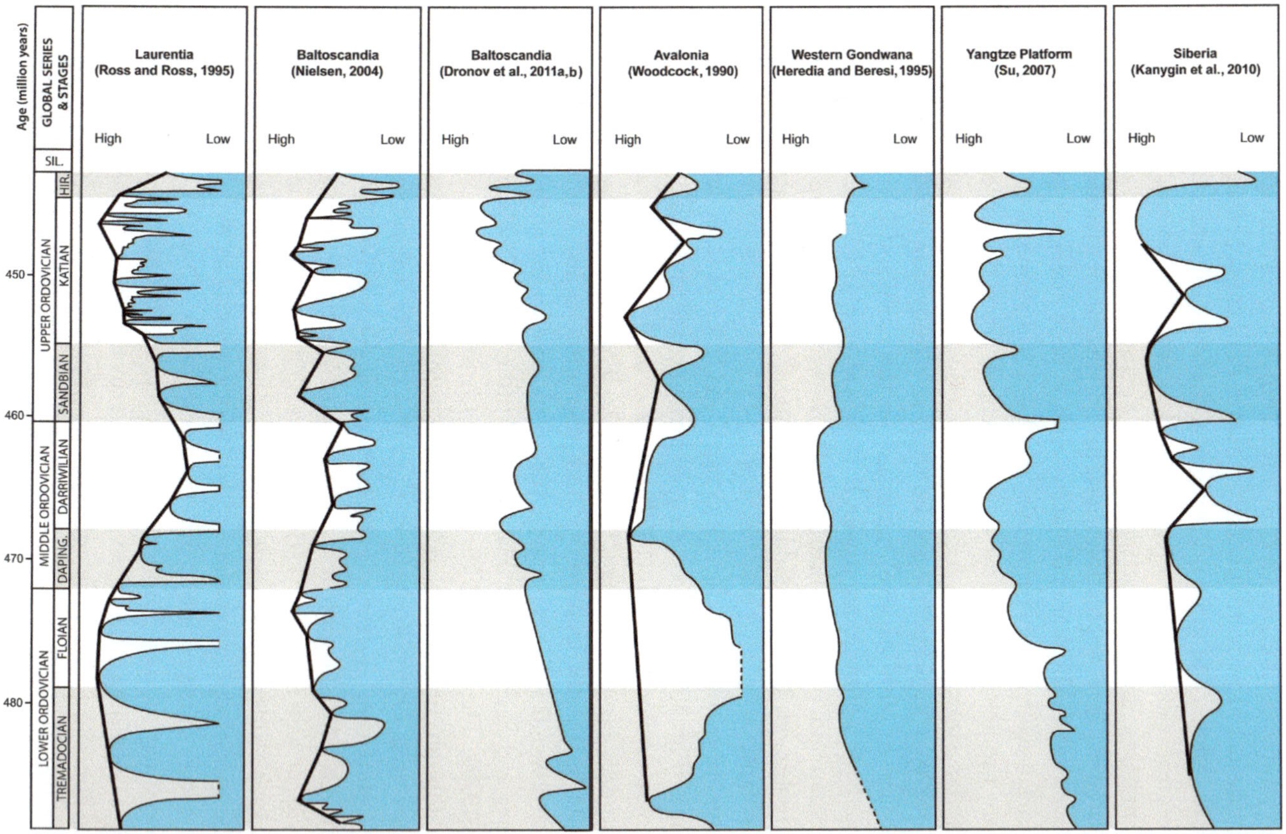

Fig. 16 Compilation of different regional sea-level curves for the Ordovician. *After Munnecke, A., Calner, M., Harper, D.A.T., Servais, T., 2010. Ordovician and Silurian sea-water chemistry, sea-level, and climate: a synopsis. Palaeogeogr. Palaeoclimatol. Palaeoecol. 296, 289–413 with modifications.*

term lithological changes and paleogeographical and geological patterns constitute one semi-global domain. The rest of the Ordovician platforms united under the name "Gondwanan platforms," which also share sea-level patterns, long-term lithological changes, and paleogeographical and geological patterns, constitute the other semi-global domain.

The differences between the two domains could be explained by the global distribution and the relative movements of the Ordovician lithospheric plates. Laurentia and Siberia seem to belong to the same plate bounded by one continuous subduction zone (see Fig. 3) similar to the modern-day subduction zone beneath the Cordilleras and Andes along the eastern margin of North and South America. The "Gondwanan platforms" have been rifted from Gondwana and were moving toward this subduction zone during the Ordovician. In the Middle Ordovician the Gondwanan platforms underwent subsidence because of subduction of this plate under the Laurentia–Siberian plate. At the same time the eastern margin of Laurentia and the western margin of Siberia (in present-day orientation) were uplifted because of the same tectonic process (Dronov, 2012).

At the moment the existence of synchronous global eustatic sea-level fluctuations for all of the Ordovician period seems not to be firmly established. However, if such fluctuations existed, which seems to be probable, the general shape of the sea-level curve for the entire Ordovician seems to be affected by semi-global plate tectonic movements. As a result instead of one global sea-level curve for the Ordovician (Haq and Schutter, 2008), it would be probably more correct to suggest two semi-global curves for the two major tectonic regions, one of which includes the Siberian and North American platforms and the other combines the Baltica and Gondwanan platforms. This subdivision probably reflects the position of the main Ordovician lithospheric plates.

6. CONCLUSIONS

1. The pattern of long-term lithological changes in the Ordovician of the Siberian Tungus basin demonstrates striking similarities with contemporaneous basins of Laurentia. On both platforms, the Ordovician succession starts with tropical stromatolite-bearing carbonates, which abruptly change to siliciclastic deposits (Baykit Sandstone in Siberia

and Eureka Sandstone in North America, respectively), and terminates with cool-water carbonates. This kind of succession is unique for Laurentia and Siberia and differs markedly from Baltica or any other platform rifted and drifted from Gondwana.

2. Both Siberia and Laurentia were situated in tropical to subtropical latitudes during the entire Ordovician, and a wide distribution of cool-water *heterozoan* carbonates on the two paleocontinents during the Upper Ordovician indicates almost simultaneous upwelling of cool-water oceanic waters and penetration of them into the corresponding epicontinental basins. The similarity between the Ordovician successions of Siberia and Laurentia is emphasized also by numerous K-bentonite beds associated with cool-water carbonates on the western (in present-day coordinates) margin of the Tungus basin. The timing of volcanism in the Ordovician of Siberia is surprisingly close to the period of volcanic activity of the Taconic Arc near the margin of Laurentia which suggests the existence of a single Taconic-Yenisei Arc along the united margin of Laurentia and Siberia.
3. Despite the obvious difficulties in the precise worldwide chronostratigraphic correlation of the Ordovician depositional sequences the shape of the sea-level curve for the Ordovician of Siberia demonstrates its general similarity with the sea-level curve for Laurentia. The sea-level curve for the Ordovician of the North American Platform assumes a prominent sea-level drop at the base of the Middle Ordovician and a long-term lowstand during all the Dapingian and Darriwilian (80–100 m lower than in the Lower and Upper Ordovician). The sea-level curve for the Ordovician of the Siberian Platform looks roughly the same.
4. The sea-level curves for the Ordovician of the Gondwanan platforms (North Africa, Yangtze Platform, South America, Avalonia) seem to share different patterns. The Middle Ordovician represents here a rather highstand interval. There are two different sea-level models for Baltica, but the shallow-water model seems to fit better to the platforms rifted from the Gondwanan paleocontinent. As a result instead of one global sea-level curve for the Ordovician, it would probably be more suitable to suggest two semi-global curves for two major tectonic regions one of which includes the Siberian and North American platforms and the other combines Baltica and the Gondwanan platform. This subdivision probably reflects the position of the main Ordovician lithosphere plates.

ACKNOWLEDGMENTS

This chapter represents a contribution to the IGCP 653 Project "The onset of the Great Ordovician Biodiversification Event." Financial support for this research was provided by the Russian Foundation for Basic Research Grant No. 16-05-00799. I like to thank Warren Huff for his helpful review and linguistic improvements.

REFERENCES

Aigner, T., 1985. Storm depositional systems. Lecture Notes in Earth Sciences, vol. 3 Springer-Verlag, Berlin.

Barnes, C.R., Fortey, R.A., Williams, S.H., 1996. The pattern of global bio-events during the Ordovician Period. In: Walliser, O.H. (Ed.), Global Events and Event Stratigraphy in the Phanerozoic. Springer-Verlag, New York, pp. 139–172.

Beauchamp, B., Desrochers, A., 1997. Permian warm- to very cold-water carbonates and cherts in northwest Pangea. In: James, N.P., Clarke, J. (Eds.), Cool-Water Carbonates, pp. 327–347. Special Publications, Soc. Sediment. Geol. 56.

Bergström, S.M., Chen, X., Gutiérrez-Marco, J.C., Dronov, A., 2009. The new chronostratigraphic classification of the Ordovician System and its relations to major series and stages and to δ13C chemostratigraphy. Lethaia 42, 97–107.

Bgatov, V.I., 1973. Litologo-geokhimicheskie zakonomernosti osadkonakoplenija v ordovike i silure Sibirskoi platformy. [Lithological-Geochemical Trends in the Ordovician–Silurian Sedimentation in the Siberian Platform]. Krasnoyarsk Knizh. Izd., Krasnoyarsk (in Russian).

Brandley, R.Y., Krause, F.K., 1997. Thermoclines and wave-sweeping on an equatorial carbonate ramp: Lower Carboniferous strata of Western Canada. In: James, N.P., Clarke, J. (Eds.), Cool-Water Carbonates, pp. 365–390. Special Publications, Soc. Sediment. Geol. 56.

Brookfield, M.E., 1988. A mid-Ordovician temperate carbonate shelf—the Black River and Trenton Limestone Groups of southern Ontario, Canada. Sediment. Geol. 60, 137–153.

Cocks, L.R.M., Torsvik, T.H., 2005. Baltica from the late Precambrian to mid-Paleozoic times: the gain and loss of terrain's identity. Earth Sci. Rev. 72, 39–66.

Cocks, L.R.M., Torsvik, T.H., 2007. Siberia, the wandering northern terrane, and its changing geography through the Paleozoic. Earth Sci. Rev. 82, 29–74.

Cooper, R.A., Sadler, P.M., 2012. The Ordovician Period. In: Gradstein, F.M., Ogg, J.G., Schmitz, M., Ogg, G. (Eds.), The Geologic Time Scale 2012. Elsevier, Boston, USA, pp. 489–523.

Dronov, A.V., 2001. Otlichitel'nye osobennosti netropicheskikh karbonatov na primere nizhne-sredneordovikskikh otlozhenij Pribaltiki. [Distinctive features of non-tropical carbonates exemplified by the Lower-Middle Ordovician deposits of the Baltic region]. Proceedings of the 2nd Russian Lithological Conference and 8th Russian Symposium on Fossil Coral and Reefs, Geoprint, Syktyvkar, Russia, pp. 18–19, (in Russian).

Dronov, A.V., 2012. Traces of volcanic eruptions in the Upper Ordovician of the Siberian Platform and their significance for palaeotectonic reconstructions. In: Kuznetsov, N.B. (Ed.), Sedimentary basins and geological prerequisites for prognosis of new objects for oil and gas exploration, pp. 119–122. Proceedings of the XLIV Tectonic Meeting, Moscow, GEOS. (in Russian).

Dronov, A. 2013. Late Ordovician cooling event: evidence from the Siberian Craton. Palaeogeogr. Palaeoclimatol. Palaeoecol. V.389, 1, p. 87–95.

Dronov, A.V., Holmer, L.E., 1999. Depositional sequences in the Ordovician of Baltoscandia. In: Kraft, P., Fatka, O. (Eds.), Quo vadis Ordovician? Short Papers of the 8th International Symposium on the Ordovician System, pp. 133–136. Acta Universitis Carolinae. Geologica 43 (½), Praha.

Dronov, A., Holmer, L., 2002. Ordovician Sea-Level Curve: Baltoscandian View. The Fifth Baltic Stratigraphical Conference, Geological Survey of Lithuania, Vilnius, Lithuania, pp. 33–35.

Dronov, A., Rozhnov, S., 2007. Climatic changes in the Baltoscandian basin during the Ordovician: sedimentological and palaeontological aspects. Acta Palaeontol. Sin. 46 (Suppl), 108–113.

Dronov, A.V., Kanygin, A.V., Timokhin, A.V., Tolmacheva, T.J., Gonta, T.V., 2009. Correlation of eustatic and biotic events in the Ordovician paleobasins of the Siberian and Russian platforms. Paleontol. J. 43 (11), 1477–1497.

Dronov, A.V., Huff, W.D., Kanygin, A.V., Gonta, T.V., 2011a. K-bentonites in the Upper Ordovician of the Siberian Platform. In: Gutiérrez-Marco, J.C., Rábano, I., García-Bellido, D. (Eds.), Ordovician of the World, pp. 135–141. Cuadernos del Museo Geominero, Instituto Geológico y Minero de España, Madrid.

Dronov, A.V., Ainsaar, L., Kaljo, D., Meidla, T., Saadre, T., Einasto, R., 2011b. Ordovician of Baltoscandia: facies, sequences and sea-level changes. In: Gutiérrez-Marco, J.C., Rábano, I., García-Bellido, D. (Eds.), Ordovician of the World, pp. 143–150. Cuadernos del Museo Geominero, Instituto Geológico y Minero de España, Madrid.

Ettensohn, F.R., 2010. Origin of the Late Ordovician (mid-Mohawkian) temperate-water conditions on southeastern Laurentia: glacial or tectonic? In: Finney, S.C., Berry, W.B.N. (Eds.), The Ordovician Earth System, pp. 163–175. Geological Society of America Special Paper 466.

Fortey, R.A., 1984. Global earlier Ordovician transgressions and regressions and their biological implications. In: Bruton, D. (Ed.), Aspects of the Ordovician System, pp. 37–50. Palaeontological Contributions from the University of Oslo, Universitetsforlaget, Oslo, Norway.

Haq, B.U., Schutter, S.R., 2008. A chronology of Paleozoic sea-level changes. Science 322, 64–68.

Harris, M.T., Sheehan, P.M., Ainsaar, L., Hints, L., Männik, P., Nõlvak, J., Rubel, M., 2004. Upper Ordovician sequences of western Estonia. Palaeogeogr. Palaeoclimatol. Palaeoecol. 210, 135–148.

Henrich, R., Freiwald, A., Betzler, C., Bader, B., Schafer, P., Samtleben, C., Brachert, T.C., Wehrmann, A., Zankl, H., Kuhlmann, D.H.H. 1995. Controls on modern carbonate sedimentation on warm-temperate to arctic coasts, shelves and seamounts in the northern hemisphere: implications for fossils counterparts. Facies, v. 32, p. 71–108.

Heredia, S., Beresi, M., 1995. Ordovician events and sea-level changes on the western margin of Gondwana: the Argentine Precordillera. In: Cooper, J.D., Droser, M.L., Finney, S.C. (Eds.), Ordovician Odyssey: Proceedings of 7th International Symposium on the Ordovician System, pp. 315–318. Pacific Section. SEPM, California, Fullerton.

Herrmann, A.D., Haupt, B.J., 2010. Toward identifying potential causes for stratigraphic change in subtropical to tropical Laurentia during the Mohawkian (early Late Ordovician). In: Finney, S.C., Berry, W.B.N. (Eds.), The Ordovician Earth System, pp. 29–35. Geological Society of America Special Paper 466.

Herrmann, A.D., Patzkowsky, M.E., Pollard, D., 2004. The impact of paleogeography, pCO2, poleward ocean heat transport and sea level change on global cooling during the Late Ordovician. Palaeogeogr. Palaeoclimatol. Palaeoecol. 206, 59–74.

Holland, S.M., Patzkowsky, M.E., 1996. Sequence stratigraphy and long-term paleoceanographic change in the Middle and Upper Ordovician of the eastern United States. In: Witzke, B., Ludvigson, G., Day, J. (Eds.), Paleozoic Sequence Stratigraphy: Views From the North American Craton, pp. 117–129. Geological Society of America. Special Paper 306, Boulder, CO.

Huff, W., Bergström, S.M., Kolata, D.R., 2010. Ordovician explosive volcanism. In: Finney, S.C., Berry, W.B.N. (Eds.), The Ordovician Earth System, pp. 13–28. Geological Society of America Special Paper 466.

Huff, W.D., Dronov, A., Sell, B., Kanygin, A.V. & Gonta, T.V. 2014. Traces of explosive volcanic eruptions in the Upper Ordovician of the Siberian Platform. Est. J. Earth Sci. V. 64, No 4, pp. 244-250.

Jaanusson, V., 1982. Introduction to the Ordovician of Sweden. In: Bruton, D.L., Williams, S.H. (Eds.), Field Excursion Guide. IV International Symposium on the Ordovician System, pp. 1–10. Paleont. Contr. Univ. Oslo, 279.

James, N.P., 1997. The cool-water carbonate depositional realm. In: James, N.P., Clarke, J.A.D. (Eds.), Cool-Water Carbonates, pp. 1–20. SEPM. Special Publication, 56.

Kanygin, A.V., Moskalenko, E.A., Yadrenkina, A.G., 1982. Ordovik Sibirskoi platformy. Opornij razrez po reke Kulyumbe. [The Ordovician of the Siberian Platform. The Key Section at the Kulumber River]. Nauka, Moscow, Russia (in Russian).

Kanygin, A.V., Moskalenko, E.A., Divina, T.A., Matukhina, V.G., Yadrenkina, A.G., 1984. Ordovik zapadnoi chasti Irkutskogo amphiteatra. [The Ordovician of the Western Irkutsk Amphitheater]. Nauka, Moscow, Russia (in Russian).

Kanygin, A.V., Timokhin, A.V., Sychev, O.V., Yadrenkina, A.G., 2006. Osnovnye etapy evolutsii i biofacial'noe raionirovanie ordovikskogo paleobasseina Sibirskoi platformy. [The main Ordovician evolutionary stages and biofacies of the Siberian platform basin]. Proceedings of the International Symposium. GEO, Novosibirsk, Russia, pp. 22–25 (in Russian).

Kanygin, A.V., Yadrenkina, A.G., Timokhin, A.V., Moskalenko, T.A., Sychev, O.V., 2007. Stratigraphija neftegazonosnykh basseinov Sibiri. Ordovik Sibirskoi platformy. [Stratigraphy of the Oil- and Gas-Bearing Basins of Siberia. The Ordovician of the Siberian Platform]. GEO, Novosibirsk, Russia (in Russian).

Kanygin, A., Dronov, A., Timokhin, A., Gonta, T., 2010a. Depositional sequences and palaeoceanographic change in the Ordovician of the Siberian craton. Palaeogeogr. Palaeoclimatol. Palaeoecol. 296 (3–4), 285–294.

Kanygin, A.V., Koren, T.N., Yadrenkina, A.G., Timokhin, A.V., Sychev, O.V., Tolmacheva, T.Y., 2010b. Ordovician of the Siberian platform. In: Finney, S.C., Berry, W.B.N. (Eds.), The Ordovician Earth System, pp. 105–117. Geological Society of America Special Paper 466.

Kazarinov, V.P., Bgatov, V.I., Budnikov, V.I., Divina, T.A., Zhabin, V.V., Kazansky, Y.P., Matukhin, R.G., Rodin, R.S., 1969. Kontinental'nye pereryvy i kory vyvetrivanija Sibirskoi platformy. [Regional unconformities and weathering crusts in the Siberian platform]. Transactions of SNIIGGiMS Issue 98, Nauka, Novosibirsk, pp. 5–16 (in Russian).

Kreisa, R.D., 1981. Storm-generated sedimentary structures in subtidal marine facies with examples from the middle and upper Ordovician of southwestern Virginia. J. Sediment. Petrol. 51, 823–848.

Lashkov, E.M., Paškevičius, J., 1989. Stratigraphicheskie probely i sedimentatsionnye pereryvy v razreze ordovika zapadnogo kraja Vostochno-Evropeiskoi platformy. [Stratigraphic gaps and discontinuities in the Ordovician succession of the western margin of the East-European platform]. In: Nauchnye trudy Vyshikh uchebnykh zavedeniy Litovskoi SSR, pp. 12–37. Geologija 10. (in Russian).

Lavoie, D., 1995. A Late Ordovician high-energy temperate-water carbonate ramp, southern Quebec, Canada: implications for Late Ordovician oceanography. Sedimentology 42, 95–116.

Lavoie, D., Asselin, E., 1998. Upper Ordovician facies in the Lac Saint-Juan outlier, Quebec (eastern Canada): palaeoenvironmental significance for Late Ordovician oceanography. Sedimentology 45, 817–832.

Lees, A., Buller, A.T., 1972. Modern temperate-water and warm-water shelf carbonate contrasted. Mar. Geol. 13, 67–73.

Lindström, M., 1984. The Ordovician climate based on the study of carbonate rocks. In: Bruton, D. (Ed.), Aspects of the Ordovician System, pp. 81–88. Palaeontological Contributions from the University of Oslo, Universitetsforlaget, Oslo, Norway.

Loutit, T.S., Hardenbol, J., Vail, P.R., Baum, G.R., 1988. Condensed sections: the key to determination and correlation of continental margin sequences. In: Wilgus, C.K., Hastings, B., Kendall, C., Posamentier, H., Ross, C., Van Wagoner, J. (Eds.), Sea-Level Changes: An Integrated Approach, pp. 183–213. Society of Economic Paleontologists and Mineralogists Special Publication 42.

Männil, R.M., 1966. Istorija razvitija baltijskogo basseina v ordovike [The Ordovician History of the Baltic Basin]. Valgus, Tallinn (in Russian).

Markov, E.P., 1970. Ordovik i ranny Silur jugo-zapada Tungusskoi sineclizy [Ordovician and Early Silurian of the South-west of Tungus Sineclise]. Nedra Publishing House, Leningrad. 144 pp. (in Russian).

Markov, E.P., Markova, L.G., Matukhina, V.G., 1971. Pereryvy osadkonakoplenija I kory vyvetrivanija v ordovike prieniseiskoj chasti Sibirskoi platformy [Breaks in sedimentation and weathering crusts in the Ordovician of the Yenisei region of the Siberian platform]. Transactions of SNIIGGiMS. Issue 126, Nauka, Novosibirsk, pp. 66–69 (in Russian).

Munnecke, A., Calner, M., Harper, D.A.T., Servais, T., 2010. Ordovician and Silurian seawater chemistry, sea-level, and climate: a synopsis. Palaeogeogr. Palaeoclimatol. Palaeoecol. 296, 289–413.

Nelson, C.S., 1988. An introductory perspective on non-tropical shelf carbonates. Sediment. Geol. 60, 3–14.

Nelson, C.S., Keane, S.L., Head, P.S., 1988. Non-tropical carbonate deposits on the modern New Zealand shelf. Sediment. Geol. 60, 71–94.

Nestor, H., Einasto, R., 1997. Ordovician and Silurian carbonate sedimentation basin. In: Raukas, A., Teedumäe, A. (Eds.), Geology and Mineral Resources of Estonia. Estonia Academy Publishers, Tallinn, pp. 192–204.

Nielsen, A.T., 2004. Ordovician sea-level changes: a Baltoscandian perspective. In: Webby, B.D., Paris, F., Droser, M.L., Persival, G. (Eds.), The Great Ordovician Biodiversification Event. Columbia University Press, New York, pp. 84–93.

Oradovskaya, M.M., 1988. Ordovician and Silurian Biostratigraphy and Facies in the Northeast USSR. Nedra, Moscow. 160 pp. (in Russian).

Pedley, M., Carannante, G., 2006. Cool-water carbonate ramps: a review. In: Pedley, H.M., Carannante, G. (Eds.), Cool-Water Carbonates: Depositional System and Palaeoenvironmental Controls, pp. 1–9. Geological Society, London, Special Publications 225.

Ross, J.R.P., Ross, C.A., 1992. Ordovician sea-level fluctuations. In: Webby, B.D., Laurie, J.R. (Eds.), Global Perspectives on Ordovician Geology. Balkema, Rotterdam, pp. 327–335.

Ross, C.A., Ross, J.R.P., 1995. In: Cooper, J.D., Droser, M.L., Finney, S.C. (Eds.), North American depositional sequences and correlations. Ordovician Odyssey: Short Papers for the Seventh International Symposium on the Ordovician System. Fullerton, pp. 309–313.

Schlager, W., 2007. Carbonate sedimentology and sequence stratigraphy. In: Crossey, L.J. (Ed.), Concepts in Sedimentology and Paleontology No. 8. SEPM (Society for Sedimentary Geology), 200 pp.

Schutter, S.R., 1996. The Glenwood Shale as an example of a Middle Ordovician condensed section. In: Witzke, B., Ludvigson, G., Day, J. (Eds.), Paleozoic Sequence Stratigraphy: Views From the North American Craton, pp. 55–65. Geological Society of America Special Paper 306, Boulder, CO.

Sokolov, B.S., Tesakov, Y.I., 1975. Ordovician Stratigraphy of the Siberian Platform. Nauka, Novosibirsk. 255 pp. (in Russian).

Su, W., 2007. Ordovician sea-level changes: evidence from the Yangtze Platform. Acta Palaeontol. Sin. 46 (Suppl), 471–476.

Tolmacheva, T.Y., and Abaimova, G.P. 2009. Late Cambrian and Early Ordovician conodonts from the Kulumbe River section, northwest Siberian Platform. Mem. Assoc. Australas. Palaeontol., v.37, p.427–451.

Vail, P.R., Mitchum, R.M., Thompson, S., 1977. Seismic stratigraphy and global changes of sea-level, part 4: Global cycles of relative changes in sea-level. In: AAPG Memoir, pp. 83–96. Tulsa, OK.

Van Wagoner, J.C., Posamenier, H.W., Mitchum, R.M., Vail, P.R., Sarg, J.F., Loutit, T.S., Hardenbol, J., 1988. An overview of the fundamentals of sequence stratigraphy and key definitions. In: Wilgus, C.K., Hastings, B., Kendall, C., Posamentier, H., Ross, C., Van Wagoner, J. (Eds.), Sea-Level Changes: An Integrated Approach, pp. 39–45. Society of Economic Paleontologists and Mineralogists Special Publication 42.

Videt, B., Paris, F., Rubino, J.-L., Boumendjel, K., Dabard, M.-P., Loi, A., Ghienne, J.-F., Marante, A., Gorini, A., 2010. Biostratigraphical calibration of third order Ordovician sequences on the northern Gondwana platform. Palaeogeogr. Palaeoclimatol. Palaeoecol. 296, 359–375.

Woodcock, N.H., 1990. Sequence stratigraphy of the Paleozoic Welsh Basin. J. Geol. Soc. Lond. 147, 537–547.

Yu, B., Chen, J., Lin, C., 2001. Cambrian-Ordovician sequence stratigraphy of the northern Tarim Platform and its correlation with Yangtze Platform and North China Platform. Sci. China D 44 (4), 373–384.

CHAPTER SIX

Upper Devonian Biostratigraphy, Event Stratigraphy, and Late Frasnian Kellwasser Extinction Bioevents in the Iowa Basin: Western Euramerica

Jed Day[*,1], **Brian J. Witzke**[†]
[*]Illinois State University, Normal, IL, United States
[†]University of Iowa, Iowa City, IA, United States
[1]Corresponding author: e-mail address: jeday@ilstu.edu

Contents

Stratigraphy & Timescales, Volume 2
ISSN 2468-5178
http://dx.doi.org/10.1016/bs.sats.2017.08.002

Abstract

Upper Devonian (Upper Frasnian) strata in the Iowa area (central United States) provide one of the best records in North America of Late Frasnian sea-level history and the associated Kellwasser Extinction Events within a detailed conodont biostratigraphic framework. Rich conodont faunas span Frasnian Zones 11 through 13c. Diverse benthic faunas of the Lime Creek Formation provide an unparalleled paleontologic record to examine the effects of the Kellwasser extinctions of benthic faunas in the tropical seaway. These strata preserve an onshore-to-offshore transect of depositional facies ranging from shallow-water carbonate shelf to basinal deeper water shale facies in an epeiric ramp system. Late Frasnian sedimentation was initiated above a deeply incised and karstified surface on underlying Givetian–Mid Frasnian carbonates. Regional stratigraphic relations require a minimal sea-level rise of at least 65–90 m to onlap this older surface. The Lime Creek Formation includes two stratigraphic sequences correlated with the global eustatic transgressive–regressive (T-R) cycles IId-1 and IId-2. The Lime Creek records a major extinction of benthic faunas through the Lower Kellwasser Extinction Event (LKE) associated with T-R cycle IId-2 deepening (upper part Zone 12) and onset of a major positive $\delta^{13}C_{carb}$ isotope excursion that spans the LKE and post-LKE recovery interval. The change from brachiopod-dominated to mollusk-dominated benthic associations is evidence of significant reorganization of benthic paleocommunity structure in the post-LKE survival-and-recovery interval. The near total extinction of the Late Frasnian conodont fauna during the Upper Kellwasser Event was coincident with regional expansion of Grassy Creek Shale organic-rich dysoxic–anoxic shale facies in southeast Iowa.

1. INTRODUCTION

This report summarizes the current state of knowledge of the integrated conodont and brachiopod biostratigraphy, $\delta^{13}C_{carb}$ chemostratigraphy, sea level, and Kellwasser extinction history of the Late Devonian (Late Frasnian–Early Famennian) of the Iowa Basin of central North America (Figs. 1–3). Significant new integrated biostratigraphic data from surface and subsurface localities in eastern and northern Iowa refine local brachiopod and conodont ranges, revise zonal correlations, confirm the timing of Late Devonian sea-level fluctuations, and clearly identify Late Frasnian Kellwasser extinction intervals in shallow subtidal platform, off-platform deep ramp, and

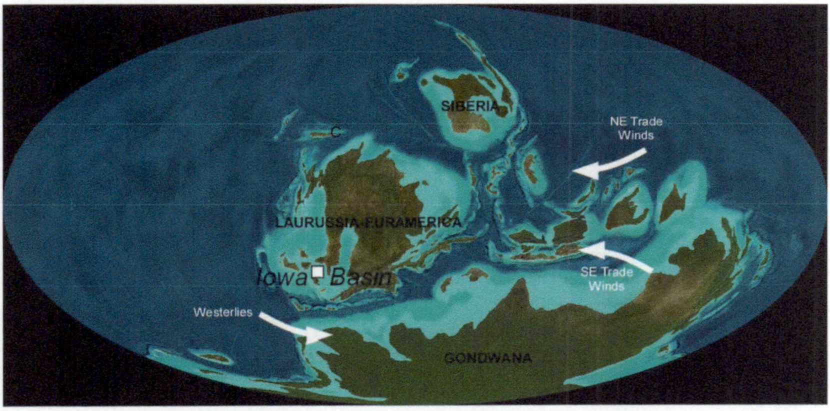

Fig. 1 Global Late Devonian (Frasnian) paleogeographic map showing the location of the subtropical epeiric Iowa Basin in western Laurussia–Euramerica. *Arrows* indicate predicted atmospheric surface wind systems. *Courtesy of and adapted from R. Blakey Deep Time Maps™ produced by Colorado Plateau Geosystems, Inc.*

basinal sections in the Iowa Basin (Fig. 3). Deep-ramp facies of the Amana Beds and North Liberty Beds of the Lime Creek Formation and basinal deposits of the Sweetland Creek and Grassy Creek Shales feature conodont assemblages of the offshore *Palmatolepis* biofacies (Figs. 2 and 3, localities PRC, NLB, TSC). Directly associated offshore conodonts, brachiopods and ammonoids in the Amana Beds Member of the Lime Creek provide an exceptional record of the Lower Kellwasser Event (LKE) in the deep-ramp environment in eastern Iowa (Fig. 2, locality PRC; Fig. 3). That record is directly comparable to the shallow subtidal record documented by conodont and brachiopod sequences in the Owen Member of the Lime Creek in northern Iowa (Fig. 3).

The type section of the Sweetland Creek Shale (Figs. 2 and 3) is the principle central North American reference section for the Late Frasnian and Early Famennian (Klapper, in Johnson and Klapper, 1992; Over, 2002, 2006) and was used in the development of the Frasnian Composite Standard for graphic correlation (Klapper and Kirchgasser, 2016). New data from our samples from the type Sweetland Creek Shale permit revisions of local species ranges and zonal boundary positions as reported by Klapper (in Johnson and Klapper, 1992) and Over (2002, 2006, table 1) and locate positions of the LKE and Upper Kellwasser Extinction (UKE) intervals below the Frasnian–Famennian (F—F) stage boundary in Muscatine County of eastern Iowa. Investigation of subsurface sections in Des Moines and Lee counties in

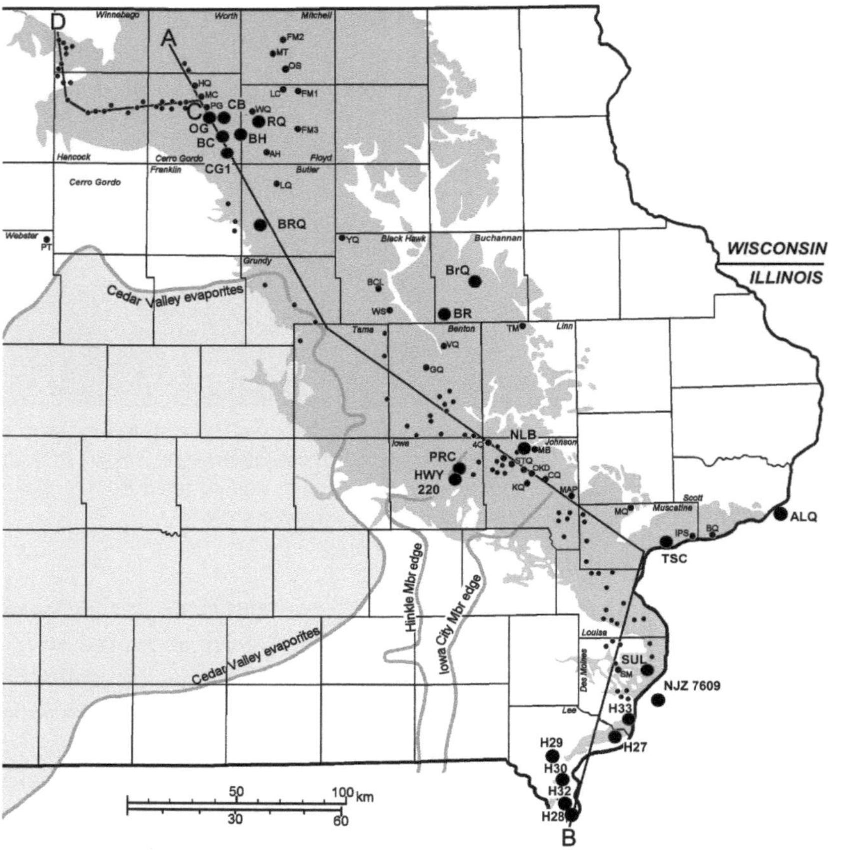

Fig. 2 Map of the eastern half of the state of Iowa showing the locations of core, well, and outcrop sections of Middle–Upper Devonian (Eifelian–Famennian) strata in the Iowa Basin. *Dark shading* = Devonian outcrop belt, light shading shows extent of Givetian–Lower Frasnian evaporites within the Cedar Valley Group (after Witzke et al., 1989). The distal margins of the Hinkle and Iowa City Members (*labeled lines*) mark the general southeastward extent of Late Givetian inner shelf facies within the Cedar Valley Group as shown in Fig. 4. *Large dots* = locations of core, outcrop, and quarry sections discussed in this study. *Medium dots* = locations of core and outcrop sections (see Bunker et al., 1986; Witzke and Bunker, 1997; Witzke et al., 1989 for descriptions and precise locations); *small dots* = well sections based on well cuttings only. County boundaries are outlined, and relevant counties are labeled. Locations of well and outcrop sections available in the IGS Geosam database (www.igsb.uiowa.edu/webapps/geosam). *Modified from fig. 1 of Witzke, B.J., Bunker, B. J., 2006. Middle shelf facies of the Cedar Valley Group (Devonian) and their stratigraphic relationships in eastern Iowa. In: Day, J.E., Luczaj, J., Anderson, R. (Eds.), New Perspectives and Advances in the Understanding of Lower and Middle Paleozoic Epeiric Carbonate Depositional Systems of the Iowa and Illinois Basins. Iowa Geological Survey Guidebook 25, pp. 23–46; and fig. 1 of Day, J., 2008.*

southeastern Iowa reveals complex depositional patterns in that part of the basin.

The Sweetland Creek Shale is highly condensed in the Iowa Geological Survey's (IGS) Sullivan Slough Core in Des Moines County (Fig. 2, locality SUL). Despite extreme condensation, the position of the LKE and the lower part of the UKE intervals are identifiable. We also identify a stratigraphic leak of Late Frasnian conodonts hosted in Early Frasnian Lithograph City Formation carbonates and reworked Late Frasnian faunas in Early Famennian Grassy Creek Shale. Reworked faunas identify positions of submarine erosional surfaces in that section (Fig. 7; Table 3). New data permit zonal correlations of Frasnian Zones (F.Z.) 11, 12, and Subzone 13a in the Sweetland Creek and Grassy Creek shale sections in the IGS H-30 core in Lee County of southeastern Iowa (Figs. 2 and 7; Table 4).

The conodont fauna described from the Lime Creek Formation in central and northern Iowa by Anderson (1966), Metzger (1989), and Day (1990) is characteristic of the shallow-water Frasnian *Polygnathus* biofacies (see Klapper and Lane, 1985, 1989). Intercorrelation between conodont successions developed in the shallow-water *Polygnathus* and offshore *Palmatolepis* biofacies have been difficult because shallow platform deposits generally lack species that define global Frasnian zonal boundaries in the offshore successions, which are generally rare or absent in platform deposits (Day, 1990; Klapper and Lane, 1985, 1989). Remarkable new occurrences of diagnostic offshore species of *Ancryognathus*, *Palmatolepis*, and *Polygnathus* in platform sections (Fig. 2, localities CG-1, BH, BQ, BRQ) now clearly identify key Frasnian global zonal boundary positions and the LKE interval within the lower and middle parts of the Owen Member of the Lime Creek Formation in northern Iowa. Resampling of the IGS CG-1 core yields new occurrences of *Polygnathus samueli* and *Polygnathus planarius* whose restricted ranges identify Frasnian Zone 12. *Icriodus altnernatus* above *P. samueli* and *P. planarius* are within Subzone 13a. New samples with *Palmatolepis foliacea* and *Pa. winchelli* in the upper Cerro Gordo and lower Owen Members at Bird Hill and the Bruns Quarry (Fig. 2, localities BH, BRQ) confirm a

Stop 7—Rockford Brick and Tile Quarry—Floyd County Fossil and Prairie Center and Park Preserve. In: Groves, J.R. Walters, J.C., Day, J. (Eds.), Carbonate Platform Facies and Faunas of the Middle and Upper Devonian Cedar Valley Group and Lime Creek Formation, Northern Iowa. Iowa Geological Survey Guidebook 28, pp. 89–98; Day, J., 2013. Stop 5. Rockford Brick and Tile Quarry—Floyd County Fossil and Prairie Center and Park Preserve. In: Day, J. Witzke, B.J., Lundy, S. (Eds.), Aspects of the Paleozoic History of Epeiric Seas of the Iowa Basin. Iowa Geological Survey Guidebook 29, pp. 109–118.

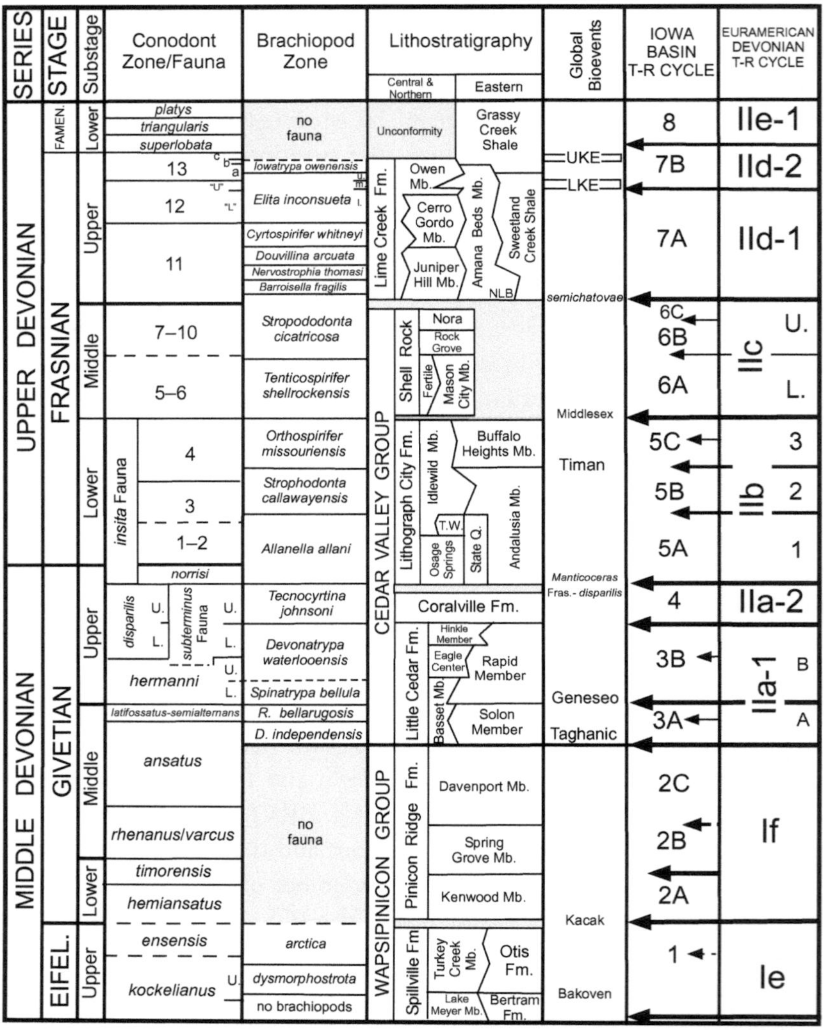

Fig. 3 Chronostratigraphy of the Middle–Late Devonian (Late Eifelian–Early Famennian) strata of the northern shelf facies of the Iowa Basin. Devonian lithostratigraphy after Witzke et al. (1989), Bunker and Witzke (1992), Witzke and Bunker (1996), Day (1997, 2006), Day et al. (2008, 2013), and this study. Qualitative Euramerica Devonian eustatic T-R cycles of Johnson et al. (1985, 1996), Johnson and Klapper (1992), and subdivisions proposed by Day et al. (1996a, 2008, 2013), Day (1998), and Whalen and Day (2008). Iowa Devonian T-R cycles of Witzke et al. (1989), Bunker and Witzke (1992), Witzke and Bunker (1997, 2006), Day (2006), and Day et al. (2008, 2013). Devonian conodont biostratigraphy follows Witzke et al. (1985, 1989), Klapper, in Johnson and Klapper (1992), Bunker and Witzke (1992), Witzke and Bunker (1997), Day (1990, 1992, 2006, Day, 2010a,b), Rogers (1998), Over (2002, 2006). Frasnian conodont zones after Klapper and Kirchgasser

position high within Zone 12. The occurrence of *Palmatolepis bogartensis*, *Polygnathus breviarina*, and *O. dissimilis* in the Owen Member of the Lime Creek is correlated with Frasnian Subzone 13a, constrain the position of the LKE, and support Metzger's (1989) correlation of part of the Upper Lime Creek (undivided) in the subsurface of Central Iowa with Alberta very Late Frasnian Faunal Interval 8 of Klapper and Lane (1989). These new data also indicate that strata deposited during the very Latest Frasnian (UKE interval) are absent across much of the peritidal and shallow subtidal Lime Creek platform in northern Iowa (Figs. 3, 5, and 6).

New sampling and recent taxonomic revisions of Lime Creek atrypide, spirideride, and lingulide brachiopod taxa (Day, 1998; Day and Copper, 1998; Ma and Day, 2000, 2003) necessitate revision to local species range zones and renaming of the basal Lime Creek *Lingula fragila* Zone of Day (1996a). Associated conodonts and brachiopods in the Upper Lime Creek Formation fix the position the base of the *Iowatrypa owenensis* Zone within Subzone 13a of Klapper and Kirchgasser (2016) above the LKE interval (Fig. 3). The base of that zone was not confidently known in terms of the conodont sequence and has been positioned high in Zone 12 or at or just above the base of Zone 13 in older reports of Day (1989, 1995, 1996a, 1998) and Day et al. (2008, 2013). The Lime Creek brachiopod sequence in eastern and northern Iowa is reevaluated based on recognition of the LKE interval within the middle *Elita inconsueta* Zone in surface and subsurface surface study sites (Figs. 2 and 3). Based on revisions herein, the post-LKE survivor and recovery shelly fauna characterize the upper *E. inconsueta* and *I. owenensis* Zones, respectively.

The reported occurrence of the ammonoid *Crickites lindneri* (Baker et al., 1986, Preslicka et al., 2013) from the Buseman quarry in Butler County

(2016) as adapted from Montagne Noire Frasnian zones of Klapper (1989), and Girard et al. (2006). Famennian global standard zonation after Spalletta et al. (2017). Devonian brachiopod biostratigraphy from Day (1989, 1992, 1996b, 1997, 1998, this study). Abbreviations: *B.* = *Barroisella*; *D.* = *Desquamatia*; *Fm.* = Formation; *l* = lower; *L* = Lower; *m* = middle; *M* = Middle; *Mb.* = Member; *NLB* = North Liberty Beds; *R.* = *Rhyssochonetes*; *T.W.* = Thunder Woman Member; *u* = upper; *U* = Upper. *Modified from fig. 3 of Day (2006) and Day, J., 2008. Stop 7—Rockford Brick and Tile Quarry—Floyd County Fossil and Prairie Center and Park Preserve. In: Groves, J.R. Walters, J.C., Day, J. (Eds.), Carbonate Platform Facies and Faunas of the Middle and Upper Devonian Cedar Valley Group and Lime Creek Formation, Northern Iowa. Iowa Geological Survey Guidebook 28, pp. 89–98; Day, J., 2013. Stop 5. Rockford Brick and Tile Quarry—Floyd County Fossil and Prairie Center and Park Preserve. In: Day, J. Witzke, B.J., Lundy, S. (Eds.), Aspects of the Paleozoic History of Epeiric Seas of the Iowa Basin. Iowa Geological Survey Guidebook 29, pp. 109–118.*

(Fig. 2, locality BUS) is likely from the post-LKE interval in the middle-upper Owen Member in northern Iowa. In Iowa County of eastern Iowa, directly associated conodont, ammonoid, and brachiopod faunas in the Amana Beds Member of the Lime Creek are correlated with the upper part of Zone 12 and the lower part of Subzone 13a (Fig. 2, locality PRC; Fig. 3). Stepped extinctions of brachiopods and ammonoids in the Amana Beds in the upper part of F.Z. 12 comprise a deep-ramp record of the LKE (Fig. 1, locality PRC; Figs. 2, 3, 5, and 10).

Johnson and Klapper (1992, pp. 131–132, fig. 2) aligned the base of Euramerica Devonian Transgressive–Regressive (T-R) cycle IId within Montagne Noire Zone 9 of Klapper (1989) in the US midcontinent at the base of the Sweetland Creek Shale in Muscatine County (Fig. 1, locality TSC). Those samples from the lower Sweetland Creek outcrop and in the subsurface of eastern Iowa are correlated with the lower part of Zone 11 (Fig. 3) based on ranges of species in terms of the current Frasnian Composite Standard (Klapper and Kirchgasser, 2016, table 2). The start of marine flooding that initiated deposition of Devonian T-R cycle IId-2 of Day (1998, p. 212, figs. 4 and 9; Day and Copper, 1998, fig. 2) is within the upper part of F.Z. 12 coincident with the onset of the LKE within the lower TST interval of T-R cycle IId-2 (Figs. 2 and 8–10).

2. METHODS AND DATA

New biostratigraphic and whole-rock geochemical data were collected from Lime Creek Formation in northern Iowa (Fig. 2, localities BH, BSQ, BRQ, and CG-1) and surface and subsurface offshore deep subtidal facies of the Sweetland Creek Shale and Lower Grassy Creek Shale in eastern and southeastern Iowa (Fig. 2, localities TSC, SUL, H-30, and H-32). The principle reference section for the subsurface Lime Creek Formation in north-central Iowa is in the IGS CG-1 core (see reports by Day, 1989, 1990). We sampled a continuous split of the CG-1 Core for whole-rock $\delta^{13}C_{carb}$ and bulk magnetic susceptibility (MS). Samples weighing 8–20 g were taken at 10 cm intervals prior to sampling for conodonts. The remainder of the core split was sampled for conodonts every 30 cm (sample weights varied from 57 to 225 g). Brachiopods where collected directly from the core and recovered from conodont samples after crushing during their preparation.

The type section of the Sweetland Creek Shale in Muscatine County, Iowa, was resampled continuously at 10-cm intervals by J. Day and

A. Wallenberg in June of 2016 to supplement data reported in by Klapper (in Johnson and Klapper, 1992, table 1) and Over (2002, 2006). A continuous split of the IGS Sullivan core was sampled for conodonts at 30-cm intervals through the Lower Frasnian Cedar Valley Group (Lithograph City Formation) and then at 2–5 cm through the highly condensed Sweetland Creek Shale and Lower Grassy Creek Shale to document the positions of the LKE and UKE intervals and Frasnian–Famennian boundary. We sampled splits of the IGS H-30 and H-32 cores of the Sweetland Creek and Lower Grassy Creek Shales continuously at 10-cm intervals for conodonts with sample masses ranging from 60 to 190 g. Core chips weighing 10–30 g were cut from the core split of the H-32 core by D. De Vleeschouwer at 5-cm intervals for whole-rock $\delta^{13}C_{carb}$ in the Sweetland Creek Shale and for $\delta^{13}C_{org}$ in the Grassy Creek Shale. Preliminary bulk MS data from the CG-1 and H-32 cores are reported by Day et al. (2014, 2015a, 2016) and De Vleeschouwer et al. (2014, 2016).

All brachiopods reported from the CG-1 core by Day (1989) are in the collections of the University of Iowa (SUI) Paleontology Repository. All specimens from 51 additional samples will be reposited there. All conodonts reported by Day (1990) are in the SUI Repository. All new samples from the ISG CG-1, H-30, H-32, and Sullivan cores in Day's collection at the ISU Paleontology Lab. Identifications of conodonts from the IGS Sullivan Core (Figs. 1, 7, and 9) were made by G. Klapper, and those from the CG-1 and H-30 cores by J. Day.

All conodont and macrofossils sampled from of Amana Bed Member from Price Creek reported below were collected during visits to that locality starting in 1995, and most recently in April of 2016. All brachiopods and conodonts in collections from the Rockford Quarry, Clay Banks, Beaver Creek, Bird Hill, Bruns and Buseman quarries, and Owens Grove locality were identified by J. Day. All carbonate and calcareous shale samples were digested in buffered 10% formic acid for 24 h in the ISU Paleontology laboratory by J. Day and ISU undergraduate research assistants (2 kg each from Bird Hill; 3.5–4.5 kg from the Bruns Quarry). All new conodonts and brachiopod collections reported here will be transferred to the SUI Paleontology repository.

In situ geochemical samples from the Lime Creek Formation in the Bruns Quarry were collected every 25 or 30 cm. One-centimeter slabs were cut from each sample and polished. Micrites on polished surfaces were drilled to generate powders for carbon isotopic analysis. All samples were analyzed by A. Morse in B. Finney's Laboratory at the Center for Ecological

Research and Education (CERE) at Idaho State University. Sample analyses were performed on a Finnigan GasBench II interfaced to a Delta V advantage mass spectrometer through the ConFlo IV system. Dried powdered carbonate samples were loaded into Labco 12 mm Exetainer glass vials and reacted with phosphoric acid on a heating block at 72 °C until the reaction was complete. $\delta^{13}C_{carb}$ and $\delta^{18}O_{carb}$ isotopic values are reported as ‰ values relative to the V-PDB scale. Four in-house standards (Carrara, Fisher, Merck, and Calcite 1), which are directly calibrated using international standards (NBS-18, NBS-19, and LSVEC), were used to create a two-point calibration curve to correct the raw data and to monitor the accuracy of the data. Precision for both $\delta^{13}C$ and $\delta^{18}O$ is better than ±0.2‰. Only $\delta^{13}C_{carb}$ values of sample are reported below.

3. IOWA BASIN UPPER DEVONIAN LITHOSTRATIGRAPHY

The Eifelian to Early Famennian chronostratigraphy of the Iowa Basin is shown in Fig. 3. The Upper Frasnian deposits in this region overlay a complex and deeply incised and karsted erosional disconformity (Figs. 4 and 5). This complex paleotopographic surface formed during a profound regional relative sea-level lowstand during the latter part of Middle Frasnian that led to emergence and erosion of the older Cedar Valley Group carbonate platform throughout the Iowa and adjacent parts of the western Illinois basins (Figs. 3–5; Witzke, in Day et al., 2008, 2013). Fig. 6 shows our generalized depositional model and cross-section of Late Frasnian deposits in the region. Upper Frasnian and Lower Famennian age Sweetland Creek and Grassy Creek Shales are deeper water facies interpreted to have accumulated in dysoxic and anoxic parts of the density stratified Late Devonian seaway in Scott and Muscatine counties and in the subsurface throughout southeastern Iowa (Figs. 3, 6, and 7–9). Equivalents of the Lower Sweetland Creek Shale in Johnson County are the North Liberty Beds (Figs. 2 and 3). Mixed carbonate-clastic platform facies that accumulated in shallow inner to mid-shelf oxygenated to dysoxic parts of the Late Frasnian seaway are included in the Lime Creek Formation (Figs. 3, 5, and 10–15).

The Independence Shale of Iowa (Cooper, 1967; Cooper and Dutro, 1982; Müller and Müller, 1957; Stainbrook, 1945) includes stratigraphic leaks of Late Frasnian Shales and minor carbonates that fill karst cavities hosted in older Cedar Valley Group carbonates in eastern Iowa and western Illinois (see Fig. 4). Most Independence Shale leak deposits in eastern Iowa yield microfossils (abundant conodonts). Macrofossils in Independence

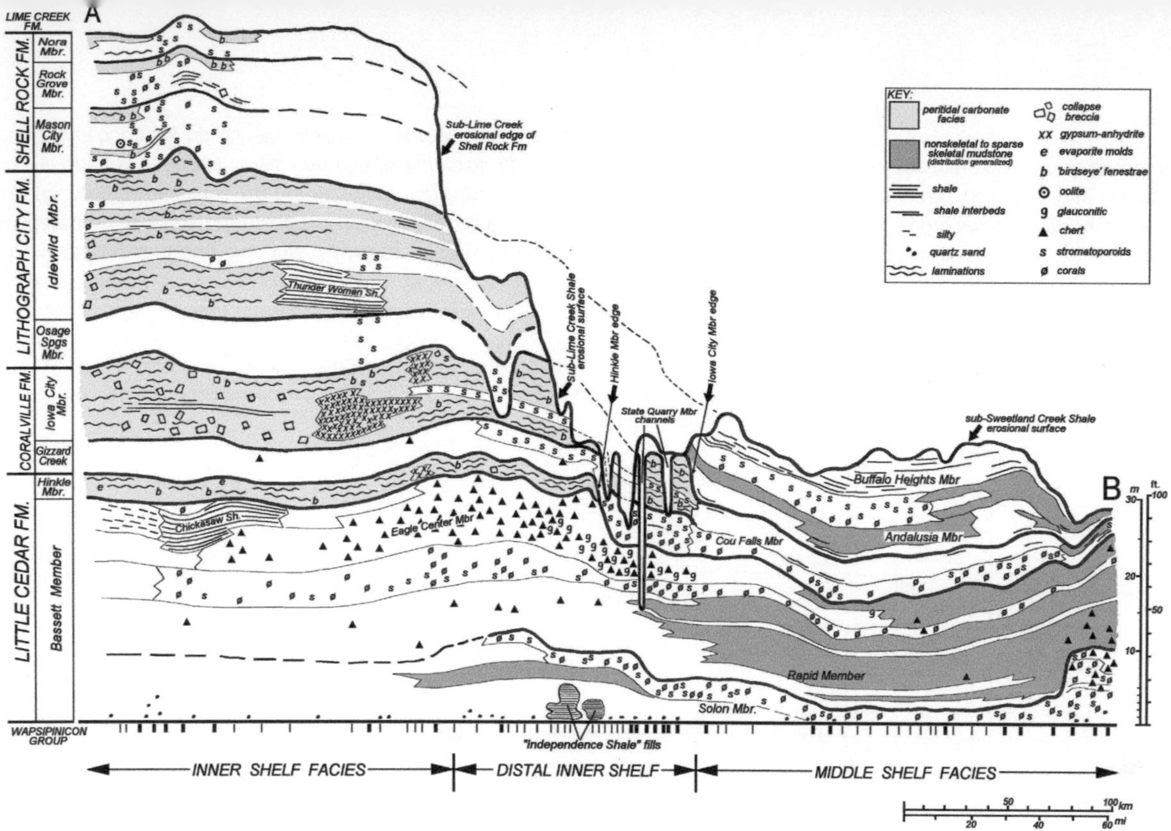

Fig. 4 Northwest to southeast stratigraphic cross-section (Fig. 2, cross-section line A–B) of the Cedar Valley Group across north-central and eastern Iowa. Significant Late Givetian (pre-*norrisi* Zone) post-Coralville and Middle Frasnian sub-Lime Creek/Sweetland Creek erosion have truncated Cedar Valley strata, especially in the distal inner shelf area. Independence Shale fills represent stratigraphic leaks of Late Frasnian Lime Creek Formation or Sweetland Creek Shale within Cedar Valley Group, Wapsipinicon Group, or Silurian-hosted karst networks and openings. Datum is disconformity at top of Wapsipinicon Group. *After fig. 2 of Witzke, B.J., Bunker, B. J., 2006. Middle shelf facies of the Cedar Valley Group (Devonian) and their stratigraphic relationships in eastern Iowa. In: Day, J.E., Luczaj, J., Anderson, R. (Eds.), New Perspectives and Advances in the Understanding of Lower and Middle Paleozoic Epeiric Carbonate Depositional Systems of the Iowa and Illinois Basins. Iowa Geological Survey Guidebook 25, pp. 23–46.*

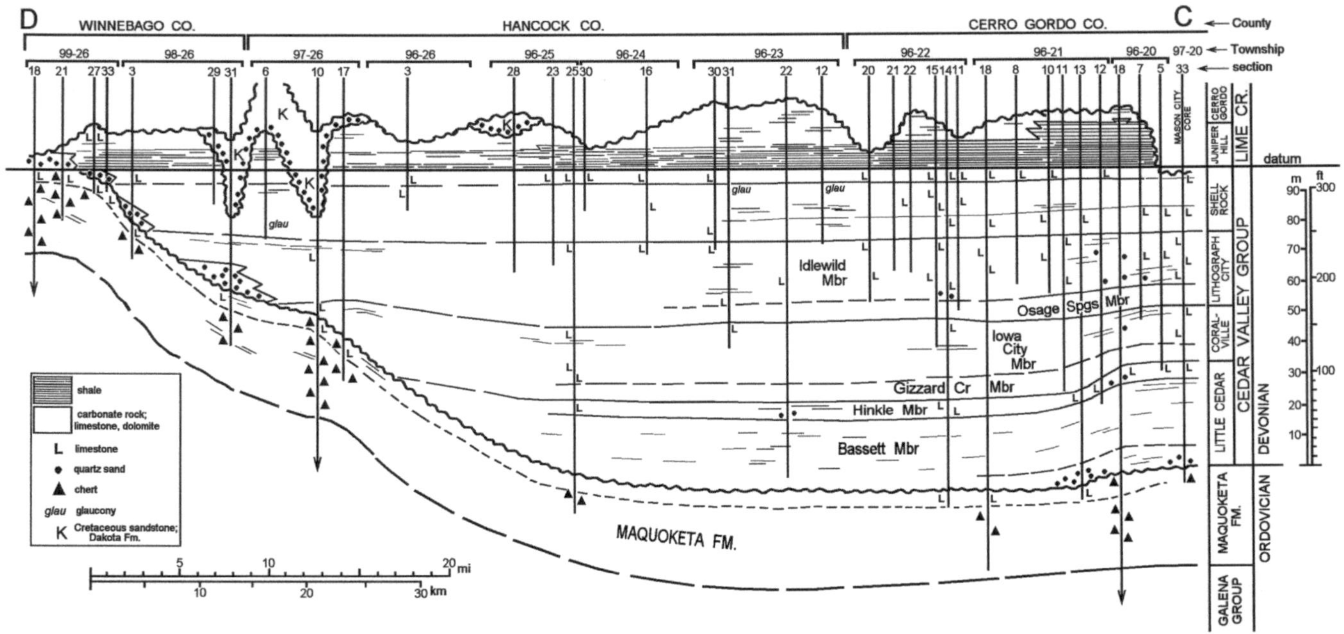

Fig. 5 Stratigraphic cross-section showing northwestward onlap of units within the Cedar Valley Group and Late Frasnian Lime Creek Formation across the Devonian–Ordovician unconformity in north-central Iowa. Datum is base of Lime Creek Formation (interpolated where truncated by Cretaceous strata). See Fig. 2 for line of cross-sections D–C and well points used in the cross-section (see well descriptions on Geosam database, http://gsbdata.igsb.uiowa.edu/geosam/). *After fig. 3 of Witzke, B.J., Bunker, B. J., 2006. Middle shelf facies of the Cedar Valley Group (Devonian) and their stratigraphic relationships in eastern Iowa. In: Day, J.E., Luczaj, J., Anderson, R. (Eds.), New Perspectives and Advances in the Understanding of Lower and Middle Paleozoic Epeiric Carbonate Depositional Systems of the Iowa and Illinois Basins. Iowa Geological Survey Guidebook 25, pp. 23–46, and fig. 7 of Day et al. (2013).*

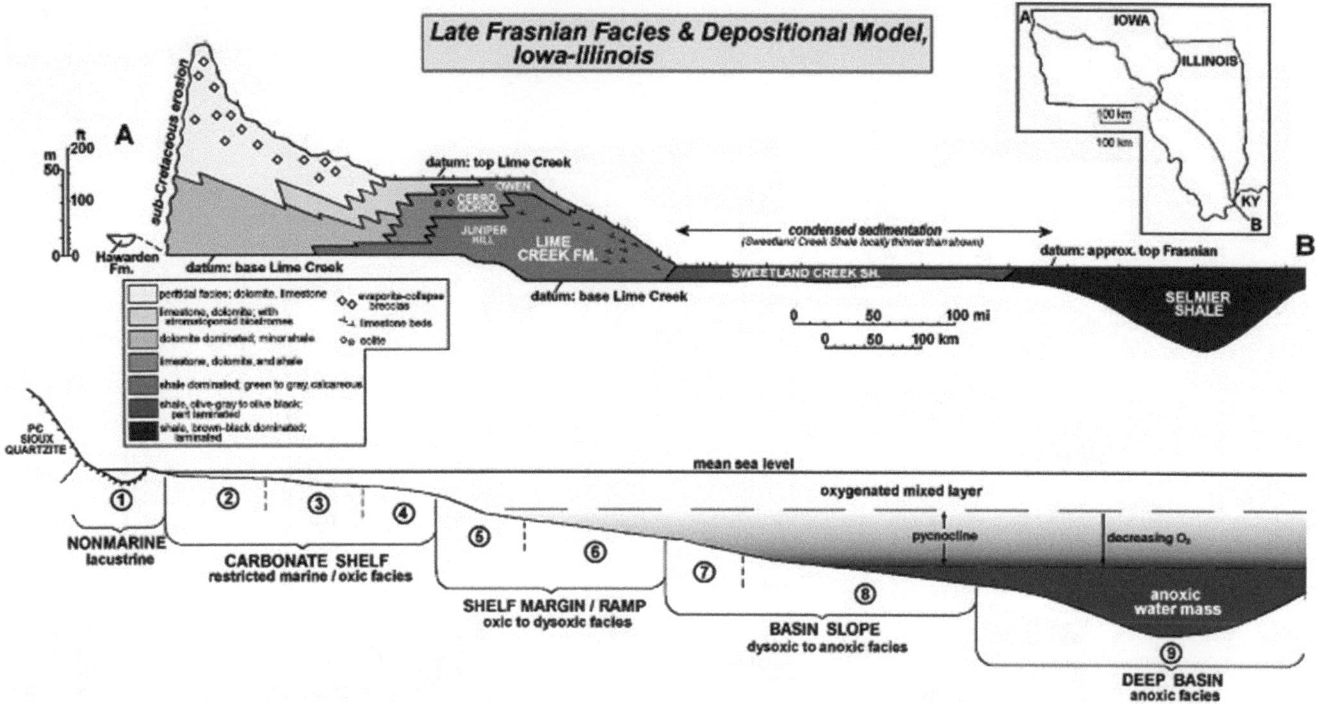

Fig. 6 Cross-section of Late Frasnian rocks in the central Midcontinent and interpreted bathymetric profile illustrating seaway stratification during widespread development of organic-rich anoxic basinal facies during the Late Frasnian (vertically exaggerated). Numbered facies record deposition in: (1) nonmarine lacustrine, (2) and (3) mixed peritidal mudflat, restricted marine, shallow subtidal, (4) and (5) subtidal platform with diverse benthic shelly faunas (type area of Lime Creek Formation), (6) mixed oxic and dysoxic deep ramp (Amana Beds Member), (7) dysoxic (North Liberty Beds facies), (8) and (9) mixed dysoxic and anoxic (Sweetland Creek and Grassy Creek shales). *Modified from fig. 5 of Witzke, B.J., 1987. Models for circulation patterns in epicontinental seas applied to Paleozoic facies of the North America craton. Paleoceanography, 2, 229–248.*

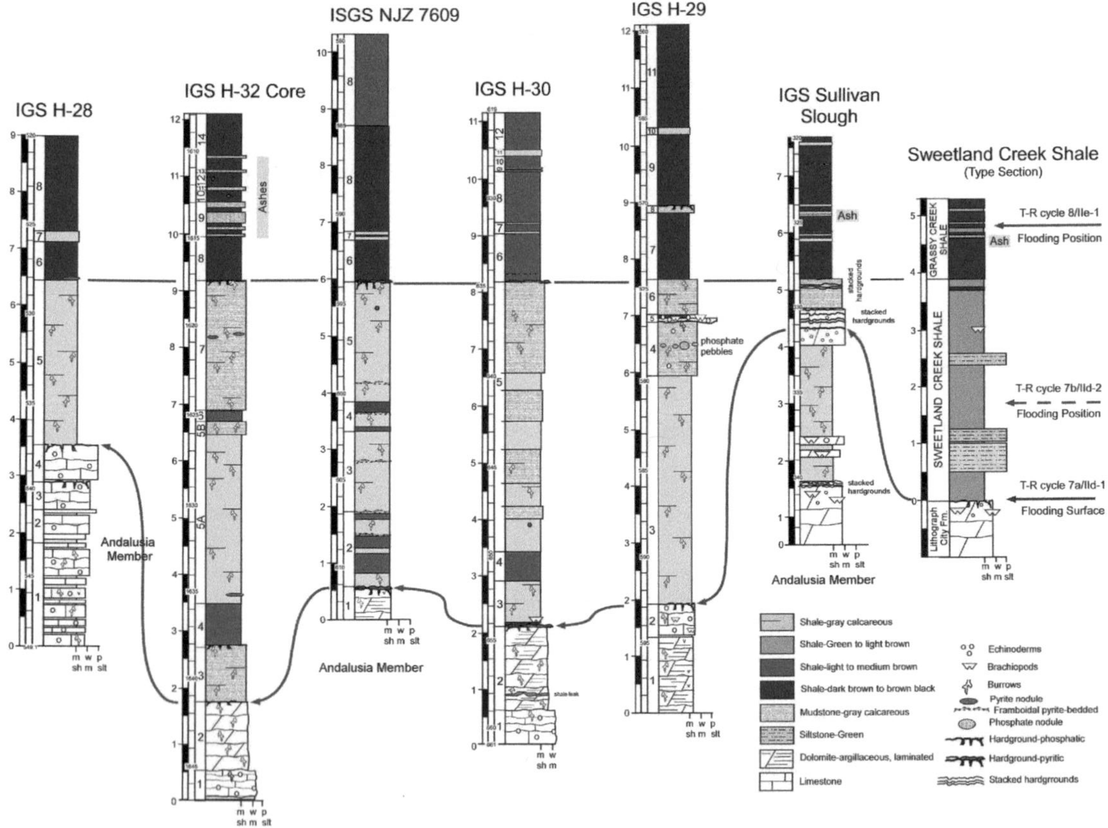

Fig. 7 See legend on opposite page.

Shale leaks occur only in Buchanan County (Fig. 2, localities BR and BrQ). The Amana Beds Member of the Lime Creek Formation consists of 13–14 m of shales with thin carbonates overlying Cedar Valley Group carbonates in Iowa County in eastern Iowa (Figs. 2, 3, and 10, localities PRC, HWY-220). The brachiopod fauna of the Amana Beds is similar to the fauna described from the Independence Shale leaks in Buchanan County and was included in the Independence Shale by Stainbrook (1945). The Independence Shale in Buchanan County and Amana Beds represents a distinct deeper water offshore Lime Creek benthic biotope (see the Appendix and discussion below).

In much of the Iowa Basin, platform emergence led to erosional truncation or nondeposition of uppermost Frasnian and Lower Famennian deposits where the Lime Creek is overlain by Middle Famennian deposits of the Sheffield Formation (Day, 1996a, 1998; Day et al., 2008, 2013; Witzke and Bunker, 1997; Witzke et al., 1989). Subtidal conditions persisted in southeastern Iowa where outcrop and subsurface sections of the Late Frasnian Sweetland Creek Shale are conformably or disconformably overlain by the uppermost Frasnian and Lower Famennian organic-rich Grassy Creek Shale (Figs. 3 and 6–9). Cryptic submarine disconformities are developed within the Grassy Creek Shale in the IGS Sullivan Core as identified by Late Frasnian conodonts reworked with Early Famennian taxa in the Sullivan Core section in Des Moines County, Iowa (Figs. 7 and 9).

Fig. 7 Cross-section of Late–Latest Frasnian and Early Famennian strata that onlap older Cedar Valley Group (Lithograph City Formation) carbonate rocks in southeastern Iowa. See Fig. 2 for locations of all core and outcrop (type Sweetland Creek = TSC) sections. The datum is the top of the Sweetland creek Shale. Key subsurface Sweetland Creek reference sections in southeastern Iowa are the IGS H-30 and H-32 cores. All Iowa cores housed at the Iowa Geological Survey's Oakdale Core Repository, and the NJZ 7609 core is at the Illinois State Geological Survey's core library, University of Illinois. Rock texture abbreviations: *m* = mudstone; *p* = packstone; *sh* = shale; *slt* = siltstone; *w* = wackestone. *Vertical scale of left-hand column* in all sections is in meters, with well depth footage given in the adjacent column. Descriptions of IGS H-28, H-29, H-30, Sullivan cores by J. Day, IGS H-32 core by J. Day and D. De Vleeschouwer in 2013, and ISGS NJZ-7609 core by J. Day in 2013. *Type section of Sweetland Creek Shale adapted from fig. 2 of Over, D.J., 2006. Frasnian-Famennian boundary at the type Sweetland Creek Shale locality. In: Day, J., Luczaj, J., Anderson, R. (Eds.), New Perspectives and Advances in the Understanding of Lower and Middle Paleozoic Epeiric Carbonate Depositional Systems of the Iowa and Illinois Basins. Iowa Geological Survey Guidebook 26, pp. 74–79 and J. Day's section descriptions in 1990 and 2016.*

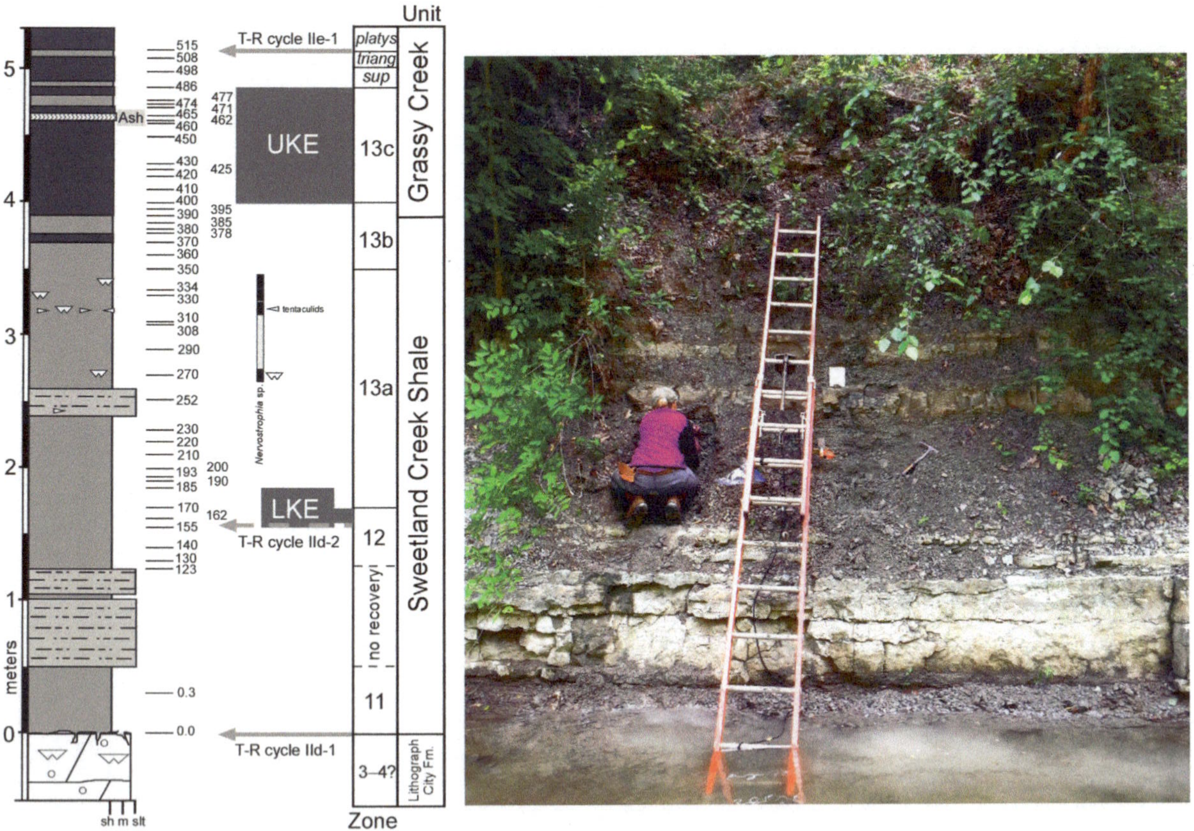

Fig. 8 See legend on opposite page.

3.1 Sweetland Creek Shale

The type section of the Sweetland Creek Shale is located on the east cutbank of Sweetland Creek in Muscatine County, Iowa (Fig. 2, locality TSC; Fig. 7, Type Section; see fig. 1 of Over, 2006). At the type section, and throughout the subsurface in eastern and southern Iowa, the Sweetland Creek Shale overlies a disconformity surface (phosphatic hardground) developed on Lower Frasnian carbonates of the Andalusia or Buffalo Heights Members of the Lithograph City Formation (Figs. 3 and 7–9). Dark-brown organic-rich shales of the Grassy Creek Shale conformably overlay the Sweetland Creek at the type section (Figs. 7 and 8; see Over, 2006, fig. 2). In the subsurface IGS H-core sections (Fig. 7) a pyritic–phosphatic hardground caps the Sweetland Creek Shale at its contact with the Grassy Creek (Figs. 7 and 9). The capping hardground formed as a result of sediment starvation and onset of basinal anoxia in southeastern Iowa (Figs. 2, 6, and 7).

The thickness of the Sweetland Creek ranges from 3.9 m at the type section to 7.5 m in the IGS H-32 core section in Lee County in southeastern Iowa (Figs. 2 and 7). In the IGS Sullivan core (Figs. 2, 7, and 9) an extremely condensed 85 cm thick section of the Sweetland Creek consists of thin dolomite, limestone, and calcareous mudstone with closely spaced hardground surfaces in both the basal dolomite and upper mudstone (see Fig. 9) with a highly condensed conodont sequence shown in Table 3 (samples 40–61).

Over (2006, fig. 2) illustrated the type section where it is 3.9 m thick (Fig. 8). The type section consists of a basal 50 cm light green-brown shale;

Fig. 8 Type section of the Sweetland Creek Shale exposed on Sweetland Creek, Muscatine County, Iowa in June of 2016 (location shown in Fig. 2, see location in appendix 1 of Klapper and Kirchgasser, 2016, p. 28). Photograph was taken in June of 2016 by J. Day 100 m north of the exposure sampled by Klapper (in Johnson and Klapper, 1992), Day (unpublished), and Over (2002, 2006) that is now covered by slumping. The stratigraphic column is adapted from Over (2002, p. 159, fig. 6b; 2006, p. 77, fig. 2) and J. Day's field notes of 1990. Conodont sample positions listed in cm above the base of the Sweetland Creek Shale. Composite of all samples reported by Klapper (in Johnson and Klapper, 1992, p. 132, fig. 2), Over (2002, p. 159, fig. 6b; 2006, p. 77, fig. 2, p. 78, table 1), and unpublished data from samples collected by Day in 1990 with J. Over, G. Klapper, and W. Kirchgasser. Frasnian and Famennian conodont zonal boundaries and Lower and Upper Kellwasser Extinction interval (LKE and UKE) positions based on correlations in Table 2. Numbered zones 3–4, 11–13c are Frasnian Zones or Subzones of Klapper and Kirchgasser (2016). Early Famennian zones: *platys* = *Pa. delicatula platys*; *sup* = *Palmatolepis subperlobata*; *triang* = *Pa. triangularis* zones of Spalletta et al. (2017). Lithologic or rock texture abbreviations and symbols as in Fig. 7.

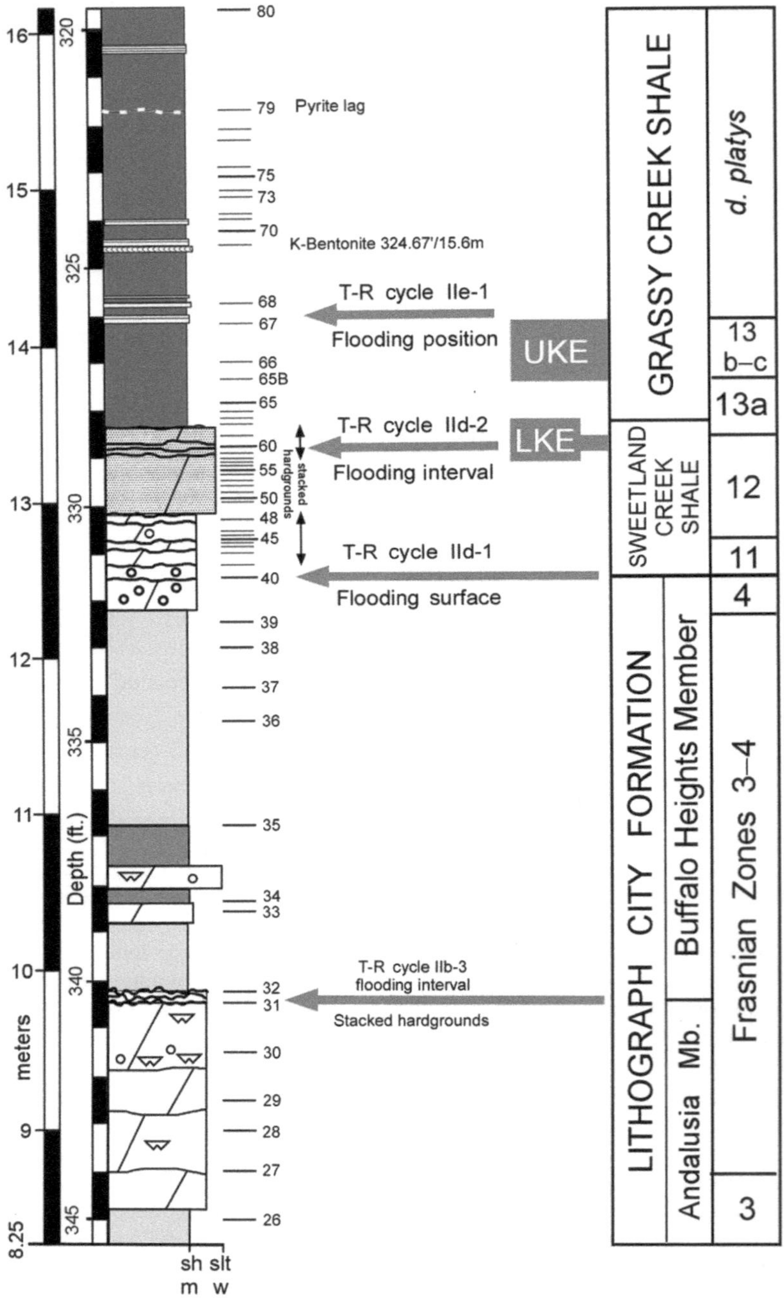

Fig. 9 Early Frasnian to Early Famennian stratigraphy, conodont correlations, positions of marine flooding surfaces or intervals, and the LKE and UKE intervals in the IGS Sullivan Slough Core, Des Moines County, Iowa. Positions of conodont sample intervals (bases marked by *labeled black horizontal lines*) to the right of the stratigraphic column (see

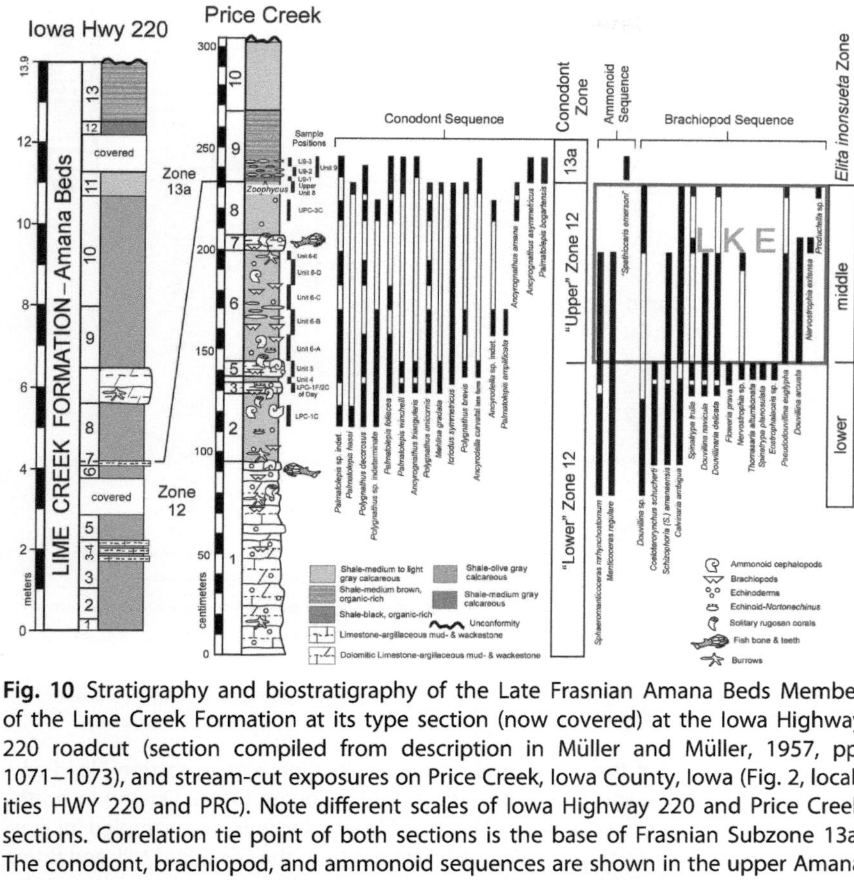

Fig. 10 Stratigraphy and biostratigraphy of the Late Frasnian Amana Beds Member of the Lime Creek Formation at its type section (now covered) at the Iowa Highway 220 roadcut (section compiled from description in Müller and Müller, 1957, pp. 1071–1073), and stream-cut exposures on Price Creek, Iowa County, Iowa (Fig. 2, localities HWY 220 and PRC). Note different scales of Iowa Highway 220 and Price Creek sections. Correlation tie point of both sections is the base of Frasnian Subzone 13a. The conodont, brachiopod, and ammonoid sequences are shown in the upper Amana Beds composite section at Price Creek are compiled from Hanson and Preslicka (1996), and Day's 1995 and 2016 field notes, brachiopod, and conodont collections. Ranges of taxa shown as *vertical columns*, with labeled *black rectangles* to the right of the column mark positions of macrofossil and conodont samples. The Lower Kellwasser Extinction interval is enclosed by a *gray rectangle* labeled LKE and identified by brachiopod and ammonoid extinctions in "upper" Zone 12 within the middle *Elita inconsueta* Zone, immediately below the onset of apparent benthic anoxia indicated by the *thin dark-gray shale* at the base of unit 9 (=base of Subzone 13a).

Table 3 for plotted conodont sequence). Numbered zones 3–4 and 11–13c are Frasnian zones or subzones of Klapper and Kirchgasser (2016). Lithologic or rock texture abbreviations and symbols as in Fig. 7. *Vertical scales* are in meters above base, or total depth in feet, of the core. Sample 28 yields mixed Zone 11 and Zone 4 taxa, and reworked Late Frasnian Zone 13 taxa are admixed with Early Famennian Middle *triangularis* Zone taxa in samples 67 and 79 as shown in Table 3.

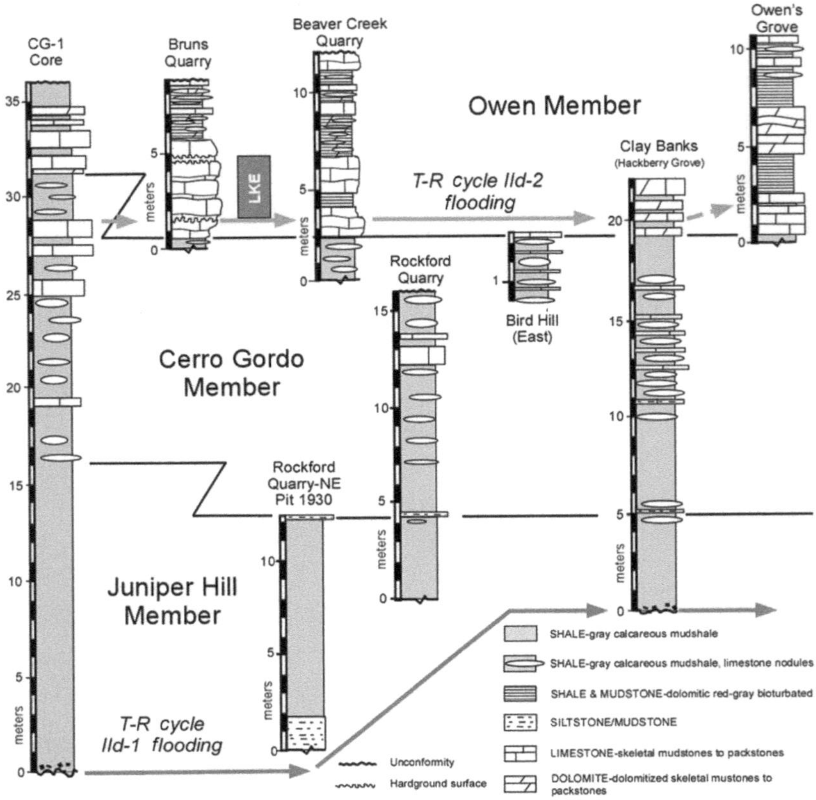

Fig. 11 Stratigraphy of the Lime Creek Formation its type area in eastern Floyd and Cerro Gordo counties, and new exposures in Butler County (Bruns Quarry) in north-central Iowa. The CG-1, Clay Banks, and Rockford Quarry sections were sampled for conodonts by Day (1990). Data from additional samples from the CG-1, Bird Hill, and Bruns localities are shown in Tables 6 and 7 reported here. Brachiopods were sampled at the Rockford Quarry by C. Belanski (unpublished samples in the Belanski Collection 1918–1930) and Day (1989, 1995, 1998, 2008, 2013; Day and Copper, 1998; Ma and Day, 2000, 2003; new samples since those reports taken in 2014 to March of 2017). See Day (1990, figs. 3, 6, 7, 8, and appendix) for sample positions and occurrence of conodonts reported at that time from the IGS CG-1 core, Rockford Quarry, and Clay Banks sections. Geochemical data from analyses of Late Frasnian conodont apatite and brachiopod calcite from specimens collected the CG-1, Rockford Quarry, Bird Hill, and Clay Banks sections reported in Joachimski et al. (2004, 2009) and Van Geldern et al. (2006). Positions of the marine flooding events that initiated Iowa Devonian T-R cycles 7A and 7B coincide with positions of Euramerican Devonian T-R cycles IId-1 and IId-2, respectively (Fig. 3). The position of the Late Frasnian Lower Kellwasser Extinction interval coincides with the T-R cycle IId-2 flooding interval (TST and lower MFI of Fig. 14) at the Bruns Quarry and IGS CG-1 Core. Locations of all section localities are shown in Fig. 2 (*BH* = Bird Hill; *BRQ* = Bruns Quarry; *CB* = Clay Banks; *CG-1* = IGS Cerro Gordo Project Hole #1 core; *OG* = Owen's Grove; *RQ* = Rockford Quarry. *Modified from fig. 14 of Day, J., 2013. Stop 5. Rockford Brick and Tile Quarry—Floyd County Fossil and Prairie Center and Park Preserve. In: Day, J. Witzke, B.J., Lundy, S. (Eds.), Aspects of the Paleozoic History of Epeiric Seas of the Iowa Basin. Iowa Geological Survey Guidebook 29, pp. 109–118.*

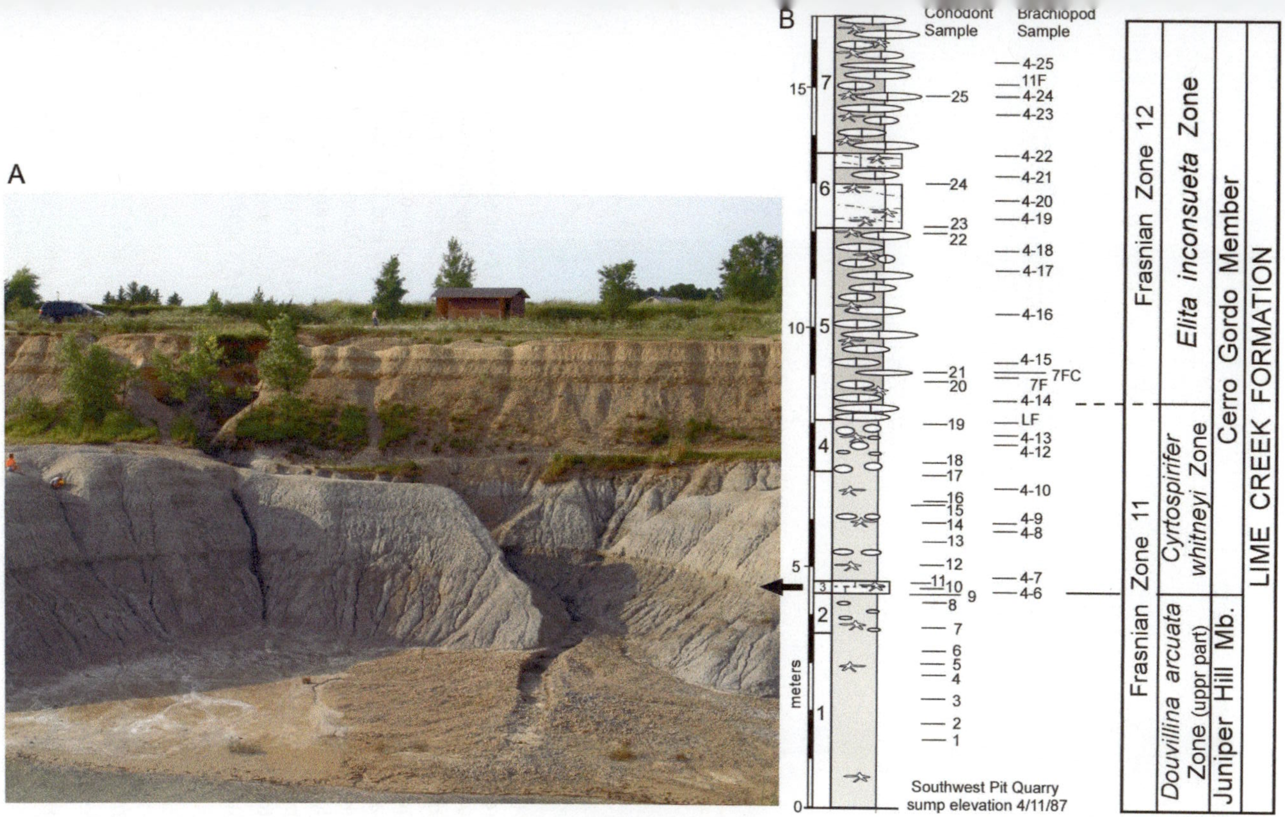

Fig. 12 Late Frasnian stratigraphy and biostratigraphy of the Lime Creek Formation at the historic Rockford Brick and Tile Quarry (Fig. 2, locality RQ). (A) Photograph of the exposure of the exceptionally fossiliferous shales and argillaceous platform carbonate strata of the middle part of the Lime Creek Formation at the Rockford Brick and Tile Quarry (Floyd County Fossil and Prairie Preserve) taken by J. Day in June of 2014 (see fig. 1 of Day, 2013). (B) Stratigraphic section modified from Day (1990, p. 617, fig. 3; adapted from fig. 2 of Day, 2013). The *arrow* on the left side of stratigraphic column identifies the contact of the Juniper Hill and Cerro Gordo Members in Fig. 12A. Conodont sequence originally reported by Day (1990) and in Table 7. Conodont zones are Frasnian Zones of Klapper and Kirchgasser (2016). Brachiopod sample positions and zonation of Day (1989). This is the stratotype section of the lower boundaries of *Cyrtospirifer whitneyi* and *Elita inconsueta* zones as defined by Day (1989).

Fig. 13 Exposures of the Upper Lime Creek Formation spanning the Lower Kellwasser Extinction interval at the Bruns Quarry in Butler County, Iowa, operated by Bruening Construction Company of Decorah, Iowa. Composite stratigraphic section of the high-wall and quarry sump exposures shown in Fig. 14. (A) Exposure of most of the Owen Member along the east high wall of the quarry sampled by J. Day and B. Witzke in June of 2014, and J. Day with students E. Menzier, K. Murphy, R. Olson, and N. Weidman in July of 2014. Additional sampling by J. Day in April of 2016 and March of 2017. (B) Exposure of the uppermost Cerro Gordo Member and its contact with the basal limestones of the Owen Member in the quarry sump pit 200 m west of the high-wall exposures shown in (A). *Photographs by J. Day, July of 2014. Locality BRQ as shown in Fig. 2.*

65–90 cm of green bedded siltstones (no conodonts reported); and 250 cm of slightly calcareous green-brown shales with a 30-cm bed of green siltstone–mudstone from 2.85 to 3.15 m. In the IGS H cores of southeastern Iowa the Sweetland Creek is comprised of medium- to light-gray calcareous shales, mudstones, rare thin skeletal lags with echinoderm plates and rare

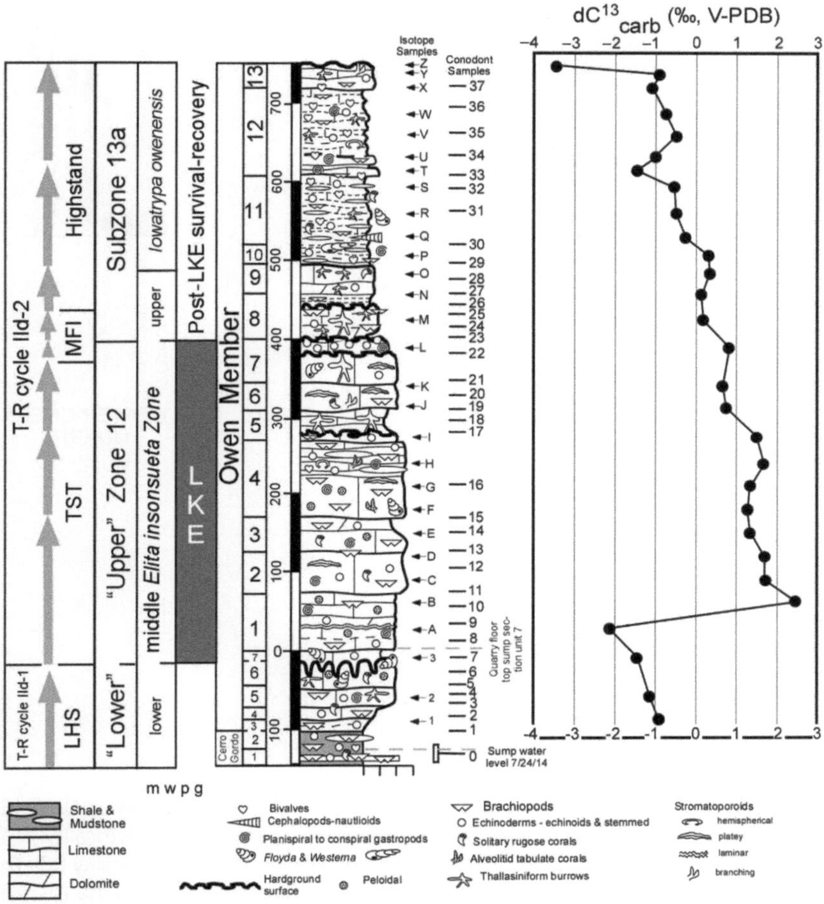

Fig. 14 Section of the Upper Lime Creek Formation at the Bruns Quarry in Butler County, Iowa (Fig. 2, locality BRQ). The Bruns Quarry composite section consists of exposures in the quarry sump pit, and high-wall ramp section described on east side of the quarry (measured and described by J. Day and B. Witzke, June and July, 2014, resampled in April of 2015, 2016, 2017). Bruns Quarry conodont sequence shown in Table 7. *Vertical gray arrows* on the left-hand side denote small-scale fourth-order shallowing-up cycles (note hardground bounding flooding surfaces) in the Owen Member. The Lower Kellwasser Extinction interval (=LKE) is identified in the uppermost part of Frasnian Zone 12 (samples 17–22) immediately below the base of Zone 13a at position of sample 23. Abbreviations: *g*=grainstone; *LHS*=Late Highstand System tract; *m*=mudstone; *MFI*=maximum flooding interval; *p*=packstone; *TST*=Transgressive System Tract; *w*=wackestone. *Vertical scale* is in cm above (base to top of high-wall section) and below (depth of sump pit exposure) quarry floor position.

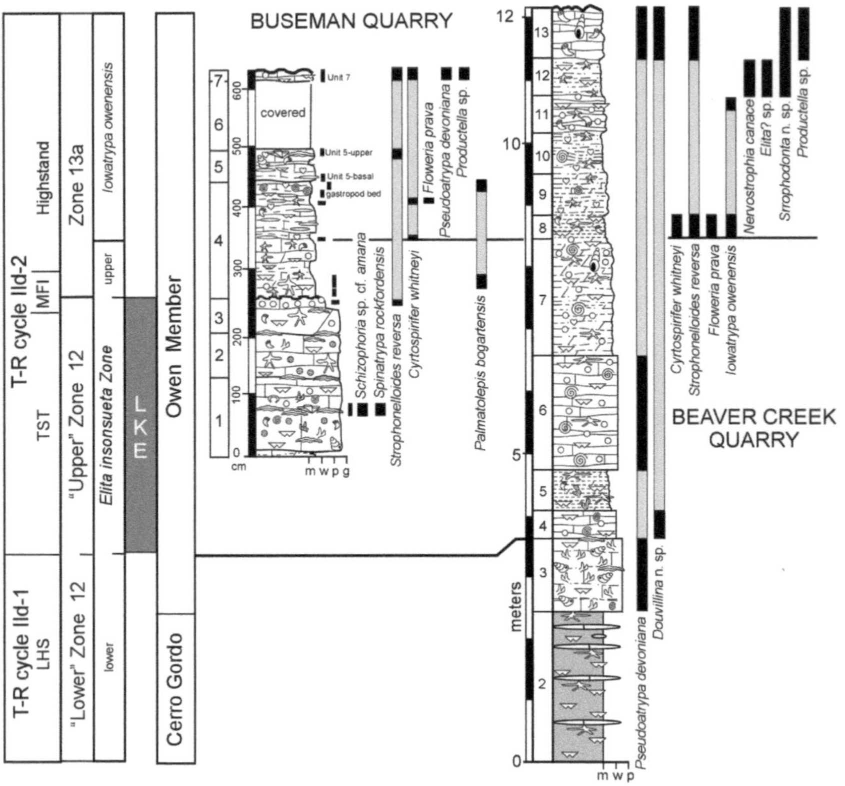

Fig. 15 Section of the Upper Lime Creek Formation exposed at the Buseman Quarry in Butler County (Fig. 2, locality BRQ), and the Beaver Creek Quarry (=Belanski locality 17, in Strimple and Leverson, 1969) in southeastern Cerro Gordo County (Fig. 2, locality BC). The upper datum is the base of the *Iowatrypa owenensis* Zone. The lower datum is the T-R cycle IId-2 flooding surface. The now flooded Buseman Quarry is located 0.8 km east of the active Bruns Quarry (Fig. 14). Abbreviations: *g*=grainstone; *LHS*=Late Highstand System tract; *m*=mudstone; *MFI*=maximum flooding interval; *p*=packstone; *TST*=Transgressive System Tract; *w*=wackestone.

brachiopods (see Fig. 7, H-29 core), thin brown shale interbeds, capped by pyritic hardgrounds, and overlain by the dark-brown organic-rich Grassy Creek Shale (Fig. 7).

Sweetland Creek macrofossils reported by Müller and Müller (1957, p. 1074) include carbonized plant fragments, fish scales and bone fragments, decalcified ammonoid aptychi, and phosphatic linguloid brachiopod shell fragments. Pyritized shell fragments of a species of *Nervostrophia* sp. occur in four of Day's samples from the lower part of the type section (Fig. 8) associated with *Pa. winchelli* and *Pa. bogartensis* (Subzone 13a). In northern Iowa, *Nervostrophia canace* occurs in a comparable position in the middle-upper Owen Member with Subzone 13a conodonts at the Bruns and Beaver Creek quarries (Figs. 13–15; Table 10).

3.2 Grassy Creek Shale

The Grassy Creek Shale is a thick shale succession (40–50 m thick) in eastern Iowa consisting of laminated dark-brown organic-rich shales and green-gray shales of uppermost Frasnian to mid-Famennian age in surface and subsurface sections in eastern and southeastern Iowa (Figs. 2 and 7–9). The bulk of the Grassy Creek succession regionally correlates with Famennian T-R cycles IIe-1 and lower IIe-2 of Johnson and Klapper (1992), although its basal interval includes deposits of upper cycle IId-2 and spans the UKE. Only the Lower Famennian part of the Grassy Creek is shown in Figs. 7–9. The Grassy Creek overlies the Sweetland Creek Shale throughout eastern and southern Iowa and accumulated under anoxic conditions in the deeper part of the Late Frasnian and Early Famennian seaway (Figs. 6 and 7). At all sections in the subsurface (Figs. 7 and 9) the Sweetland Creek–Grassy Creek contact is a burrowed pyritic–phosphatic hardground representing a starved surface that likely formed during the highstand phase following the maximum flooding interval (MFI) of Devonian T-R cycle IId-2. Where preserved, mid-Upper Famennian gray and gray-green shales of the Saverton Shale conformably overlie the Grassy Creek throughout the subsurface of southeastern Iowa (see fig. 2 of Witzke and Bunker, 2002).

Thin light- to medium-gray shales and K-bentonite clay-rich altered volcanic ash beds are interbedded with organic-rich brown shales in the lower part of the Grassy Creek Shale at the Sweetland Creek Shale type Section (Figs. 7 and 8), the Sullivan Core (Figs. 7 and 9), and the IGS H-32 core (Fig. 7). The ash bed in the type Sweetland Creek (Figs. 7 and 8) was sampled for zircons by J. Over in 1995. Samples of seven ashes from the IGS H-32 core (Fig. 7) were processed for zircons by M. Schmidt at Boise State University in 2013. Neither site yielded datable zircons.

3.3 North Liberty Beds

The North Liberty Beds consist of 1.5 m (5 ft.) of blue-gray calcareous shales overlying carbonates of the Cedar Valley Group 2.4 km northeast of the town of North Liberty in Johnson County, Iowa (Müller and Müller, 1957, pp. 1075–1076; Youngquist, 1947, p. 95; Figs. 2 and 3).

IGS geologists (B. Witzke and B. Bunker) revisited this locality in the late 1980s and sampled the underlying Cedar Valley Group carbonates. Those contain breviconic nautiloids and the brachiopod *Allanella allani*. The latter permits an assignment to the uppermost Givetian to Lower Frasnian State Quarry Member of the Lithograph City Formation (Fig. 3, *norrisi* to Zones 1–3, *A. allani* Zone of Day, 1989). Important conodonts from the North Liberty Beds identified by Youngquist (1947, pl. 24 and 25) include *Palmatolepis semichatovae, P. proversa,* and *Ancyrognathus triangularis* from the basal 30 cm. Müller and Müller (1957, pl. 142, figs. 9, 6, and 12) also illustrated *Pa. semichatovae* with *P. proversa* from their North Liberty Beds samples. As shown in Table 1, the global FAD of *Palmatolepis proversa* defines the base of Frasnian Zone 10, and the FADs of both *Ag. triangularis* and *Pa. semichatovae* are low in Zone 11 of Klapper and Kirchgasser (2016, see their table 2). Hence, the North Liberty Beds are no older than Zone 11.

3.4 Lime Creek Formation

Following the usage of the IGS, the Lime Creek Formation includes four members. Type sections of three members are located in eastern Cerro Gordo County in north-central Iowa (Figs. 2 and 11). These are, in ascending order, the Juniper Hill, Cerro Gordo, and Owen Members (Figs. 3, 5, 11, and 12). The Amana Beds Member's type section is the roadcut on the north side of Iowa Highway 220 immediately west of Middle Amana in Iowa County (Figs. 2, 3, and 10). The type section is covered, although small (2–3 m thick) exposures of the Amana Beds occur along nearby Price Creek (Fig. 10).

The Lime Creek is the surface bedrock unit in the type area in northern Iowa where the Holocene erosional surface truncates the upper part of the formation in the outcrop area in Floyd, Cerro Gordo, and Butler counties east of the Lime Creek erosional edge. The Lime Creek Formation is up to 43 m thick in Cerro Gordo and Floyd counties (Figs. 2 and 11–15). It overlies the Middle Frasnian Nora Member of the Shell Rock Formation above a pronounced erosional disconformity (Figs. 3 and 4). Where preserved, the Owen Member is overlain by Middle Famennian Shales and minor carbonates of the Sheffield Shale in Butler and Franklin counties to the south of the type area (Fig. 2), and throughout the subsurface of central and western Iowa and eastern Nebraska (see fig. 2 of Metzger, 1989). In Iowa and Johnson counties in eastern Iowa, the Amana Beds and North Liberty Beds overlie older Upper Givetian or Lower Frasnian Coralville or Lithograph City strata above a complex erosional unconformity (Fig. 3).

Northern and Eastern Iowa, and in the Offshore Facies of the Sweetland Creek and Lower Grassy Creek Shales in Eastern and Southeastern Iowa

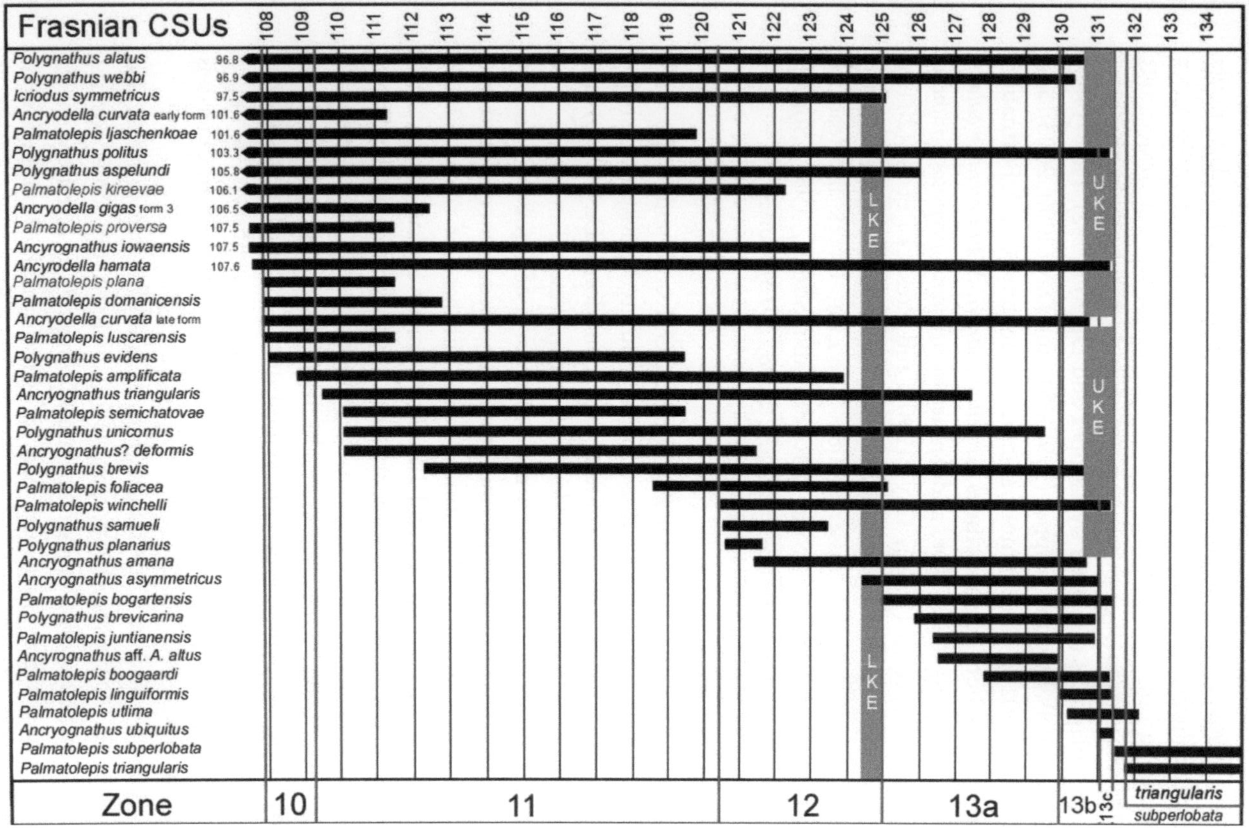

Numbered Frasnian Zones and Subzones are those of Klapper and Kirchgasser (2016), as adapted from Klapper (1989). Standard zones for the Famennian and Frasnian–Famennian boundary follow Spalletta et al. (2017). Plotted ranges are scaled according to the current Frasnian Composite Standard (CS) developed through graphic correlation of all sections used to date to compile the Frasnian CS as reported by Klapper and Kirchgasser (2016, see table 1). Species ranges are plotted based on their lowest bases and highest tops in Composite Standard Units (CSUs) of the current Frasnian CS. Lowest bases in CSU values of ranges of the first 11 taxa listed to the left of *plotted range bars*. Species ranges compiled from appendix 1 of Klapper (1997, 2007a, appendix 1; 2007b, 2009), and table 2 of Klapper and Kirchgasser (2016). Positions of the Lower (LKE) and Upper Kellwasser Extinction (LKE) intervals are based on reports by Girard et al. (2006).

The well-documented brachiopod fauna of Lime Creek Formation (including the Independence Shale) includes at least 73 species (Day, 1989, 1995, 1996a; Appendix). Most of these taxa are widespread elements of the cosmopolitan Late Frasnian fauna known from similar middle-outer shelf facies in carbonate platform and deeper subtidal deposits throughout the Late Devonian tropics of Euramerica (western North America and eastern Eurasia–Poland) as discussed by Cooper and Dutro (1982), Day (1989, 1995, 1996a, 1998), Day and Copper (1998), Ma and Day (2000, 2003), and Day et al. (2008, 2013). The ammonoid fauna of the Lime Creek Formation of eastern and northern Iowa includes *Crickites lindneri*, *Manticoceras regulare*, *Tornoceras* (*Tornoceras*) *uniangulare*, and *Sphaeromanticoceras rhynchostomum* as reported by Baker et al. (1986), Day (1990), and Preslicka et al. (2013). These are comparable to coeval species described from eastern North America in House and Kirchgasser (2008). The Lime Creek also features cosmopolitan tropical rugose coral (McLean and Sorauf, 1989; Sorauf, 1998) and stromatoporoid (Stock, 1984a,b, 2008, 2013) faunas.

3.4.1 Juniper Hill Member

The basal 10–15 cm of the Juniper Hill in the CG-1 core and at the Clay Banks section (Figs. 2 and 11) consists of an indurated phosphatic shaley lag deposit with abundant placoderm fish plates, conodonts (Day, 1990), lingulide brachiopods (*Barroisella fragila*), and green algae *Tasmanites*. The remainder of Juniper Hill (Figs. 3, 11 and 12) consists of green-gray to gray silty calcareous shale or bioturbated mudstone. The occurrence of scattered calcareous nodules or bedded siltstones marks the base of the overlying Cerro Gordo Member (Figs. 11 and 12). The Juniper Hill is 16.1 m thick in the CG-1 core and is10.5 m in thickness at its type section at Clay Banks (Fig. 11). Belanski (Belanski Register) described 11.8 m of Juniper Hill Shales in the old north pit of the Rockford Quarry (Fig. 11; see Day, 2008, stop 7, fig. 3). The entire Juniper Hill in the North Pit was mined for brick and tile production in the first half of the 20th century. The upper 4.3 m of the Juniper Hill Member shales exposed in the Rockford Quarry is shown in Fig. 12.

At its type section at Clay Banks (Figs. 2 and 11) the Juniper Hill yields conodonts (Anderson, 1966; Day, 1990) charophytes, scolecodonts, dacryonarids, foraminifers (*Nanicella* sp. and *Multiseptida* sp., see Day, 1990, fig. 10), and green algae *Tasmanites*. Macrofossils from the Juniper Hill at the Rockford Quarry (Figs. 11 and 12) reported by Webster (1908), Thomas (1922), and Belanski (Belanski Collection, University of Iowa) include the brachiopods *Douvillina* and *Nervostrophia*, carbonized logs and branches, hexactinellid sponges, and conularids. Twenty species of brachiopods and other shelly fossils were recovered from 39 samples taken in the CG-1 core section by Day (1989, fig. 10, table 5). All known brachiopod and macrofossils recovered in all 82 samples in the CG-1 core section are listed in Table 8. The Juniper Hill in the CG-1 core (Table 8) section is the type section for the *Barroisella fragila*, *Nervostrophia thomasi*, and *Douvillina arcuata* zones as defined by Day (1989, 1996a).

The basal condensed phosphatic lag deposits in the lower 2.2 m of the Juniper Hill Member (Fig. 11; Tables 7 and 8) with the lingulide brachiopod *B. fragila* appear to reflect initial anoxic–dysoxic benthic conditions in a highly condensed Transgressive System Tract (TST) interval above the basal Lime Creek Transgressive Surface (TS). A transition to fully oxygenated benthic conditions is indicated by the increase in numbers of species FADs and alpha diversity of brachiopod taxa in the overlying lower part of the Juniper Hill during the early highstand phase of Lime Creek T-R cycle IId-1 (Figs. 3 and 11; Table 8).

3.4.2 Cerro Gordo Member

The Cerro Gordo Member consists of fossiliferous calcareous shales with intervals of intensely bioturbated nodular shaley limestone and beds of argillaceous packstone and wackestones (Figs. 11 and 12). The upper 3–5 m Cerro Gordo consists of nodular calcareous shales with thin (2–5 cm) skeletal shell beds (packstone) as seen at the Bird Hill and Bruns Quarry sections (Figs. 2, 11, 13B, and 14). They are interpreted as tempestite or storm generated shell lags that accumulated as the Lime Creek shelf aggraded above local storm wave base during the highstand phase of Iowa Devonian T-R cycle 7A deposition (Figs. 2 and 3; Bird Hill section in Fig. 11). It ranges in the thickness from 10.3 m at its type section at Clay Banks, to up to 16 m in exposures at the Rockford Quarry and the nearby Bird Hill localities (Figs. 11 and 12). In Cerro Gordo and Floyd counties, it is overlain by thin to medium bedded *Amphipora*-rich packstones of the basal Owen Member (Figs. 11 and 15). In Butler County (Figs. 1, 2, 13, and 14) shales interbedded with thin nodular wackestones and skeletal packstones of the upper Cerro Gordo are conformably overlain by Owen Member skeletal wackestones with large archaeogastropods *Floyda gigantea* and *Westerna gigantea* (see Day, 1987), abundant bivalves (*Paracyclus* sp. and *Edmondia* sp.), and gastropods, with a low diversity brachiopod fauna (Table 10).

The Cerro Gordo contains a diverse and abundant invertebrate fauna described in older reports dating to James Hall's (1858) study. The Cerro Gordo brachiopod fauna consist of at least 42 species listed in Appendix and Tables 8–10. Day (1989, fig. 10, p. 319) defined the *Cyrtospirifer whitneyi* and *E. inconsueta* zones in the brachiopod sequence at the Rockford Quarry section in the Floyd County Fossil and Prairie Preserve (Figs. 11 and 12). The peak of Lime Creek faunal diversity (see brachiopod alpha diversity in Table 9) is documented in the middle and upper parts of Cerro Gordo Member reflecting continued migrations of cosmopolitan taxa into the region during the prolonged highstand interval of the Iowa Devonian T-R cycle 7A (Day, 1996a; Figs. 3 and 11, Euramerican T-R cycle IId-1). Other fossils common or abundant in the Cerro Gordo include green algae (*Tasmanites*, charophytes); calcareous benthic foraminifera (Cushman and Stainbrook, 1943; Day, 1990; Metzger, 1989); tentaculids; hexactinellid sponges; stromatoporoids (Stock, 1984a,b, 2008, 2013); tabulate corals, solitary, and colonial rugose corals (Sorauf, 1998); bryozoans (cyclostomes, cryptostomes, trepostomes); bivalves (largely undescribed); gastropods (Day, 1987, 1990); cephalopods (*M. regulare*, Baker et al., 1986; Preslicka et al., 2013); serpulid worm tubes; scolecodonts; echinoderms (crinoids, echinoids); ostracodes (Gibson, 1955); conodonts (Anderson, 1966; Metzger, 1989; Day, 1990, this report); and fish material (placoderms, sharks). The Cerro Gordo conodont and brachiopod successions are shown in Tables 6–10.

3.4.3 Owen Member

Owen Member carbonate facies overlay shales and thin-interbedded limestones of the upper Cerro Gordo Member in the type area in Cerro Gordo County and to the south in Butler County (Figs. 2, 11, and 13–15). In southeastern Cerro Gordo County, the equivalents of lower Owen carbonates seen elsewhere are Cerro Gordo Shales in the CG-1 core (Fig. 11) and, hence, the lower Owen and upper Cerro Gordo Members share regional facies relationships. Conodont correlations outlined below also demonstrate that the Amana Beds Member at Price Creek (Fig. 10) are deep-ramp equivalents of the lower-middle Owen Member. The type section at Owens Grove is largely covered (Fig. 2, locality OG; Fig. 11). Belanski (Belanski Register, University of Iowa Repository) described 10 m of Owen carbonates and shales at its type section, and 9.5 m are exposed in the Beaver Creek Quarry in eastern Cerro Gordo County (Fig. 2, locality BC and BRQ;

Figs. 11 and 15). It is 8 m thick at the Bruns Quarry in Butler County (Fig. 2, locality BRQ; Figs. 11, 13, and 14).

The Owen is characterized by fossiliferous limestone, dolomitic limestone, dolomite, interbedded with calcareous and dolomitic shales (Figs. 11 and 13–15). Oolitic and peloidal pack- and grainstones occur in the lower part of the Owen Member at a number of sections in Cerro Gordo County as described by Lynn (1978). Those facies reflect deposition of the lower Owen in a shallow subtidal setting above fair weather wave base during the late highstand and regression of Iowa T-R cycle 7A and initial transgressive phase of Iowa T-R cycle 7B (Fig. 3; Day et al., 2008, 2013).

Fenton and Fenton (1924) divided the Owen into three intervals in Cerro Gordo County. These are: (1) a basal bed with abundant digitate stromatoporoids *Amphipora* (called the "*Idiostroma* Zone"); (2) a thick interval of fossiliferous dolomitic limestones and shales ("*Floydia* Zone"); and (3) an upper interval of dolomitic limestones, dolomite, and dolomitic shales characterized by an abundance of corals and stromatoporoids (the so-called *Acervularia* [= *Hexagonaria*] Zone). Stock (1984a,b, 2008, 2013) reported a low diversity stromatoporoid fauna from the upper interval. The latter two intervals share a partial lateral facies relationship across the outcrop belt of Cerro Gordo, Franklin, and Butler counties (Figs. 2, 11, and 13–15).

The lower 10–20 cm of the Owen Member exposed at Bird Hill (Figs. 2 and 11) consists of *Amphipora* packstones with a sparse brachiopod fauna including *Pseudoatrypa devoniana* and *C. whitneyi*. At Clay Banks (Figs. 2 and 11) the Owen consists of basal *Amphipora* packstone overlain by pelleted skeletal wackestones with large archaeogastropods *W. gigantea* and *F. gigantea*. The only well-exposed sections available for study at present are in Butler County at the Bruns and Buseman quarries (Figs. 2, 11, and 13–15). Facies typical of the Fenton's upper two intervals makes up the Owen in those localities in Butler County (Figs. 13–15). Basal Owen lithologies in the Bruns Quarry sump section (Figs. 11, 13B, and 14, units 3–7) comprise a shallowing-up succession of thin-bedded argillaceous skeletal mudstone and wackestone (Fig. 14, units 3 and 4) and medium-bedded packstones with large *F. gigantea* and *W. gigantea* (see Day, 1987, 1990) and the brachiopods *P. devoniana* and *Rigauxia orestes*. This basal interval is capped by a pronounced sculpted hardground flooding surface (Fig. 14, units 5 and 6). The basal Owen package is interpreted as the edge of the shallow subtidal to supratidal inner shelf facies tract that prograded over deeper subtidal calcareous shales of the Cerro Gordo Member in the upper

Highstand System Tract (HST) interval of Iowa Devonian T-R cycle 7A (Fig. 3, Euramerican T-R cycle IId-1; Figs. 11 and 14).

The hardground capping the first Lime Creek depositional cycle is the transgressive surface of the second Lime Creek sequence package. It is overlain by three hardground-bound packages of coral and stromatoporoid-bearing skeletal mudstones, wackestones, and packstones (Figs. 13A and 14, highwall section units 1 to lower 7). These packages become progressively thinner upsection and record repeated flooding events and initial sediment starvation marked by hardground surfaces followed by accumulation in the TST of Iowa T-R cycle 7B (Figs. 3, 11, and 14). The TST interval is overlain by a hardground-bound crinoidal wackestone (Fig. 14, upper unit 7) and a dolomitic wackestone (lower unit 8) representing the MFI of the upper Lime Creek sequence. The same MFI was also observed in the nearby Buseman Quarry section in the upper 10–15 cm of unit 3 and lower part of unit 4 (Fig. 15). The remainder of the upper Owen Member (Figs. 13B and 14, Bruns Quarry units upper 8–13; Fig. 15, Buseman Quarry units 4–7; Beaver Creek Quarry units 7–13) consists of packages of very silty dolomitic and calcareous shales, mudstones, and dolomitic skeletal mudstones and wackestones interpreted as the HST of the upper Lime Creek depositional sequence.

The base of Subzone 13a in the Bruns Quarry is at the base of sample 23 (Fig. 14; Table 7), confirming that the underlying TST and MFI deposits (with Zone 12 conodonts) are within the LKE interval. Unlike the faunas seen in underlying strata, the post-LKE benthic fauna of the upper Owen is mollusk-dominated. Bivalves (*Edmondia* sp., *Paracyclus* sp., other taxa) and gastropods (*Westerna* sp. and other indeterminate high-spire, low-spire, and planispiral taxa) are common, but scattered to common brachiopods also occur (Fig. 14; Table 10; Appendix: OU = upper Owen). Solitary and colonial rugose corals (Sorauf, 1998), tabulate corals (*Alveolites* sp.), and laminar to branching stromatoporoids (see Stock, 2008, 2013) are present, generally increasing in abundance upward in the succession. The ammonoids *C. lindneri* and *S. rhynchostomum*? are reported (Preslicka et al., 2013, figs. 9 and 10) from the Owen Member from the Buseman Quarry locality (Figs. 2 and 15). In the Frasnian of New York, *C. lindneri* occurs above the Pipe Creek Shale horizon (House and Kirchgasser, 2008) above the LKE interval that is correlated with a position in Frasnian Subzone 13a (Klapper and Kirchgasser, 2016). Owen ammonoids have only been found as float specimens at the Buseman Quarry (Fig. 15), but likely would be derived from the post-LKE interval in the upper Owen Member correlated with Subzone 13a (Figs. 14 and 15).

3.4.4 Amana Beds Member

The name Amana Beds was proposed by Stainbrook (1945, p. 7) to include shales, thin-interbedded limestones, and dolomites exposed in roadcut exposures on the north side of Iowa Highway 220 immediately west of the village of Middle Amana (Figs. 2 and 10). The only published description of those exposures is that of Müller and Müller (1957, pp. 1071–1073). Float collections in the UI Paleontology Repository made by M. Stainbrook and C.H. Belanski are from the lower 3–4 m of the type locality (Day, 1995, 1998). In the early 1950s, the Highway 220 roadcut section consisted of 12–14 m of gray and dark olive gray shales with minor thin-bedded and nodular argillaceous limestones and dolomitic limestones in the lower 4–5 m, overlain by 7–8 m of dark olive gray and brown organic-rich shales (Fig. 10). Conodonts including *Ancryognathus calvini* reported in Müller and Müller (1957) indicate that the lowest Frasnian Subzone 13a fauna is in their unit 8 permitting correlation with the nearby Price Creek section (see correlation tie point in Fig. 10).

New Amana Beds exposures were discovered in stream cuts along the banks of Price Creek about 2 miles (3–3.2 km) north–northeast of the type section after the major flooding in 1993 (Fig. 2, locality PRC; Fig. 10). These outcrops yield well-preserved ammonoids (*Manticoceras* and *Sphaeromanticoceras*) in direct association with brachiopods, echinoderms (*Nortonechinus*), corals, and conodonts. Day (1995) correlated these new exposures with the *E. inconsueta* Zone of the type Lime Creek based on typical Zone 12 conodonts (sampled by G. Klapper, J. Day, and J. Preslicka in 1994) associated with the brachiopods *Calvinaria ambigua, Coeloterorhynchus schucherti*, and *Spinatrypa trulla*. All brachiopods known from the Amana Beds are listed in Appendix (AM). The brachiopod sequence in the Price Creek outcrops is summarized in Fig. 10. The Amana Beds Member in eastern Iowa is a deeper water facies of the proximal platform deposits of most of the Owen Member Lime Creek in its type area. The brachiopod assemblage in the lower part of the Member comprises an offshore biofacies characteristic of deeper parts of the ramp system. The majority of Amana Beds species are restricted to those deposits or the stratigraphic leak deposits of the Independence Shale in Buchannan County (Appendix; Figs. 3 and 10). See further comments on the Amana Beds brachiopod fauna in the discussion of the Independence Shale later.

A conodont sample collected by J. Preslicka from the nodular brown shale interval of unit 9 of the Price Creek section yields Subzone 13a conodonts (Fig. 10, sample unit 9, lowest *Pa. bogartensis*) confirming the hypothesis of Preslicka et al. (2013, p. 67; see Fig. 4) that the upper part of the Price Creek section spanned the LKE interval. Four additional samples collected through unit 9 in 2016 confirm a Subzone 13a correlation and constrain the underlying strata high within "upper" Zone 12 (discussed below) and the LKE interval as shown in Fig. 10.

3.5 Independence Shale

As now restricted, the Independence Shale includes karst fillings (stratigraphic leaks) of Lime Creek or Sweetland Creek Shale hosted in older Cedar Valley Group, Wapsipinicon Group, or underlying Silurian carbonates discovered in various locations in eastern Iowa (Figs. 4 and 9). Lime Creek stratigraphic leaks are common in karst and fracture fills over a wide geographic area of eastern Iowa and western Illinois. Wherever samples have been processed, these green-gray and gray shale fills have yielded well-preserved Lime Creek-Sweetland Creek conodont faunas. Independence Shale leaks with brachiopods, corals, and other macroinvertebrates only occur in the type area near the towns of Independence and Brandon in central and southwestern Buchanan County (Fig. 2, localities BR and BrQ). Independence Shale leaks in the type area with macrofauna are contained within large, cavern-like karst openings in older Devonian limestone strata, but none of those original exposures are available for modern investigation. A number of brachiopod species first described by Stainbrook (1945) from Independence Shale have been identified in the Juniper Hill and Cerro Gordo Members of the Lime Creek by Day (1989, 1995; see Appendix), although most are restricted to the Buchannan County Independence Shale leaks and the Amana Beds Member in Iowa County (Appendix).

Nearly all Independence Shale species have been redescribed in reports by Cooper and Dutro (1982), Copper and Chen (1995), Day and Copper (1998), and Ma and Day (2003). Conodonts associated with brachiopods in Independence deposits in Buchannan County (Müller and Müller, 1957, pp. 1073–1074) include *Pa. winchelli* with *Pa. foliacea*, indicating a correlation with Zone 12 of Klapper and Kirchgasser (2016), and correlate at a position within the lower to middle *E. inconsueta* Zone of Day (1989). A majority of species in the Independence Shale in the type area, and the Amana Beds assemblage, comprise a deeper water Lime Creek brachiopod biofacies (Appendix; Figs. 3 and 10). These include the orthides *Schizophoria (S.) amanaensis, Skenidium independense*; the strophomenides *Pseudodouvillina euglyphea, Douvillinaria perversa, Nervostrophia extensa*; the productoid *Leptolosia inexpectans*; the terebratulide *Cranaena amana*; the atrypide *S. trulla*; the spiriferides *Acutatheca propria*, *Cyrtospirifer placitus*; and the distinctive rhynchonellides *Hypothyridina emmonsi*, *C. schucherti*, and *C. ambigua*. Species common to faunas of the Independence and Lime Creek Formation of northern Iowa (Day, 1989, 1995) include *Stainbrookia infera*, *D. arcuata*, *N. thomasi*, *Strophonelloides reversa*, *Sulcatostrophia camerata*, *Productella rugatula*, *P. thomasi*, *Retichonetes brandonensis*, *Floweria prava*, *Thomasaria altumbonata*, *Spinatrypa planosulcata*, and *Atryparia* (*Costatrypa*) *varicostata* (Appendix; see Tables 8 and 9).

Other conodont-bearing leak deposits with Late Frasnian offshore species typical of the Lime Creek Formation and Sweetland Creek Shale occur at numerous localities in eastern Iowa and western Illinois in karst and fracture fills within underlying Devonian and Silurian strata (Witzke and Bunker, 2006). Several of these localities are identified in Fig. 2 including the IGS IPSCO cores in Muscatine County (locality, IPS), the Sullivan Core section in Des Moines County (locality SUL, Figs. 7 and 9), and the Allied Quarry in Rock Island County in western Illinois (locality ALQ).

4. CONODONT FAUNA AND BIOSTRATIGRAPHY

The sequence of conodont species in the offshore sections of the Sweetland Creek and Grassy Creek Shale (see Figs. 3 and 7–9) is shown in Tables 2–4. The sequence of conodont taxa in subtidal platform facies of the Lime Creek Formation (see Figs. 3 and 10–15) is shown in Tables 6–8 and Figs. 10 and 15. Correlations shown therein are with the current Frasnian zonation of Klapper and Kirchgasser (2016) and the Famennian standard zonation of Spalletta et al. (2017).

The Frasnian Stage is currently subdivided into 12 zones and 3 subzones based on the 13-fold Frasnian zonation first proposed by Klapper (1989). Klapper's (1989) sections in the Montagne Noire region served as the initial Standard Reference Section for graphic correlation of sections throughout the Frasnian tropics where it has been widely duplicated (Klapper and Kirchgasser, 2016, see cited studies on p. 8). Girard et al. (2006, pp. 189–190) subdivided Frasnian Zone 13 of Klapper (1989) into three subzones. Subzones 13b and 13c coincide with the lower and upper parts of the *linguiformis* Zone as proposed by Sandberg et al. (2002). Klapper and Kirchgasser (2016) suggested that the use of Montagne Noire Frasnian zones of Klapper (1989) is referred to as F.Z. 1–12 and Subzones 13a–13c. The Late Frasnian spans the interval of most of Zone 11 to the base of the

Table 2 Late Frasnian and Early Fammenian Conodont Sequence and Biostratigraphy of the Type Section of the Sweetland Creek Shale and Lower Grassy Creek Shale on the North Bank of Sweetland Creek, Muscatine County, Iowa

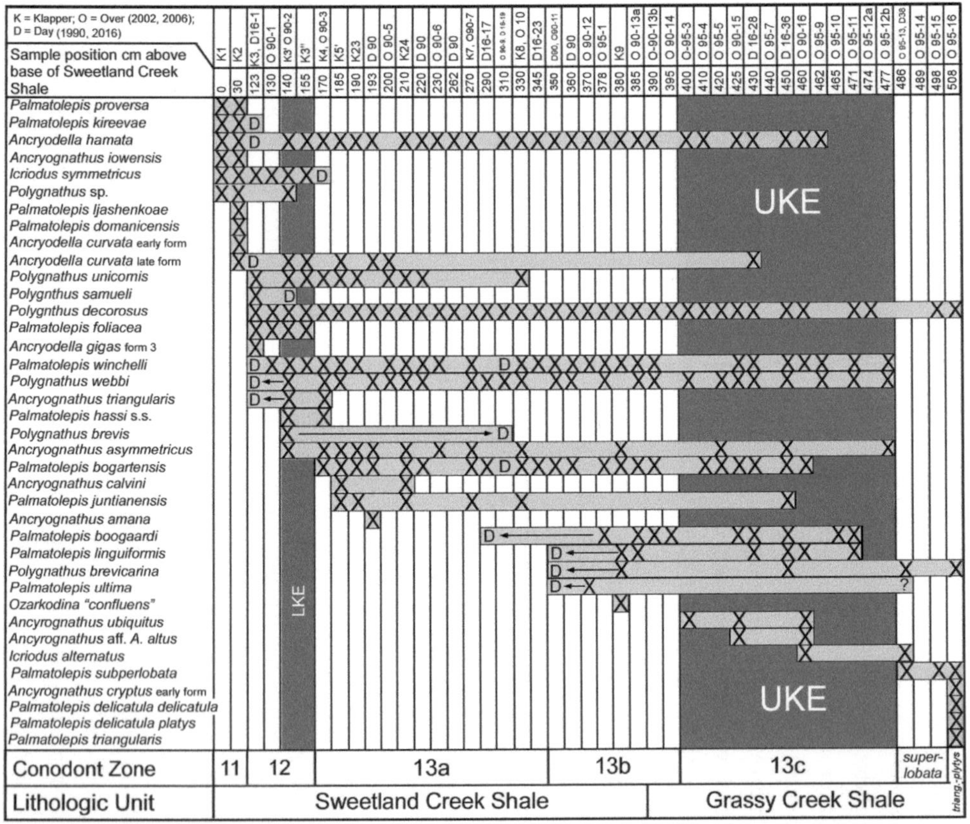

Conodont sample composite from published sample data of Klapper (in Johnson and Klapper, 1992, p. 132, fig. 2), Over (2002, 159, fig. 6b; 2006, p. 77, fig. 2, p. 78, table 1), and J. Day's unpublished samples taken with J. Over, W. Kirchgasser, and G. Klapper in 1990, and 40 new continuous samples collected with A. Wallenberg in 2016. Sample identified by collector: K = Klapper (in Johnson and Klapper, 1992); O = Over's samples collected in 1990 (=O90) and 1995 (=O95); D = Day's (1990) (=D90) and 2016 samples (=D16). Conodont occurrences in samples marked by *Xs*, and *gray horizontal rows* specify local ranges in the section. Sample occurrences marked by (D) are new local range extensions or occurrences for *Ancryognathus asymmetricus*, *Ag. triangularis*, *Icriodus symmetricus*, *Polygnathus samueli*, *P. brevicarina*, *P. brevis*, *P. webbi*, *Palmatolepis boogardi*, *Pa. linguiformis*, and *Pa. ultima*. Frasnian Zones after Klapper and Kirchgasser (2016). Occurrences of *Pa. triangularis?* identified in the report by Over (2006, table 1) are suspected occurrences of *Pa. ultima*. Famennian standard zones after Spalletta

Table 5 Upper Devonian (Early Frasnian–Early Famennian) Conodont Sequence in the IGS Sullivan Slough Core, Des Moines County

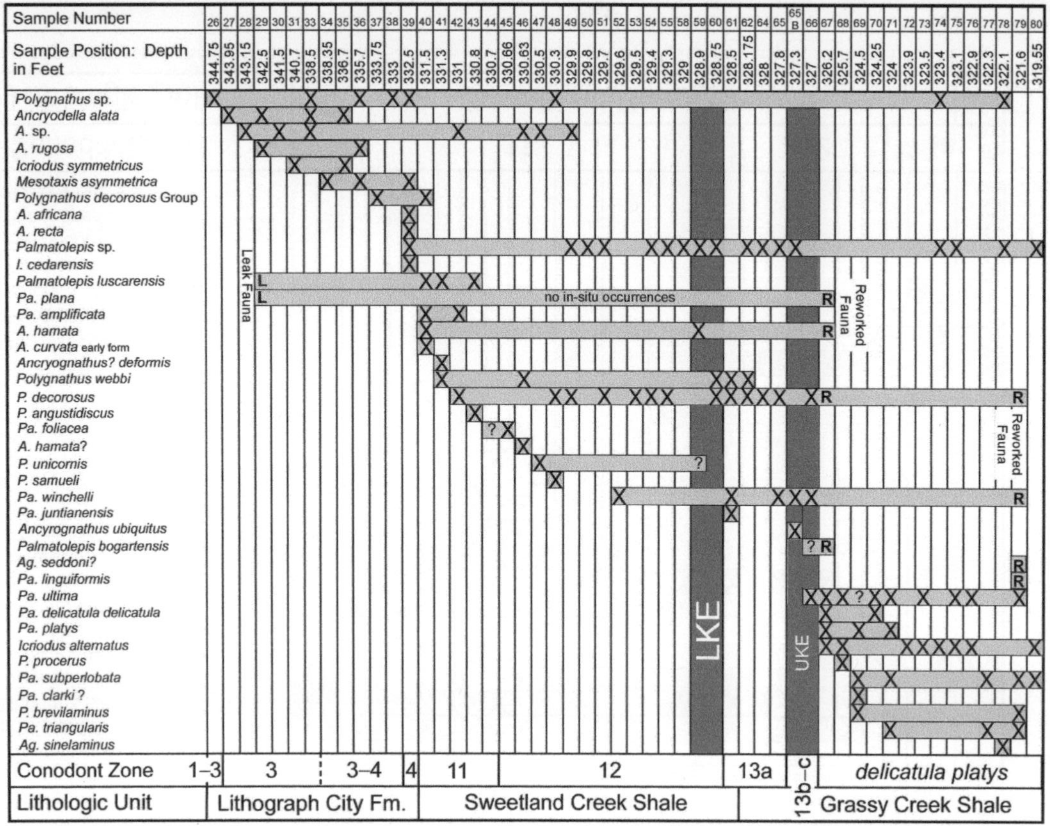

Note the condensed Late Frasnian conodont sequence in the Sweetland Creek Shale and Lower Grassy Creek Shale, with Early Frasnian Zones 3–4 faunas in the Lithograph City Formation separated by disconformity spanning Middle Frasnian Zones 5–10. Sample 29 contains admixed Late Frasnian Zone 11 (occurrences labeled as L) and Zones 3–4 taxa beneath lowest in situ Zone 11 fauna in sample 40 representing an Independence Shale stratigraphic leak. Reworked Zone 13 taxa (occurrences labeled as R) identified in the Lower Grassy Creek Shale samples 67 and 79 admixed with *Palmatolepis delicatula platys* Zone (see Spalletta et al., 2017) taxa identify submarine disconformities at the base or within those sample intervals. Positions of samples shown in Fig. 9, and core location shown in Fig. 2. Lower and Upper Kellwasser Extinction intervals indicated by *vertical dark-gray columns* labeled LKE and UKE, respectively as in Tables 1 and 2. The UKE interval truncated by the submarine erosional surface disconformity at the base of sample position 67. Sampled by J. Day in 2009 with H. Rowe, S. Rimmer, and S. De LaRue, resampled by J. Day in 2011. Conodonts identified by G. Klapper in 2009 and 2010, and J. Day.

Table 4 Frasnian Conodont Sequence in the Sweetland Creek Shale and Lower Grassy Creek Shale (=G.C.) in the IGS H-30 Core in the Subsurface of Lee County, Iowa

Sample Number	1	2	3	4	5	6	7	8	9	10	11	12	13	14	15	16	17	18	19	20	21	22	23	24	25	26	27	28	29	30
Cm above base of Sweetland Creek Shale	−5 to +10	10–30	50–70	70–90	90–112	112–132	132–150	150–170	190–215	235–255	275–295	295–315	315–335	335–355	355–375	375–395	415–435	435–455	455–475	475–495	495–515	515–535	535–555	555–575	575–595	615–635	635–655	655–668	676–685	685–694
Palmatolepis amplificata	X						X																							
Pa. kireevae	X	X				X	X	X		X	X		X	X				X						X						
Icriodus symmetricus	X	X				X	X	X	X										X	X										
Ancryodella hamata	X	X					X			X																				X
Polygnathus decorosus	X	X				X	X	X		X	X		X												X			X		X
Ancryognathus triangularis	X																													
Polygnathus sp.			X		X																		X							
Palmatolepis sp.				X								X			X		X		X	X	X	X	X							
P. unicornis						X			X					X																
Pa. semichatovae								X		X																				
Pa. sp. cf. *Pa. domanicensis*								X		X																				
Ozarkodina sp.								X																						
Mehlina gradata									X			X																		
Ancryognathus iowensis										X																				
P. brevis																X														X
Ag. sp.																X														
P. webbi																				X								X	X	X
Pa. winchelli																									X			X		X
Pa. bogartensis																										X				X
Icriodus alternatus																											X			X
Pa. linguiformis																														X
Pa. boogaardi																														X
P. brevicarina?																														X
Ag. asymmetricus																														X
Conodont Zone	11																			12						13a				13b
Lithologic Unit	Sweetland Creek Shale																													G.C.

Numbered Frasnian conodonts zones and subzones as in Tables 1–3. Core location shown in Fig. 2, and stratigraphic section shown in Fig. 7.

Palmatolepis subperlobata Zone and begins with the eustatic *semichatovae* marine transgression (Fig. 3) that initiated Devonian T-R cycle IId (Becker, in Gradstein et al., 2012, p. 520).

The main Late Frasnian and Early Famennian reference section in eastern and southeastern Iowa is the type section of the Sweetland Creek Shale. That succession consists of condensed record of Frasnian Zones 11 to Subzone 13c, the LKE and UKE intervals, and the Lower Famennian *Pa. subperlobata* to *Palmatolepis delicatula platys* Zone (Figs. 2, 3, 7, and 8; Table 2). The Platform sections of the Lime Creek Formation span the interval of Zones 11 to Subzone 13a (Figs. 3 and 11–15; Tables 5–7).

Klapper (2007a, 2009, p. 406) redefined the boundary of the basal Famennian Lower *triangularis* Zone and revised the closely related very Late Frasnian and Early Famennian species *Pa. ultima* Ziegler (1958) and *Palmatolepis triangularis* Sanneman (1955). In his revision, the FAD of *Palmatolepis superlobata* defined the lower zonal boundary of his revised lower *triangularis* Zone. Spalletta et al. (2017) subdivided the lower *triangularis* Zone (as revised by Klapper, 2007a), into a lower *P. superlobata* and overlying *Pa. triangularis* Zone whose lower boundaries are defined by the FADs of their nominal species (Table 1). Spalletta et al. (2017, pp. 34–35) renamed the Middle *triangularis* Zone after the species whose global FAD defines its base (now the *Pa. delicatula platys* Zone).

Global ranges of cosmopolitan Frasnian conodonts species in terms of Composite Standard Units (CSUs) of the Frasnian Composite Standard (CS) are well known (see Klapper and Kirchgasser, 2016, tables 1 and 2). Ranges of important Frasnian and Early Famennian conodont taxa (in terms of Frasnian CSUs) known from the Iowa Frasnian and Lower Famennian strata are plotted in Table 1. The durations of the LKE and UKE intervals are approximated in Tables 1–4 given that their lower boundaries are not precisely calibrated in terms of Frasnian CSUs in the reports by Girard et al. (2006), Klapper (2007a, 2009), and Klapper and Kirchgasser (2016). Authorship and current synonymies for Frasnian and Early Famennian conodont species listed in Tables 1–7 can be found in the reports by Klapper et al. (2004), Girard et al. (2006), Klapper (2007a,b), and Klapper and Kirchgasser (2016).

Klapper and Kirchgasser (2016) revised a number of widespread species of the Frasnian genus *Ancyrodella* Ulrich and Bassler (1926). Klapper and Kirchgasser (2016, pp. 13–15) reassigned *A. lobata* Branson and Mehl (1934), *A. buckeyensis* Stauffer (1938), and *A. nodosa* (as restricted in their synonymy, p. 15) as synonyms of *A. hamata* Ulrich and Bassler (1926).

Table 5 Late Frasnian Conodont Sequence in the Lime Creek Formation at the type section of the Juniper Hill and Cerro Gordo Members at the Cerro Gordo County Clay Banks Nature Preserve (Hackberry Grove of the Pre-1980s literature) in Cerro Gordo County in North-Central Iowa

Sample Number	1	2	3	4	5	6	7	8	9	10	11	12	13	14	15	16	17	18	19	20	21	22	23	24	25	26	27	28	29
Meters above base Lime Creek Formation	0.0	0.50	0.75	0.90	1.25	3.50	3.70	4.30	4.70	5.06	5.70	6.40	7.20	7.70	8.20	8.70	9.20	9.70	10.00	10.35	10.55	11.10	11.95	12.70	13.50	14.20	15.20	16.65	17.5
Palmatolepis kireevae	X					X												X											
Icriodus symmetricus	X		X		X																								
Ancryodella curvata early form	X				X	X																							
Ancyrognathus? aff. *A. deformis*	X							X																					
Polygnathus aff. *P. evidens*	X	X	X	X	X		X	X																					
I. "subterminus"						X	X	X			X	X		X	X	X													
Pa. semichatovae									X	X	X	X	X	X	X	X	X	X	X	X									
P. alatus									X					X				X	X					X	X	X	X		X
P. unicornis									X		X	X					X	X		X	X			X		X	X		X
P. evidens											X		X	X		X	X	X	X	X	X	X				X			
Ag.? deformis												X						X		X	X					X			
A. curvata late form													X		X			X					X						
P. brevis													X						X						X	X	X	X	X
P. politus													X						X										X
P. webbi															X													X	
A. hamata																X		X											
A. curvata																X	X		X	X	X								
Conodont Zone	11																									12			
Lithostratigraphy	Juniper Hill Member																		Cerro Gordo Member										

Section location shown in Fig. 2, and section shown in Fig. 11. Abbreviations: *A.* = *Ancyrodella*; *Ag.* = *Ancryognathus*; *I.* = *Icriodus*; *Pa.* = *Palmatolepis*; *P.* = *Polygnathus*. Modified from fig. 6 of Day, J., 1990. The Upper Devonian (Frasnian) conodont sequence in the Lime Creek Formation of north-central Iowa and comparison with Lime Creek Ammonoid, brachiopod, foraminifer, and gastropod sequences. J. Paleontol. 64, 614–628.

Table 6 Late Frasnian Conodont Sequence in the Lime Creek Formation in the Iowa Geological Survey Cerro Gordo Project #1 (CG-1) Core Section From Southeastern Cerro Gordo County, North-Central Iowa (Fig. 2, locality CG-1; Fig. 11, section CG-1)

Sample Number	1	2	3	4	5	6	7	8	9	10	11	12	13	14	15	16	17	18	19	20	21	22	23	24	25	26	27	28	29	30	31	32
Sample Position: meters above base Lime Creek Fm.	0	0.35	0.53	1.06	1.37	1.67	1.98	2.28	2.44	2.59	2.90	3.05	3.35	3.5	3.96	4.11	4.39	4.72	5.03	5.34	5.67	5.95	6.1	6.55	6.86	7.17	7.47	7.78	7.95	8.08	8.39	8.84
Icriodus symmetricus	X		X		X				X			X																				
Palmatolepis sp.	X			X		X				X	X		X																			
Polygnathus sp.	X	X	X	X	X																			X	X				X			
Pa. kireevae			X				X	X		X																				X		
P. unicornis					X		X								X		X		X	X			X					X			X	
Pa. semichatovae					X		X	X				X			X		X	X	X	X		X										
Ancryodella sp.								X				X																				
P. sp. cf. *P. evidens*								X																								
I. sp.								X																							X	
P. evidens									X	X	X	X	X		X		X			X	X	X	X	X		X	X			X	X	X
P. alatus												X			X					X												
P. aspelundi												X							X													
P. brevis													X								X							X				
I. "subterminus"													X																			
Ancyrognathus triangularis														X																		
P. sp. cf. *P. politus*																		X														
Ag. sp.																				X												
A. curvata late form																							X									
P. sp. cf. *P. webbi*																							X									
P. sp. cf. *P. equalis*																											X					
A. hamata																														X		
P. politus																																
Mehlina gradata																																
Ag.? deformis																																
A. gigas																																
P. planarius																																
P. webbi																																
P. samueli																																
Ag. asymmetricus																																
Icriodus alternatus																																
Conodont Zone																																
Lithologic Unit	Juniper Hill Member																															

(Continued)

Day (1990) reported *Ancryodella nodosa* and *A. lobata* in the Lime Creek Formation (see p. 619, fig. 6, samples 16 and 17; p. 620, fig. 7, sample 19; appendix, pp. 627–628). Klapper (in Johnson and Klapper, 1992) reported the same taxa from the Sweetland Creek Shale (p. 132, fig. 2, samples 1 and 2, 3′ to 7, 8 and 9). Over (2006, p. 77, table 1, occurrences therein) reported *A. nodosa* and *A. buckeyensis* from the type section of the Sweetland Creek Shale. Ziegler et al. (2000, p. 612) designated *Polygnathus pacificus* Savage and Funai (1980) as a junior synonym of *Polygnathus politus* Ovnatanova (1969). Reported occurrences of *P. pacificus* from the Lime Creek Formation by Day (1990, p. 619, fig. 6, samples 13, 19, and 29; p. 620, fig. 7, samples 10, 17, and 18) and Metzger (1989, p. 508, fig. 10, samples 31–41 and 43; p. 520, fig. 15.9) are reassigned to *P. politus* as shown in Tables 5–7.

33	34	35	36	37	38	39	40	41	42	43	44	45	46	47	48	49	50	51	52	53	54	55	56	57	58	59	60	61	63	64	65	66	67	68	69	70	71	72	73
9.18	9.3	9.6	9.91	10.37	11.28	11.59	11.9	12.05	12.35	12.65	12.73	13.02	13.26	13.57	13.88	14.18	14.49	14.79	15.1	15.4	15.95	16.0	16.1	16.32	16.62	16.93	17.23	17.54	18.15	18.45	18.76	19.06	19.37	19.67	19.98	20.28	20.65	20.98	21.26
					X			X	X																														
			X																																				
															X		X	X		X												X							
X															X		X																						
					X		X	X						X				X					X	X	X	X			X							X			
																					X									X									
X	X	X	X	X	X	X		X	X			X	X	X		X	X	X	X		X		X		X		X	X		X	X	X					X		
															X										X	X				X									
										X																X													
																													X										
								X									X		X																				
		X			X																																		
									X	X			X																										
													X																				X					X	
																			X	X			X	X					X				X				X		X
																							X		X														

11

(Continued)

The type section of the Sweetland Creek Shale (Figs. 2, 3, 7, and 8; Table 2) is the principal reference section for the Late Frasnian, Frasnian–Famennian Boundary, and Early Famennian interval in the Iowa Basin. Previous taxonomic and biostratigraphic investigations of the Sweetland Creek conodont fauna include Miller and Youngquist (1947); Youngquist and Miller (1948); Müller and Müller (1957); Klapper and Furnish (1963); Klapper (1990); Klapper, in Johnson and Klapper (1992); Klapper and Foster (1986, 1993); and Over (2002, 2006). Klapper (in Johnson and Klapper, 1992, fig. 2) correlated the Sweetland Creek conodont sequence with Montagne Noire zones 9–13 of Klapper (1989). The faunas in their lowest two samples correlated with Zones 9 and 10 we reassign to Zone 11 as discussed below (see Fig. 8; Table 2). Over (2006, fig. 2) identified

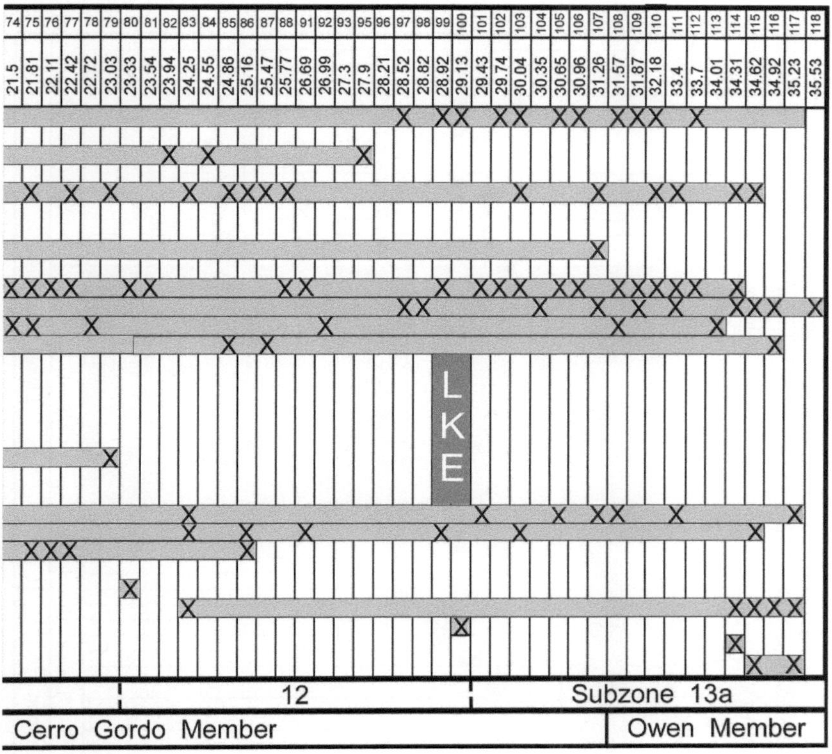

Day (1990, fig. 8) published data from 25 CG-1 core samples. The position of the LKE interval shown as LKE placed at the base of climbing limb of the Lower Kellwasser carbon-isotope excursion identified in the CG-1 core δC^{13}_{carb} isotope stratigraphy by Day et al. (2015a,b, 2016). The base of Subzone 13a is immediately above the occurrence of *Polygnathus samueli* (sample 100) based on its restricted range in F.Z. 12 as shown in Table 1. Abbreviations of genera as in Table 2, *Mb.* = Member.

the bases of Montagne Noire Subzones 13a, 13b, 13c and the lower and middle *triangularis* zones. We reposition the base of Frasnian Subzone 13b in the type Sweetland Creek based on a lower FAD of *Palmatolepis linguiformis* in our samples as shown in Table 2. All data from the subsurface Late Frasnian strata in IGS Sullivan and H-30 cores (Figs. 7 and 9; Tables 3 and 4) are new.

4.1 Revisions to Iowa Frasnian Conodont Species Ranges

New data on conodont occurrences in the type section of Sweetland Creek Shale (Figs. 2, 7, and 8; Table 2) supplement those from reports by Klapper (in Klapper and Johnson, p. 132, fig. 2) and Over (2006, pp. 76–77, fig. 2,

Table 7 Late Frasnian Conodont Composite Sequence in the Lime Creek Formation in Floyd (Rockford Quarry and Bird Hill) and Bulter (Bruns Quarry) Counties in North-Central Iowa (Fig. 2, localities RQ, BH, BRQ)

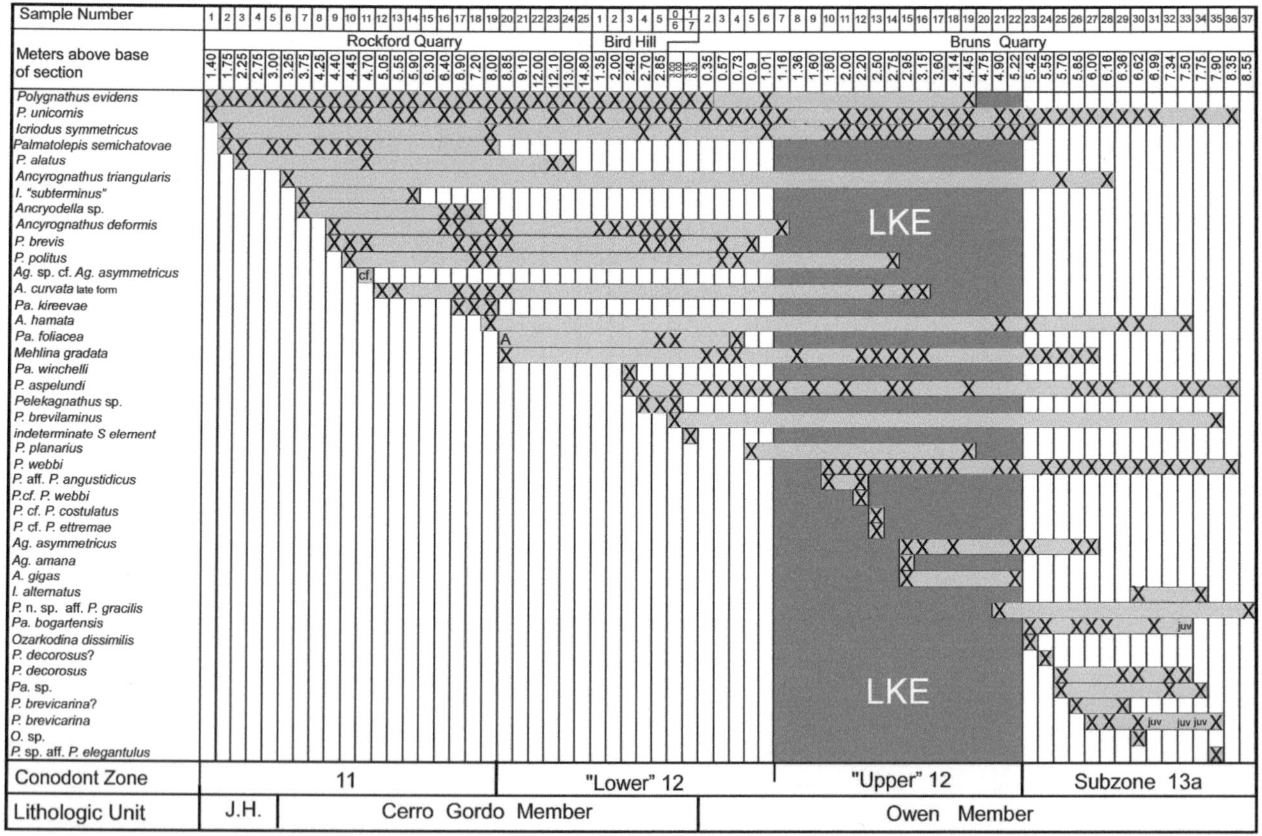

The Lower Kellwasser Event Interval (LKE) coincides with onset and peak values of Lower Kellwasser δC^{13}_{carb} excursion below the base of Frasnian Zone 13a in the lower-middle part of the Owen Member. Stratigraphic sections for the Rockford Quarry and Bird Hill localities shown in Figs. 11 and 12. The Bruns Quarry composite section is shown in Figs. 13 and 14. Abbreviations of genera as in Table 2, *juv.* = juvenile; *O.* = *Ozarkodina*.

table 1). These new data allow for local ranges extensions of: *Palmatolepis boogardi*, *Pa. linguiformis*, *Pa. ultima*, *Pa. winchelli*, *Polygnathus brevicarina*, *Polygnathus brevis*, *P. samueli*, and *Icriodus symmetricus* as shown in Table 2. Eighty new samples from the IGS CG-1 core, seven from Bird Hill, and 38 from the Bruns Quarry in (Figs. 2, 11, 13, and 14) include important new occurrences of offshore conodont species in the upper Cerro Gordo and Owen Members of the Lime Creek Formation in northern Iowa. These include *Ancryognathus asymmetricus*, *Ag. amana*, *Icriodus alternatus*, *Pa. bogartensis*, *Pa. foliacea*, *Pa. winchelli*, *P. brevicarina*, *P. planarius*, and *P. samueli*. Prior to this investigation only *Ag. asymmetricus* and *Pa. foliacea* were reported by Anderson (1966) and Day (1990) from the Lime Creek Formation of northern Iowa.

The occurrence of *I. symmetricus* at in our sample at 170 cm in the type Sweetland Creek Shale (Table 2) moves its local Last Appearance Datum (LAD) from the uppermost part of Frasnian Zone 12 through the LKE interval into the lowest part of Frasnian Subzone 13a, consistent with its global LAD in the Frasnian CS (Table 1). *P. samueli* in the type Sweetland Creek sample at 140 cm moves its LAD upward to base of the LKE interval (Table 2). As shown in Table 1, its global LAD plots below that position at CSU 123.5 as cited by Klapper and Kirchgasser (2016, p. 7, table 1). As revised, the range of *P. samueli* would span most of Zone 12. The local LAD of *P. brevis* in the type Sweetland Creek Shale is moved upward from high in Zone 12 into the lower part of Subzone 13a (Table 2, sample at 310 cm). In the IGS H-30 Core *P. brevis* ranges into Subzone 13b where it occurs in a diverse assemblage with *Pa. linguiformis* in sample 30 (Fig. 7; Table 4). The global LAD of *P. brevis* is very high in Subzone 13b and is extinct during the onset of the UKE crisis interval in the upper part of Zone 13b (Table 1). In eastern Iowa, the first occurrence of *P. brevicarina* is lowered to 350 cm within the upper Sweetland Creek Shale where it occurs with *Pa. linguiformis* and *Pa. ultima* at the base of Subzone 13b (Fig. 8; Table 2, sample at 350 cm). Over (2006, table 1, samples 95-13 and 90-19) reported *P. brevicarina* ranging into the Lower Famennian (lower part of the *P. delicatula platys* Zone = Middle *triangularis* Zone). Those occurrences are problematic in that *P. brevicarina* is not thought to range into the Lower Famennian based on species ranges outlined in fig. 2 of Spalletta et al. (2017). In the Bruns Quarry in northern Iowa, *P. brevicarina* occurs in the upper Owen Member of the Lime Creek in Subzone 13a with its FAD above that of *Pa. bogartensis* (Figs. 13 and 14; Table 7).

The local FAD of *Pa. boogardi* in the Sweetland Creek Shale is lowered from 378 cm (Subzone 13b) to 290 cm to a position within the mid-upper part of Subzone 13a (Fig. 8; Table 2) and is consistent with its known global range in the Frasnian CS (Table 1). The local FAD of *Pa. ultima* is lowered to 350 cm with respect to Over's reported FAD at 370 cm (Table 2). The new occurrence of *Pa. linguiformis* in our sample D90 at 350 cm lowers the base of Subzone 13b to that position from the higher level in sample 095–1 at 378 cm reported by Over (2002, 2006, table 1). New occurrences of *Polygnathus webbi* and *Pa. winchelli* in sample D16-1 at 123 cm place their local FADs in the lowest Zone 12 position in the type Sweetland Creek Shale (Table 2).

The occurrence of *Palmatolepis amplificata* in unit 6B in the Amana Beds at the Price Creek section in Iowa County (Figs. 2 and 10) is within the LKE interval in an "upper" Zone 12 position close to the lower boundary of Frasnian Subzone 13a. The LAD of *Pa. amplificata* plots at CSU 123.9 in the Frasnian CS (Table 1; Klapper and Kirchgasser, 2016, table 2). The range of this species may require upward extension based on where the base of the LKE interval falls precisely in the Frasnian CS.

P. planarius is reported here from the Frasnian of Iowa for the first time. We identify it in sample 80 from the Cerro Gordo Member in the CG-1 core section (Fig. 11; Table 6) in Cerro Gordo County, and samples 5 and 19 (Table 7) from the Owen Member at the Bruns Quarry in Dumont County. Its occurrence in the CG-1 core appears to be low in Zone 12 (Table 6), whereas the two occurrences in the Owen Member are just below and within the LKE interval in the upper part of Zone 12 (Table 7). The present range in the Frasnian CS is restricted to the lower part of Zone 12 (Table 1; Klapper and Kirchgasser, 2016, table 2). Its local LAD within the LKE interval in northern Iowa indicates that its global LAD be adjusted to a position within upper Zone 12 (Table 1).

4.2 Frasnian Zone 11

The global FAD of *Palmatolepis feisti* Klapper (2007a,b) defines the base of Frasnian Zone 11, with its upper boundary marked by the global FAD of *Pa. winchelli* (Klapper and Kirchgasser, 2016, table 1). In the absence of *Pa. feisti*, the global FADs of *Ag. triangularis*, *Ag.? deformis*, *Pa. semichatovae*, and *Polygnathus unicornis* characterize the lower part of Zone 11 (Table 1). *Pa. feisti* has not been recovered in the Upper Frasnian strata in Iowa.

As mentioned earlier, the basal 30 cm of gray shales of the North Liberty Beds, overlying Lower Frasnian Lithograph City Formation carbonates, in Johnson County in eastern Iowa yields *Pa. semichatovae*, *Pa. proversa*, and *Ag. triangularis* (see Youngquist, 1947, pl. 24 and 25). The FADs of both *Ag. triangularis* and *Pa. semichatovae* are low in Zone 11 (Table 1).

The basal skeletal lag bed of the Sweetland Creek Shale immediately above the basal hardground in the IGS H-30 core in Lee County (Figs. 2 and 7; Table 4, sample 1) contains *Ag. triangularis*, with longer-ranging *A. hamata*, *Polygnathus decorosus*, *Pa. amplificata*, and *Palmatolepis kireevae*. *Ag. triangularis* places the base of the Sweetland Creek in the IGS H-30 core in the lower part of Zone 11 (Table 1). *Pa. semichatovae* enters the sequence in sample 8 (Table 4). The upper boundary of Zone 11/base of Zone 12 in the IGS H-30 core is marked by the lowest *Pa. winchelli* in sample 20 (Table 4).

Klapper's (in Johnson and Klapper, 1992, table 1) lowest two samples at 0 and 30 cm above the base of the type Sweetland Creek Shale do not feature characteristic Zone 11 taxa and were correlated with Montagne Noire Frasnian zones 9–10 in their report. Klapper reported *I. symmetricus*, *A. hamata*, *Pa. proversa*, *Ancryognathus iowensis*, and *Pa. kireevae* in his sample 1 in the basal 30 cm (Table 2, =sample K1). The local FADs of *Ancryodella curvata* early form, *Pa. ljashenkoae*, *Pa. domanicensis*, and *A. curvata* late form are in the overlying sample (Table 2, sample K2) below the 73 cm thick siltstone (Fig. 8). The global FADs of *Pa. proversa*, *Pa. domanicensis*, and *A. curvata* late form define and characterize the base of Zone 10 constraining the maximum age of the type Sweetland Creek Shale. All species recovered in the lower two samples are long ranging taxa with their global FADs in older Middle Frasnian or even Upper Givetian strata (Tables 1 and 2). All of those species ranges overlap in the interval of Zone 11 in the Frasnian CS. The global LADs of *A. curvata* early form, *Pa. domanicensis*, and *Pa. proversa* are all in the lower part of Zone 11 (Table 1).

The basal conodont-rich skeletal lag of the Sweetland Creek Shale in the IGS Sullivan core in Des Moines County (Figs. 2, 7, and 9; Table 3, sample 40) includes *Palmatolepis luscarensis* and *Pa. plana*, both with global FADs at the base of Zone 10, and *Pa. amplificata* with its global FAD high in Zone 10 (Table 1). Five centimeter above, sample 41 yields *Ag.*? *deformis* correlated with Zone 11 (Tables 1 and 3).

Day (1990, figs. 3, 6, and 8) reported conodonts from basal Lime Creek Formation in northern Iowa at its type section at Clay Banks in the IGS CG-1 core (Figs. 2 and 11; Tables 5 and 6, respectively). The fauna in the lowest sample (basal skeletal lag deposit) at Clay Banks (Table 5, sample 1) includes *I. symmetricus*, *Pa. kireevae*, *A. curvata* early form, *Ancyrognathus*? aff. *Ag.*? *deformis*, and *Polygnathus* sp. aff. *Polygnathus evidens*. Nearly all of those are long ranging forms, except for *Ag.*? sp. aff. *Ag.*? *deformis* which is an endemic taxon and presumed ancestor of *Ag.*? *deformis*, The latter has its global FAD low in Zone 11. *Pa. semichatovae* and *P. unicornis* have their local FADs in sample 9 at 470 cm above the base of the Juniper Hill at Hackberry Grove (Table 5).

I. symmetricus, *Palmatolepis* sp., and *Polygnathus* sp. make up the lowest Lime Creek fauna in the IGS CG-1 core (Table 6, sample 1), followed by local FADs of *Pa. kireevae* at 53 cm, and the Zone 11 indicators *Pa. semichatovae* and *P. unicornis* at 137 cm above the base of the Lime Creek (Table 6, samples 3 and 5, respectively). CG-1 core samples 1–79 (Table 6) document an expanded Zone 11 interval, as is the case for the surface sections at Clay Banks and the Rockford Quarry (Figs. 11, 12, and 16).

The collective conodont data from the basal Sweetland Creek, North Liberty Beds, and lower Lime Creek Formation above the regional karstified paleotopographic surface on the Cedar Valley Group carbonate platform strata (Figs. 4–11) indicate correlation of those units low in Zone 11 (Fig. 3). The marine flooding that initiated Late Frasnian sedimentation in Iowa coincides to the *semichatovae* transgression of Sandberg et al. (2003) otherwise known as Euramerican Devonian T-R cycle IId of Johnson et al. (1985, 1989) and Johnson and Klapper (1992) or T-R cycle IId-1 of Day (1998).

4.3 Frasnian Zone 12

The lower boundary of Frasnian Zone 12 is defined by the FAD of *Pa. winchelli* and its upper boundary is at the FAD of *Pa. bogartensis* (Klapper, 1989; Klapper and Kirchgasser, 2016; Table 1). The global Fads of *P. planarius*, *P. samueli*, and *Ag. amana* are all in the lower part of Zone 12 (Table 1). The Lower Kellwasser Extinction interval is high in Zone 12 (Girard et al., 2006; Table 1).

For the purposes of further discussion below, we informally divide Frasnian Zone 12 into "lower" and "upper" intervals. "Lower" Zone 12 spans the interval from the lower boundary at the FAD of *Pa. winchelli* and spans most of the zonal interval as currently understood in the Frasnian CS. As used here, the "upper" Zone 12 interval begins with the onset of marine flooding of Devonian T-R cycle IId-2 and LKE interval as seen in the Amana Beds Member in Iowa County in eastern Iowa (Figs. 2 and

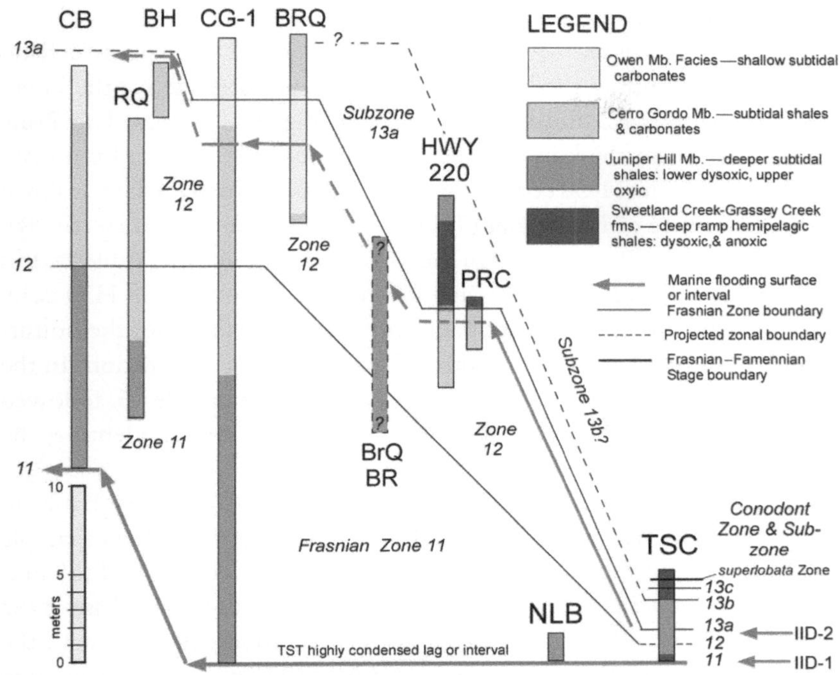

Fig. 16 Chronostratigrahic cross-section of the Late Frasnian strata in the Iowa Basin (to scale in m) showing expanded intervals of Frasnian Zones 11, 12, and Subzone 13a (in part) in the carbonate platform sections of northern Iowa, and stratigraphic condensation into the deeper part of the ramp–shelf system in southeastern Iowa. Platform sections at localities CB, BH, CG-1, BRQ shown in Figs. 11–15, deep-ramp (BrQ and BR leaks, HWY 220, PRC), and basinal sections (NLB, TSC) shown in Figs. 7, 8, and 10. Positions of Frasnian zonal boundaries in the type Sweetland Creek Shale section (locality TSC, see Figs. 3, 7, and 8) and Table 2. Positions of Frasnian zonal boundaries in all other sections as outlined in this report. Basal datum is the base of Frasnian Zone 11 as marked by the *gray arrow* coinciding to the Euramerican Devonian T-R cycle IId-1 flooding surface. *Dark-gray arrows* are Euramerican Devonian T-R cycles IId-1 and IId-2 flooding surfaces or intervals.

10), and the lower Owen Member in sections in Cerro Gordo and Dumont counties in northern Iowa (Figs. 2 and 13–15). At the Bruns Quarry (Figs. 13 and 14), this interval coincides with the onset of the 4%**0** Lower Kellwasser $\delta^{13}C_{carb}$ isotope excursion that occurs within the lower part of the Owen Member in the interval of the middle *E. inconsueta* Zone (Fig. 14; Table 9). The local FAD of *Ag. asymmetricus* also occurs within this interval in northern Iowa (Fig. 14; Table 7, conodont sample 15) and its global FAD plots at the approximated base of the LKE interval as shown in Table 1.

4.3.1 "Lower" Zone 12

The first *Pa. winchelli* with *Pa. foliacea* and *P. samueli* in the type Sweetland Creek section identifies the base of Zone 12 immediately above the lower siltstone in samples K3 and D16-1 at 123 cm (Fig. 8; Table 2). The Zone 12 interval is 47 cm thick in that section (Fig. 8; Table 2). The local FAD of *Pa. winchelli* in the H-30 core is high in the Upper Sweetland Creek above the last occurrence of *Pa. kireevae* (Table 4, samples 24 and 25, respectively). The global LAD of *Pa. kireevae* is in the lower part of Zone 12 (Table 1). It is likely that the base of Zone 12 is at a lower position in the IGS H30 core. Poor conodont recovery in many samples probably accounts for the unusually high position of the first definitive Zone 12 taxa in that section. In the IGS Sullivan core the first appearance of *Pa. foliacea* in sample 45, followed by the local FAD of *P. samueli* in sample 47 (Fig. 9; Table 3), identifies the lower part of Zone 12 in that section.

The base of Zone 12 in the Lime Creek type section at Clay Banks in Cerro Gordo County (Figs. 2 and 11; Table 5) is likely at the base of sample 26 at 14.3 m. This placement is based on the lowest occurrence of the brachiopod *E. inconsueta* (see Day, 1995, p. 35, fig. 8) identifying the base of the *E. inconsueta* Zone. The lower boundary of that zone coincides with the lower part of Zone 12 at the nearby Rockford Quarry section in western Floyd County (Figs. 2, 11, and 12; Table 8).

The occurrence of *P. planarius* in sample 80 marks the lowest Zone 12 fauna at 23.33 m above the base of the Lime Creek in the CG-1 core in the subsurface of Cerro Gordo County (Fig. 11; Table 5). The range of *P. planarius* plots in the lower part of Zone 12 in the Frasnian CS (Table 1). The occurrence of *P. samueli* in CG-1 core sample 100 identifies the upper part of Zone 12 (Table 5). That species is restricted to Zone 12 (Table 1).

At the Rockford Quarry section (Figs. 11 and 12; Table 7) the base of Zone 12 has been placed at the lowest occurrence of *Pa. foliacea* in the Corro Gordo Member based on the report by Anderson (1966). Klapper and Kirchgasser (2016, table 2) now plot the global FAD of *Pa. foliacea* high in Zone 11 (Table 1). In the Rockford Quarry section (Figs. 11 and 12; Table 7) *Pa. foliacea* enters the succession in sample 20 immediately above the highest occurrences of *Pa. kireevae* and *Pa. semichatovae* in sample 19 (Table 7). *Pa. semichatovae* has its global LAD high in Zone 11, and *Pa. kireevae* in the lower part of Zone 12 (Table 1). We correlate sample 19 with a position high in Zone 11 based on the co-occurrence of those two species within the highest part of the range of *Pa. semichatovae*

Table 8 Late Frasnian Benthic Invertebrate (Primarily Brachiopod) Sequence in the Lime Creek Formation in the IGS Cerro Gordo Project #1 (CG-1) Core

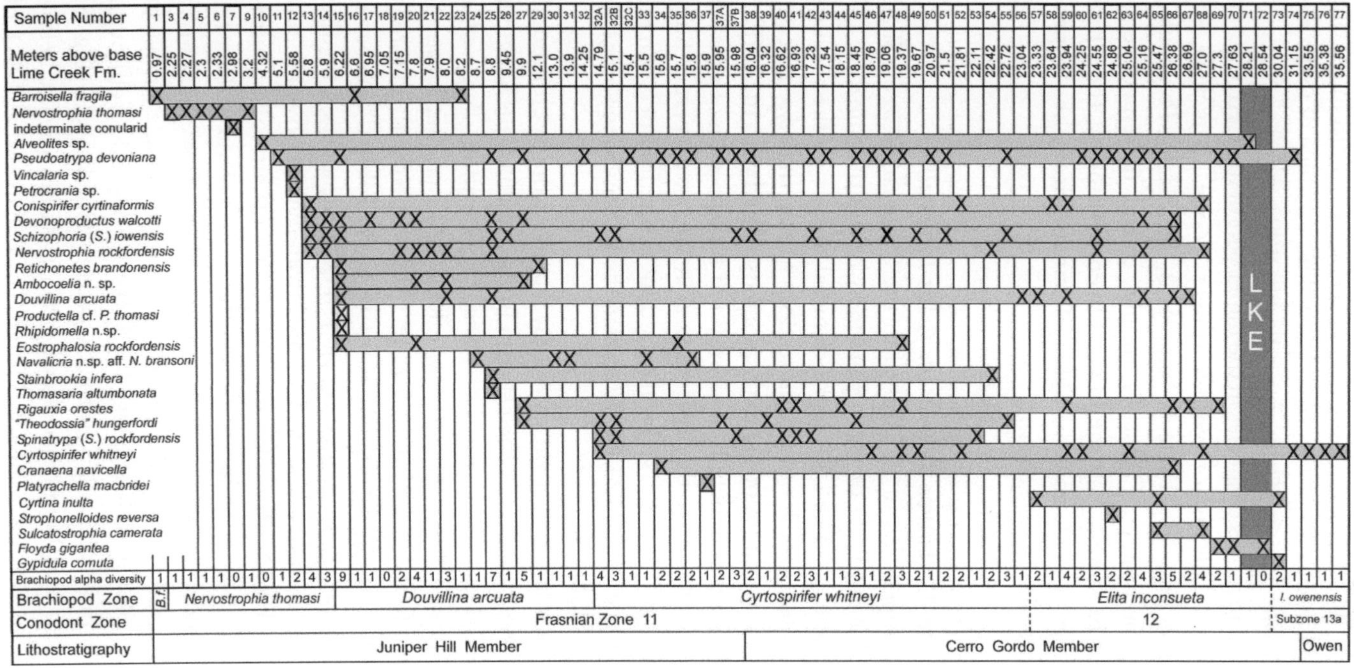

New data from resampling of the CG-1 core are 51 additional samples numbered 32–77. Position of the LKE interval approximated from conodont sequence shown in Table 6. Core location shown in Fig. 2 (locality CG-1) and lithostratigraphic section shown in Fig. 11.

Adapted and modified from fig. 9 of Day, J., 1989. The brachiopod succession of the late Givetian-Frasnian of Iowa. In: McMillan, N.J., Embry, A.F., Glass, D.J., (Eds.), Devonian of the World: Proceedings of the Second International Symposium on the Devonian System. Canadian Society of Petroleum Geologists Memoir 14(3), pp. 303–325.

(Table 7). Prior to the range extension of *Pa. foliacea* downward into Zone 11 by Klapper and Kirchgasser (2016, table 1), the known range of that form was restricted to Zone 12 (see Klapper and Foster, 1993, fig. 2).

Notable new occurrences of species in the upper part of the Cerro Gordo Member in the Bird Hill section within "lower" Zone 12 include *Pa. foliacea*, *Pa. winchelli* (Fig. 11; Table 7, samples 3, 5, and 6), *Polygnathus aspelundi*, and *P. brevilaminus* (Table 7, samples 3 and 6, respectively). *P. planarius* occurs in sample 5 in the lower Owen Member at the Bruns Quarry section (Figs. 11, 13, and 14; Table 7), ranging into the "upper" Zone 12 LKE interval as noted above.

4.3.2 "Upper" Zone 12 (Lower Kellwasser Crisis Interval)

The base of the "upper" Zone 12 interval in the Sweetland Creek Shale type section starts at the FAD of *Ag. asymmetricus* below the Subzone 13a position (Fig. 8; Table 2). The LKE had a minimal impact on the offshore *Palmatolepis*-dominated conodont fauna in sections in eastern and southeastern Iowa. The only notable conodont extinctions associated with the LKE in the Sweetland Creek section include *P. samueli* and *I. symmetricus*. The latter has its LAD immediately above the LKE interval in the lowest part of Subzone 13a (Figs. 7 and 8; Table 2). No local conodont extinctions occur in the highly condensed upper Zone 12 LKE interval in the IGS Sullivan Core section (Figs. 7 and 9; Table 3).

The LKE had a local impact on conodont faunas in the deep subtidal mid-ramp part of the Lime Creek shelf system in the Amana Beds Member of the Lime Creek (Fig. 10). There *Ag. amana*, *I. symmetricus*, *Polygnathus brevis*, *Po. unicornus*, and *Mehlina gradata* have local LADs at that location. Most or all of those forms range into Subzone 13a in the Sweetland Creek Shale in Muscatine County in eastern Iowa (Fig. 8; Table 2). What is remarkable at Price Creek is the total extinction of the associated benthic shelly fauna (Fig. 10). In equivalent strata of the lower-middle Owen Member of the Lime Creek at the Bruns Quarry section in northern Iowa a number of conodont taxa have their local LADs in the LKE interval in "upper" Zone 12 (Figs. 13 and 14; Table 7). These include *Ag.? deformis*, *A. curvata* late form, *P. politus*, and *P. planarius*. *Ag. amana* occurs in a single sample in the LKE interval in Northern Iowa (Fig. 14; Table 7, sample 15), although its regional LAD is in Subzone 13a in the Sweetland Creek Shale (Fig. 8; Table 2, sample D90 at 193 cm). Its global LAD is within or just below the UKE interval in the Frasnian CS (Table 1). The regional LAD of *Ag.? deformis* in the lower Owen Member permits extension of its range into the LKE interval within “upper” Zone 12 in northern Iowa (Fig. 14; Table 7, sample 7). Overall, the LKE appears to have had a larger impact on species populations in the deeper mid-ramp and shallow subtidal carbonate platform settings in the Iowa Basin.

4.4 Frasnian Subzone 13a

The FAD of *Palmatolepis bogartensis* (Table 1) defines the lower boundary of Frasnian Subzone 13a (Klapper, 1989; Klapper and Kirchgasser, 2016). The upper boundary of this zone is at the FAD of *Pa. linguiformis* (Table 1). Additional taxa with FADs in the lower part or within Subzone 13a include *Pa. boogardi*, *Pa. juntianensis*, and *P. brevicarina* (Table 1). *Ag. calvini* (not shown in Table 1) is also restricted to the lower part of Subzone 13a based on its known range in the type Sweetland Creek Section (Fig. 8; Table 2, Klapper's samples K5′ and K24). It was reported by Müller and Müller (1957) from the Amana Beds under the name *Ancyroides princeps* Miller and Youngquist (1947) from bed 8 of the type section (Fig. 10). Ziegler (1981, p. 8) synonymized both *Ancryognathus princeps* and *Ag. uddeni* under *Ag. calvini* as discussed in the report by Klapper (1990, pp. 1006 and 1010).

Regionally the base of Subzone 13a constrains the position of the Lower Kellwasser Extinction bioevent interval in “upper” Zone 12 (Figs. 3, 10, 14, and 15). In the Sweetland Creek Shale the base of Subzone 13a occurs at 170 cm in the type section (Table 2, samples K4 and O90-3), followed by the entry of *Ag. calvini* and *Pa. juntianensis* in the next sample at 185 cm (Table 2, sample K5). *Pa. boogardi* enters the succession at 290 cm in the upper part of the subzone (Table 2, samples D16-17). In Des Moines County, the base of Subzone 13a is at the position of sample 60 based on the local FAD of *Pa. juntianensis* in the IGS Sullivan Core (Figs. 2, 7, and 9; Table 3).

The base of Subzone 13a is at lowest sample from unit 9 in the Price Creek Section of the Amana Beds Member of the Lime Creek Formation in Iowa County in eastern Iowa (Figs. 2 and 10, sample U9-1). As mentioned earlier, the lowest *Ag. calvini* is reported from bed 8 in the Highway 222 roadcut exposures (Müller and Müller, 1957, p. 1072). Beds 8–13 of the type Amana Beds Member (Fig. 10) represent an expanded Subzone 13a record in the mid-ramp setting of the Lime Creek Shelf (Fig. 16).

In the IGS CG-1 core (Figs. 2 and 11; Table 6) the base of Subzone 13a is provisionally placed immediately above the highest occurrence of *P. samueli*. That occurrence appears to be in the upper part of Zone 12, based on its range in the Sweetland Creek Shale in Muscatine Iowa (Fig. 8; Table 2) and its global range in the Frasnian CS shown in Table 1. Above the *P. samueli*, *Ag. asymmetricus* appears followed by *I. alternatus* in samples 114 and 115, respectively. All other Frasnian occurrences of *I. alternatus* in Iowa are within Subzones 13a and 13b in the IGS H30 core in Lee County (Fig. 7, Table 4, samples 27 and 30), and within Subzone 13c and the *Pa. superlobata* Zone in the type Sweetland Creek Shale in Muscatine County, Iowa (Table 2).

Equivalent strata in northern Iowa include platform carbonates and dolomitic shales and mudstones in the upper half of the Owen Member (Figs. 3, 11, and 13–15). In the Bruns Quarry in Dumont County (Figs. 13 and 14, tables 7 and 10) samples document the local FADs of *Pa. bogartensis* with *Ozarkodina dissimilis* in sample 23 at 5.42 m, followed by the local FAD of *P. brevicarina* in sample 27 (Table 7). *Pa. bogartensis* first appears at a similar position in the nearby Buseman Quarry (Fig. 15, just above base of unit 4). *P. decorosus* also has its local FAD in sample 25 at the Bruns Quarry (Table 7). That species ranges through most of the offshore section in the type Sweetland Creek Shale (Figs. 7 and 8; Table 2) and is a typical element of the *Palmatolepis* biofacies fauna. *Ag. asymmetricus* ranges into and has its local LAD the lower part of the Subzone 13a (Table 7, samples 23, 26, and 27). These remarkable occurrences reflect a return of deeper water conditions and expansion of the offshore conodont fauna typical of the *Palmatolepis* biofacies onto the Lime Creek platform edge during the sea-level highstand phase of Devonian T-R cycle IId-2 (Figs. 3 and 14).

4.5 Frasnian Subzones 13b and 13c (Upper 13b–13c UKE Interval)

The base of Subzone 13b is at the FAD of *Pa. linguiformis* (Tables 1 and 2). The lower boundary of Subzone 13c is at the FAD of *Ancryognathus ubiquitus* (Tables 1 and 2). The global record of the UKE is marked by the onset of extinctions of most typical Frasnian conodont taxa within the upper part of Subzone 13b, extending through the LKE crisis interval of Subzone 13c (Table 1). All significant Frasnian species extinctions occurred within the LKE crisis interval of Subzone 13c (Fig. 8; Table 2) in the Iowa Basin.

4.6 *Pa. superlobata* Zone (Frasnian–Famennian Stage Boundary: Post-UKE Interval)

The base of the Famennian is the lower boundary of the *Pa. subperlobata* Zone in the global standard conodont zonation of the Famennian Stage (Spalletta et al., 2017). This boundary is exactly the same proposed by

Klapper (2007b) in his revision to the definition of the lower *triangularis* Zone as suggested in the report by Klapper et al. (2004, pp. 375–377) and later adopted by the International Commission on Stratigraphy-Subcommission on Devonian Stratigraphy (see Becker et al., 2012, p. 570). The FAD of *Pa. subperlobata* coincides with the flood occurrence of *Pa. ultima* in association with the nominal species immediately above the top of the UKE interval (Table 1).

At the type section of Sweetland Creek Shale the base of the *subperlobata* Zone is clearly identified by the FAD of *Pa. subperlobata* in Over's (2006) sample O95-13 at 486 cm above the base of the of the Sweetland Creek Shale (Figs. 2, 7, and 8). Hynes et al. (2014) reported *Pa. superlobata* in the Lower Grassy Creek Shale immediately below the first volcanic ash bed in the H-32 core from the subsurface in Lee County (Figs. 2 and 7).

5. BRACHIOPOD BIOSTRATIGRAPHY AND LKE RECORD

Day (1989) defined five of six Late Frasnian Lime Creek brachiopod zones (Fig. 3). Day (1996a) revised brachiopod identifications and local species ranges and proposed the "*Lingula*" *fragila* Zone to span the interval of lower Juniper Hill Member below the base of the *N. thomasi* Zone. The six brachiopod zonal intervals (taxon range or assemblage zones) are in the ascending order: the *B. fragila*, *N. thomasi*, *D. arcuata*, *C. whitneyi*, *E. inconsueta*, and *I. owenensis* zones (Fig. 3). The original basis for these is the brachiopod succession documented in key surface sections and the CG-1 core section in the type area of the Lime Creek in Cerro Gordo and Floyd counties (Figs. 2, 11, and 12; Tables 8 and 9). New data reported here are from the Amana Beds Member outcrop in Iowa County (Figs. 2 and 10), the IGS CG-1 core and Bird Hill sections in Cerro Gordo County (Fig. 11, tables 8 and 9), and the Bruns and Buseman quarries in Butler County (Figs. 13–15; Table 10). Appendix lists all species and authorships of Late Frasnian brachiopod taxa from northern and eastern Iowa.

Integrated brachiopod and conodont data now identify the LKE interval within the "middle" *E. inconsueta* Zone in eastern and northern Iowa (Figs. 10–15; Tables 8–10). Three informal subzone intervals are discussed later that include lower *E. inconsueta* Zone = interval from the base of the zone to the base of the LKE interval; middle *inconsueta* Zone = the LKE interval (crisis interval); and upper *inconsueta* Zone = post-LKE survivor interval from the top of the LKE interval to the base of the *I. owenensis* Zone

Table 9 Late Frasnian Brachiopod Sequence in Cerro Gordo and Owen Members of the Lime Creek Formation at the Rockford Quarry (Belanski Locality 4), Bird Hill, and Owen's Grove (Belanski Locality 9, Owen Mb. Type Section) Sections in Floyd and Cerro Gordo Counties, North-Central Iowa (see Day, 1989, fig. 10; 1995, fig. 6; 2008, 2013)

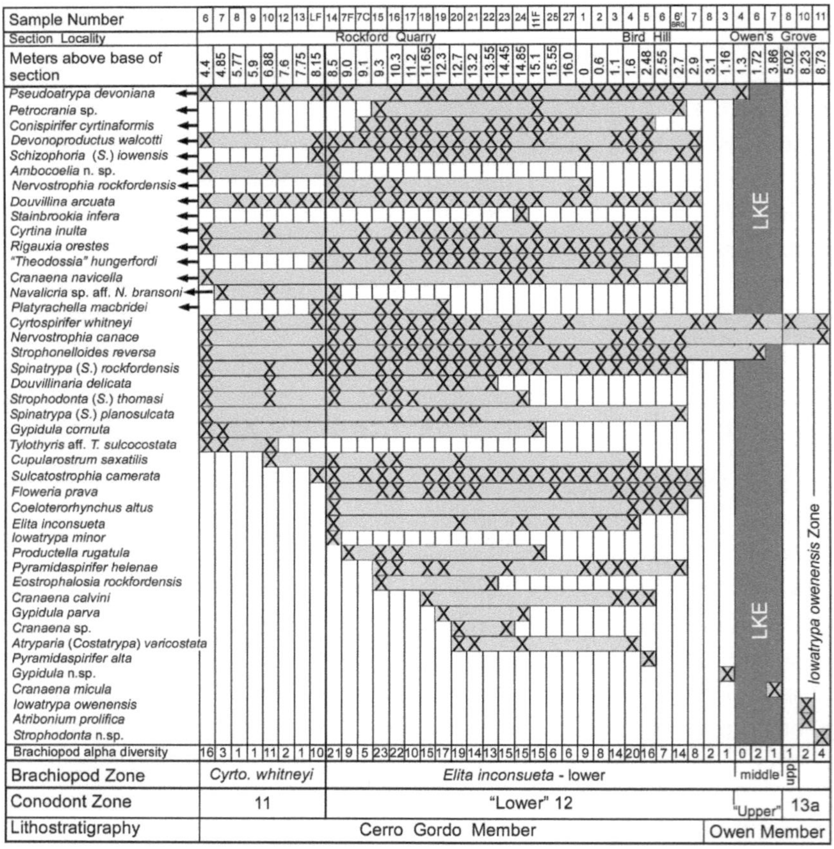

Arrows to left of listed brachiopod species indicate local First Appearance Datums (FADs) in the IGS CG-1 Core section (see Table 8) in the Juniper Hill Member (see Figs. 11 and 12). Bird Hill sample 6′ coincides with the lowest sample at the Bruns Quarry (=BR0) in Bulter County, Iowa (see Fig. 14; Table 10). *LKE*=Lower Kellwasser Extinction interval.

(Figs. 8 and 10; Tables 5 and 6). The overlying *I. owenensis* Zone represents the post-LKE recovery interval.

5.1 Brachiopod Range Revisions

New data, and reports published after those of Day (1989, 1995, 1996a), require revisions of ranges of a number of the Lime Creek brachiopod taxa.

Table 10 Brachiopod Sequence in the Uppermost Cerro Gordo and Owen Members of the Lime Creek Formation at the Bruns Quarry, Butler County, Iowa (Fig. 1, Locality BRQ; Figs. 9 and 10)

Sampled Unit	1	2	3	4	5	6	1	2	4	5	5	6	8	8	9	10	11	12	13
	Sump Section						South High-wall Section												
Meters above base of section	0.0	0.2	0.43	0.57	0.73	1.01	0.0	0.75	2.61	2.73	3.0	3.11	4.35	4.5	4.85	4.95	5.2	6.09	7.14
Devonoproductus walcotti ←	X																		
Petrocrania sp. ←	X																		
Douvillina arcuata ←	X																		
Floweria prava ←	X																		
Cranaena navicella ←	X																		
Sulcatostrophia camerata ←	X									LKE									
Pyramidaspirifer helenae ←	X																		
Coeloterorhynchus altus ←	X																		
Spinatrypa (S.) planosulcata ←	X																		
Schizophoria (S.) iowensis ←	X	X	X																
Rigauxia orestes ←	X	X			X	X					X								
Strophonelloides reversa ←	X															X	X	X	
Pseudoatrypa devoniana ←	X							X	X									X	
Cyrtospirifer whitneyi ←	X	X				X							X	X	X	X	X	X	X
Nervostrophia canace ←	X																		X
Conispirifer cyrtinaformis ←																			X
Productella rugatula ←																	X		X
Spinatrypa (S.) rockfordensis ←	X								X		X					X		X	
Cyrtina inulta ←	X									X							X		
Iowatrypa owenensis																X			
Douvillina n.sp.										LKE							X		
Rhipidomella sp.																			X
Gypidula n.sp.																			X
Brachiopod alpha diversity	17	3	1	0	1	2	0	1	2	1	2	0	1	1	1	4	5	4	6
Brachiopod Zone	lower *Elita inconsueta*						middle						upper			*Iowatrypa owenensis*			
Conodont Zone	"Lower" 12						"Upper" 12						Subzone 13a						
Lithostratigraphy	C.G.		Owen Member																

Arrows to left of listed brachiopod species indicate local First Appearance Datums (FADs) in the Juniper Hill and Cerro Gordo Members in the CG-1 core or in the Rockford Quarry sections (Figs. 2, 11, and 15). *LKE* = Lower Kellwasser Extinction interval.

Day (1998) and Day and Copper (1998) reported *A.* (*C.*) *varicostata* from sample position 24 in the Cerro Gordo Member at the Rockford Quarry section (Fig. 11; Table 9). Restudy of Belanski's collections identify *A.* (*C.*) *varicostata* in samples 20 and 21 at the Rockford Quarry (Fig. 12; Table 9), and from Day's sample 4 in the Bird Hill section (Fig. 11; Table 9). These occurrences fix its range in the Iowa Basin within in the lower *E. inconsueta* Zone, within "lower" Zone 12, as shown in Table 9.

This species' range in the Sly Gap Formation of southern New Mexico is from near the base of the *Nervostrophia geniculata* Zone into the upper part of the *Iowatrypa rara* Zone of Day (1988, 1989) spanning Frasnian conodont zones 11–12 as shown in Day (1998, figs. 9–11). The North American LAD of *V.* (*C.*) *varicostata* is in Subzone 13b in the upper Simla Member (Southeask Formation) at Ancient Wall in the Alberta Rocky Mountains of western Alberta (Day, 1998; Day and Whalen, 2006, 2008). The FAD of *P. devoniana* is in sample 11 in the upper part of the *N. thomasi* Zone in the CG-1 Core (Fig. 11; Table 8), with its LAD in unit 13 of the upper Owen Member at the Beaver Creek Quarry in the upper *I. owenensis* Zone (Fig. 15). The FAD of *Spinatrypa rockfordensis* remains unchanged in sample 32A in the CG-1 core section (Table 8), and in sample 6 in the Rockford Quarry (Fig. 12; Table 9). Its LAD is now fixed in unit 13 in the Bruns Quarry section (Fig. 14; Table 10) in the upper *I. owenensis* Zone. *I. owenensis* has its FAD in unit 10 in the Bruns Quarry section (Fig. 14; Table 10) fixing the base of the *I. owenensis* Zone in that section above the base of Subzone 13a (Figs. 3 and 14), and at the same level in the zonal stratotype in the Beaver Creek Quarry section in unit 8 (Fig. 15). Its LAD is in the upper part of unit 11 at the latter locality (Fig. 15). In western North American carbonate platforms, nearly all species of *Iowatrypa* are restricted to strata spanning Frasnian Zones 11 and 12 and are extinct prior to or during the LKE (Day, 1998). The occurrence of *I. owenensis* represents a late migration of *Iowatrypa* into North America during the post-LKE survivor recovery interval in the highstand interval of Euramerican Devonian T-R cycle IId-2 prior to the extinction of all atrypide taxa during the UKE. Additional post-LKE atrypide migrants into North American platforms at this time include *Spinatrypina* and *Desquamatia* (*D.*) in western Canadian Simla Member (Southesk Formation) platform in western Alberta and the Kakisa Formation (Members 1 and 2) Platform in the southern NWT (Day, 1998; Sullivan and Day, 2008).

Ma and Day (2000, see synonymy therein) reassigned most Lime Creek specimens of *Tenticospirifer cyrtinaformis* to the genus *Conispirifer* (*C. cyrtinaformis*, Tables 8–10). As reported by Day (1989, fig. 9) its FAD is 5.8 m above the base of the Lime Creek Formation in sample 13 in the CG-1 core section (Fig. 11; Table 8). Its LAD is in the upper part of the *I. owenensis* Zone in the uppermost Owen Member in unit 13 at the Bruns Quarry (Figs. 13 and 14; Table 10). Lime Creek specimens illustrated as *Platyrachella cyrtinaformis hellenae*, *P. alta* and *P. a. latimarginata* by the Fenton and Fenton (1924, pp. 164–167, pl. 29) serve as the types for two species of the genus *Pyramidaspirifer* of Ma and Day (2000). In situ specimens of the genotype *P. alta* occur in sample 5 in the Bird Hill East section in westernmost Floyd County (Figs. 2 and 11, locality BH; Table 9). Fenton and Fenton (1924) reported this species as ranging throughout their "*Spirifer*" Zone = upper Cerro Gordo Member = *Elita inconsueta* Zone of Day (1989) in the now covered exposures of the classic Bird Hill section in easternmost Cerro Gordo County. The FAD of *P. helenae* is in sample 15 in the Rockford Quarry section (Fig. 12B; Table 9), and its LAD is in the upper Cerro Gordo Member in unit 1 of the Sump Section at the Bruns Quarry (Figs. 13 and 14; Table 10) in the uppermost part of the "lower" *E. inconsueta* Zone.

Ma and Day (2003) redescribed two species of *Cyrtospirifer* assignable to different species groups from the Late Frasnian of the Iowa Basin. *C. whitneyi* (Tables 8–10) was assigned by Ma and Day (2003, p. 272) to their *verneuiliformis* group. This species is common in subtidal middle and shallow inner shelf Lime Creek facies in northern Iowa. The Global FAD of *C. whitneyi* is in the Middle Frasnian Member C of the Hay River Formation in the southern Northwest Territories of western Canada (see Ma and Day, 2003, pp. 279–281). During the Late Frasnian it dispersed from northwestern Canada into carbonate shelves in the western, southwestern and central US and elsewhere in western Canada. *C. placitus* (Appendix) is a member of the of *syringothyriformis–verneuili* group of Ma and Day (2003, p. 272) and is restricted to deeper water facies preserved as Independence Shale stratigraphic leak deposits in Buchannan County. This species appears to have been restricted to offshore deep oxic–dysoxic subtidal ramp biotope in the Lime Creek seaway. Associated conodonts reported by Müller and Müller (1957) indicate a "lower" Zone 12 age for the Independence Shale fauna as discussed above.

5.1.1 B. fragila *Zone*

Day (1996a, p. 280) defined base of the *L. fragila* Zone in the CG-1 core at its entry in conodont sample 1 (Table 8, sample 1) just above the base of the Juniper Hill Member. *L. fragila* is reassigned to *Barroisella* Hall and Clarke (1892). It is in no way comparable to species of the genus *Lingula* Bruguière (1797). That genus was described from fossil Quaternary and Holocene specimens collected from modern coastal settings in the western Pacific and Indian oceans. The thickness of the *B. fragila* zonal interval in the CG-1core is 2.25 m. *B. fragila* is the only brachiopod with its FAD at the base of this interval. Conodonts of Zone 11 with their local FADs in this interval include *I. symmetricus*, *Pa. kireevae*, *Pa. semichatovae* and *P. unicornis* (Table 7, samples 3–8).

5.1.2 N. thomasi *Zone*

The FAD of *N. thomasi* marks the base of this zone 2.25 m above the base of the Juniper Hill Member in the CG-1 section (Day, 1989; Fig. 11; Table 8, sample 3). Seven other stepped species FADs occur in the upper part of the Zone (see Table 8, samples 11–13) reflecting additional migrations or increasingly favorable environmental conditions in deep subtidal shelf in this region. These include *Psuedoatrypa devoniana* (Table 8, sample 11); *Vincalaria* sp. and *Petrocrania* sp. (Table 8, sample 12); and *Conispirifer cyrtinaformis*, *Schizophoria iowaensis*, and *Nervostrophia rockfordensis* (Table 8, sample 13). Associated Zone 11 conodonts *Pa. semichatovae* and *Pa. kireevae* range into this interval, with the local FAD of *P. evidens* just above the lower zonal boundary (Table 6, sample 9), followed by *P. aspelundi*, *P. alatus*, and *P. brevis*, *Ag. triangularis*, and *A. curvata* late form (Table 6, samples 12–14 and 23).

5.1.3 D. arcuata *Zone*

The type stratum and base of this assemblage zone (Day, 1989) is sample 15 in the CG-1 core section 6.22 m above the base of the Lime Creek (Fig. 11; Table 8). Day (1989) characterized the base of the zone by the species FADs of *Retichonetes brandonensis*, *Eostrophalosia rockfordensis*, *Productella* sp. cf. *P. thomasi*, *Ambocoelia* sp., and *Rhipidomella* sp. Additional brachiopod FADs within this zone include "*Theodossia*" *hungerfordi*, *R. orestes*, *Thomasaria altumbonata*, *Stainbrookia infera*, and *Navalicria* sp. aff. *N. bransoni*. Species of conodonts associated with brachiopods in this interval of the Juniper Hill are all Zone 11 taxa ranging up from the underlying interval of the *N. thomasi* Zone (Table 6).

R. brandonensis, *S. infera*, and *T. altumbonata* were originally described from the Independence Shale fauna by Stainbrook (1945). At this time, FADs of those taxa in the Iowa Frasnian brachiopod sequence in the CG-1 core section are within Zone 11 (Fig. 9; Table 6). As discussed earlier the Juniper Hill Member Shales accumulated in deeper water than overlying facies of the Cerro Gordo Member and the shared occurrences of the above taxa suggest similar relative depth and oxygenated seafloor settings for Juniper Hill and deeper ramp sediments that sourced the stratigraphic leaks of the Independence Shale in Buchannan County and the Amana Beds Member in Iowa County (Figs. 2, 4, and 10).

5.1.4 C. whitneyi *Zone*

Day (1989) defined the base of this assemblage zone by the FAD of *C. whitneyi* in the basal Cerro Gordo Member at the Rockford Quarry in sample 6 of Day (1989, =Belanski samples 4–6, fig. 10; see Fig. 12B; Table 9, sample 6). Other taxa with their FADs at the base of this zone are the atrypides *S. rockfordensis* and *S. planosulcata*; the strophomenides *Strophonelloides reversa, Douvillinaria delicata, N. canace, Strophodonta* (*S.*) *thomasi*; and the spiriferide *Tylothyris* sp. aff. *T. sulcocostata* (Table 9). Other brachiopods with FADs above the base of this zone are *Cupularostrum saxatilis* and *Sulcatostrophia camerata* in the Rockford Quarry section (Table 9), and the spinocyrtid *Platyrachella macbridei* in the middle part of the zone in the CG-1 core (Table 8, sample 37). Frasnian Zone 11 conodonts are associated with brachiopods in this interval (Tables 8 and 9). Conodonts with local FADs within the *C. whitneyi* Zone in the CG-1 core section include *P. politus* (Table 6, sample 42) and *Mehlina gradata* (Table 6, sample 46). Both *Pa. semichatovae* and *Pa. kireevae* have their local LADs in the highest part of the *C. whitneyi* Zone in the Rockford Quarry section (Figs. 11 and 12B; Table 7, sample 19) immediately below the FAD of *E. inconsueta* and *Pa. foliacea* (Tables 7 and 9). The FADs of both *C. whitneyi* and *S. rockfordensis* in sample 32A identify the base of the *C. whitneyi* Zone in the IGS CG-1 core in the upper Juniper Hill Member at 14.79 m above the base of the Lime Creek Formation (Fig. 11; Table 8).

5.1.5 E. inconsueta *Zone*

As mentioned earlier, the LKE crisis and post-LKE survivor intervals fall in the middle and upper parts of this zonal interval in eastern and northern Iowa (Figs. 10, 14, and 15). The lower part of the zone is characterized by numerous species FADs reflecting continued migrations of cosmopolitan taxa into the Iowa Basin during the prolonged sea-level highstand phase of Devonian T-R cycle IId-1 (Day, 1989, 1996a). Only one species FAD is noted (*Gypidula micula*; Table 9) within the middle (LKE interval) and upper parts of the *E. inconsueta* Zone in the Iowa Basin.

The base of this zone (Day, 1989) is defined at the FAD of the nominal species in sample 14 of Day (1989, fig. 10, =Belanski Samples 4–14; =Table 9, sample 6) in the middle Cerro Gordo Member at the Rockford Quarry (Figs. 11 and 12; Table 9). Additional brachiopod FADs that characterize the base of the zone include *Floweria prava, Coeloterorhynchus altus, N. canace*, and *Iowatrypa minor*. *I. minor* was described by Fenton and Fenton (1924, pp. 140–141, pl. 27, figs. 17 and 18) on the basis of two cotypes and later redescribed by Day and Copper (1998). That species may be a synonym of *Iowatrypa americana* of comparable age described from the Independence Shale leak deposits in Buchanan County by Stainbrook (1945). Additional species with FADs within the lower *E. inconsueta* Zone (Table 9) are *P. rugatula*, *Eostrophalosia rockfordensis*, *Pyramidaspirifer alta*, *P. helenae*, *A.* (*C.*) *varicostata*, *Cranaena calvini*, *C. macula*, C. sp., *Gypidula parva*, and *Gypidula* n. sp. The FADs of both *E. inconsueta* (Table 9, sample 6) and the conodont *Pa. foliacea* (see Anderson, 1966; Table 7, sample 20) are coincident at the Rockford Quarry aligning the bases of the *E. inconsueta* Zone and Zone 12 in northern Iowa. In the CG-1 core the base of the *E. inconsueta* Zone is placed at the local FAD of *Cyrtina inulta* at 23.33 m in sample 57 (Table 8) in the upper Cerro Gordo Member. The first occurrence of the conodont *P. planarius* with *C. inulta* in the CG-1 core (Fig. 11, Tables 6 and 8) indicates an assignment to the lower part of Zone 12 based on the restricted range of *P. planarius* (Table 1).

Alpha diversity of the Cerro Gordo Member brachiopod fauna peaks at 23 species, remains relatively high through the Cerro Gordo in the pre-LKE interval of the lower *inconsueta* Zone, and abruptly falls to one to two species per sample in the *Amphipora* packstones in the basal Owen Member in the Bird Hill East section (Table 9). The abrupt reduction in alpha diversity in the Bird Hill section in Floyd County reflects rapid shallowing during the late highstand phase of Iowa T-R cycle 7A. The same pattern is seen in the Bruns Quarry composite section (Fig. 14; Table 10) where alpha diversity falls abruptly from 17 species in the uppermost Cerro Gordo Member (Table 10, sump section, unit 1) to 3 then 1 to 2 in the lower Owen Member (Fig. 14; Table 10, sump section, units 2–6). In coeval deep-ramp facies of the Amana Beds in Iowa County, extinctions of *C. schucherti*, *S. planosulcata*, *T. altumbonata*, *Eostrophalosia* sp., and *F. prava* in the uppermost part of the lower *E. inconsueta* Zone signal the onset of the LKE crisis in the Price Creek section (Fig. 10). Numerous species have their LADS in the upper part of the lower *E. inconsueta* Zone including all species of the rhynchonellide *Coeloterorhynchus* (*C. altus* and *C. schucherti*), two of three species of the atrypide *Spinatrypa* (*S. planosulcata* and *S. trulla*) and the spiriferides *T. altumbonata* and "*Theodossia*" *hungerfordi*. Thus far no species of the distinctive spiriferide *Thomasaria* are reported from the post-LKE interval in North America.

At the Bruns Quarry (Fig. 14; Table 10) the LKE crisis interval of the middle *E. inconsueta* Zone is in the lower-middle Owen Member where brachiopod alpha diversity ranges from 0 to 2 in sampled intervals (Tables 9 and 10). The fauna recovered in the LKE interval includes *C. whitneyi*, *C. inulta*, *P. devoniana*, *S. rockfordensis*, *S. reversa*, *Schizophoria* sp. cf. *S. amana*, and the only known occurrence of *G. micula*. In the Buseman Quarry and Owen Grove sections (Figs. 11, 14, and 15; Tables 9 and 10) brachiopod diversity through the LKE interval ranges from 1 to 2 and then totally collapsed to 0 within the MFI of Iowa Devonian T-R cycle 7B (Figs. 11 and 14; Table 6). The majority of taxa that characterize the diverse Cerro Gordo brachiopod fauna of the lower *E. inconsueta* Zone are extinct, and brachiopod species LADs in the LKE interval of the middle *E. inconsueta* Zone include *G. micula* in the Owens Grove (Table 9) and *Rigauxia orestes* in the Bruns Quarry (Fig. 14; Table 10) sections. *C. whitneyi* is the only brachiopod that ranges into the upper *E. inconsueta* Zone post-LKE survivor interval in the northern Iowa (Figs. 11, 14, and 15; Tables 9 and 10).

After the initial brachiopod extinctions in the uppermost part of the lower *E. inconsueta* Zone in the Amana Beds section, all other brachiopods and two of three ammonoid taxa become extinct within the middle *E. inconsueta* Zone (LKE interval) at that location (Fig. 10). Notable among those are *Calvinaria ambigua*, *P. euglyphea*, and *N. extensa*. These extinctions may mark North American or global LADs of the genera *Calvinaria* and *Pseudodouvillina*. In southern New Mexico, *N. extensa* has its regional FAD in sample 24 at Lava Gap section the Upper Sly Gap Formation in the San Andres Mountains at a similar position high in Frasnian Zone 12 (Day, 1988, table 8; 1998, fig. 10).

Brachiopod assemblages in the post-LKE survivor interval of the upper *E. inconsueta* Zone yield one species *C. whitneyi* (Figs. 14 and 15; Tables 9 and 10). Additional carryover taxa reappear in the succession in younger deposits of the recovery interval in the *I. owenensis* Zone.

5.1.6 **I. owenensis** *Zone (Post-LKE Recovery Interval)*

Day (1989, p. 319) originally defined the base of this zone on the FAD of the nominal species reported in unit 8 in the lower part of the Owen Member at the Beaver Creek section (now covered) of the Upper Lime Creek Formation (Figs. 2, 11, and 15). Because of the absence of data at that time, the alignment of the base of the *I. ownensis* Zone within the conodont sequence was uncertain when first defined by Day (1989) and has been aligned high in Zone 12 or above the base of Frasnian Subzone 13 in older reports by Day (1989, 1995, 1996a, p. 290). Our new data from the Bruns Quarry document the first occurrence of *I. owenensis* in unit 10 in the post-LKE recovery interval (Fig. 15; Table 10) above the lowest *Pa. bogartensis* marking the base of Frasnian Subzone 13a (Tables 7 and 10). Other species with FADs within the *I. ownensis* Zone in the Beaver Creek and Bruns Quarry sections (Figs. 14 and 15; Tables 9 and 10) include *Douvillina* n. sp., *Rhipidomella* sp., and *Gypidula* sp. At the Owen's Grove type section of the Owen Member (Fig. 5; Table 1), *Strophodonta* (*Strophodonta*) n. sp. also has its FAD in that zone.

The benthic shelly fauna from the interval of the *I. owenensis* Zone is dominated by molluscs including the archaeogastropods *W. gigantea*, *Westerna* sp. and bivalves *Edmondia* sp., *Paracyclus* sp., *Spathella* sp. with subordinate brachiopods (Sanks et al., 2016). Assemblages also include common solitary and rare colonial rugose corals *Tabulophyllum expansum* and *Pachyphyllum dumonti* (described by Sorauf, 1998). The low-moderate diversity brachiopod fauna in this interval (Day, 1989, 1996a; Tables 9 and 10) consists of the small numbers of new occurrences such as *I. owenensis* and LKE survivor taxa known from the lower *E. inconsueta* Zone deposits of the upper Cerro Gordo Member. These carryover taxa include the strophomenides *S. reversa* and *N. canace*; the atrypides *P. devoniana* and *S. rockfordensis*; the productoid *P. rugatula*; the spiriferides *C. whitneyi* and *Conispirifer cyrtinaformis*; and the spiriferinide *C. inulta*. Brachiopod alpha diversity ranges from 2 to 6 species in sampled assemblages from this interval (Tables 9 and 10; Fig. 15) reflecting the major collapse of brachiopod alpha diversity from peak values as high as 17 to 20 species in the pre-LKE interval of the Cerro Gordo Member (Table 9). Restructuring of the benthic community in the Owen subtidal platform is evidenced by the increased abundance and dominance of archaeogastropods and infaunal bivalves in the benthic fauna of the *I. owenensis* Zone.

6. IOWA BASIN LATE FRASNIAN DEPOSITIONAL SETTING AND HISTORY

Current stratigraphic, biostratigraphic, and sea-level frameworks for Middle–Late Devonian (Eifelian–Early Famennian) rocks of the Iowa Basin are shown in Fig. 3. Fig. 6 is the modified version of Witzke's (1987) depositional model and regional cross-section for the Lime Creek–Sweetland Creek seaway in the Iowa and Illinois basins. The depositional model presented herein differs from earlier versions especially in the recognition of thicker carbonate shelf facies (50–100+ m) of the Lime Creek Formation in the subsurface of central and western Iowa. These thicker sections record widespread progradation of peritidal mudflat facies, and notable evaporite solution collapse breccias in the upper part.

The regional depositional history and setting of the Lime Creek Formation of northern Iowa and its equivalents in eastern Iowa have been outlined in Witzke (1987), Witzke et al. (1989), Witzke and Bunker (1997), Day (1990, 1995, 1996a, 1998), and Day et al. (2008, 2013). The Lime Creek and equivalents have been interpreted to comprise a single third-order depositional sequence designated as Iowa Devonian Transgressive–Regressive Cycle 7, representing the regional record of the global Euramerican Devonian eustatic T-R cycle IId of Johnson et al. (1985) in the older reports by Witzke (1987), Witzke et al. (1989), Day (1990, 1995, 1996a), Johnson and Klapper (1992), and Witzke and Bunker (1997). Day (1998) proposed the subdivision of T-R cycle IId into T-R cycles IId-1 and IId-2 based on recognition of an additional third-order depositional sequence and sequence boundary or boundary intervals in the upper part of T-R cycle IId interval in Iowa, New Mexico, and the western Canadian sedimentary basin (Alberta and the NWT). Correlations outlined above identify T-R cycle IId-2 deposits in the Iowa Basin represented by most of the Owen Member of the Lime Creek Formation (Figs. 3, 10, and 13–16), and equivalent deep water facies of the Sweetland Creek Shale at its type section in southeastern Iowa (Figs. 3 and 7–9). Widespread erosional truncation of uppermost Lime Creek platform strata in northern Iowa, overlain by Middle Famennian Sheffield Shale, indicates that a very Latest Frasnian or Early Famennian sea-level lowstand led to platform emergence and erosion in much of the Iowa Basin at that time.

6.1 Frasnian Zone 10 Regression-Relative Sea-Level Lowstand

The Upper Cedar Valley Group is deeply eroded beneath overlying strata of the Late Frasnian Lime Creek–Sweetland Creek–Grassy Creek formations in the Iowa and western Illinois basins (Figs. 4 and 5; Day et al., 2008, 2013; Witzke and Bunker, 2006; Witzke et al., 1989). This surface developed during a profound sea-level lowstand, leading to platform emergence and associated subaerial erosion during the latter part of the Middle Frasnian. During that time the Shell Rock Formation was sharply truncated across the distal inner shelf area (Figs. 3, 4, and 7) and older Cedar Valley Group units were beveled and truncated across Iowa. The presence of middle-shelf Shell Rock facies in southeastern Iowa would also be predicted from the facies patterns seen in older Cedar Valley T-R sequences (Figs. 3 and 4), and the

Shell Rock Formation was most likely deposited across the middle shelf. However, the Shell Rock Formation was entirely eroded across this region during the upper Middle Frasnian erosional episode (Witzke and Bunker, 2006).

The total erosional relief developed on the Cedar Valley Group exceeds the total thickness of the Shell Rock Formation (Figs. 3 and 4), and the Lime Creek Formation and equivalents are known to overlie all formations of the Cedar Valley Group (Fig. 4) at various localities in eastern Iowa and northeastern Missouri (see Figs. 4 and 7–9; Over, 2002, 2006). Considering the general southeastward depositional slope and deepening of all Cedar Valley carbonate ramps (Fig. 4) the actual bathymetric relief across the inner and middle shelf was likely even greater than 65 m. The development of Late Frasnian Independence Shale stratigraphic leaks filling karst cavities hosted in Older Wapsipinicon and Cedar Valley groups carbonates (Fig. 4, Independence Shale) or older Silurian carbonates in central and eastern Iowa and western Illinois (Bunker et al., 1985, p. 53) illustrates widespread emergence and a major lowering of erosional base levels at that time. Independence Shale leaks are documented in solutional cavities up to 90 m stratigraphically below the highest parts of the regional sub-Lime Creek erosional surface across the inner shelf (Fig. 4). It appears likely that sub-Lime Creek erosional base level fell enough (in excess of 125 m) to develop karst systems through the entire thickness of the Wapsipinicon and Cedar Valley groups, and Silurian strata in eastern Iowa during the Middle Frasnian.

Lowering of erosional base levels (to develop channel incisions and karst systems) in the stable cratonic interior during the Paleozoic was likely driven by sea-level fall associated with the withdrawal of interior seaways, and such changes have been interpreted to reflect global eustasy (Day et al., 2008, 2013; Witzke and Bunker, 1997, 2006). The influence of epeirogenic structural movements associated with Acadian tectonism may have contributed to platform uplift, although the stable configuration of facies and thicknesses across the Iowa Basin suggests relative tectonic stability during the Eifelian–Frasnian in the region. Furthermore, the relatively short geologic time frame for such large-scale sea-level changes is not easily accommodated by any known tectonic mechanism. If the sub-Lime Creek base-level changes were primarily driven by global eustasy, then base-level changes of 65–90 m (potentially up to 125 m) during the upper part of Frasnian Zone 10 indicate that very large eustatic sea-level changes occurred over a relatively short time span (less than one conodont zone) in the Late Devonian. These changes are similar in magnitude to glacial-eustatic sea-level changes in the Quaternary, Carboniferous, and Ordovician–Silurian driven by waxing and waning of continental glaciers in Gondwana. Identifying mechanisms for large Middle and Late Devonian eustatic sea-level changes remains problematic given that there is no direct clear-cut evidence of continental glaciers in Gondwana prior to the Late Famennian at this time.

6.1.1 Devonian T-R Cycle IId-1 (semichatovae *Event)*

In Iowa, T-R cycle IId-1 deposits make up the most of the Lime Creek Formation in north-central and eastern Iowa (Figs. 3 and 11–16), and the lower portion of the Sweetland Creek Shale at its type section (Figs. 7 and 8) in Muscatine County, Iowa, and significant part of the Sweetland Creek Shale in the subsurface of southeastern Iowa (Figs. 7 and 9). In its type area, the Lime Creek Formation is comprised of two shallowing upward successions of shales and carbonate rocks that accumulated during two significant third-order transgressive–regressive cycles during the Late Frasnian of the Late Devonian (Fig. 3). The Late Frasnian sea-level rise (Fig. 3) that initiated Lime Creek–Sweetland Creek Shale deposition in the Iowa Basin is designated as Iowa Devonian T-R cycle 7 of Witzke et al. (1989) and 7A of Day et al. (2008, 2013) corresponding to most of the interval of Devonian eustatic T-R cycle IId of Johnson et al. (1985) and Johnson and Klapper (1992) as shown in Fig. 3.

In its type area the Lime Creek Formation was deposited in inner-mid-shelf environments near the margin of an immense carbonate platform dominated by peritidal carbonates and evaporates in the basin interior in central and western Iowa (Figs. 3 and 6). During the late highstand phase of the initial and largest sea-level rise of the Devonian, shallow-water inner-platform and platform margin facies of the Owen Member of the Lime Creek prograded from the west and north to the current vicinity of the type area of the Lime Creek in northern Iowa (Figs. 11 and 14–16). Diversity of the cosmopolitan subtidal platform benthic shelly faunas of the Lime Creek Formation reached their maximum during the prolonged relative sea-level highstand during the interval of Frasnian Zones 11 (upper part) and 12. Seaward of the platform margin, deeper water facies of the Independence Shale, Amana Beds Member (Fig. 10) accumulated in deep subtidal ramp settings above the regional oxycline. Deep water hemipelagic shales of the North Liberty Beds, Sweetland Creek, and Grassy Creek shales accumulated where bottom conditions (either anoxia or temperature) generally excluded most stenotypic benthic invertebrates including nearly all calcareous brachiopods.

6.1.2 Devonian T-R Cycle IId-2

In their original synthesis of Devonian sea-level records of Euramerican sedimentary basins, Johnson et al. (1985, fig. 12) noted that their T-R cycle IId was comprised of a pair of widely recognized transgressions. In Iowa and elsewhere in western North America (Day, 1998; Whalen and Day, 2008, 2010; Whalen et al., 2016) the second of their transgressions is widely recognized as the rapid deepening event that began in upper Zone 12 as evidenced by carbonate platform drowning and backstepping known to be associated with the LKE (Fig. 3). In the Appalachian Foreland Basin, the same event led to expansion of the Pipe Creek Shale (Johnson et al., 1985, p. 581, fig. 6) over older Upper Frasnian deposits in that region and the onset of the first episode of brachiopod extinctions (Bush et al., 2015; Dutro, 1981; Johnson et al., 1985; Klapper and Kirchgasser, 2016). This event was designated as Devonian T-R cycle IId-2 by Day (1998).

No distinct lithologic signature recording a deepening event is evident in offshore sections of the Sweetland Creek Shale in eastern and southeastern Iowa (Figs. 7–9). In that part of the basin the position of the T-R cycle IId-2 flooding surface or flooding interval is approximated by position below lowest Frasnian Subzone 13a conodont faunas. The very Late Frasnian T-R cycle IId-2 transgression has a distinct lithologic signature in mid-ramp and shallow subtidal platform sections in Iowa County (Fig. 10) and the type area of the Lime Creek Formation (Figs. 3 and 13–15). In Iowa County, brown shales overlaying gray shales immediately above the LKE interval mark the deepening event in the Amana Beds section at Price Creek (Fig. 10, unit 9). This seems to coincide with the expansion of anoxic bottom waters over this region of the Lime Creek ramp associated with regional relative sea-level rise. As discussed earlier, this deepening event is recognized within the lower to middle part of the Owen Member at the Bruns and Buseman quarries recorded by the series of prominent flooding surface/hardground-bound facies packages comprising parasequences making up the TST, MFI, and HST intervals of Latest Frasnian depositional T-R cycle IId-2 (Figs. 14 and 15).

Subsequently the Late Frasnian Lime Creek platform in the Iowa Basin experienced a significant period of emergence that eroded Latest Frasnian and Early Famennian deposits across most of the Iowa region. Deposition continued during the very Late Frasnian and Early Famennian in southeastern Iowa as recorded by the Grassy Creek Shale where we can recognize the F–F boundary in surface and subsurface localities (Figs. 2 and 7–9). The UKE event interval is characterized by the regional onset and expansion of pronounced organic-rich brown shale deposition likely associated with pronounced anoxia in that part of the basin where the type Sweetland Creek accumulated (Figs. 7 and 8).

7. SUMMARY AND CONCLUSIONS

The Late Devonian Lime Creek and Sweetland Creek formations in the Iowa area provide one of the best records of Late Frasnian history and events available anywhere on the Euramerican paleocontinent. Biostratigraphic, paleontologic, and chemostratigraphic data in the area augment the regional lithostratigraphy, sequence stratigraphy, and sea-level history of central North America. Of particular note, these strata preserve the Late Frasnian Kellwasser Extinction bioevents which can be tightly constrained within a biostratigraphic and chemostratigraphic framework, and the rich paleontologic record further documents changes in the benthic invertebrate faunas across these events. Salient observations and interpretations of Lime Creek–Sweetland Creek stratigraphy and deposition are summarized below.

7.1 Sub-Lime Creek Surface

Regional Lime Creek–Sweetland Creek sedimentation was initiated above a widespread eroded and deeply karstified surface developed on underlying Devonian Cedar Valley Group strata during the Mid-Frasnian. Flooding of this surface during the Late Frasnian led to the infilling of karstic openings by fossiliferous shales (the Independence Shale).

7.2 Regional Upper Frasnian Lithostratigraphy

A northwest-to-southeast profile of Upper Frasnian strata across Iowa (Fig. 6) shows major changes in lithofacies patterns and thickness. The thickest strata (50–100+ m) are located in central and northwestern Iowa and are dominated by a complex of shallow-water carbonate facies included in the Lime Creek Formation. These carbonate facies interfinger with fossiliferous subtidal carbonate and shale facies in the type area of the Lime Creek Formation in northern Iowa (35–50 m thick), where the formation includes three members: a basal Juniper Hill Shale, a middle highly fossiliferous mixed shale-limestone interval (Cerro Gordo Member), and an upper carbonate-dominated Owen Member (Figs. 11 and 16). Lime Creek strata progressively thin to the southeast (20–40 m thick) across east-central Iowa (Fig. 16), an area characterized by shale and mixed shale'carbonate facies that includes several important fossiliferous localities (Independence Shale, Amana Beds, North Liberty Beds). Equivalent strata in southeastern Iowa and adjoining Illinois are assigned to the Sweetland Creek Shale, a greatly thinned interval (0.85–7.5 m) of gray to green-brown shale and siltstone units. Uppermost Frasnian strata of the basal Grassy Creek Shale overlie the Sweetland Creek in that area, characterized by a thin interval of organic-rich brown shale (Figs. 7 and 16).

7.3 Late Frasnian Depositional Profile

The transect of Upper Frasnian strata across the outcrop belt and shallow subsurface from northern Iowa to southeastern Iowa conveniently parallels an onshore-to-offshore depositional profile (Figs. 6 and 16). A relatively thick succession of carbonate facies in central and northwestern Iowa forms a broad carbonate shelf or platform that includes subtidal facies in the lower part, stromatoporoid-rich biostromal facies in the middle part, and prograding peritidal/mudflat facies with evaporite-collapse breccias in the upper part. The type Lime Creek area of northern Iowa is located along the margin of this platform, where an entirely subtidal succession of fossiliferous carbonate facies interfingers with more offshore shale facies (Fig. 11). Proceeding southeastward from the platform margin across east-central Iowa, the Lime Creek Formation was deposited across a broad and gently sloping ramp, and deposition was in progressively deeper environments in that direction. Differences in faunal compositions are evident in these progressively deeper water facies. Further offshore to the southeast, equivalent strata in southeastern Iowa and Illinois (Sweetland Creek-basal Grassy Creek) are significantly condensed (Fig. 16) reflecting decreased sediment accumulation in deeper basinal dysoxic to anoxic bottom environments. Shelly benthic faunas in these generally inhospitable environments were sparse to absent.

7.4 Frasnian Sea-Level History

Because the Iowa area was situated in the relatively tectonically stable continental interior of North America during the Devonian, it is a particularly good place to evaluate sea-level changes that most likely are dominantly of eustatic origin. The sub-Lime Creek Mid-Frasnian erosional and karstified surface marks a profound and relatively short-lived sea-level lowstand event in the Mid-Frasnian. Based on regional stratigraphic relations, this lowstand event appears to have lowered erosional base levels 65–90 m (or more). Such large-scale eustatic sea-level changes may suggest a glacial-eustatic mechanism, although there is, as yet, no evidence of Mid-Frasnian continental glaciations in Gondwana. The overlying Lime Creek Formation was deposited during the Late Frasnian T-R cycle IId of Johnson and Klapper (1992). This cycle is divided into two constituent T-R cycles, IId-1 and IId-2 (Iowa T-R cycles 7A and 7B), each forming a stratigraphic sequence. Most of the Lime Creek is included within IId-1; its base is marked by a transgressive phosphatic lag, overlain by a highstand succession (Juniper Hill, Cerro Gordo Members) that shallows upward to an interval of stromatoporoid packstones or oolitic–peloidal carbonates at the top (lower Owen Member). T-R cycle IId-2 encompasses most of the Owen Member in northern Iowa, most of the type Amana Beds in eastern Iowa, and the upper half of the Sweetland Creek Shale and Lower Grassy Creek Shale in southeastern Iowa (Fig. 16). Its complex TST includes stromatoporoidal limestone units bounded by hardgrounds (sediment-starved surfaces); maximum flooding is marked by a cluster of hardgrounds in crinoidal lithologies. Overlying highstand deposits are dominated by deeper water subtidal argillaceous skeletal mudstones, and the presence of deeper water *Palmatolepis* conodont biofacies faunas in these strata further supports a sea-level highstand. The interval shallows upward and was erosionally beveled at the top during the succeeding sea-level lowstand that subaerially exposed the Owen carbonate platform in northern Iowa.

The sea-level history for the Late Frasnian succession in eastern Iowa (Amana and North Liberty beds) and southeastern Iowa (Sweetland Creek-basal Grassy Creek) is more difficult to interpret, as these strata were deposited in deeper water ramp and basinal settings where decameter-scale sea-level changes would not necessarily significantly affect depositional processes. A transgressive hardground surface marks the base of succession, and additional complex stacked hardgrounds units and phosphatic lags in some sections likely reflect episodes of sediment starvation in the basin during T-R cycle IId-1. The base of T-R cycle IId-2 is also marked by a transgressive hardground or starved surface in most, but not all, sections. Furthermore, T-R cycle IId-2 includes anoxic–dysoxic organic-rich shale facies of the basal Grassy Creek Shale, and this shift in deposition probably reflects sea-level deepening with increased seaway stratification and/or changes in seaway circulation. Organic-rich shales interfinger upslope in the lower-ramp succession of the Amana Beds during cycle IId-2.

7.5 Biostratigraphic and Paleontologic Information

The benthic faunas of the Lime Creek Formation are among the most diverse and best documented in the Upper Frasnian of Euramerica, and the brachiopod-dominated associations of the Lime Creek are particularly rich and informative. The succession of brachiopods is used to delineate a series of brachiopod biozones based on the first appearance of key species and associated taxa, and revisions and additions for many taxa in these biozones are summarized herein. This zonation also has utility in areas of North America outside of Iowa (Day, 1989, 1998). The brachiopod ranges within the lithostratigraphic and depositional framework further help document the Lower Kellwasser bioevent/extinction in deep ramp (Fig. 10) and shallow subtidal shelf settings (Figs 11, 14, and 15; Tables 9 and 10) and environmental changes in the succession.

The stratigraphic succession, brachiopod biozones, and associated history of sea-level changes and bioevents are constrained within the standard Frasnian conodont biostratigraphic zonation (Figs. 3, 10, and 14; Tables 8–10). Further, zonally significant ammonoids identified in Lime Creek strata of northern and eastern Iowa provide further biostratigraphic control (Fig. 10). Extensive sampling of the diverse Lime Creek and Sweetland Creek/Grassy Creek conodont faunas provides a refined biostratigraphic basis that enables global correlation of these strata (Figs. 7–16; Tables 2–7). Revisions and additions to the Lime Creek–Sweetland conodont faunas, as presented herein, provide a more detailed biostratigraphic framework for these strata. Further, conodont data from the Iowa Frasnian succession will necessitate some range adjustments and extensions in the Frasnian Composite Standard (Klapper and Kirchgasser, 2016); correlation of the new Sweetland Creek conodont ranges can improve the efficacy of the composite.

The base of the Lime Creek–Sweetland Creek interval is now correlated with Frasnian Zone 11, not Zone 9 as previously reported, and the flooding event at the base of T-R cycle IId is thereby drawn at this position (the "*semichatovae* Event"). As interpreted herein, T-R cycle IId-1 (Iowa T-R cycle 7A) spans Zone 11 to middle Zone 12, which primarily includes the Juniper Hill and Cerro Gordo Members in northern Iowa. T-R cycle IId-2 (Iowa T-R cycle 7B) spans the upper Zone 12 through Subzone 13c interval in southeastern Iowa, but the upper interval is truncated across northern Iowa, where the Owen Member contains conodont faunas of upper Zone 12 and Subzone 13a. The discovery of conodont faunas with species typical of deeper water *Palmatolepis* biofacies in mid-Owen strata of northern Iowa supports the idea of maximum depositional deepening during Zone 13a. Proceeding further down the Lime Creek paleo-ramp into deeper depositional settings in the Amana Beds and Independence Shale of eastern Iowa, conodont faunas are entirely within the *Palmatolepis* biofacies. Dark-gray and brown organic-rich shales are found in Zone 13a of the Amana Beds, also suggestive of maximum depositional deepening in that zone (Fig. 16).

The greatly thinned Sweetland Creek–basal Grassy Creek Shale succession occupies the deepest depositional setting in the Late Frasnian of Iowa, entirely within *Palmatolepis* biofacies. T-R cycle IId-1 includes conodonts of Zone 11 and lower Zone 12. Although the Sweetland Creek Shale is a relatively thin and condensed interval, biostratigraphic information from multiple sections indicates significant variations in the stratigraphic thicknesses of these zones across southeast Iowa, thinner in the northern sections (<2 m) and thicker to the south (up to 8 m). Succeeding strata of T-R cycle IId-2 are also significantly condensed in southeast Iowa. Succeeding conodont biozones of T-R cycle IId-2 are best developed in southeastern Iowa, also marked by significant sediment condensation (strata <1–3.5 m). Grassy Creek organic-rich shale deposition expanded across this area during Subzones 13a and 13b.

The upper part of this cycle correlates with Subzone 13c where nearly all typical Late Frasnian conodont taxa become extinct and are succeeded by Early Famennian species of the *Pa. subperlobata* and *Pa. triangularis* zones in the Lower Grassy Creek Shale in southeastern Iowa (Fig. 16).

7.6 Kellwasser Extinction Events

The Upper and Lower Kellwasser Extinction Events (LKE, UKE) are identified within the Upper Frasnian succession of Iowa, and the contained brachiopod and other benthic faunas are well constrained within the standard conodont zonal scheme. The LKE is recognized in "upper" Zone 12, associated with the transgressive systems tract of T-R cycle IId-2. In northern Iowa, this interval in the Owen Member includes multiple hardground surfaces that mark punctuated episodes of sediment starvation, probably recording several parasequence-scale transgressive pulses (Fig. 14). The associated carbonate strata display low alpha-diversity brachiopod faunas (zero to two species) within a coral-stromatoporoid biofacies (Figs. 11, 14, and 15; Tables 8–10). The decline in brachiopod diversity is especially noteworthy when compared to the rich and diverse brachiopod faunas in the Cerro Gordo Member below (Table 9). Much of this decline may be related to progressively shallowing benthic environments through upper Cerro Gordo–lower Owen deposition. Some of the Cerro Gordo taxa survived the LKE, but the interval marked a crisis and extinction interval for most of the Cerro Gordo brachiopod fauna (and other benthic taxa). Even though deeper water depositional environments (*Palmatolepis* biofacies) returned above the LKE interval, the benthic paleocommunities have a notably different faunal composition in the middle to upper Owen Member (the post-LKE survival-and-recovery interval). Instead of the brachiopod-rich benthic associations seen in the Cerro Gordo Member, the post-LKE faunas in northern Iowa are dominated by molluscan taxa (bivalves, gastropods), and a general reorganization of benthic community structure seems to have occurred during and following the LKE crisis across the carbonate shelf of northern Iowa. Nevertheless, eight species of Cerro Gordo brachiopods survived the LKE, and these survivors along with the appearance of several new species show a general upward increase in alpha diversity through the recovery interval.

Compared to many other sections of the LKE elsewhere in the world, the interval in the Owen Member of northern Iowa is notably thicker; this stratigraphic expansion enables finer-scale details to be identified within the LKE interval. This is particularly evident in the $\delta^{13}C$ carbon-isotope excursion in the LKE interval identified in the Owen Member (Fig. 14). A large positive shift (+4.5‰) is identified in the basal LKE, coincident with transgressive deepening of the T-R cycle IId-2. The falling limb of the excursion (−6‰) is recorded through the overlying 7 m of section, beginning in the upper LKE interval and continuing through the survival-and-recovery interval of the upper Owen; this is coincident with highstand deposition. The large fluctuations in $\delta^{13}C$ underscore the idea that major changes in carbon recycling/burial and productivity were occurring in the global ocean–atmosphere system during and after the LKE.

The LKE interval is also preserved in a deeper depositional setting along the paleo-ramp in eastern Iowa (Amana Beds, Price Creek section; Fig. 10). The highest diversity of brachiopods is found immediately below the LKE interval (12 species), and the ranges of at least five of these species terminate at the base of the LKE. Brachiopod alpha diversity is notably reduced through the LKE interval (five to six species). Survival and recovery above the LKE cannot be evaluated in this area, as significant depositional deepening at the base of Frasnian Subzone 13a largely eliminated brachiopod biofacies from the deeper ramp. Overall, the Iowa LKE record across both platform and ramp facies records the regional extinction of 20 of

32 brachiopod genera, 24 of 36 brachiopod species, 4 of 6 coral taxa, and 4 of 5 stromatoporoid genera.

The LKE and UKE intervals also are preserved in the stratigraphically condensed Sweetland Creek–Lower Grassy Creek interval of southeast Iowa. Benthic faunas are largely absent in this area, and so diversity and extinction effects on the benthos cannot be evaluated through the Kellwasser events. The associated pelagic conodont faunas do not appear to show any notable changes in diversity or extinction across the LKE interval in the mid- and deep-ramp settings in eastern and southeast Iowa. The LKE and UKE in these condensed basinal facies are associated with $\delta^{13}C$ carbon-isotope excursions (in preparation). The UKE shows a large-scale expansion of organic-rich shale deposition, likely associated with a rise in sea level and changes in seaway circulation/stratification that resulted in the regional expansion of an anoxic–dysoxic water mass.

ACKNOWLEDGMENTS

The authors thank former ISU students who participated in field work and assisted with conodont sampling in the IGS's CG-1 and H cores, sample processing and picking including J. Clark, L. Crause, J. Foote, K. Garber, D. Haas, D. Kimple, A. Martins, E. Meinzer, W. Mohan, K. Murphy, M. Mustain, R. Olson, J. Rappe, T. Stieglitz, S. Taha, R. Thompson, S. Travers, A. Trinski, A. Wallenberg, and N. Weidman. Special thanks go to G. Klapper (University of Iowa Emeritus & Northwestern University) who identified conodonts from the Sullivan Quarry core section, and J. Preslicka (Iowa City) for providing his collections of brachiopods and conodont samples to J. Day from the Price Creek locality of the Amana Beds, Iowa County, Iowa. Mr. J. Volz kindly permitted access to the Price Creek locality on numerous occasions since 1993, most recently in April of 2016. We also thank Mr. G. Bohling for sharing his special beef jerky and permission to sample the type section of the Sweetland Creek Shale in Muscatine County, Iowa in June of 2015 and 2016. M. Root, J. Dohlman, and K. Bruening Construction Company-Skyline Quarries of Decorah, Iowa (owners of Greene Limestone Company) have kindly permitted access to their properties in northern Iowa for the past 30 years, including the Bruns Quarry in Butler County, Iowa since 2014. The IGS Sullivan Slough Core was sampled by J. Day with H. Rowe (Texas Bureau of Economic Geology, Austin, Texas), S. Rimmer (Southern Illinois University-Carbondale), and S. De Larue (Chicago, Illinois). Funding to J. Day in support of this work was provided by grants from: the Donors of the Petroleum Research Fund of the American Chemical Society; Illinois State University-College of Arts and Sciences Faculty Research grant program; and an United States Geological Survey research contract (P. Emsbo) for the support from 2009 to 2016 for studies of the IGS Sullivan and H cores, and surface outcrops of the Lime Creek Formation at Bird Hill, Floyd County, and the Bruns Quarry, Butler County, Iowa.

APPENDIX. EXPLANATION

Table A.1 Late Frasnian Brachiopod Fauna Known From the Iowa Basin of Central North America

Occurrence	SW	AM	IS	JH	CG	O
Lingulida–Lingulata						
Barroisella fragila (Fenton and Fenton, 1924)		AM		JH	CG	OL
Craniforma–Craniata						
Petrocrania famelica (Hall and Whitfield, 1873)			IS	JH	CG	OL
P. sp.				JH	CG	
Philhedra sheldoni (White, 1862)					CG	OL
Strophomenata–Strophomenida						
Nervostrophia canace (Hall and Whitfield, 1873)					CG	OU
N. calvini Stainbrook (1945)			IS			
N. extensa (Cooper and Dutro, 1982)		AM				
N. rockfordensis (Fenton and Fenton, 1924)					CG	
N. thomasi Stainbrook (1945)			IS	JH		
N. sp.	SW					
Douvillina arcuata (Hall, 1858)		AM	IS			
D. navicula Stainbrook (1945)			IS			
D. n. sp.						OU
Douvillinaria delicata (Fenton and Fenton, 1924)					CG	
D. variabilis (Calvin, 1878)		AM	IS			
Pseudodouvillina euglyphea Stainbrook (1945)		AM	IS			
Strophonelloides reversa (Hall, 1858)			IS		CG	OU
S. reversa deeringeri Stainbrook (1945)			IS			
Sulcatostrophia camerata (Fenton and Fenton, 1924)			IS		CG	
S. calvini (Miller, 1883)			IS			
Strophodonta (Strophodonta) thomasi (Fenton and Fenton, 1924)					CG	

Table A.1 Late Frasnian Brachiopod Fauna Known From the Iowa Basin of Central North America—cont'd

Occurrence	SW	AM	IS	JH	CG	O
S. (*S.*) *umbonata* Cooper and Dutro (1982)					CG	
S. (*S.*) n. sp.						OU
Orthotetida						
Floweria prava (Hall, 1858)		AM			CG	OL
F. sp.						OU
Productida–Productidina						
Productella rugatula (Stainbrook, 1945)					CG	OU
P. thomasi (Stainbrook, 1945)			IS	JH		
Devonoproductus walcotti (Fenton and Fenton, 1924)				JH	CG	OL
Strophalosiidina						
Eostrophalosia rockfordensis (Hall and Clarke, 1892)				JH	CG	
Leptalosia inexpectans Stainbrook (1945)			IS			
Chonetidina						
Retichonetes brandonensis (Stainbrook, 1945)			IS	JH		
Rhynchonellata–Rhynchonellida						
Atribonium proifica (Belanski, 1928a,b)						OU
Calvinaria ambigua Stainbrook (1945)		AM	IS			
Navalicria n. sp. aff. *N. bransoni*				JH	CG	
Vincalaria sp.				JH		
Hypothyridina emmonsi (Hall and Whitfield, 1872a,b)			IS			
Coeloterorhynchus altus(Thomas and Stainbrook, 1922)					CG	
C. schucherti (Stainbrook, 1945)		AM	IS			
"*Cupularostrum*" *saxatilis* (Hall, 1858)					CG	
Orthida						
Cariniferella iowensis Stainbrook (1945)			IS			
Rhipidomella n. sp. 1				JH		

Continued

Table A.1 Late Frasnian Brachiopod Fauna Known From the Iowa Basin of Central North America—cont'd

Occurrence	SW	AM	IS	JH	CG	O
Rhipidomella n. sp. 2						OU
Schizophoria (*Schizophoria*) *amanaensis* Stainbrook (1945)		AM	IS			
Schizophoria (*S.*) *iowensis* Hall (1858)				JH	CG	OL
Skenedium independense Stainbrook (1945)			IS			
Stainbrookia infera (Calvin, 1878)				JH	CG	
Pentamerida						
Gypidula cornuta (Fenton and Fenton, 1924)			IS		CG	
Gypidula cornuta stainbrooki Cooper and Dutro (1982)			IS			
Gypidula munda Calvin (1878)			IS		CG	
Gypidula parva (Fenton and Fenton, 1924)					CG	
Atrypida						
Atryparia (*Costatrypa*) *varicostata* (Stainbrook, 1945)			IS		CG	
Pseudoatrypa devoniana (Webster, 1921)		AM		JH	CG	OU
Iowatrypa americana (Stainbrook, 1945)			IS			
I. minor (Fenton and Fenton, 1924)					CG	
I. ownensis (Webster, 1921)						OU
Spinatrypa (*Spinatrypa*) *planosulcata* (Webster, 1888)		AM			CG	
Spinatrypa (*S.*) *rockfordensis* (Fenton and Fenton, 1924)					CG	OU
Spinatrypa (*S.*) *trulla* (Stainbrook, 1945)		AM	IS			
Spiriferida						
Cyrtospirifer whitneyi (Hall, 1858)					CG	OU
C. placitus (Stainbrook, 1945)			IS			
Conispirifer cyrtinaformis (Hall and Whitfield, 1872a)				JH	CG	OU
Pyramidaspirifer helenae (Fenton, 1918)					CG	

Table A.1 Late Frasnian Brachiopod Fauna Known From the Iowa Basin of Central North America—cont'd

Occurrence	SW	AM	IS	JH	CG	O
P. alta (Fenton and Fenton, 1924)					CG	
Platyrachella macbridei (Calvin, 1878)				JH	CG	
Ambocoelia n. sp.				JH	CG	
"*Theodossia*" *hungerfordi* (Hall, 1858)				JH	CG	OL
Elita inconsueta (Fenton and Fenton, 1924)					CG	
Thomasaria altumbona Stainbrook (1945)		AM	IS	JH		
Acutatheca propira Stainbrook (1945)			IS			
Rigauxia orestes (Hall and Whitfield, 1872a,b)				JH	CG	OL
Tylothyris sp. aff. *T. sulcocostata* Cooper and Dutro (1982)					CG	
Spiriferinida						
Cyrtina inulta Stainbrook (1945)				JH	CG	OU
Terebratulida						
Cranaena amana Stainbrook (1945)		AM				
C. calvini (Hall and Whitfield, 1872a,b)					CG	
C. micula (Belanski, 1928b)						OL
C. navicella (Hall, 1858)				JH	CG	OL
C. sp.					CG	

Authors of all described species listed are cited in the bibliography. Occurrences by stratigraphic unit abbreviated as follows: *AM*, Amana Beds; *CG*, Cerro Gordo Member; *IS*, Independence Shale; *JH*, Juniper Hill Member; *O*, Owen Member; *OL*, lower Owen Pre-LKE; *OU*, upper Owen Post-LKE; *SW*, Sweetland Creek Shale.

REFERENCES

Anderson, W.I., 1966. Upper Devonian conodonts and the Devonian-Mississippian boundary of north-central Iowa. J. Paleontol. 40, 395–415.

Baker, C., Glenister, B.F., Levorson, C.O., 1986. Devonian ammonoid *Manticoceras* from Iowa. Proc. Iowa Acad. Sci. 93, 7–15.

Becker, R.T., Gradstein, F.M., Hammer, O., 2012. The Devonian Period. In: Gradstein, F.M., Ogg, J.G., Schmitz, M.D., Ogg, G.M. (Eds.), The Geologic Time Scale. Geological Society of Iowa, Iowa City, IA, pp. 559–601.

Belanski, C.H., 1928a. Peentameracea of the Devonian of Northern Iowa. University of Iowa Studies, University of Iowa, Iowa City, IA, vol. 7 (7), 34 pp.

Belanski, C.H., 1928b. Terebratulacea of the Devonian of northern Iowa. University of Iowa Studies, University of Iowa, Iowa City, IA, vol. 7 (8), 32 pp.

Branson, E.B., Mehl, M.G., 1934. Conodonts From the Grassy Creek Shale of Missouri. University of Missouri Studies, University of Missouri, Columbia, MO, vol. 8, pp. 171–259 [date of imprint, 1933].

Bruguière, J.G., 1797. Vers, Coquilles, Mollusques et Polypiers. In: Tableau encyclopédique et méthodiquedes trois règnes de la nature, 2. Agasse, Paris, pp. 96–314.

Bunker, B.J., Witzke, B.J., 1992. An upper Middle through lower Upper Devonian lithostratigraphic and conodont biostratigraphic framework of the Midcontinent Carbonate Shelf area, Iowa. In: Day, J., Bunker, B.J. (Eds.), Iowa Geological Survey Guidebook. vol. 16, pp. 3–26.

Bunker, B.J., Ludvigson, G.A., Witzke, B.J., 1985. The Plum River Fault Zone and the structural and stratigraphic framework of eastern Iowa. Iowa Geological Survey, Technical Information Series. vol. 13. 126 pp.

Bunker, B.J., Witzke, B.J., Day, J.E., 1986. Upper Cedar Valley Stratigraphy, North-Central Iowa, Lithograph City Formation. GeologicalSociety of Iowa Guidebook, vol. 44, 41 pp.

Bush, A.W., Csonka, D., DiRenzo, G.V., Over, D.J., Beard, J.A., 2015. Revised correlation of the Frasnian–Famennian boundary and Kellwasser Events (Upper Devonian) in shallow marine paleoenvironments of New York State. Palaeogeogr. Palaeoclimatol. Palaeoecol. 433, 233–246.

Calvin, S., 1878. On some dark shale recently discovered below the Devonian limestones, at Independence, Iowa; with a notice of its fossils and description of new species. US Geol. Surv. Bull. 4, 25–730.

Cooper, G.A., 1967. Correlation of the Devonian sedimentary Formations of North America. Geol. Soc. Am. Bull. 53, 1792–1794.

Cooper, G.A., Dutro, J.T., 1982. Devonian brachiopods of New Mexico. Bull. Am. Paleontol. 82–83, 215.

Copper, P., Chen, Y., 1995. *Invertina*, a new Middle Devonian atrypid brachiopod genus from south China. J. Paleontol. 69, 251–256.

Cushman, J.A., Stainbrook, M.A., 1943. Some foraminifera from the Devonian of Iowa. Cushman Lab. Foram. Res. Contr. 19 (4), 73–79.

Day, J., 1987. Revision, distribution and extinction of the Middle and Upper Devonian Palaeotrochid gastropod genera *Floyda* and *Turbonopsis*. J. Paleontol. 61, 960–973.

Day, J., 1988. Stratigraphy, Biostratigraphy, and Depositional History of the Givetian and Frasnian Strata in the San Andres and Sacramento Mountains of Southern New Mexico. 238 pp. PhD dissertation, University of Iowa, Iowa City.

Day, J., 1989. In: McMillan, N.J., Embry, A.F., Glass, D.J. (Eds.), The brachiopod succession of the late Givetian-Frasnian of Iowa. Devonian of the World: Proceedings of the Second International Symposium on the Devonian System, pp. 303–325. Canadian Society of Petroleum Geologists Memoir 14(3).

Day, J., 1990. The Upper Devonian (Frasnian) conodont sequence in the Lime Creek Formation of north-central Iowa and comparison with Lime Creek Ammonoid, brachiopod, foraminifer, and gastropod sequences. J. Paleontol. 64, 614–628.

Day, J., 1992. Middle-Upper Devonian (late Givetian-early Frasnian) brachiopod sequence in the Cedar Valley Group of central and eastern Iowa. In: Day, J., Bunker, B.J. (Eds.), The Stratigraphy, Paleontology, Depositional and Diagenetic History of the Middle-Upper Devonian Cedar Valley Group of Central and Eastern Iowa. Iowa Department of Natural Resources Guidebook, vol. 16, pp. 53–105.

Day, J., 1995. The brachiopod fauna of the Upper Devonian (late Frasnian) Lime Creek Formation of north-central Iowa, and related deposits in eastern Iowa. In: Bunker, B.J. (Ed.), Geology and Hydrogeology of Floyd-Mitchell Counties, pp. 21–40. Geological Society of Iowa Guidebook 62.

Day, J., 1996a. Faunal signatures of Middle-Upper Devonian depositional sequences and sea level fluctuations in the Iowa Basin: U.S. Midcontinent. In: Witzke, B.J.,

Ludvigson, G.A., Day, J. (Eds.), Paleozoic Sequence Stratigraphy, Views From the North American Craton, pp. 277–300. Geological Society of America Special Paper 306.

Day, J., 1996b. Faunal signatures of Middle-Upper Devonian depositional sequences and sea level fluctuations in the Iowa Basin: US mid-continent. In: Witzke, B.J., Ludvigson, G.A., Day, J. (Eds.), Paleozoic Sequence Stratigraphy, Views From the North American Craton, 277–300. Geological Society of America Special Paper 306.

Day, J., 1997. Phylogeny and biogeography of *Tecnocyrtina* (Brachiopoda-Spiriferinida) in the Devonian (Givetian-Frasnian) of North America. In: Klapper, G., Murphy, M.A., Talent, J.A. (Eds.), Paleozoic Sequence Stratigraphy, Biostratigraphy, and Biogeography: Studies in Honor of J. Granville ("Jess") Johnson. Geological Society of America Special Paper 321, pp. 245–261.

Day, J., 1998. Distribution of latest Givetian-Frasnian Atrypida (Brachiopoda) in central and western North America. Acta Palaeontol. Pol. 43, 205–240.

Day, J.E., 2006. Overview of the Middle-Upper Devonian sea level history of the Wapsipinicon and Cedar Valley Groups, with discussion of new conodont data from the subsurface Cedar Valley Group of southeastern Iowa. In: Day, J.E., Luczaj, J., Anderson, R. (Eds.), New Perspectives and Advances in the Understanding of Lower and Middle Paleozoic Epeiric Carbonate Depositional Systems of the Iowa and Illinois Basins. Iowa Geological Survey Guidebook, vol. 25, pp. 3–21.

Day, J., 2008. Stop 7—Rockford Brick and Tile Quarry—Floyd County Fossil and Prairie Center and Park Preserve. In: Groves, J.R., Walters, J.C., Day, J. (Eds.), Carbonate Platform Facies and Faunas of the Middle and Upper Devonian Cedar Valley Group and Lime Creek Formation, Northern Iowa, pp. 89–98. Iowa Geological Survey Guidebook 28.

Day, J., 2010a. New Developments and Refinements to Lower Cedar Valley Group Conodont Biostratigraphy in Eastern Iowa. Geological Society of Iowa Guidebook, vol. 87, pp. 93–104.

Day, J., 2010b. Subtropical record of Upper Devonian (late Frasnian–early Famennian) sea level events and Kellwasser extinction bioevents, Southern Ouachita Margin–Western Laurussia (Arizona–New Mexico, Central and Northern Mexico). Geol. Soc. Am. Abstr. Programs 42 (2), 89.

Day, J., 2013. Stop 5. Rockford Brick and Tile Quarry—Floyd County Fossil and Prairie Center and Park Preserve. In: Day, J., Witzke, B.J., Lundy, S. (Eds.), Aspects of the Paleozoic History of Epeiric Seas of the Iowa Basin, pp. 109–118. Iowa Geological Survey Guidebook 29.

Day, J., Copper, P., 1998. Revision of latest Givetian and Frasnian Atrypida (Brachiopoda) from central North America. Acta Palaeontol. Pol. 43, 155–204.

Day, J., Whalen, M.T., 2006. Extinction patterns of western Laurussian tropical benthic faunas suggest intensification of latitudinal temperature gradient and rapid climate change as major cause of Late Frasnian Kellwasser extinctions. Geol. Soc. Am. Abstr. Programs 38 (7), 267.

Day, J., Whalen, M.T., 2008. Record of Late Frasnian Kellwasser extinction bioevents, terminal Frasnian Upper Kellwasser Crisis Fauna, and Early Famennian post-extinction recovery: Alberta Rocky Mountains. Geol. Soc. Am. Abstr. Programs 40 (6), 435.

Day, J., Uyeno, T.T., Norris, A.W., Witzke, B.J., Bunker, B.J., 1996. Middle–Upper Devonian relative sea-level histories of North American cratonic interior basins. In: Witzke, B.J., Ludvigson, G.A., Day, J. (Eds.), Paleozoic Sequence Stratigraphy: Views From the North American Craton, pp. 259–276. Geological Society of America Special Paper 306.

Day, J., Witzke, B.J., Bunker, B., 2008. Overview of Middle and Upper Devonian Cedar Valley Group and Lime Creek Formation Carbonate Platform Facies, Faunas, and Event

Stratigraphy of Northern Iowa. In: Groves, J.R., Walters, J.C., Day, J. (Eds.), Carbonate Platform Facies and Faunas of the Middle and Upper Devonian Cedar Valley Group and Lime Creek Formation, Northern Iowa, pp. 15–39. Iowa Geological Survey Guidebook 28.

Day, J., Witzke, B.J., Bunker, B.J., 2013. Overview of the Middle and Upper Devonian northern shelf facies in the Iowa Basin. In: Day, J., Witzke, B.J., Lundy, S. (Eds.), Aspects of the Paleozoic History of Epeiric Seas of the Iowa Basin, pp. 1–40. Iowa Geological Survey Guidebook, 29.

Day, J., De Vleeschouwer, D., Witzke, B.J., Preslicka, J., Whalen, M.T., Claeys, P., 2014. Late Frasnian Kellwasser extinctions constrained by integrated conodont-brachiopod-ammonoid biostratigraphy, isotope chemostratigraphy and astronomical cyclostratigraphic calibration: Iowa Basin-Western Laurussia. Geol. Soc. Am. Abstr. Programs 46 (6), 756.

Day, J., De Vleeschouwer, D., Witzke, B.J., Whalen, M.T., Claeys, P., 2015a. Astrochronological calibration of Late Frasnian Sea Level and Bioevents from integrated bio-, chemo- and cyclostratigraphy: Iowa Basin, USA. In: Programs and Abstracts for the Second International Conference on Stratigraphy. STRATI 2 Conference, Graz, Austria.

Day, J., Witzke, B.J., Preslicka, J., 2015b. Basin-to-platform records of Kellwasser extinctions: Iowa Basin: Central North America. In: Program and Abstracts for the Second International Conference on Stratigraphy. STRATI 2 Conference, Graz, Austria.

Day, J., De Vleeschouwer, D., Witzke, B.J., Whalen, M.T., Claeys, P., 2016. Astrochronological calibration of Late Frasnian Sea Level and Bioevents: Iowa Basin, USA. Geol. Soc. Am. Abstr. Programs, 48 (5). http://dx.doi.org/10.1130/abs/2016NC-275682.

De Vleeschouwer, D., Day, J.E., Whalen, M.T., Claeys, P., 2014. A high-resolution chronology for the late Frasnian constructed by integrated bio-, chemo- and cyclostratigraphy. In: Kido, E., Waters, J.A., Ariunchimeg, Y.A., Sersmaa, G., Da Silva, A.C., Whalen, M., Suttner, T.J., Konigshof, P. (Eds.), Abstract Volume for the IGCP 596 & IGCP 580 Joint Meeting and Field-Workshop: International Symposium in Mongolia. Berichtedes Institutes für Erdwissenschaften, vol. 19. Karl-Franzens-Universität Graz, p. 22.

De Vleeschouwer, D., Da Silva, A.-C., Day, J.E., Whalen, M., Claeys, P., 2016. A global cyclostratigraphic framework constrains the timing and pacing of environmental changes over the Late Devonian mass extinction (Frasnian–Famennian). In: European Geosciences Union General Assembly 2016-Geophysical Research Abstracts. 18, EGU2016-8193-1.

Dutro Jr., J.T., 1981. Devonian brachiopod biostratigraphy of New York State. In: Oliver Jr., W.A., Klapper, G. (Eds.), Devonian Biostratigraphy of New York, Part 1, Text. International Union of Geological Sciences, Subcommission on Devonian Stratigraphy, Washington, DC, pp. 67–82.

Fenton, C.A., 1918. Some new brachiopods and gastropods from the Devonian of Iowa. Am. Midl. Nat. 5, 213–224.

Fenton, C.A., Fenton, M.A., 1924. The Stratigraphy and Fauna of the Hackberry Stage of the Upper Devonian. Contributions From the Museum of Geology. University of Michigan Museum of Paleontology, University of Michigan, Ann Arbor, Michigan, vol. 1, 260 pp.

Gibson, L.B., 1955. Upper Devonian ostracodes from the Cerro Gordo Formation of Iowa. Bull. Am. Paleontol. 35 (154), 38.

Girard, C., Klapper, G., Feist, R., 2006. Subdivision of the terminal Frasnian *linguiformis* conodont zone, revision of the correlative interval of Montagne Noire Zone 13, and discussion of stratigraphically significant associated trilobites. In: Over, D.J., Morrow, J.R., Wignall, P.B. (Eds.), Understanding Late Devonian and Permian-Triassic Biotic and Climatic Events: Towards an Integrated Approach, pp. 181–198. Developments in Palaeontology and Stratigraphy, vol. 20.

Gradstein, F.M., Ogg, J.G., Schmitz, M.D., Ogg, G.M. (Eds.), 2012. The Geologic Time Scale. Elsevier, Amsterdam, pp. 559–601.

Hall, J., 1858. Report on the geological survey of the State of Iowa. Part2, Chapter 8, Paleontology of Iowa, vol. 1, pp. 473–724. 29 pls.

Hall, J., Clarke, J.M., 1892. An introduction to the study of the genera of Palaeozoic Brachiopoda. Natural History of New York. Paleontology 8 (2), 217.

Hall, J., Whitfield, R.P., 1872a. Description of New Species of Fossils From the Devonian Rocks of Iowa. New York State Cabinet Natural History, Annual Report, 23, pp. 223–243.

Hall, J., Whitfield, R.P., 1872b. Description of New Fossils From the Vicinity of Louisville, Kentucky, and the Falls of the Ohio. New York State Museum Annual Report, vol. 24, pp. 181–200.

Hall, J., Whitfield, R.P., 1873. Descriptions of New Species of Fossils From the Devonian Rocks of Iowa. New York State Cabinet Natural History Annual Report, 23, pp. 223–243.

Hanson, D.W., Preslicka, J.E., 1996. Upper Devonian (Frasnian) Fauna from the Amana Beds at Price Creek, Iowa. Unpublished independent study thesis. University of Iowa, University of Iowa Department of Geoscience Repository.

House, M.R., Kirchgasser, W.T., 2008. Late Devonian Goniatites (Cephalopoda, Ammonoidea) from New York State. Bull. Am. Paleontol. 374, 288.

Hynes, A.R., Over, J., Day, J., 2014. Conodont biostratigraphy of the Sweetland Creek Shale and Grassy Creek Shale, Upper Devonian, Southeastern Iowa. Geol. Soc. Am. Abstr. Programs 46 (6), 420.

Joachimski, M.M., Van Geldern, R., Breisig, S., Buggisch, W., Day, J., 2004. Oxygen isotope evolution of biogenic calcite and apatite during the Middle and Late Devonian. Int. J. Earth Sci. 93, 542–553.

Joachimski, M.M., Breisig, S., Buggisch, W., Talent, J.A., Mawson, R.M., Gereke, M., Morrow, J.R., Day, J., Weddige, K., 2009. Devonian climate and reef evolution: insights from oxygen isotopes in apatite. Earth Planet. Sci. Lett. 284, 599–609.

Johnson, J.G., Klapper, G., 1992. North American Midcontinent Devonian T-R cycles, Norman, Oklahoma. In: Chaplin, J.R., Barrick, J.E. (Eds.), Special Papers in Paleontology and Stratigraphy: A Tribute to Thomas W. Amsden, pp. 127–135. Oklahoma Geological Survey Bulletin 145.

Johnson, J.G., Klapper, G., Sandberg, C.A., 1985. Devonian eustatic fluctuations in Euramerica. Geol. Soc. Am. Bull. 96, 567–587.

Johnson, J.G., Klapper, G., Elrick, M., 1996. Devonian transgressive–regressive cycles and biostratigraphy, northern Antelope Range, Nevada, establishment of reference horizons for global cycles. Palaios 11, 3–14.

Klapper, G., 1989. The Montagne Noire Frasnian (Upper Devonian) conodont succession. In: McMillan, N.J., Embry, A.F., Glass, D.J. (Eds.), Devonian of the World. Paleontology, Paleoecology, Biostratigraphy, pp. 449–468. Canadian Society Petroleum Geology Memoir 14 (3).

Klapper, G., 1990. Frasnian species of the Late Devonian conodont genus *Ancyrognathus*. J. Paleontol. 64, 998–1025.

Klapper, G., 1997. Graphic correlation of Frasnian (Upper Devonian) sequences in Montagne Noire, France, and western Canada. In: Klapper, G., Murphy, M.A., Talent, J.A. (Eds.), Paleozoic Sequence Stratigraphy, Biostratigraphy, and Biogeography: Studies in Honor of J. Granville ("Jess") Johnson, pp. 113–129. Geological Society America Special Paper 321.

Klapper, G., 2007a. Frasnian (Upper Devonian) conodont succession at Horse Spring and correlative sections, Canning Basin, Western Australia. J. Paleontol. 81, 513–537.

Klapper, G., 2007b. Conodont taxonomy and the recognition of the Frasnian/Famennian (Upper Devonian) stage boundary. Stratigraphy 4, 67–76.

Klapper, G., 2009. Upper Devonian conodonts in the Canning Basin, appendix 1. In: Playford, P.E., Hocking, R.M., Cockbain, A.E. (Eds.), Devonian Reef Complexes of the Canning Basin, Western Australia, pp. 405–413. Geological Survey of Western Australia Bulletin, 145.
Klapper, G., Foster, C.T., 1986. Quantification of outlines in Frasnian (Upper Devonian) platform conodonts. Can. J. Earth Sci. 23, 1214–1222.
Klapper, G., Foster, C.T., 1993. Shape analysis of Frasnian species of the Late Devonian conodont genus Palmatolepis. J. Paleontol. 32, 1–35.
Klapper, G., Furnish, W.M., 1963. Conodont zonation of the Early Upper Devonian in eastern Iowa. Proc. Iowa Acad. Sci. 69, 400–410.
Klapper, G., Kirchgasser, W.T., 2016. Frasnian Late Devonian conodont biostratigraphy in New York: graphic correlation and taxonomy. J. Paleontol. 90, 525–554.
Klapper, G., Lane, H.R., 1985. Upper Devonian (Frasnian) conodonts of the *Polygnathus* biofacies, N. W. T., Canada. J. Paleontol. 59, 904–951.
Klapper, G., Lane, H.R., 1989. Frasnian (Upper Devonian) conodont sequence at Luscar Mountain and Mount Haultain, Alberta Rocky Mountains. In: McMillan, N.J., Embry, A.F., Glass, D.J. (Eds.), Devonian of the World, Calgary, pp. 469–478. Canadian Society of Petroleum Geologists Memoir 14(3).
Klapper, G., Uyeno, T.T., Armstrong, D.K., Telford, P.G., 2004. Conodonts of the Williams Island and Long Rapids formations (Upper Devonian, Frasnian-Famennian) of the Onakawana B drillhole, Moose River Basin, northern Ontario, with a revision of Lower Famennian species. J. Paleontol. 78, 371–387 (Correction in 78, p. 1221.).
Lynn, R.J., 1978. Petrology of the Upper Devonian Owen Member of the Lime Creek Formation. University of Iowa, Iowa. Unpublished M.S. thesis, 83 pp.
Ma, X., Day, J., 2000. Revision of *Tenticospirifer* Tien, 1938 and similar spiriferid brachiopods from the Middle-Late Devonian (Givetian-Frasnian) of North America, Eurasia, and Australia. J. Paleontol. 74, 444–463.
Ma, X., Day, J., 2003. Revision of North American and selected Eurasian Late Devonian (Frasnian) species of *Cyrtospirifer* and *Regelia* (Brachiopoda). J. Paleontol. 77, 267–292.
McLean, R.A., Sorauf, J.E., 1989. The distribution of Rugose Corals in Frasnian outcrop sequences of North America. In: McMillan, N.J., Embry, A.F., Glass, D.J. (Eds.), Devonian of the World, pp. 379–396. Canadian Society of Petroleum Geologists Memoir 14(3).
Metzger, R., 1989. Upper Devonian (Frasnian-Famennian) conodont biostratigraphy in the subsurface of north-central Iowa and southeast Nebraska. J. Paleontol. 63, 503–524.
Miller, S.A., 1883, Catalogue of American Paleozoic Fossils. 2nd Edition, 298 p.
Miller, A.K., Youngquist, W., 1947. Conodonts from the type section of the Sweetland Creek Shale in Iowa. J. Paleontol. 21, 501–517.
Müller, K.J., Müller, E.M., 1957. Early Upper Devonian (independence) conodonts from Iowa. J. Paleontol. 31, 1069–1108.
Over, D.J., 2002. The Frasnian-Famennian boundary in the Appalachian Basin, Michigan Basin, Illinois Basin, and southern continental margin, central and eastern United States. Palaeogeogr. Palaeoclimatol. Palaeoecol. 181 (1–3), 153–170.
Over, D.J., 2006. Frasnian-Famennian boundary at the type Sweetland Creek Shale locality. In: Day, J., Luczaj, J., Anderson, R. (Eds.), New Perspectives and Advances in the Understanding of Lower and Middle Paleozoic Epeiric Carbonate Depositional Systems of the Iowa and Illinois Basins, pp. 74–79. Iowa Geological Survey Guidebook 26.
Ovnatanova, N.S., 1969. Some new conodonts from the Upper Frasnian and Lower Famennian deposits of the Central Russian Platform and South Timan. Transactionof All-Union Research Geological Prospecting Oil Institute (VN1GNI), vol. 93, pp. 139–141. in Russian.

Preslicka, J.E., Newsom, C.R., Blume, T.E., 2013. Ammonoids from the Devonian of Iowa. In: Day, J., Witzke, B.J., Lundy, S. (Eds.), Aspects of the Paleozoic History of Epeiric Seas of the Iowa Basin, pp. 63–72. Iowa Geological Survey Guidebook 29.

Rogers, F.S., 1998. Conodont biostratigraphy of the Little Cedar and lower Coralville formations of the Cedar Valley Group (Middle Devonian) of Iowa and significance of a new species of *Polygnathus*. J. Paleontol. 72, 726–737.

Sandberg, C.A., Morrow, J., Ziegler, W., 2002. Late Devonian sea-level changes, catastrophic events, and mass extinctions. In: Koeberl, C., MacLeod, K.G. (Eds.), Catastrophic Events and Mass Extinctions: Impacts and Beyond, pp. 473–487. Geological Society of America Special Paper 356.

Sandberg, C.A., Morrow, J.R., Poole, F.G., Ziegler, W., 2003. Middle Devonian to Early Carboniferous event stratigraphy of Devils Gate and Northern Antelope Range sections, Nevada, U.S.A. Cour. Forschungsinst. Senck. 242, 187–207.

Sanks, K., Day, J., Witzke, B.J., 2016. Latest Frasnian (Upper Devonian) Lower Kellwasser Extinction (LKE) subtropical carbonate platform survivor-recovery fauna-central North America. Geol. Soc. Am. Abstr. Programs 48 (7). http://dx.doi.org/10.1130/abs/2016AM-285521.

Sanneman, N.D., 1955. Beitrag zur Untergliederung des Oberdevons nach Conodonten. Neues Jarhbuch Geol. Paläontol. Abhandlungen 100, 324–331.

Savage, N.M., Funai, C.A., 1980. Devonian conodonts of probable early Frasnian age from the Coronados Islands of southeastern Alaska. J. Paleontol. 54, 806–813.

Sorauf, J.E., 1998. Frasnian (Upper Devonian) rugose corals from the Lime Creek and Shell Rock Formations of Iowa. Bull. Am. Paleontol. 113 (355), 1–159.

Spalletta, C., Perri, M.C., Over, D.J., Corradini, C., 2017. Famennian (Upper Devonian) conodont zonation: revised global standard. Bull. Geosci. 92 (1), 31–57.

Stainbrook, M.A., 1945. Brachiopoda of the Independence Shale of Iowa. Geol. Soc. Am. Mem. 14, 74.

Stauffer, C.R., 1938. Conodonts of the Olentangy Shale. J. Paleontol. 12, 411–443.

Stock, C.W., 1984a. Upper Devonian Stromatoporoidea of north-central Iowa: redescription of the type species of Hall and Whitfield (1873). J. Paleontol. 58, 773–788.

Stock, C.W., 1984b. The distribution of stromatoporoids in the Upper Devonian of north-central Iowa. In: Anderson, W.I. (Ed.), General Geology of North-Central Iowa. Guidebook for the 48th Annual Tri-State Geological Field Conference, pp. 125–129.

Stock, C.W., 2008. Stromatoporoid biostratigraphy of the Iowa Frasnian. In: Groves, J.R., Walters, J.C., Day, J. (Eds.), Carbonate Platform Facies and Faunas of the Middle and Upper Devonian Cedar Valley Group and Lime Creek Formation, Northern Iowa. Iowa Geological Survey Guidebook 28, pp. 43–52.

Stock, C.W., 2013. Stromatoporoid Biostratigraphy of the Iowa Givetian and Frasnian. In: Day, J., Witzke, B.J., Lundy, S. (Eds.), Aspects of the Paleozoic History of Epeiric Seas of the Iowa Basin, pp. 53–62. Iowa Geological Survey Guidebook 29.

Strimple, H.L., Leverson, C.O., 1969. Catalogue of type specimens of the Belanski Collection. Bull. Am. Paleontol. 56 (251), 259–271.

Sullivan, N., Day, J., 2008. Very Late Frasnian (Upper Devonian) Brachiopod Fauna of the *Theodossia Scopulorum* Zone in the Paleoequatorial Mackenzie Shelf (Northwest Territories-Canada)–Western Laurussia. Geol. Soc. Am. Abstr. Programs 40 (6), 475.

Thomas, A.O., 1922. Some new Paleozoic glass-sponges from Iowa. Proc. Iowa Acad. Sci. 24, 85–91.

Thomas, A.O., Stainbrook, M.A., 1922. Status of certain rhynchonellid brachiopods from the Devonian of Iowa. Iowa Acad. Sci. Proc. 29, 93–99.

Ulrich, E.O., Bassler, R.S., 1926. A classification of the toothlike fossils, conodonts, with descriptions of American Devonian and Mississippian species. Proc. U. S. Natl. Mus. 68 (12), 61.

Van Geldern, R., Joachimski, M.M., Day, J., Jansen, U., Alaverez, F., Yolkin, E.A., Ma, X.-P., 2006. Carbon, oxygen, and strontium isotope records of Devonian brachiopod shell calcite. Palaeogeogr. Palaeoclimatol. Palaeoecol. 240, 47–67.

Webster, C.L., 1888. Lime Creek Brachiopods. Am. Nat. 22, 1100–1104.

Webster, C.A., 1908. Fauna below the Hackberry Group at Rockford Iowa. Contrib. Sci. 1, 5–8.

Webster, C.A., 1921. Notes on the genus *Atrypa*, with description of new species. Am. Midl. Nat. 7, 13–20.

Whalen, M.T., Day, J.E., 2008. Magnetic susceptibility, biostratigraphy, and sequence stratigraphy: insights into Devonian carbonate platform development and basin infilling, Western Alberta. In: Lukasik, J., Simo, J.A. (Eds.), Papers on Phanerozoic Reef Carbonates in Honor of Wolfgang Schlager, pp. 291–314. Society for Sedimentary Geology, Special Publication 89.

Whalen, M.T., Day, J., 2010. Cross-basin variations in magnetic susceptibility influenced by changing sea level, paleogeography, and paleoclimate; Upper Devonian, Western Canada Sedimentary Basin. J. Sediment. Res. 80, 1109–1127.

Whalen, M.T., De Vleeschouwer, D., Payne, J.H., Day, J.E., Over, D.J., Claeys, P., 2016. Pattern and timing of the late Devonian biotic crisis in western Canada: insights from carbon isotopes and astronomical calibration of magnetic susceptibility data. In: New Advances in Devonian Carbonates: Outcrop Analogs, Reservoirs, and Chronostratigraphy. SEPM Special Publication 107, p. 17.

White, C.A., 1862. Description of new species of fossils from the Devonian and Carboniferous rocks of the Mississippi Valley. Boston Soc. Nat. History Proc. 9, 8–33, 5 fig.

Witzke, B.J., 1987. Models for circulation patterns in epicontinental seas applied to Paleozoic facies of the North America craton. Paleoceanography 2, 229–248.

Witzke, B.J., Bunker, B.J., 1997. Sedimentation and stratigraphic architecture of a Middle Devonian (late Givetian) transgressive-regressive carbonate-evaporite cycle, Coralville Formation, Iowa area. In: Klapper, G., Murphy, M.A., Talent, J.A. (Eds.), Paleozoic Sequence Stratigraphy, Biostratigraphy, and Biogeography: Studies in Honor of J. Granville ("Jess") Johnson: Geological Society of America, Special Paper 321, pp. 67–88.

Witzke, B.J., Bunker, B.J., 2002. Bedrock Geology in the Burlington Area, Southeastern Iowa. Iowa Geological Survey Guidebook 23, pp. 23–52.

Witzke, B.J., Bunker, B.J., 2006. Middle shelf facies of the Cedar Valley Group (Devonian) and their stratigraphic relationships in eastern Iowa. In: Day, J.E., Luczaj, J., Anderson, R. (Eds.), New Perspectives and Advances in the Understanding of Lower and Middle Paleozoic Epeiric Carbonate Depositional Systems of the Iowa and Illinois Basins, pp. 23–46. Iowa Geological Survey Guidebook 25.

Witzke, B.J., Bunker, B.J., Klapper, G., 1985. Devonian stratigraphy in the Quad Cities area, eastern Iowa-northwestern Illinois. In: Hammer, W.R., Anderson, R.C., Schroeder, D.A. (Eds.), Devonian and Pennsylvanian Stratigraphy of the Quad-Cities Region, Illinois-Iowa. Society of Economic Paleontologists and Mineralogists, Great Lakes Section, 15th Annual Field Conference Guidebook, pp. 19–64.

Witzke, B.J., Bunker, B.J., Rogers, F.S., 1989. Eifelian through Lower Frasnian stratigraphy and deposition in the Iowa area, Midcontinent, U.S.A. In: McMillan, N.J., Embry, A.F., Glass, D.J. (Eds.), Devonian of the World, pp. 221–250. Canadian Society of Petroleum Geologists, Memoir 14 (1).

Youngquist, W.L., 1947. A new Upper Devonian conodont fauna from Iowa. J. Paleontol. 21, 95–112.

Youngquist, W.L., Miller, A.K., 1948. Additional conodonts from the Sweetland Creek Shale of Iowa. J. Paleontol. 22, 440–450.
Ziegler, W., 1958. Conodontenfeinstratigraphische Untersuchungen an der Grenze Mitteldevon/Oberdevon und in der Adorfstufe. Notizbl. Hess. Landesamt Bodenforsch. 87, 7–77.
Ziegler, W., 1981. Ancyrognathus bifurcatus, A. calvini, A. irregularis, A. symmetricus, A. triangularis. In: Ziegler, W. (Ed.), Catalogue of Conodonts, IV: 1–14. E. Schweizerbart'sche Verlagsbuchhandlung, Stuttgart, pp. 17–25.
Ziegler, W., Ovnatanova, N., Kononova, L., 2000. Devonian polygnathids from the Frasnian of the Rheinisches Schiegergebirge, Germany, and the Russian Platform. Senckenb. Lethaea 80, 593–645.

FURTHER READING

Anderson, W.I., Furnish, W.M., 1987. The Lime Creek Formation of north-central Iowa. In: Biggs, D.L. (Ed.), North-Central Section. Geological Society of America Centennial Field Guide, 3, pp. 89–92.
Bassler, R.S., 1925. Classification and stratigraphic use of the conodonts. Geol. Soc. Am. Bull. 36, 218–220.
Becker, R.T., House, M.R., 1997. Sea-level changes in the Upper Devonian of the Canning Basin, Western Australia. Cour. Forschungsinst. Senck. 199, 129–146.
Becker, R.T., Feist, R., Flajs, G., House, M.R., Klapper, G., 1989. Frasnian–Famennian extinction events in the Devonian at Coumiac, southern France. C. R. Acad. Paris Sér. II 309, 259–266.
Becker, R.T., House, M.R., Kirchgasser, W.T., Playford, P.E., 1991. Sedimentary and faunal changes across the Frasnian/Famennian boundary in the Canning Basin of Western Australia. Hist. Biol. 5, 183–196.
Brice, D., 1988. Brachiopodes du Dévonien de Ferques (Boulonnais–France). In: Brice, D. (Ed.), Le Dévonien de Ferques. Bas-Boulonnais (N. France). Biostratigraphie du Paléozoïque, vol. 7, pp. 323–395.
Calvin, S., 1883. On the fauna found at Lime Creek, Iowa, and its relation to other geological Faunas. Am. J. Sci. 25, 432–436.
Calvin, S., 1897. Geology of Cerro Gordo County. Iowa Geol. Surv. Annu. Rep. 7, 117–195.
Copper, P., 1973. New Siluro-Devonian atrypoid brachiopods. J. Paleontol. 47, 484–507.
Copper, P., 1978. Devonian atrypoids from western Canada. In: Stelck, C.R., Chatterton, B.D.E. (Eds.), Western and Artic Canadian Biostratigraphy, pp. 289–331. Geological Association of Canada Special Paper 18.
Copper, P., 2002a. Reef development at the Frasnian/Famennian mass extinction boundary. Palaeogeogr. Palaeoclimatol. Palaeoecol. 181, 27–65.
Copper, P., 2002b. Silurian and Devonian reefs: 80 million years of global greenhouse between two ice ages. In: Kiessling, W., Flügel, E., Golonka, J. (Eds.), Phanerozoic Reef Patterns. Society for Sedimentary Geology (SEPM) Special Publication 72, pp. 181–238.
Day, J., Whalen, M.T., 2001. Temporal and geographic extinction patterns of Late Devonian (Late and Latest Frasnian) platform shelly faunas: Kellwasser mass extinction bioevents: western Laurussia. Geol. Soc. Am. Abstr. Programs 33 (4), 8.
Day, J., De Vleeschouwer, D., Witzke, B.J., Whalen, M.T., and Claeys, P., 2016. Astrochronological calibration of late Frasnian sea level and bioevents: Iowa Basin, USA. Geol. Soc. Am. Abstr. Programs, 48(5). http://dx.doi.org/10.1130/abs/2016NC-27568.
De Vleeschouwer, D., Whalen, M.T., Day, J.E., Claeys, P., 2012. Cyclostratigraphic calibration of the Frasnian (Late Devonian) time scale (western Alberta, Canada). Geol. Soc. Am. Bull. 124, 928–942.

Fenton, C.A., 1919. Hackberry Stage of the Upper Devonian. Am. J. Sci. 48, 355–376.
Fenton, C.A., 1930. Studies on the genus *Atrypa*. Am. Midl. Nat. 12, 1–13.
Fenton, C.A., 1931. Studies of the Evolution of the Genus *Spirifer*. Wagner Free Institute of Science II. 436 pp.
Fenton, C.A., 1932. Alate shell lamellae and spines in the genus *Atrypa*. Am. Midl. Nat. 8, 203–221.
Fenton, C.A., 1935. Atrypae described by Clement L. Webster and related forms (Devonian, Iowa). J. Paleontol. 9, 369–384.
Glenister, B.F., Klapper, G., 1966. Upper Devonian conodonts from the Canning Basin, Western Australia. J. Paleontol. 440, 777–842.
Gouwy, S., Bultynck, P., 2000. Graphic correlation of Frasnian sections (Upper Devonian) in the Ardennes, Belgium. Bull. Inst. R. Sci. Nat. Belg. Sci. Terre 70, 25–52.
Gouwy, S., Haydukiewicz, J., Bultynck, P., 2007. Conodont-based graphic correlation of upper Givetian–Frasnian sections of the eastern Anti-Atlas (Morocco). Geol. Q. 51, 375–392.
Hall, J., 1867. Descriptions of the fossil Brachiopoda of the Upper Helderberg, Hamilton, Portage, and Chemung groups: New York Natural History Survey. Paleontology 4 (1), 422.
House, M.R., 1962. Observations on the ammonoid succession of the North American Devonian. J. Paleontol. 36, 247–284.
House, M.R., Becker, R.T., Feist, R., Flajs, G., Girard, C., Klapper, G., 2000. The Frasnian/Famennian GSSP at Coumiac, southern France. In: Bultynck, P. (Ed.), Subcommission on Devonian Stratigraphy: Recognition of Devonian Series and Stage Boundaries in Geological Areas, pp. 59–75. Courier Forschungsinstitut Senckenberg 225.
Huddle, J.W., 1968. Redescription of Upper Devonian Conodont Genera and Species Proposed by Ulrich and Bassler in 1926. US Geological Survey Professional Paper, 578: I–IV, p. 55.
Huddle, J.W., 1970. Revised descriptions of some Late Devonian polygnathid conodonts. J. Paleontol. 44, 1029–1040.
Joachimski, M.M., Buggisch, W., 2002. Conodont apatite delta signatures indicate climatic cooling as a trigger of the Late Devonian (F-F) mass extinction. Geology 30, 711–714.
Klapper, G., 2000. Species of Spathognathodontidae and Polygnathidae (Conodonta) in the recognition of Upper Devonian stage boundaries. In: Bultynck, P. (Ed.), Subcommission on Devonian Stratigraphy: Fossil Groups Important for Boundary Definition, pp. 153–159. Courier Forschungsinstitut Senckenberg, 220.
Klapper, G., Becker, T.R., 1999. In: Serpagli, E., Coradini, C. (Eds.), Comparison of Frasnian (Upper Devonian) conodont zonations. Studies on Conodonts: Proceedings of the Seventh European Conodont Symposium, Bologna-Modena 23–25 June, 1998, pp. 339–348. Bolletin. Italian Society Paleontology, vol. 37(2–3).
Klapper, G., Feist, R., Becker, R.T., House, M.R., 1994. Definition of the Frasnian/Famennian stage boundary. Episodes 16, 433–441(imprint, 1993).
Klapper, G., Kirchgasser, W.T., Baesemann, J.F., 1995. Graphic correlation of a Frasnian (Upper Devonian) composite standard. In: Mann, K.O., Lane, H.R. (Eds.), Graphic Correlation, pp. 177–184. Society for Sedimentary Geology Special Publication 53.
Klapper, G., Kuz'min, A.V., Ovnatanova, N.S., 1996. Upper Devonian conodonts from the Timan-Pechora region, Russia, and correlation with a Frasnian composite standard. J. Paleontol. 70, 131–152.
McLean, R.A., Klapper, G., 1998. Biostratigraphy of Frasnian (Upper Devonian) strata in western Canada, based on conodonts and rugose corals. Bull. Can. Petrol. Geol. 46, 515–563.

Miller, A.K., 1936. In: Devonic ammonoids of Iowa. Proceeding of the Iowa Academy of Sciences, Fiftieth Meeting, April 3, pp. 336–338.
Morrow, J.R., 2000. Shelf-to-basin lithofacies and conodont paleoecology across Frasnian–Famennian (F-F, mid-Late Devonian) boundary, central Great Basin (Western U.S.A.). Cour. Forschungsinst. Senck. 219, 1–57.
Morrow, J.R., Sandberg, C.A., 2003. Late Devonian sequence and event stratigraphy across the Frasnian–Famennian (F-F) boundary, Utah and Nevada, Chapter 10. In: Harries, P.J. (Ed.), Approaches in High-Resolution Stratigraphic Paleontology. Springer Publishers. Topics in Geobiology.
Over, D.J., 1997. Conodont biostratigraphy of the Java Formation (Upper Devonian) and the Frasnian-Famennian boundary in western New York State. In: Klapper, G., Murphy, M.A., Talent, J.A. (Eds.), Paleozoic Sequence Stratigraphy, Biostratigraphy, and Biogeography: Studies in Honor of J. Granville ("Jess") Johnson. Geological Society of America Special Paper 321, pp. 161–177.
Over, D.J., Baird, G.C., Kirchgasser, W.T., 2013. Middle–Upper Devonian strata along The Lake Erie shore, western New York. In: Deakin, A.K., Lash, G.G. (Eds.), New York State Geological Association Field Trip Guidebook, SUNY Fredonia. Fredonia, New York, pp. 182–219.
Owen, D.D., 1852. Report of a Geological Survey of Wisconsin, Iowa, and Minnesota. Lippencott Gambo and Company, Philadelphia. 636 pp.
Sandberg, C.A., Ziegler, W., Dreesen, R., Butler, J.L., 1988. Late Frasnian mass extinction: conodont event stratigraphy, global changes, and possible causes. Cour. Forschungsinst. Senck. 102, 263–307.
Sandberg, C.A., Poole, F.G., Johnson, J.G., 1989a. Upper Devonian of Western United States. In: McMillan, N.J., Embry, A.F., Glass, D.J. (Eds.), Devonian of the World. Regional Syntheses, pp. 183–220. Canadian Society of Petroleum Geologists Memoir 14 (1). [imprint, 1988].
Sandberg, C.A., Ziegler, W., Bultynck, P., 1989b. New standard conodont zones and early *Ancyrodella* phylogeny across Middle-Upper Devonian boundary. In: Walliser, O.H., Ziegler, W. (Eds.), Contributions to Devonian Palaeontology and Stratigraphy: Part I, Chinese—German Collaboration, Part II: Various Devonian Topics, pp. 195–230. Courier Forschungsinstitut Senckenberg, 110.
Sandberg, C.A., Morrow, J.R., Warme, J.E., 1997. Late Devonian Alamo impact event, global Kellwasser events, and major eustatic events, eastern Great Basin, Nevada and Utah. Brigham Young Univ. Geol. Stud. 42 (1), 129–160.
Stock, C.W., 2005. Devonian stromatoporoid originations, extinctions, and paleobiogeography: how they relate to the Frasnian-Famennian extinction. In: Over, D.J., Morrow, J.R., Wignall, P.B. (Eds.), Understanding Late Devonian and Permian-Triassic Biotic and Climatic Events: Towards an Integrated Approach. Elsevier, Amsterdam, pp. 71–92.
Stock, C.W., Turner, M.S., 2006. Stromatoporoids in the Idlewild Member of the Lithograph City Formation in Iowa: how they reflect global paleobiogeographic trends of the Middle-Late Devonian. In: Day, J.E., Luczaj, J., Anderson, R. (Eds.), New Perspectives and Advances in the Understanding of Lower and Middle Paleozoic Epeiric Carbonate Depositional Systems of the Iowa and Illinois Basins, pp. 81–87. Iowa Geological Survey Guidebook Series 25.
Thomas, A.O., 1920. Echinoderms of the Iowa Devonian. Proc. Iowa Acad. Sci. 56, 387–550.
Walliser, O.H., 1996. Global events in the Devonian and Carboniferous. In: Walliser, O.H. (Ed.), Global Events and Event Stratigraphy in the Phanerozoic. Springer, Berlin, pp. 225–250.

Webster, C.A., 1909. Illustration and description of some fossil species from the Hackberry Group. Iowa Nat. 5, 1–11.
Whalen, M.T., Śliwinski, M.G., Payne, J.H., Day, J.E., Chen, D., Da Silva, A.-C., 2015. Chemostratigraphy of the Late Devonian Frasnian-Famennian transition in Western Canada and Southern China: implications for carbon and nutrient cycling and mass extinction. In: Da Silva, A.C., Whalen, M.T., Hladil, J., Chadimova, L., Chen, D., Spassov, S., Boulvain, F., Devleeschouwer, X. (Eds.), Magnetic Susceptibility Application: A Window Onto Ancient Environments and Climatic Variations, pp. 37–72. Geological Society, London, Special Publication 414.
Witzke, B.J., Day, J., Bunker, B., 2009. Middle-Upper Devonian (Middle Givetian-Early Famennian) record of relative sea level and climate change in the Iowa and Western Illinois Basins, Western Laurussia. Over, D.J. (Ed.), Paleontographica Americana, vol. 64, pp. 229–231.
Youngquist, W.L., 1945. Upper Devonian conodonts from the Independence Shale (?) of Iowa. J. Paleontol. 19, 355–367.
Ziegler, W., Sandberg, C.A., 1984. Palmatolepis-based revision of the upper part of the standard Late Devonian conodont zonation. In: Clark, D.L. (Ed.), Conodont Biofacies and Provincialism, pp. 179–194. Geological Society of America, Special Paper 196.
Ziegler, W., Sandberg, C.A., 1990. The Late Devonian standard conodont zonation. Cour. Forschungsinst. Senck. 121, 115.

CHAPTER SEVEN

Challenging Darwin: Evolution of Triassic Conodonts and Their Struggle for Life in a Changing World

Ali M. Kılıç*[,1], Pablo Plasencia[†], Jean Guex[‡], Francis Hirsch[§]
*Balıkesir University, Balıkesir, Turkey
[†]Universidad de Valencia, Valencia, Spain
[‡]University of Lausanne, Lausanne, Switzerland
[§]University of Naruto, Naruto, Japan
[1]Corresponding author: e-mail address: alimurat@balikesir.edu.tr

Contents

Stratigraphy & Timescales, Volume 2
ISSN 2468-5178
http://dx.doi.org/10.1016/bs.sats.2017.08.003

Abstract

The phylogeny and distribution of Triassic conodonts reveal many aspects of their natural history. Conodonts incorporate the morphologic response to temperature as well as to eustatic cycles. Speciation, radiation, and extinction are not fortuitous and evolution uses heterochrony (progenesis and neoteny) in response to stress-generating events. Proteromorphosis (reappearance of ancestral morphs) and paedomorphosis (retention of juvenile traits) is a reaction to sublethal environmental stress. This often follows radiation of fully developed forms in the recovery stage after extinction that timely matches transgressions. Evolutionary retrogradation (neoteny) during eustatic high stands often precedes extinction. The 51-Ma long Triassic Period consists of the Induan–Olenekian first 5 Ma with a post Permian extinction recovery rate of 13 speciations/Ma. The next 10 Ma of the Anisian–Ladinian saw a drop in speciations to 8 S/Ma, but during the remaining Carnian–Rhaetian 36 Ma, the rate was down to below 2 S/Ma, until the end of the Triassic. Roughly a dozen Permian survivors became extinct during the Induan. Soon after, Gondolellidae saw a revival with the *Neospathodus–Kashmirella* lineage. The *Paragondolella–Misikella* lineage emerged in the Aegean (Early Anisian) and dominated the Triassic scene for the next 45 Ma until extinction. In pelagic environments Gladigondolellidae ranged from Late Spathian until Mid-Carnian. *Neogondolella* became extinct by the Late Ladinian, and *Paragondolella* was replaced by *Metapolygnathus* during Julian times. The evolutionary trends of a free blade, platform adornment, widening of the posterior end, and bifurcation of the basal cavity were initiated sporadically during Late Ladinian and became the norm, reaching their peak in Early Norian *Ancyrogondolella*. Strongly ornamented *Mazzaella* and *Carnepigondolella* marked the Julian Pluvial Event and the Late Tuvalian LST (lowstand systems tract). The atavistic (neotenic) *Norigondolella* came in around the Tuvalian–Lacian boundary and a generalized posterior acumination set in with Alaunian (Middle Norian) *Mockina*. Catastrophes of various origin, such as global LST events and temperature rises, precipitated retrogradation with Dienerian *Neospathodus*, Pelsonian *Nicoraella*, Illyrian *Pseudofurnishius*, Cordevolian *Mosherella*, Tuvalian *Neocavitella*, and Sevatian–Rhaetian *Misikella*. Bithynian

Kamuellerella–Ketinella–Gedikella in the Istanbul Zone define a North Tethyan Prepontic faunal subprovince. Late Anisian early forms of *Pseudofurnishius* initiated a Ladinian Sephardic faunal Province along the northern Gondwanian shelf. A stepwise decrease in speciation linked Triassic conodont evolution in unison to climatic–eustatic events.

En science on peut aisément donner des nouveaux noms,
à condition de ne pas créer d'idées nouvelles.
*(In science one can easily give new names,
Provided one does not create new ideas)*

Rodolphe Truempy

1. INTRODUCTION

In the second half of the 20th Century Louis Dollo's law of irreversibility (1893) was widely taught, with the result that the young paleontology student was well insulated from Neo-Lamarckian heresy.

Evolution was Darwinian, and the only innovation might have been the theory of punctuated equilibria as an alternative to phyletic gradualism. The significance of "gaps" in the fossil record was a recurrent source for proposals of spontaneous generation in alternation to the painful incompleteness of the fossil record. The wonderful books of Stephen J. Gould such as Ontogeny and Phylogeny (1977) have illustrated these principles of evolution and have generated interest in evolutionary research in the most eloquent way, even if contested by subsequent research.

This brought the tragedies of Darwin's life closer and gave us the satisfaction of living in the 20th century, well armed with the tools of radiometry on the one hand and the world of Mendelian genetics on the other. Revolutions were below the horizon with the improvements of such geochemical skills as $\delta^{13}C$ and $\delta^{18}O$. The world of oil exploration with its thousands of kilometers of drilling records started to draw a number of rules concerning the tracing of eustatic sea level changes and led to the famous eustatic sea level curve of Haq et al. (1988). This graphic representation of the coming and going of transgressions and regressions was an eye opener for the senior author, when trying to put the succession of taxa into a phylogenetic tree (Hirsch, 1994b). The revelation was that phylogeny seemed to depend on the eustatic curve and those certain aberrant taxa constantly reappeared at lows, to disappear again with the next transgression. The interplay of measurements, such as carbon and oxygen excursions plotted into the sea level curve, would soon give a sense to the evolutionary aberrations observed in the fossil record.

Also a full literature exists about so-called Lazarus, Elvis, and zombie taxa.

Observations of the nonconventional behavior in the development of fossils are not new. Paleozoic and Mesozoic strata abound with examples, such as the unrolling of ammonites. It was Guex who pointed to such simplifications in Jurassic ammonites and who coined the term proteromorphosis for the phenomenon.

Gould tried to demonstrate the absence of a global increase in complexity during the evolution of living organisms in his famous book, "Full House," which was published in 1996. Guex (2016) considered Gould's argument invalid because size decreases are always related to environmental stress, a phenomenon known since the beginning of the 20th century (Shimer, 1908; see also Mancini, 1978; after Guex, 2016).

When examining the conodont phylogeny, their taxonomy incorporates the morphologic riposte to temperature as well as to eustatic cycles, expressed in speciation, radiation, and extinction as these are not fortuitous and evolution uses diverse strategies such as heterochrony (progenesis and neoteny) in response to stress-generating events. Proteromorphosis (reappearance of ancestral morphs) and paedomorphosis (retention of juvenile traits) is a reaction to sublethal environmental stress and is often followed by radiation of fully developed forms, in the recovery stage after extinction, timely matching transgressions. Evolutionary retrogradation (neoteny) during eustatic high stands often precedes extinction. This was the case of the Alaunian *Mockina* whereafter the ultimate proteromorph *Misikella* brought no post-Rhaetian recovery. The cryptic gondolellid features, encoded in "neospathid" proteromorphs permitted the conodont survival throughout the entire Triassic, signaling Dienerian, Anisian, Ladinian, Carnian, and Norian crises, but ultimately to no avail during the terminal Rhaetian. The Permian–Triassic evolution of platform conodonts (Gondolellidae) consists mainly in the development of the carina and the platform. During the sublethal environmental stress conditions subsequent to the Permian–Triassic extinction, the Wuchiapingian–Griesbachian *Clarkina* lineage is replaced by the primitive looking platform-lacking Dienerian–Aegean *Neospathodus kummeli–Kashmirella timorensis* lineage. It is assumed that, above *Jinogondolella denticulata,* end of the Capitanian *Jinogondolella* lineage, "*Neospathodus*" *arcucristatus*, an atavistic blade-like homeomorph that lacks a platform, underlies *Protoclarkina crofti*, of the base of the anagenetic *Clarkina* lineage. These primitive-looking forms are derived from their immediate ancestors by retrograde evolution, a phenomenon that has been described as proteromorphosis. Such events suggest that proteromorphosis occurs during periods of crisis, with the sudden reappearance of homeomorphic forms that are atavistic representatives of the clade. The phenomenon is further substantiated by several additional retrogradations that pace the Triassic, a period prone to such events (Kiliç et al., 2016).

The Late Triassic represents an extremely long time span of 37 Ma, which is 70% of the total length of the period. Evolutionary rebounds after quasi extinction of subfamily Neogondolellinae took place by radiation out of the single surviving genus *Paragondolella*. First by Julian genera *Metapolygnathus* and *Mazzaella*, thereafter by the Tuvalian–Lacian genera *Quadralella–Carnepigondolella–Ancyrogondolella*. The survival of the clade throughout Alaunian and Sevatian time took place by successive retrogradations (proteromorphosis) of the Alaunian *Mockina* and Sevatian–Rhaetian *Misikella*, bringing no ultimate post-Rhaetian recovery.

2. METHODS

2.1 Triassic Conodont Databases (De Renzi et al., 1996; Plasencia, 2009; Plasencia et al., 2013)

They reveal the impact of periodic environmental stress, from the immediate aftermath of the end Permian mass extinction to subsequent sea level rises, cyclic changes, sudden regressions, and the gradual sea level drop that lasted 27 Ma until the end of the Triassic Period. The study of these evolutionary criteria allows to assess the history of their development.

2.2 Graphic Representation of the Relationships Between Stress, Time, and Evolutionary State

Thom's (1972) catastrophe theory is a domain of differential topology and aims at building the simplest continuous dynamic model which can generate a morphology, given empirically, or a set of discontinuous phenomena (Guex, 2016).

2.3 Stratigraphical and Environmental Studies

Information on environmental factors impacting the evolution of conodonts was gained from the stratigraphical and environmental studies, often included in conodont papers. Finally, during reconstruction of the phyletic lineages, several events were discovered.

3. STRATIGRAPHY

Stratigraphy studies sedimentary sequences layer by layer. Some were deposited rapidly, such as clastic sediments like sandstone, or slow for biochemical limestones, to extremely slowly like deep sea radiolarian oozes. The sedimentation rate can be negative in the case of erosion or denudation. The nature and environment of deposition are always revealed by the content, while fossil traces of life provide relative time constraints. Darwin was familiar with the relative stratigraphic concept of time and its subdivision, but the unavailability of precise timing caused him great discomfort. The discovery of isotopes and the technology of radiometry appeared after Darwin's death. Evolution was not linked with genetics, and Mendel's name was not attached to the basic laws of genetics until the XXth Century. Darwin, as Galton (2015) puts it, was de facto unaware of Mendel (1865). However, Hoffmann's (1869) book, with Darwin's handwritten marginal notes on pages 50–51 and 53–55, facing a long excerpt from Mendel (1865) on p. 52, is kept in the Cambridge University Library. Did Darwin skip over the passage on Mendel's paper, or did he not appreciate its significance? (Ramirez, 2015).

The Triassic system is of particular interest because geologists were greatly misled about its length. The time span of each ammonoid Zone being interpreted as relatively constant, their higher frequency during the Lower Triassic resulted in considering this epoch as relatively long, in comparison to the Middle and Late Triassic. Lately, however, radiometric ages reveal a mere 5 Ma long Early Triassic and a 10-Ma long Middle Triassic. These added together represent less than a third of the total Triassic 51 Ma time span. The interesting aspect lies in what caused such discrepancy and conodonts provide a number of clues, especially in relation with the impact of succeeding ecological crises of the marine ecosystems. Characteristically Pangea marks the start of the Triassic period (i), coinciding with the aftermath of the P–T mass extinction (ii), the probable absence of an equatorial humid belt during most of the Triassic time span (Truempy, 1982) (iii), low relative sea levels, although absolutely ice free (iv), and being one of the hottest periods of Earth history (Shaviv and Veizer, 2003) (v).

Triassic warming events have been documented by $\delta^{13}C$ curves (Payne et al., 2004), showing perturbations of the carbon cycle during recovery from end-Permian extinction, and the $\delta^{18}O$ composite curve for surface waters of the Tethyan subtropics (Trotter et al., 2015) that show long-term cycles of warm episodes W1–W3 and subsequent Scythian–Ladinian warming events (Fig. 1).

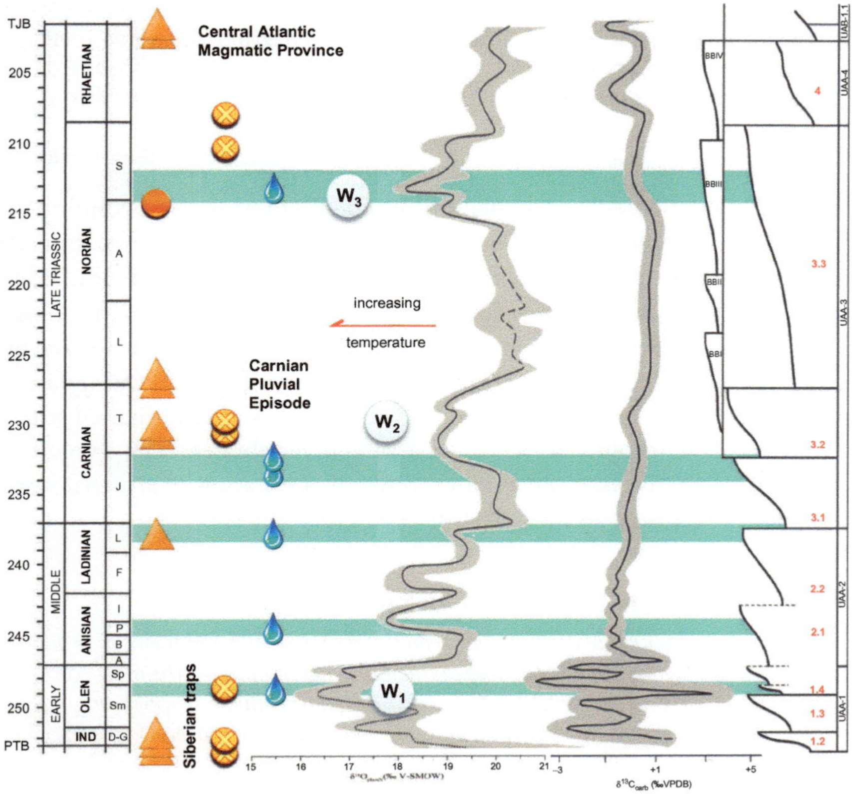

Fig. 1 Triassic events. The Triassic Sequences by Haq et al. (1988) are complemented by Embry's (1997); in Orchard et al., 2001) BB I–IV sequences. The graph, adapted for the ICS time scale (2014), runs between Siberian traps and CAMPS (Central Atlantic Magmatic Province). $\delta^{13}C$ curve, after Payne et al. (2004), showing perturbations of the carbon cycle during recovery from end-Permian extinction. The $\delta^{18}O$ composite curve for surface waters of the Tethyan subtropics (Trotter et al., 2015) shows long-term cycles of Triassic climate change, warm episodes W1–W3, volcanism, oceanic anoxia, humid phases, and craters. Key: *A*, Aegean; *A*, Alaunian; *B*, Bithynian; *D*, Dienerian; *F*, Fassanian; *G*, Griesbachian; *IND*, Induan; *J*, Julian; *L*, Lacian; *L*, Longobardian; *OLEN*, Olenekian; *P*, Pelsonian; *S*, Sevatian; *Sm*, Smithian; *Sp*, Spathian; *T*, Tuvalian.

4. RETROGRADE EVOLUTION DURING MAJOR EXTINCTION CRISES

Evolutionary trends of conodonts concern more or less gradual geometrical and ornamental transformations occurring over long periods of ecologically stable periods. By contrast, major evolutionary jumps in several invertebrate groups occur during massive extinction periods, which are characterized by the appearance of primitive forms resembling remote ancestors of their immediate progenitors. These are defined as atavistic forms. Homeomorphic species generated during sublethal environmental stress can be separated from the ancestral group by several millions of years.

Guex (2016)

Guex (2016) presented a new theoretical model of retrograde evolutionary changes during sublethal environmental stress and analyzed the evolutionary patterns of some planktonic foraminifera, radiolarians, nautiloids, conodonts, corals, and ammonoids during major extinction periods. "In ecologically stable periods, the transformations of the skeletons are characterized by an increase of shell curvature, corresponding to an increase in the apparent geometrical complexity. During periods of sublethal environmental stress, rapid retrograde evolution occurs in many invertebrates. The evolution of silicoflagellates is discussed as an example of application of artificial stress to modern organisms" (Guex, 2016).

4.1 Sublethal Environmental Stress

The concept of sublethal environmental stress denotes specific conditions that are critical to the survival or normal development of living organisms. The most common cases of environmental perturbations include pollution or poisoning by chemicals such as toxic gases, nutriment shortage, large scale sea level falls, major climatic changes, hydric stress, acid rains, marine anoxic events, etc. The origins of environmental stress can easily be identified and quite naturally the particular kind of stress can be hostile to some organisms and favourable to others.

Guex (2016)

4.2 Timing of the End Triassic Extinction

The End Triassic Extinction has long been suspected to be related to the onset of the Central Atlantic Magmatic Province (CAMP) volcanism but it is only recently that U-Pb ages measured on zircons have allowed a precise correlation with the relative timescale based on the evolution of the ammonites. That correlation has been established based on very detailed stratigraphic research in the American Cordilleras (Northern Peru and Nevada, USA) and on the discovery of ash beds deposited in the same levels as age diagnostic ammonites.

Guex (1995), Schoene et al. (2010), after Guex (2016)

According to De Renzi et al. (1996), extinction and origination rate averages for lineages of the Phanerozoic marine invertebrates are not coherent with the dynamics of conodonts. De Renzi et al. (1996) considered the possibility of random extinction. "Their extinction rate diminished throughout

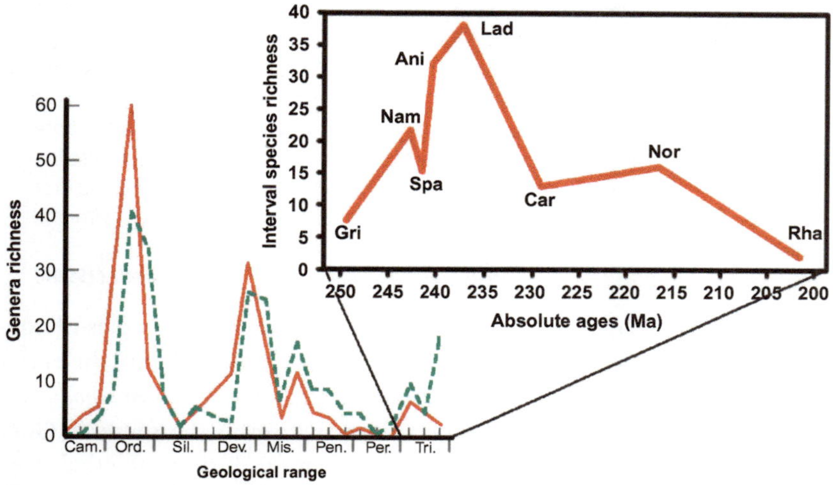

Fig. 2 Genera (based on discrete elements) richness for conodonts in their complete geological range. Number of genera appeared in each period (*solid line*) compared with total number extinct during the same period (*dashed line*). *Right upper figure*: Conodont lineage richness during the Triassic. Richness is given at the middle point of each stage as its total number of lineages. Key: *Ani*, Anisian; *Car*, Carnian; *Gri*, Griesbachian; *Lad*, Ladinian; *Nam*, Nammalian; *Nor*, Norian; *Rha*, Rhaetian; *Spa*, Spathian. *Modified after De Renzi, M., Budurov, K., Sudar, M., 1996. The extinction of conodonts—in terms of discrete elements—at the Triassic-Jurassic boundary. Cuad. Geol. Iber. 20, 347–364.*

the Lower Triassic to the Upper Triassic, but the lineage origination rate diminished since the Ladinian as well, the clade had a negative growth rate, and the probability of extinction was high (Fig. 2).

Diffuse competition with better animal designs belonging to Paleozoic and Modern evolutionary faunas and, in a more secondary way, a decrease of available marine areas and spreading of environments with abnormal Salinity in the Western Tethys are pointed out as possible causes of their extinction" (De Renzi et al., 1996).

4.3 Retrograde Evolution of Triassic Conodonts

Conodonts proliferated during the Paleozoic and the Triassic and disappeared during the end Triassic extinction (ETE). They were extensively studied because they evolved very fast and represent very good stratigraphic markers, sometimes better than ammonites because one finds them in many environments where ammonites are absent. These organisms have complex apatitic teeth which are generally well preserved in a variety of sediments.

Guex (2016)

Triassic conodonts frequently show iterative evolutions, with repetition of heterochronous homeomorph lineages. Specific examples of such evolutionary repetition, accompanied by a typical geometrical simplification during the transition between the ancestors and descendants, can be found above the Permian Triassic major mass extinction: for example the appearance of Neospathodus *at the base of two successive lineages,* Clarkina *during the upper Permian and* Kashmirella–Paragondolella *during the Lower and Middle Triassic. It is tempting to surmise that the catastrophic event, which prompted proteromorphosis, coincides with the Dienerian negative excursion.*

Hirsch (1994b)

Kiliç et al. (2016) have analyzed evolutionary trends and strategies of conodonts across and after the Permian–Triassic mass extinction until the end of the Ladinian. They note that neospathid genera such as *Neospathodus* are seen as the result of proteromorphic retrogradation in the sense of Guex (2001) and atavistic homeomorph forms appearing during sublethal stress events are often separated by several millions of years. Hypothetically, the simple and economic architecture as a potential source of renewed phylogenetic development explains such sudden appearances (Guex et al., 2014; Kiliç et al., 2016).

Evolution in Lopingian–Induan (Late Permian–Early Triassic) anagenetic gondolellid lineages is comprised of the development of the carina and the platform. Loss of platform in homeomorph "Neospathodus" arcucristatus *(Late Capitanian) and the* Neospathodus kummeli–Kashmirella timorensis *lineage (Dienerian–Aegean) took place within the* Jinogondolella–Protoclarkina *and* Clarkina–Paragondolella *intervals. Additional side branches of these lineages include those starting with Lopingian* Jinogondolella postserrata *(Wardlaw and Mei), Smithian* Borinella *and Anisian* Pridaella. *Proteromorphic retrogradations occur during periods of crisis, with the sudden appearance of homeomorphic forms that are atavistic representatives of the clade. These primitive-looking forms are derived from their immediate ancestors by retrograde evolution, a phenomenon which has been described as proteromorphosis.*

Guex (2001), Guex et al. (2014), Kiliç et al. (2016).

Conodont phylogeny under the sublethal environmental stress conditions following the Permian–Triassic extinction, was paced by proteromorphosis (reappearance of ancestral morphs). Thom's (1972) cusp catastrophe diagram illustrates this catastrophic event (Fig. 3).

Kiliç et al. (2016)

5. CLASSIFICATION

The classification employed here is modified after Dzik (1991), using multielement taxonomy for the suprageneric level.

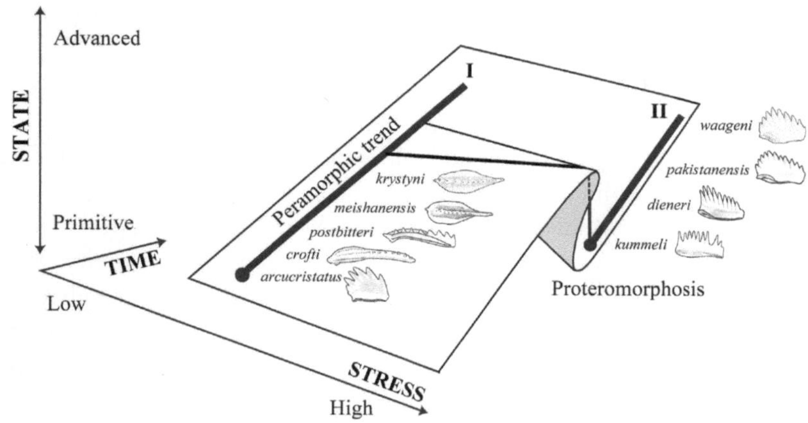

Fig. 3 The cusp catastrophe of Thom (1972): Diagrammatic representation of a punctuated anagenesis (Kiliç et al., 2016). Peramorphic trend: Anagenetic lineage of *"Neospathodus" arcucristatus, Pr. crofti, Clarkina bitteri, C. meishanensis, C. krystyni;* Proteromorphosis: Catastrophic reappearance of atavistic *Neospathodus* and stasis of *Neospathodus kummeli–K. timorensis,* and rediversification in *Pa. regale–Pa. excelsa.* The catastrophic event that prompted proteromorphosis coincides with the Dienerian negative carbon excursion and sea level high stand. The *Clarkina* peramorphic trend started after the Late Guadalupian atavistic *"Neospathodus" arcucristatus* Clark and Behnken, 1971 (Ratterman, 1976), with *Pr. crofti,* Kozur and Lucas, 1996 (Wardlaw and Mei, 1998, pp. 40–41, plate 7, figs. 1–4, 7–2) showing a reduced platform that passes into *Clarkina postbitteri* Wardlaw and Mei (1998; in Mei et al., 1994), the oldest species of the *Clarkina* lineage.

Replacing any other form of taxonomy, the system, followed among others by Hirsch (1981) and Orchard (2005), clearly enhances the numerical and morphological relationships between platform elements and accompanying bars and blade elements of the apparatus, for which the locational notation used is P_1, P_2, M, S_0, S_1, S_2, S_3, and S_4 (Purnell et al., 2000).

The platform P_1 element is the most changing feature and therefore of interest for both stratigraphy and evolution (Kiliç et al., 2016).

The clade to which Late Triassic gondolellids belong is almost exclusively the subfamily Neogondolellinae that differs from other gondolellid subfamilies by having a bifid S_3 element in its multielement apparatus. Henderson and Mei (2007) observed a number of features: (i) bifid S_3 elements as transition of *Mesogondolella* to *Jinogondolella*; (ii) changes in ontogenetic developmental timing, including lack of a platform or its reduction to a narrow rib in juvenile specimens within the development of *Clarkina* (*Neoclarkina*); (iii) the proteromorphic loss of platform in entire ontogenetic series (*Neospathodus*); (iv) the peramorphic platform development in later parts of the ontogenetic series. Thus, *Neospathodus* developed by a proteromorphic process wherein the entire ontogenetic series lacks a platform. Evolving from *Neospathodus*, platform bearing forms may have evolved by a peramorphic process.

Evolution may appear as reversed under environmental stress, giving way to "simplified" morphs, from which often totally new genera emerge. Hirsch (1994b), following Gould (1977), interpreted these "simplifications" in terms of heterochrony such as paedomorphism, progenesis, and neoteny. Since then, Guex (2001, 2006) coined the term proteromorphosis for the phenomenon that is not a simple paedomorphosis but, repeating itself, possibly goes back millions of years. In our case, *Neospathodus* is neither neotenic as this term implies reversibility from generation to generation and one should have neotenes cohabiting almost normally in the fossil record, nor is it progenetic, since it is not a truncated ontogenesis of the ancestor but homeomorphic to the ancestral ancestor. Thus, the new term proteromorphosis imposed itself.

The Carnian–Rhaetian, 37 Ma long time span, represents 70% of the Triassic Period and encompasses several evolutionary rebounds paced by the Early Julian *Mosherella*, Late Tuvalian *Neocavitella* retrogradations, each followed by anagenetic lineages. The successive Alaunian *Mockina* and Sevatian–Rhaetian retrogradations are the long fight for survival of the clade.

Triassic Gondolellidae subdivides into Neogondolellinae (Hirsch, 1994b), Sephardiellinae (Plasencia et al., 2007), and Epigondolellinae (Orchard, 2005). They differ from Gladigondolellidae by the shape of the basal cavity and the morphology of the elements of otherwise remarkably parallel apparatuses (Ishida and Hirsch, 2011; Plasencia et al., 2007).

In platform elements (P_1) taxonomic differentiation at the generic level is (i) the morphology and configuration of the basal groove, including shape, position, and size of the basal cavity or pit; (ii) evolution of the platform from monolobal to bilobal (bifid), including splitting or bifurcation of the basal groove; (iii) degree of platform reduction and development and length of a free blade; (iv) ornamentation of the platform rim, from smooth to nodose and denticulated.

At the specific level, biometric relations (platform length and width, length of the free blade, blade, size, number, and inclination of denticles of the carina), platform shape, and ornamentation, such as the development of nodes and denticles, constriction, curvature, and development of a parapet, are considered. An additional parameter is platform reticulation, the appearance of which is not clear if it is due to evolution or to certain ecological conditions.

The reappearance of monolobal platforms toward the end of generic lineages has been interpreted as paedomorphic (Hirsch, 1994b). Guex (1992) has demonstrated evolutionary jumps in ammonites, from complex to simple, caused by environmental stress. This seems to be the case with iterations of Late Anisian, Late Ladinian–Early Carnian cavital morphs as well as with the majority of Middle Norian to Rhaetian genera that reach the adult state at a juvenile ontogenic stage. The ontogeny of the P element in gondolellid genera starts as a neospathid blade with denticles, evolving into units with platforms that progressively grow toward well-developed adults (Hirsch, 1994b; March et al., 1990; Mosher, 1968; Trammer, 1971).

5.1 Natural History of the Triassic Gondolellidae

The evolutionary "simplifications" (retrogradation) are interpreted in the paedomorphic terms of neoteny and progenesis (Gould, 1977). Neogondolellinae (Hirsch, 1994b) contains the genera *Neogondolella, Neospathodus, Platyvillosus, Paragondolella, Nicoraella, Mosherella, Metapolygnathus, Mazzaella, Quadralella, Carnepigondolella, Neocavitella, Ancyrogondolella*, and *Norigondolella*. Subfamily Sephardiellinae (Plasencia et al., 2007) contains the genera *Pseudofurnishius* and *Sephardiella*. Subfamily Epigondolellinae (Orchard, 2005) encompasses *Epigondolella, Mockina*, and *Misikella*.

The P_1 element consists of a blade (carina) that toward adulthood generally develops a platform that borders a variable portion of the elongate to curved unit. The median carina is denticulated and a cusp develops above the basal cavity. The basal surface is keeled by a longitudinal groove that widens posteriorly to enclose a loop-like basal pit beneath the posterior cusp. Finely to coarsely reticulate pitting occurs on the platform. Small teeth or nodes may border the platform edge in several genera.

5.2 Genetic Memory in the Evolution of Triassic Conodonts: Trends in Evolution

The evolution of the genera in each subfamily consists of a number of reiterating trends: (i) displacement of the basal cavity from its posterior position toward the middle, accompanied by the modification of the shape from loop-like to amygdaloid; (ii) reduction of the platform, that in most Early and Middle Triassic genera of the family borders the entire unit of adult specimens, by the formation of a free blade; (iii) splitting of the monolobate basal groove into a bilobate, forked platform. The evolutionary trends are paced by sea level changes that are reflected in recurring paedomorphic neospathid morphs followed by accelerated rates of speciation. The Bithynian and Pelsonian *Nicoraella*, Early Julian *Mosherella*, Late Tuvalian *Neocavitella* and Sevatian–Rhaetian *Misikella* radiations are global LST (lowstand systems tract) events. Local events are Early Anisian *Kamuellerella–Ketinella–Gedikella* (Istanbul Terrane) and the Late Anisian *Pseudofurnishius priscus–shagami* (Levant) that initiated Sephardiellinae. Apparently Western Tethyan is the Julian Pluvial Event (*Mazzaella*). Highly ornamented forms may relate to Carnian global warming. The initial Lacian (kerri Zone) TST (transgressive systems tract) witnesses the end of genuine *Metapolygnathus* and the advent of *Ancyrogondolella* as well as the recurrence of the atavistic gondolellid *Norigondolella*. The latter may be induced after the appearance of proteromorph *Neocavitella*. During the Alaunian (rutherfordi Zone) HST (highstand systems tract), the neotenic genus *Mockina* appeared, persevering throughout Sevatian–Rhaetian times, together with *Misikella*.

Ontogeny, as well as evolution, may apparently become reversed under environmental stress, giving way to "simplified" morphs, from which often totally new genera emerge (Hirsch, 1994b). The evolutionary "simplifications" are interpreted in terms of heterochrony such as paedomorphism, progenesis, and neoteny (Gould, 1977). For Lieberman (2011: p. 35) paedomorphosis is when the descendant species is underdeveloped relatively to the ancestor, smaller in size and simpler in shape, resembling juvenile ancestors; peramorphosis is when the descendant species transcends its ancestor in terms of size and shape; and a neotenous descendant is of the same size as the adult ancestor but is underdeveloped (simpler) in terms of shape. Henderson and Mei (2007), in their cladogram of gondolellid taxa, consider the evolution of bifid S_3 elements as the transition of *Mesogondolella* to *Jinogondolella* changes in ontogenetic developmental timing. This includes lack of a platform or reduction of the platform to a narrow rib in juvenile specimens within the development of *Clarkina*, such as *Neoclarkina*. Paedomorphic loss of platform occurs in entire ontogenetic series (*Neospathodus*) and peramorphic development of platform in later parts of the ontogenetic series. Thus, *Neospathodus* developed by a paedomorphic process wherein the entire ontogenetic series lacks a platform. *Neogondolella* may have evolved from *Neospathodus* by a peramorphic process whereby ontogeny was either accelerated or additional stages were added in which a platform develops in intermediate and adult forms.

The case of the Permian–Triassic boundary is particularly interesting as conodonts were not affected too much by the extinction event at 252.6 Ma.

Located in Meishan at the contact between limestone bed 24e and volcanic clay bed 25, this event falls within the *Clarkina taylorae* Zone (Jiang et al., 2007). Above the 8-cm thick volcanic clay (bed 25) follows a black mudstone (bed 26), a calcareous mudstone (bed 27), and a second volcanic clay (bed 28), dated at 252.5 Ma. The base Triassic GSSP golden spike was driven in the middle of the 16-cm thick bed 27, where the conodont *Hindeodus parvus* appears, a rather incomprehensible choice if based on faunal criteria alone. Both the *Clarkina* and *Hindeodus* conodont lineages briskly moved through the UAA-1.1.2 LST (latest Lopingian), although bed 24e is marked by the onset of a sharp reduction in average size (from 0.63–0.69 to 0.54 mm) as well as deviation to juvenile or dwarfed size of *Neogondolella* (= *Clarkina*) (Yin et al., 2001). Spathognathodontidae, to which *Hindeodus* belongs, became extinct during the Griesbachian TST. In the course of the Late Induan, shortly before the extinction of *Clarkina*, the appearance of *Neospathodus* coincides with the Dienerian HST. As an iterative development in gondolellid conodonts, platform reduction leading to blade-like neospathid forms occurred with *Clarkina bitteri* and *Neospathodus divergens* in the Gerster Formation of the Confusion Range, Utah (Wardlaw and Collinson, 1986).

5.3 Lineages

Alternating low-, transgressive-, and highstand systems tracts pace the succeeding relative sea level Upper Absaroka A1–4 to B1 cycles and control the coming and going of the successive fossil lineages that divulge the history of life. Our subdivision follows "Eustatic Sea Level Changes" (Haq et al., 1988), while the BB (Black Bear) sequences are after Embry (Embry, 1997, in Orchard et al., 2001).

5.3.1 Late Permian–Early Triassic Lineages (Fig. 4)

Late Permian Spathognathodontidae (*Hindeodus*), Ellisoniidae (*Ellisonia*), and Gondolellidae (*Clarkina*) persisted into the earliest Triassic. *Hindeodus typicalis* and a number of Griesbachian-derived species, such as *Isarcicella isarcica*, remained limited within the Early Induan. *Ellisonia* radiated during the Olenekian into *Furnishius, Parachirognathus, Hadrodontina, Pachycladina, Parachirognathus*, and *Foliella*.

A. Late Permian–Induan: Two lineages seem to exclude each other biofacially, the pelagic *Hindeodus* lineage that spans the Late Permian–Griesbachian interval, and the shallow marine "*Neogondolella*"-type

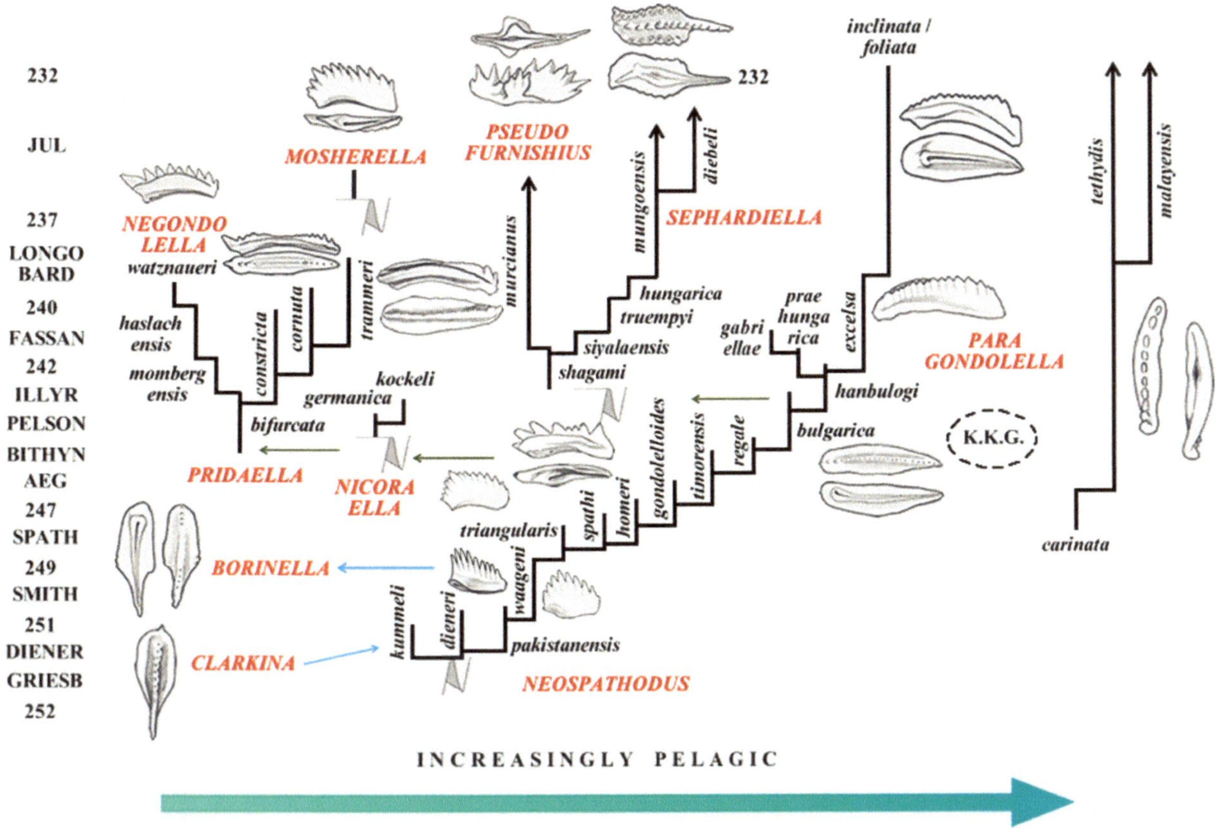

Fig. 4 See legend on next page.

bloom from the Late Permian to the Early Dienerian (*Clarkina yini, C. meishansensis, C. carinata, C. taylorae, C. planata, C. krystyni*). Although such attributions may not be hermetic, many taxa may extend to more than one habitat, and *Hindeodus* can possibly also be a shelf dweller (Hirsch, 1994a). Regarding the origin of the *Clarkina* lineage, the earliest Wuchiapingian *Pr. crofti* (Kozur and Lucas, 1996), for *Neogondolella* sp. (Croft, 1978: pp. 51–52, plate 5, figs. 1–10; Wardlaw and Mei, 1998: pp. 40–41, plate 7, figs. 1–4, 7–20) that has the Neospathid-reduced platform of the initial peramorphic *Clarkina* trend, may be preceded by the end Guadalupian "*Neospathodus*" *arcucristatus* (Clark and Behnken, 1971: in Ratterman, 1976), suggesting a Lopingian *"Neospathodus"–Protoclarkina–Clarkina* peramorphic trend.

B. Dienerian–Spathian: The ranges of *Clarkina* and *Neospathodus* overlap shortly/directly? (Mosher, 1968). The Dienerian–Spathian *Neospathodus* lineage of *N. dieneri–cristagalli–pakistanensis–waageni–conservativus–triangularis–homeri* ends with the genus *Chiosella*. In the Smithian, the resurgence of *Neogondolella* takes places in the form of the *Borinella nevadensis–B. jubata* lineage that branches into *B. nepalensis* and *Scythogondolella milleri*. Few *Neospathodus triangularis* may occur together with abundant Ellisoniids (Perri and Andraghetti, 1987).

Fig. 4 Early–Middle Triassic conodont lineages (252–237 Ma). The Scythian–Ladinian Gondolelloid lineages are represented as a function of time (Ma) and increasing pelagic habitat. *"Neogondolella"* consists of the Lopingian–Induan *Clarkina* that retrogrades into the Dienerian *Neospathodus kummeli*; Olenekian *"Neogondolella"* (*Borinella–Scythogondolella*) radiated out of *N. waageni*. The *Neospathodus–Kashmirella* lineage initiates a 45-Ma long succession of reinitialization-radiation that continues until the last Rhaetian *Misikella*. Within the range of the Scythian–Ladinian 15 Ma time span, *Pa. regale* originates from *K. timorensis*. Thom's fall-back events generate the appearances of isolated proteromorphic taxa, such as *Nicoraella* at the origin of new Anisian–Ladinian *Pridaella* and *Neogondolella* and Anisian–Early Julian *Pseudofurnishius–Mosherella*. Most pelagic is the Tethyan equatorial Late Spathian–Julian Family Gladigondolellidae and possibly the exceptional Early Anisian North-Tethyan *Kamuellerella–Ketinella–Gedikella* (KKG) group. The taxa in this interval are arranged according to their increasingly pelagic habitat. *Modified after Kiliç, A.M., Plasencia, P., Ishida, K., Guex, J., Hirsch, F., 2016. Proteromorphosis of Neospathodus (conodonta) during the Permian–Triassic crisis. Rev. Micropaléontol. 59, 33–39.*

	Induan (252.2–251.2 Ma)
UAA-1.2 TST	*Clarkina, Hindeodus*; Extinction of Spathognathodontidae
UAA-1.2 HST	Dienerian *Neospathodus*; *Clarkina carinata*
	Smithian (251.2–248.7 Ma)
UAA-1.3 LST	*Neospathodus waageni*
UAA-1.3 TST	*S. milleri, Neogondolella sibiriensis, N. nepalensiss*
UAA-1.3 HST	*N. jubata, Platyvillosus*
	Spathian (248.7–247.2)
UAA-1.4	*N. triangularis*
UAA-1.5 LST	*N. homeri*
UAA-1.5 TST	*Kashmirella gondolelloides, Gladigondolella carinata*

Summing up Early Triassic Gondolellidae, they comprise: (a) Late Permian–Griesbachian *Clarkina*; (b) the Dienerian subphase of the appearance of the *Neospathodus–Kashmirella* lineage (Dienerian–Spathian); and (c) the Smithian–Spathian subphase of *Borinella* Budurov, a genus put into uncertain subfamily (Goudemand et al., 2012) and of the taxonomically problematic Spathian genera *Icriospathodus* and *Platyvillosus*. Klets and Kopylova (2007) have discussed the Smithian appearance of *Neogondolella* in the context of the Early Olenekian of the northern latitudes that yields endemic *Borinella buurensis, B. composita, B. jakutensis, B. taimyrensis, B. sibirica,* all having characteristics of *Neogondolella* in the form of the basal cavity (Dagys, 1984). Specimens of *B. buurensis* from A. Dagys' collection were, however, referred by Kozur (1989) to *Paragondolella sweeti* "perhaps wrongly." Therein, Kozur wrote "Therefore, most probably, so-called *Neogondolella* had evolved from *Neospathodus* by the beginning of the Early Olenekian and was widely distributed in southern and northern latitudes." Hirsch (1994b) postulated the sudden Smithian (re?)appearance of "*Neogondolella*" (read *Borinella*) from *Neospathodus,* branching into a lineage of *B. nevadensis–B. jubata* and "side" branches of *B. nepalensis* and *S. milleri* during an assumed long Early Triassic time span that would have left a larger time gap between Early Dienerian extinction of *Clarkina* and Early Smithian appearance of *Borinella.*

5.3.2 Middle Triassic Lineages (Fig. 4)

The Middle Triassic is dominated by the lineage of *Paragondolella* that derived from *Kashmirella.* Conservative "*Neogondolella*" apparently dwells in shallower environments until extinction in the Late Ladinian. Initially, quasi homeomorphic *Paragondolella* dominate the more open marine scene. The Pelsonian–Illyrian neospathid *Nicoraella* and earliest *Pseudofurnishius* appear and range into the Julian with the latter developing into the Fassanian Sephardiellinae.

Anisian (247.2–242). Two new lineages emerge during the Aegean stage: *Paragondolella* and the Anisian *Neogondolella,* but the source of the latter lineage is not clear. One alternative is the continuation of the Spathian *jubata* lineage (Hirsch, 1994b), the other derived from *K. gondolelloides* over *K. timorensis*. Kozur (1989) saw "*Neogondolella*" *aegaea* as a synonym of the latter. It should be emphasized that these forms are highly homeomorphic. Also, the origin and differentiation of *Neo-* and *Paragondolella* is difficult because it is primarily based on few morphological or ornamental criteria. Their discrimination is more easily substantiated by widely divergent evolutionary trends among the respective lineages. As such, "*Neogondolella*" preserved ancestral characters during its entire range, while at an early stage *Paragondolella* incorporates the evolutionary trends that are realized in the course of the Late Ladinian and Late Triassic. Budurov and Sudar (1998) proposed the genus *Pridaella* for the lineage starting with *Pridaella constricta,* branching into a normal marine *trammeri* lineage (Peri-Tethyan) on one side, its other offshoot being the restricted marine lineage of *Neogondolella mombergensis* (Germanic facies area). Dzik and Trammer (1980) observed the gradual evolutionary increase in the latter's morphologically juvenile stages of *N. haslachensis–N. watznaueri* (Ladinian of Mid-European epicontinental environment), comparable to the *postera–bidentata–andrusovi* (Norian–Rhaetian Tethys) lineage (Trammer, 1974). The elongation of the neospathid morph (*homeri*) into *K. timorensis* from the Spathian into the Aegean (Gaetani et al., 1992; Mosher, 1968) generates a new lineage that persists until the end of the Triassic. This lineage of *Paragondolella regale–bulgarica–excelsa–foliata–inclinata,* during the Middle Triassic, consists of relatively unchanged neogondolellin morphs. A so far unidentified event during the Early Anisian precipitated the appearance of the forms *Kamuellerella–Ketinella–Gedikella,* in the central part of the North Tethys. The small size of these unique ramiform shapes, only found so far in the Turkish Istanbul Zone (Gedik, 1975; Kiliç, 2004), may suggest some extraordinary local warming event. Iterations of neospathid forms occurred from the Bithynian onward, coinciding with sea level LST through TST. These include *Nicoraella* that retrograded from early *Paragondolella* stock and out of which the *Pridaella* lineage radiated. Later, in the Pelsonian the narrow bladed *Ps. shagami* and *Ps. siyalaensis* appeared as the possible Illyrian forerunners of *Pseudofurnishius murcianus.*

In a pelagic bedded chert unit of the Chichibu Belt (South-West Japan), Ishida (1981: p. 116, plate 2, figs. 2a–c) identified as *Neogondolella* cf. *haslachensis haslachensis,* a form resembling *N.* cf. *haslachensis* that had been reported from the Early Anisian of the Kocaeli Peninsula (Gedik, 1975). Ishida (1984: p. 21) reinterpreted this form as *Neogondolella shoshonensis*, moreover precising that in Japan, this species occurs within the upper range of *Pa. bulgarica,* preceding *N.* cf. *cornuta* and *Pa.* aff. *excelsa*, being thus presumably latest Bithynian to Early Pelsonian in age and of Panthalassan low-latitude origin. The real *N. haslachensis* is characteristic in the Late Fassanian of the Germanic Basin (Dzik and Trammer, 1980).

Narkiewicz and Szulc (2004), discussing the adaptive evolution of *N. mombergensis* and the emergence of the *mombergensis–media–haslachensis–watznaueri* lineage, noted that Trammer (1971) and Zawidzka (1975) had held *N. media* (Kozur, 1968) as the juvenile stage of a "*constricta*"-type, and that Budurov and Petrunova (1998) suggested *N. media* to be the delicate early form of *N. constricta.* Consequently, if *N. haslachensis* was a descendant of *N. constricta,* its origin was related to a Tethyan lineage, while having suitable conditions for development almost exclusively in the Germanic Basin. Assuming, after Trammer (1974), that the evolution of *mombergensis* lineage was of adaptive character, it may be supposed that rapid evolution of *N. haslachensis* was controlled by continuous environmental changes in the entire Germanic Basin under conditions of restricted marine connections with the Tethys. *N. haslachensis, N. constricta, N. cornuta, N. pridaensis,* and *N. mombergensis* are present in the Russian Far East region of Sikhote-Alin (Buryi, 1996, 1997).

As the anagenetic lineage of *K. timorensis* passed into *Pa. regale,* the question remains whether the alleged "*Neogondolella*" *aegaea* branched off the *K. timorensis* lineage around the same time as *Pa. regale.* The strong homeomorphism that characterizes these Early Anisian forms is at the core of this taxonomic and phylogenetic discussion. If *aegaea* is a valid species, its origin is *timorensis*, and it may be classified under *Pridaella.* Budurov et al. (1988) held the lineage of *Neospathodus cristagalli–N. dieneri–N. pakistanensis–N. waageni–N. discretus–N. conservativus–N. zarnikovi–N. bransoni–N. triangularis–N. homeri* to be, parallel to that of the genus *Kashmirella,* comprising *K. kummeli–K. novaehollandiae–K. nepalensis–K. albertii–K. zaksi–K. spathi–K. gondolelloides,* from which the genus *Paragondolella* emerged. According to Budurov and Petrunova (in Muttoni et al., 2000: p. 233), the generic names for *Gondolella* are to be replaced by *Paragondolella* in the case of *Pa. regale, Pa. excelsa, Pa. inclinata, Pa. praehungarica, Pa. fueloepi, Pa. bulgarica,*

Pa. hanbulogi, Pa. praezaboi, Pa. bystrickyi; and *Pridaella* in the case of *Pr. trammeri, Pr. cornuta, Pr. bakalovi, Pr. longa, Pr. transita, Pr. bifurcata*; the name for *Chiosella* being *Kashmirella* and for *Budurovignathus* it is *Sephardiella.* Budurov and Petrunova (1998) also wrote that "in the coastal waters of the Peri-Tethys, the start of *Pr. constricta* marked the beginning of the development of the typical Peri-Tethyan genus *Pridaella* (with type species *Gondolella constricta*), the genus comprising *Pridaella shoshonensis, Pr. cornuta, Pr. longa*, and *Pr. bakalovi.*

	Aegean (247.2–246.2)
UAA-1.5 HST	*K. timorensis, Pa. regale*
	Bithynian (246.2–245.2)
UAA-2.1 LST	*Pa. regale, Pa. bulgarica, Nicoraella–Pr. bifurcata*
	Pelsonian (245.2–244)
UAA-2.1 TST	*Pa. bulgarica, Nicoraella kockeli, Pr. bifurcata*
	Illyrian (244–242 Ma)
UAA-2.1 HST	*Pr. constricta, Pr. cornuta, Pr. eotrammeri, Pr. trammeri, N. mombergensis, Pa. excelsa*
UAA-2.2 LST	*Ps. shagami*

Kiliç (2004) and Kiliç and Plasencia (unpublished\accepted paper) proposed that the genera *Ketinella, Kamuellerella*, and a new genus (*Gedikella*) stand for ramiform skeletal multielements, such as the hypothetical apparatus of *Cornudina.* These elements are forms that have changed very little from their Paleozoic ancestors and designate common Triassic skeletal apparatus-elements. Isolated specimens may have been found in different parts of the world (Israel, Japan, and China), but in NW Turkey they flourished during the Bithynian. There is nothing in the literature about the role of these enigmatic forms. Are these variants that resulted from some mutation within the evolution of apparatus-elements in the course of the Bithynian?–Pelsonian? Simultaneously, the very rich conodont faunas in the Kocaeli Peninsula (NW Turkey) saw a substantial modification of the *Gladigondolella tethydis* P_1 element (Kiliç, 2004). What happened during Bithynian times? Why has *G. tethydis* changed its P_1 element and why did the small K.K.G. forms flourish?

Ladinian (242–237). *Ps. murcianus* is found in stratigraphic sequence, immediately above *Ps. shagami,* from the very base of the Fassanian curionii Zone. The Fassanian–Early Julian *Sephardiella* lineage of *S. truempyi–S. hungarica–S. japonica–S. mungoensis–S. diebeli–S. mostleri* comprises, starting from adult forms of *S. mungoensis,* the trend of a slight splitting of its relatively central amygdaloid basal groove. A link between this provincial Sephardic *Pseudofurnishius* lineage *shagami–murcianus?–sosioensis* and the Late Ladinian and Early Carnian neospathid LST due iterations of *Mosherella microdus* and *M. newpassensis,* proposed by Sadeddin and Kozur (1992), should be considered as new neospathid iteration (Plasencia et al., 2015) (Fig. 4).

The main evolutionary trends during the Ladinian–Carnian interval are (a) the reduction of the platform and development of a free blade as in *Metapolygnathus tadpole* and (b) the trend of splitting the basal groove as in *Metapolygnathus communisti* (Budurov, 1977).

	Fassanian (242–239.5)
UAA-2.2 TST	*Ps. murcianus, S. truempyi, Pr. trammeri, N. mombergensis–N. haslachensis*
	Longobardian (239.5–237)
UAA-2.2 HST	*Gladigondolella malayensis, S. mungoensis*

5.3.3 Late Triassic Lineages (Figs. 5–7)

Julian (237–232). The Julian *Metapolygnathus polygnathiformis* initializes the lineage of "smooth" platforms and moderate-free blades of *Metapolygnathus intermedius–M. tadpole–M. auriformis*. The neospathid homeomorph *M. newpassensis* is transitional between *M. intermedius* and *M. tadpole* (Orchard, 2010). Issued from *M. auriformis,* the Late Julian lineage of *Mazzaella baloghi–Ma. carnica* develops strong nodes and a medium free blade. A notable sequential event is the Western Tethys salinity crisis (Gerry et al., 1990) and the faunal extinctions of bivalves and crinoids (Johnson and Simms, 1989). The Julian–Tuvalian boundary, as noted by Rigo et al. (2005), saw the highest concentration of Triassic conodont extinctions including *G. tethydis, Pa. inclinata/foliata, Ps. murcianus*, and *S. diebeli*. The typical Early Carnian forms *M. auriformis* and the lineage of *Ma. baloghi* and *Ma. carnica* are restricted to the Julian. *M. polygnathiformis* and *M. tadpole* are the only species that survive this boundary. A rapid Tuvalian radiation follows the extinction event.

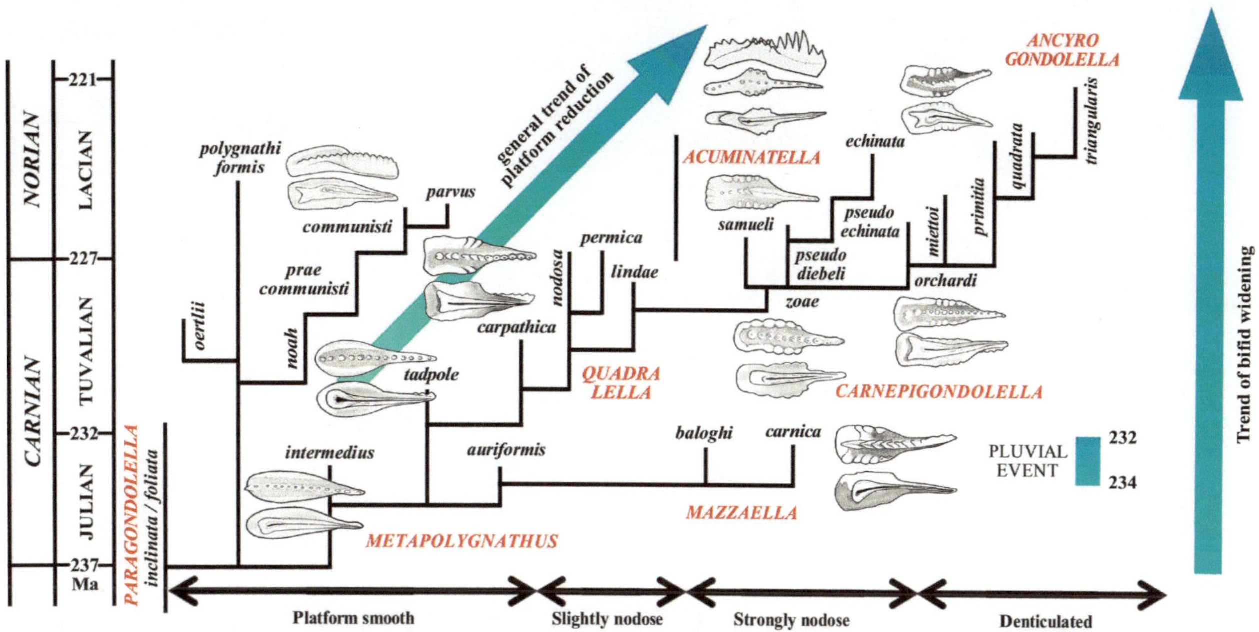

Fig. 5 Julian–Lacian conodont lineage (237–221 Ma). The Julian–Lacian Gondolellid lineages are represented in function of chronology (Ma) and evolutionary trends. These consist in the general trend of platform reduction or development of a free blade, until the remaining of a vestigial platform stub (diagonal trend), and in the trend of bifid groove-widening until the end of the Lacian (vertical trend). *Metapolygnathus* starts from conservative *Paragondolella* during the Julian with reduced platform (*intermedius, tadpole,* and *auriformis*). From the latter originates the strongly nodose Late Julian *Mazzaella* (Pluvial Event). During the Tuvalian smooth *M. polygnathiformis* persists, while general platform reduction affects the lineage of *M. noah–M. parvus.* Another Tuvalian branch consists of moderately nodose *Quadralella* (*carpathica, nodosa,* and *lindae*), from which Late Tuvalian–Lacian strongly nodose to denticulated *Carnepigondolella* derives. From the latter, the Tuvalian–Lacian *Ancyrogondolella* emerged. Thom's fall-back events of retrogradation took place during environmental stress events, such as the earliest Julian *Mosherella*, Tuvalian *Neocavitella*, and mainly Lacian *Acuminatella*.

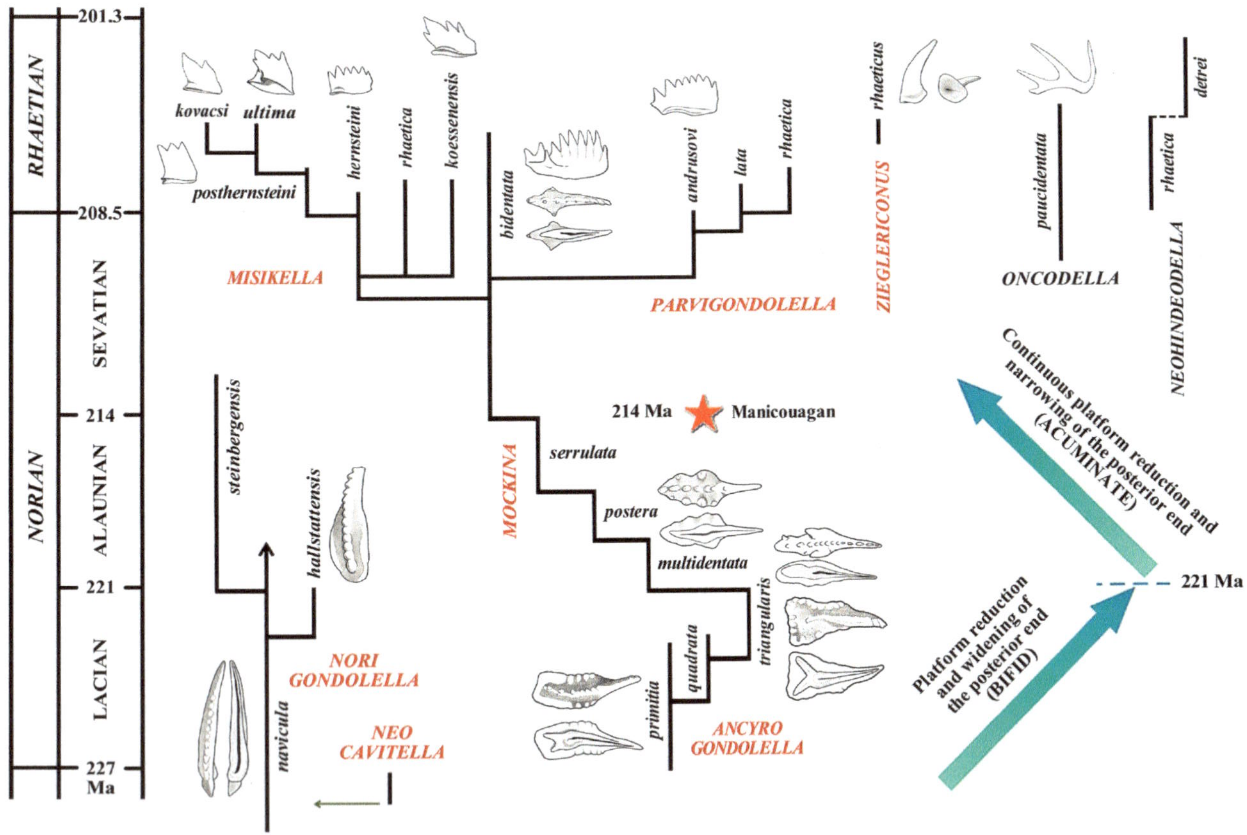

Fig. 6 See legend on opposite page.

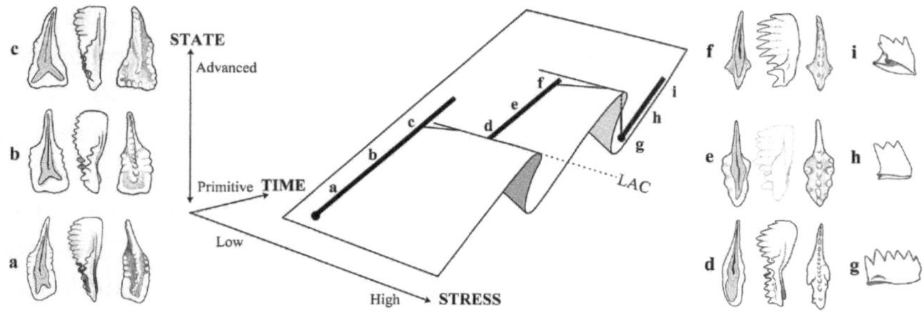

Fig. 7 Representation of Norian–Rhaetian lineages across the Lacian–Alaunian conodont crisis in Thom's (1972) cusp catastrophe diagram. (a) *A. primitia*; (b) *A. quadrata*; (c) *A. triangularis*; (d) *M. multidentata*; (e) *M. postera*; (f) *M. bidentata*; (g) *M. hernsteini*; (h) *M. posthernsteini*; (i) *M. ultima*. *LAC*, Lacian–Alaunian conodont crisis.

	Julian (237–232)
UAA-3.1 LST	*S. diebeli*, neospathid *M. microdus;* Western Tethys evaporites, extinctions: bivalves, crinoids (Johnson and Simms, 1989), conodonts
UAA-3.1 HST	Pluvial Event; *Mazzaella*

Tuvalian (232–228). Also issued from *M. polygnathiformis* is the Tuvalian *Metapolygnathus noah–M. communisti–M. parvus* lineage. These are forms with smooth platforms and gradually larger free blades (2/3 of total length in *M. parvus*). Simultaneously, a more conservative branch is

Fig. 6 Lacian–Rhaetian conodont lineage (221–201.3 Ma). During the Lacian 6 first MA of the Norian stage, the widening of the posterior end and developing of a strongly bifid lower cavity reached its maximum in the genus *Ancyrogondolella*. Simultaneously, the fall-back of *Neocavitella* possibly induced the smooth atavistic *N. navicula* that developed a free blade (*hallstattensis*) in the Late Lacian low latitudes and from the initial Alaunian, a narrowing of the platform (*steinbergensis*). The last Triassic 20 Ma of the Alaunian–Rhaetian consist entirely of atavistic genera, starting from *Mockina,* in which the evolutionary trend of platform reduction continued, in unison with the reversal from widening into the new trend of narrowing the basal groove toward a pointed or acuminate keel termination. The basal pit moves toward the middle, and its shape becomes amygdaloid (*Mockina*). As the successive fall-backs intensify, the basal cavity becomes more and more cavital (*Misikella*) and in extreme forms, only a single cone subsists (*Zieglericonus*). The youngest conodonts consist of isolated S_4 elements of an unidentified apparatus.

Metapolygnathus oertlii. M. tadpole appeared the *Quadralella carpathica–Q. nodosa–Q. lindae,* and finally *Q. permica. Q. lindae,* lineage, characterized by anterior platform - nodes and gradually reduced platforms. Most members of this lineage show strong constriction, while *Q. permica* has a shorter platform.

Increasingly strong nodes mark the edges of the entire platform of the *Carnepigondolella zoae–C. samueli/pseudodiebeli–C. echinata–C. orchardi* lineage. In the upper reaches of the *Carnepigondolella* lineage, and before the appearance of the Tuvalian–Lacian *Ancyrogondolella* lineage where denticles primarily replace all nodes. The neospathid homeomorph *Neocavitella cavitata* that appears 5 m below the Carnian–Norian boundary interval at Pizzo Mondello, in concurrent range with early *Norigondolella,* is the most probable origin of *Norigondolella.* The proteromorph *Neocavitella* appears twice in the same interval in the Black Bear Ridge section (Canada). This also takes place before the coming in of *Acuminatella,* an atavistic taxon in which the trend of widening of the platform is replaced by a narrow acuminate posterior edge. This taxon seems restricted to the cooler higher latitudes of the Ameri - Boreal Province (Hirsch, 1994a).

	Tuvalian (232–228)
UAA-3.2 LST	*M. polygnathiformis*
UAA-3.2 TST/BB-I LST	*Q. carpathica, Q. nodosa, Q. lindae, Neocavitella*
UAA-3.2 HST/BB-I TST	*M. communisti, C. samueli*

Lacian (228–221). The *Ancyrogondolella* lineage starts in the latest Tuvalian as the *A. miettoi–A. primitia* lineage, passing to the Lacian *A. quadrata–A. triangularis* lineage and several species of local significance.

The Lacian abundance of the atavistic gondolellid *Norigondolella navicula* is followed by *N. hallstattensis* that develops a free blade, as well as by the deep loop-like grooved elongated *Norigondolella steinbergensis.*

	Lacian (228–221)
UAA-3.3. LST/BB-I TST	*A. primitia*
UAA-3.3 TST/BB-I HST	*A. quadrata*
UAA-3.3 TST/BB-II LST–TST	*A. triangularis*

	Alaunian (221–214)
UAA-3.3 HST/BB-II HST	*Mockina multidentata*
UAA-3.3 HST/BB-III LST	Mockina *postera*
UUA-3.3 HST/BB-III TST	Mockina *serrulata*
	Sevatian (214–208.5)
UAA-3.3 HST/BB-III TST	Mockina *bidentata*
UUA-3.3 HST/BB-III HST	*Parvigondolella andrusovi, Misikella hernsteini*
	Rhaetian (208.5–201.3)
UAA-4/BB-IV	*Misikella posthernsteini, Misikella ultima*

Alaunian (221–214). From this stage on, all genera are atavistic. The Alaunian–Rhaetian lineage of *M. multidentata–M. postera–M. bidentata* are all characterized by a monolobal basal cavity, reduced platform and sharp denticles around the platform (Fig. 6).

Sevatian (214–208.5). The Sevatian is the acme of *M. bidentata*. The *M. hernsteini–posthernsteini* lineage emerges toward the upper part of this time span. These simple heart-shaped "cavital" basal cavities bearing earliest gondolellid ontogenetic stage-like forms are an almost perfect reiteration of the Early Triassic neospathid morphs. The Late Sevatian also sees the appearance of extremely retrograde genera, such as *Parvigondolella, Oncodella,* and *Zieglericonus* (Fig. 7).

Rhaetian (208.5–201.3). The leading form is *M. posthernsteini,* while several additional morphs form a kind of lineage ending in *M. ultima*. Some S_4 (*Neohindeodella*) elements have also been found (Fig. 6). Further retrogradations in the upper reach of *M. bidentata* generate the strongly diminished genus *Parvigondolella* and the neospathid homeomorph genus *Misikella*. During this time atavistic forms reappear such as *Zieglericonus* and *Oncodella* (Fig. 7).

5.4 Taxonomic Remarks

Wide divergences in interpreting *Paragondolella* exist between Orchard (2013) and Mazza et al. (2012): the former abandoned his earlier definition of *Metapolygnathus,* reducing it to barely more than the holotype, while Mazza et al. (2012) widened the volume of *Paragondolella* by increasing the number of species, leaving only its type species within the genus *Metapolygnathus*. Kiliç et al. (2015a,b) established the genus *Mazzaella*, and Orchard (2014) established the genus *Quadralella*.

The genus Metapolygnathus: Hayashi (1968)'s genus *Metapolygnathus* encompasses *M. communisti* (type), *M. linguiformis,* and *M. noah* to which Hayashi (1987) later added *M. echinatus, M. nodosus, M. permicus*, and *M. spatulatus.* Orchard (1983, 1991a,b) included into *Metapolygnathus* the species *nodosus, zoae, lindae, samueli, stephanae, polygnathiformis* and *communisti, reversus, primitius* and *pseudoechinatus*. Budurov (1977), Budurov and Sudar (1990), and Buryi (1996) limited the definition of *Metapolygnathus* to bifid morphs derived from evolved *Paragondolella.*

Mazza et al. (2010) described *Metapolygnathus* (type species: *M. communisti*) as follows: "the specimens of this genus are characterised on their lower side by a very narrow basal cavity, hardly visible, centrally located in primitive specimens and strongly forward shifted in advanced forms. The position of the pit is forward shifted in respect to both the platform and the keel end (Kozur, 2003). The platform is robust but reduced and generally lacks ornamentation. In some species a few nodes may be present, but they are always confined to the anterior platform or in correspondence to the geniculation point. A distinct free blade is present, with a high carina with highly fused denticles. The keel end is often, but not necessarily, bifurcated." *Metapolygnathus* has no denticles on the platform margins. Mazza et al. (2010) assign the following species to the genus, *M. communisti, M. echinatus, M. linguiformis, M. mersinensis, M. multinodosus, M. parvus.* However, 2 years later, only *M. praecommunisti* and *M. communisti* remain, all the others have been redistributed in newly defined *Paragondolella* and in *Carnepigondolella.*

The genus Carnepigondolella: Kozur (2003) defines this genus as having the basal cavity subterminal, with respect to the keel end and to the platform, or slightly forward shifted, but always distinctly behind the middle of the platform. The anterior part of the long to moderately long platform bears nodes or short node-like denticles, often extending to the entire length. The free blade, when present, is short to moderately long. The keel termination is broadly rounded; blunt or little bifurcated, but never pointed.

The genus Mazzaella: The genus *Mazzaella* Kiliç et al. (2015a,b), having as type *Epigondolella carnica.* The *Mazzaella* lineage derived from *M. auriformis,* characterized by a smooth platform with a possible constriction and a large free blade. Sharing the rather wide platform, *Mazzaella* differs from its ancestor by the strong nodes on the platform margins. *Ma. baloghi* is characterized by a platform with strong nodes and a large free blade, while *Ma. carnica* has a shorter free blade and a slightly narrower platform. *Mazzaella* appeared during the Late Julian Pluvial Event. *Mazzaella* differs from *Carnepigondolella* by having a longer free blade and a relatively broader platform without constriction, but there is no direct phylogenetic link between the Late Julian *Mazzaella* branch that derived from *M. auriformis,* and the Late Tuvalian *Carnepigondolella* lineage, beginning with *C. zoae,* whose ancestry may have derived from *Q. lindae.*

The genus Quadralella: Kiliç et al. (2015a,b) established the genus *Hayashiella* in the Tuvalian with type *M. nodosus*. Unfortunately this name was preoccupied and *Quadralella* Orchard (2014) remains the valid name. While its platform edges are smooth over most of their length, characteristic nodes occur in its anterior part. The platform varies from narrow (*Q. nodosa*) to wide and piriform (*Q. carpathica*) or narrower and constricted (*Q. lindae*). The lineage of *Quadralella* derives from *Metapolygnathus* and comprises *Q. carpathica* and *Q. nodosa* with offsprings *Q. lindae* and *Q. permica*. *Quadralella* differs from *Metapolygnathus,* only by having nodes on the anterior edge of the platform. There are direct phylogenetic links with its ancestor *Metapolygnathus* and the Late Tuvalian *Carnepigondolella*.

The genus Norigondolella: *Norigondolella* lacks a clear ancestor (Orchard, 2007). Sharing subterminal pit, lack of ornamentation, keel termination never bifurcated, *Norigondolella* may have evolved from the genus *Paragondolella* with *Neocavitella* as intermediary proteromorph (Fig. 8).

6. NATURAL HISTORY OF TRIASSIC CONODONTS

Triassic conodonts appear in three waves of development:

Wave I: From a dozen Permian–Induan survivors of the Families Spathognathodontidae, Ellisoniidae and Gondolellidae, only the latter's subfamily Neogondolellinae produced a radiating wealth of genera prior to the end of the Spathian. Most lineages that appeared during the Griesbachian and Dienerian did not persist beyond the Smithian and only one lineage persisted, thanks to repeated retrogradations, until the Rhaetian extinction. A great wealth of forms appeared and went during the first 5 Ma of Triassic times that are the recovery from the Permian–Triassic Boundary Mass Extinction. The Spathian–Anisian transition became a first environmental bottle neck through which a single lineage of Neogondolellinae could pass. Only in the pelagic realm, the Family Gladigondolellidae that appeared in the latest Spathian lasted until the end of the Julian. The first wave (Fig. 3) starts with Permian–Griesbachian survivors *Hindeodus, Ellisonia,* and *Clarkina*. The anagenetic lineage of Neogondolellinae was represented by *Clarkina* until shortly after the Griesbachian–Dienerian boundary. From then on, this lineage consists

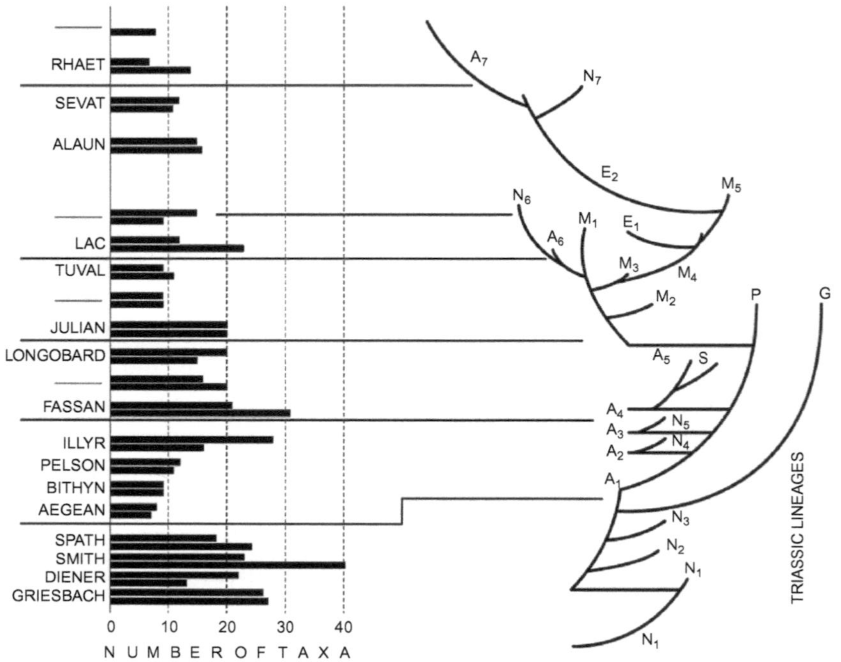

Fig. 8 Distribution of Triassic conodont lineages and number of taxa through time (Kiliç et al., 2015a,b). Taxa numbers, compiled from Martinez-Perez et al. (2013). A, *Neospathodus*; A1, *Kashmirella*; A2–3, *Nicoraella*; A4, *Pseudofurnishius*; A5, *Mosherella*; A6, *Neocavitella*; A7,*Misikella*; N, Neogondolellinae; N1, *Clarkina*; N2, *Scythogondolella*; N3, *Borinella*; N4, *Pridaella*; N5, *Neogondolella*; N6, *Norigondolella*; N7, *Parvigondolella*; P, *Paragondolella*; M1, *Metapolygnathus*; M2, *Mazzaella*; M3, *Quadralella*; M4, *Carnepigondolella*; M5, *Ancyrogondolella*; S, Sephardiellinae, *Pseudofurnishius* and *Sephardiella* lineages; E, Epigondolellinae; E1, *Acuminatella*; E2, *Mockina.*

of the atavistic retrogradations of *Neospathodus* (Dienerian–Smithian), *Kashmirella* (Dienerian–Aegean), *Smithodus* (Smithian), *Novispathodus* (Smithian–Spathian), *Eurygnathodus* (*Plativillosus*) (Smithian), and *Triassospathodus* (Spathian). Several anagenetic rediversifications appeared, such as *Borinella* (Smithian) and *Scythogondolella* (Smithian). Additional Dienerian–Spathian subfamilies include *Mullerinae* Orchard (2005) with genera *Conservatella* (Smithian), *Discretella* (Smithian) and *Guangxidella* (Smithian), Subfamily Uncertain with genus *Icriospathodus* (Spathian). The Spathian–Aegean transition formed a first bottle neck that only *Gladigondolella* and *Kashmirella* could flit through. The lineage of the Neogondolellinae consists since the Aegean of the anagenetic lineage of *Paragondolella* (Aegean–Julian), paced by atavistic retrogradations such as

Nicoraella (Bythinian–Pelsonian), *Pseudofurnishius* (Late Anisian), and anagenetic rediversifications, such as the genera *Pridaella* and *Neogondolella* and *Sephardiella*. Gladigondolellidae dwelled in pelagic seas from the Late Spathian until the end of the Julian.

Wave II: Several genera developed during Anisian and Ladinian from the single survivor lineage of Neogondolellinae, Repeated punctuated atavistic iterations in Bithynian, Pelsonian, Illyrian, and Fassanian times generated new genera with lineages all ranging until the second bottle neck of the Early Julian, 15 Ma into the Triassic. Again, only one lineage of Neogondolellinae survived, as well as the single lineage of Gladigondolellidae. The second wave consists primarily of *Gladigondolella,* which is restricted to the Tethyan pelagic realm and the lineage issued from *Kashmirella* that generated *Paragondolella* (Fig. 4). From the latter, the punctuated iterative neospathid speciations of *Nicoraella* and *Pseudofurnishius* emerged. *Nicoraella* generated *Pridaella* and from it *Neogondolella* evolved. The lineages of *Pseudofurnishius–Mosherella* and *Sephardiella* represent a subfamily characterized by the acuminate shape of the platform. Neither of these taxa ranged beyond the Early Julian and only the pelagic genus *Gladigondolella* and open-marine *Paragondolella* made it through the second bottle neck of the Early Julian, 15 Ma into the Triassic.

Wave III: The 16-Ma long Julian–Lacian time span is one of intense diversification (Fig. 5).

At least seven generic lineages resulted from three simultaneously working evolutionary trends: (i) Increased ornamentation, (ii) widening, and (iii) bifurcation of the posterior end of the unit.

The Alaunian reversal of these trends initiates the last 20 Ma of the latest Triassic, characterized by the posterior acumination of the unit (*Mockina*).

Atavistic retrogradations marked the Ladinian–Julian transition (*Mosherella*), the Late Tuvalian–Early Lacian (*Neocavitella, Norigondolella*), Sevatian (*Misikella*) and Rhaetian (*Parvigondolella, Oncodella*). Atavistic genera are the only survivors in the Rhaetian that mark the end of the third wave and signal the final extinction of conodonts. It is noticeable that *Neocavitella* sp. and *Norigondolella* sp. are found in the conodont parvus Zone of British Columbia in the radiolarian assemblage 4 (Carter and Orchard, 2013) with diminutive elements characteristic of the event.

The Julian–Lacian time span is characterized by the trend of general platform reduction (formation of a free blade) and bifid widening of the platform (splitting of the basal cavity). During the Late Julian, the surviving smooth *Paragondolella* generates *Metapolygnathus* with a smooth platform, gradually reduced by the development of a free blade. Late Julian marks the appearance of *Mazzaella* with a free blade and strong nodes bordering the platform. The Tuvalian sees an explosion of taxa: Some like *Norigondolella* are smooth, a retrogradation mimicking the Middle Triassic *Paragondolella*. But gradually genera appear with an increasing number of nodes bordering the platform. First to appear is the slightly nodose *Quadralella,* followed by the strongly nodose *Carnepigondolella* that gradually evolved to the denticulated *Ancyrogondolella,* simultaneously reaching the ultimate stage of platform widening. This ended the evolutionary trend begun in the Julian and survived only slightly into the initial Alaunian. The Alaunian reversal of this trend into platform narrowing was initiated the last 20 Ma of the Triassic. This period was dominated by *Mockina* with fewer speciations until the Late Sevatian when *Misikella* took over. Next to it, the atavistic genera *Parvigondolella, Oncodella*, and *Zieglericonus* dominate the Rhaetian and terminate the third wave in the final extinction of conodonts.

7. CAUSES AND EFFECTS

A meteorite impact at the dawn of the Sevatian may have precipitated the final extinction of conodonts. This hypothesis is substantiated by several craters at the approximate Alaunian–Sevatian boundary such as Rochechouart (France), Manicouagan and Saint Martin (Canada), Obolon (Ukraine), Red Wing (USA), and Bristol (England). The comet or asteroid was synchronously fragmented during impact (Barash, 2006). The activity of the Central Atlantic Magmatic Province (CAMP) reached its apex at 201.58 Ma (Schaltegger et al., 2008).

External causes such as paleoclimatic changes, abnormal temperature, or chemical stress could uncover morphological variants for selection to act upon (Rutherford and Lindquist, 1998). These may allow some of the rapid morphological radiations observed in the fossil record (Guex, 2001). In this context, strong activity of the Siberian Trap lasted until the end of the Griesbachian (Korte and Kozur, 2010). Also recurrent environmental changes triggering Early Triassic ecosystem instabilities such as the Middle Smithian spore spike concurring with a negative $\delta^{13}C$ anomaly, preceded a Late Smithian marine extinction event (Hermann et al., 2011). The major phylogenetic developments in Triassic conodonts seem to coincide with severe sea level changes that possibly were related to anoxic events. Correspondingly, chemical stress and high temperatures may cause inheritable resurgences of atavistic structures. Discrete, short anoxic episodes, such as an Early Dienerian shelf event on the northern Gondwanian margin that correlates in time with similar pal-oceanographic changes on the equatorial North American margin and southern Tethys, were part of the Early Triassic biotic recovery (Ware et al., 2011).

The Early Ladinian transgression caused the ascent of *Pseudofurnishius* and of the first *Sephardiella,* while the Late Ladinian–Early Carnian global regression and salinity crisis reduced niches and caused the extinction of most Neogondolellinae and Sephardiellinae, prompting the iteration of neospathodid *Mosherella.* Climatic change such as the Julian Pluvial Event may coincide with the ascent of *Mazzaella.* Tuvalian, Lacian, and Alaunian oscillations of sea level coincide with *Quadralella, Carnepigondolella, Ancyrogondolella,* and *Mockina*, respectively.

The accumulation of geologic and biological circumstances may have stressed the extinction of conodonts. Since Julian times faunal competition (fide F.D. Por, pers. comm. 1990), stressful environmental conditions, and competition from new Mesozoic biota (Martinez-Perez et al., 2013) made it difficult for conodont taxa to adapt.

Iteration of Paleozoic morphologies is interpreted as the result of retrogradation and atavistic homeomorph forms appearing during sublethal stress events, often separated by several millions of years (Guex, 2001). This is the case of the proteromorphic retrogradation in type N, shortly after the P/Tr boundary, the sudden appearance of a simple and economic architecture as a potential source for renewed phylogenetic development (Fig. 6; Guex et al., 2014).

From astronomical, sedimentological, paleogeographical, and geochemical points of view, the Triassic period was exceptionally warm. The 51-Ma long Triassic period lies between the End Permian and End Triassic mass extinctions, the first beginning during the Siberian Traps and the second ending at the height of the Central Atlantic Magmatic Province activity. Conodonts survived the former but were extinct by the end of the latter.

Three major waves characterize the evolution of Triassic conodonts:

I. The 5-Ma long Scythian recovery and intensive speciation (23 S/Ma);

II. The 10-Ma long Dinarian intensive radiation (8.5 S/Ma);

III. The 16-Ma long Julian–Lacian survival (2 S/Ma) and the 20-Ma of Alaunian–Rhaetian decline (1 S/Ma or less) until extinction.

The major causes of stress affecting the evolution of Triassic conodonts are environmental, such as marine regressions, anoxic episodes, thermal, and trophic stress (Hallam, 1978; Mancini, 1978; Valentine et al., 1994). These stressful episodes caused dwarfing and extinction of larger forms (r-selection). Smaller forms with a short life span and rapid proliferation are the result of heterochrony such as proteromorphic retrogradation (Atavism).

The disappearance of conodonts was not caused by a single mass extinction event, but the result of cumulative stress factors from hostile environments over a long time (Fig. 6).

Environmental stress such as marine regressions, anoxic episodes, and thermal or trophic periods (Hallam, 1978; Mancini, 1978; Valentine et al., 1994) may, in comparison with Cope's rule, cause decreasing size and morphological complexity, sometimes explained by elimination of the more complex forms during extinction events (Saunders et al., 1999). The larger forms being more vulnerable to extinction (r-selection), more and more decreasing small forms with short life span and rapid proliferation through proteromorphic retrogradations of type A appear during stressful episodes.

8. CONODONTS VS PALEOCURRENTS

Triassic conodonts are ecological sensors (Hirsch, 1994a,b), and their paleobiography is monitored by latitude, salinity, water depth, and geographic isolation, but some aspects of their distribution remained obscure. Within low latitude Tethys and related domains, paleocurrents provide clues to the distribution of faunal associations and eventual migratory paths. This is particularly significant for understanding the distribution of Sephardic faunal elements within deposits of Gondwana and Cimmerian terranes, while these same elements are extremely rare in environmentally equivalent deposits situated on the Eurasian side of the seaway.

The hypothesis of southern and northern westbound paleocurrents and of an eastbound counter-current may explain the distribution of Sephardic taxa like *Pseudofurnishius* south of a "Gondwana-Tethys Divide" (GTD) as well as the higher frequency of *Sephardiella* in the southern Alps than in equivalent deposits of the northern Alps. Furthermore, an eastbound counter-current would explain the allopatric migration of the Lower Ladinian *S. truempyi* from the Sardinia–Provence eastward into the Adria–Apulian domain, as far as the Izanami seamount swarms of the Izanagi Ocean (Paleo-Pacific), but not south-eastward to Africa and the Levant.

9. A BIT OF HISTORY

Triassic conodonts are a relatively young discovery. Eicher's (1947) unillustrated paper on conodonts from Egypt and Palestine was not taken seriously by the scientific community that saw them as reworked from a Paleozoic substrate, ironically very poorly represented in the region. 1956 was a good year as Tatge (1956) published her conodont paper from the German Muschelkalk and Diebel (1956) his alleged Late Cretaceous Mungo-river conodonts from Cameroon, extracted from Wurm's collection at the Humboldt Museum, ironically the very same fauna as that of Eicher. Soon Huckriede (1958) revealed the rich "Tethyan" fauna from the Alps, yielding not only elements of subfamily Neogondolellinae with the same apparatus as the widely endemic forms that dwelled in the landlocked German basin, but the entirely different clade of the fully cosmopolitan Gladigondolellidae, well differentiated by their apparatus. Less than a decade later in 1966, Van den Boogaard, in Andalusia and Hirsch in Catalonia, revealed the existence of the almost exclusively Ladinian conodonts that became the basis for the recognition of the Sephardic bioprovince that extends from Spain to the Levant (Hirsch, 1972). Moreover, the discovered fauna was identical with that of Diebel, extracted from a Catalonian Triassic ammonite, not a Cretaceous one from Cameroon (Foerster and Weddige, 1979). Hayashi (1968) published conodonts from central Japan and Mosher (1968) a comprehensive study of the conodonts of Europa and North-America. As this short and incomplete history of research reveals, in a short lapse of time Triassic conodonts from almost all existing "provinces" had been described, but not much was known about what caused their differentiation and distribution, shortly why they were where they are.

10. CONODONT FAUNAL ASSOCIATIONS

Middle and Late Triassic conodont faunal associations define paleobiogeographic realms (Hirsch, 1994a). Paleolatitude is crucial for their understanding. The low latitude Tethys realm is characterized in the Alpine-Himalayan Orogeny, from the Alps to Timor. The Pacific realm comprises terranes of mixed origin and their distribution crosses paleolatitudes, originally as well as presently as the result of plate-movements. True Tethyan low latitude taxa are found in the vicinity of both oceanic and cratonic higher latitude faunas, per example as part of displaced terranes that docked next to the North

American craton, while other taxa of high latitude origin may not be differentiated from nonterrane true craton-bound species. As such, Nevadan and NE British Columbia successions are of low and middle latitude (Orchard, 1991b). A comparison of the Middle and Upper Triassic faunal spectra from Japan with Tethys and Pacific realms reveals more Tethyan affinities, though affinities with Pacific associations are also present as well as some degree of endemism (Hirsch and Ishida, 2002).

10.1 Tethyan Associations

A large and disparate literature exists for this realm (Budurov, 1976, 1977; Budurov and Sudar, 1990; Gazdzicki et al., 1979; Kovacs and Kozur, 1980; Kozur and Mock, 1974; Krystyn, 1980, 1983; Mosher, 1968).

Anisian: *K. timorensis, Pa. regale,* and *G. carinata* mark the lowermost part of the Anisian. *G. tethydis* ranges from Anisian to Early Carnian.

Pr. constricta, Pr. bifurcata, Pr. cornuta, Pr. eotrammeri, and *Pa. bulgarica* are Anisian markers; *N. kockeli* marks a short interval at the base of the upper Anisian, and *Pa. excelsa* appears toward its upper part.

Ladinian: A number of parallel, depth-controlled lineages consist of:

(a) *Pr. trammeri, Pa. praehungarica, Pa. transita,* and *N. fuelopi* [Tethys shelf];
(b) *Pa. excelsa, Pa. foliata,* and *Pa. inclinata* [pelagic Tethys];
(c) *Ps. murcianus, S. truempyi, S. mungoensis, S. diebeli* [high saline, Sephardic];
(d) *G. tethydis* and *G. malayensis* [equatorial Tethys].

At the very top of the stage *M. microdus* appears.

Julian: The base of the Carnian still yields Sephardic migrants, e.g., *Ps. murcianus*.

Taxa are *M. polygnathiformis, Ma. carnica* and in the deeper environment *G. tethydis and G. malayensis* are found.

Tuvalian: *M. polygnathiformis, M. auriformis, M. noah, M. parvus, M. echinatus, Q. tadpole, Q. nodosa, Q. carpathica,* and *Cavitella cavitata. Carnepigondolella.*

Lacian: *M. permicus, A. quadrata, A. triangularis,* and *N. navicula*. The Tethyan character of *Ancyrogondolella spatulata* and *N. hallstattensis* was stated by Orchard (1991b).

The dominance of *A. spatulata* and its range from Lower to Middle Norian in Japan suggests a Tethyan character.

Alaunian: *Ancyrogondolella uniformis, A. spatulata, M. postera, M. multidentata,* and *N. steinbergensis*. The lineage of *Misikella* ranges from Alaunian to Rhaetian.

Sevatian: *M. bidentata* and *N. steinbergensis*.

Rhaetian: is characterized by *M. posthernsteini*.

10.2 Pacific Associations

The data on Middle–Upper Triassic associations are based on Mosher (1973), Nicora (1976), Orchard (1991a, 1991b), Bucher and Orchard (1995), Orchard and Tozer (1997), and Orchard et al. (2001).

Anisian: Besides cosmopolitan *K. timorensis* at its base, *Pridaella aldae* seems to dominate, and the craton-bound *Pr. shoshonensis* occurs.

N. kockeli is found in Nevada. Mosher (1968) mentioned *N. mombergensis,* though this probably needs revision.

Ladinian: Besides *Pr. aldae,* most taxa have a cosmopolitan character and *Pa. excelsa* appears. The Early Ladinian contains *S. truempyi*, *S. japonica* followed by the Middle Ladinian *S. hungarica*, whereas the Late Ladinian (sutherlandi Zone) is defined by *S. mungoensis,*

Julian: *M. newpassensis*.

Tuvalian: The genus *Metapolygnathus* fully dominates, among which *M. polygnathiformis*. The nodosa Zone is characterized by a wide specific spectrum of conodonts. The morphologies of species assigned here to *Metapolygnathus* have a wide range. *M. reversus* has a loop-like basic cavity and a nude platform, whereas *C. zoae* and *C. samueli* have an elongated loop-like basic cavity and nodes on the platform. A very short and rounded platform, with loop-like basic cavity, characterizes *M. stephanae,* which appears in the latest Carnian–earliest Norian primitius Zone.

Lacian: The Lacian is characterized by conodonts with moderate to strongly split basic cavities as well as sharp denticles on the platform margins.

In the Pacific realm, the predominance of *A. quadrata, A. uniformis,* and *A. triangularis* prevails over forms with affinities to *A. spatulata* seems to occur. In the uppermost part of the Lacian, "*Epigondolella*" *transitia* shows strong reduction of its split basic cavity and a narrowing of the platform.

Alaunian: The Alaunian is characterized by the development of simplified morphotypes (Hirsch, 1994b). An amygdaloid basic cavity and platform margins with sharp denticles characterizes the genus *Mockina*. The large specific diversity *of Mockina* encompasses *M. multidentata, M. elongata, M. matthewi, M. carinata, M. tozeri, M. spiculata, M. postera*, and *M. serrulata*. *M. postera* is exceptionally common in British Columbia (Orchard and Tozer, 1997). The presences in the Izanami plateau of a number of Alaunian endemic taxa define this distinctive Pacific realm.

Sevatian: The strong reduction of the platform characterizes *M. bidentata,* together with *M. englandi,* which still shows similarity with *M. postera*.

Rhaetian: *Mockina mosheri* ends the lineage of platform conodonts. *Misikella* is only known from terranes. *M. posthernsteini* is present, though not abundant in the Haida Gwaii (Canada) (Carter and Orchard, 2013).

10.3 Sephardic Associations

Ps. shagami, Ps. siyalaensis, Ps. murcianus, S. truempyi, S. mungoensis.

10.4 Environmental Conditions

Depth, salinity, and paleolatitude control faunal provincialism.

Middle Triassic *Neogondolella* is more frequent in shelf environments, whereas the ratio of *Paragondolella* increases toward the outermost shelf (Hirsch, 1994a).

In confined areas of possible abnormal salinity such as the Germanic and Sephardic provinces, specialization took place, as evidenced in the Sephardic *Pseudofurnishius*. As *Gladigondolella* (Earliest Anisian–Julian) and *N. hallstattensis* (Alaunian) are believed to represent pelagic low latitudes, it may be reasonable to assume that a high ratio of *A. spatulata*, which is dominant in the conodont associations of the Izanami plateau as well as in the bedded cherts, is low latitude related.

11. A CASE STUDY: THE MIDDLE TRIASSIC IN THE MEDITERRANEAN REGION

The Triassic period derived its name from the characteristic trilogy comprising a succession of variegated sandstone called "Buntsandstein," followed by carbonates yielding rich marine mollusk faunas called "Muschelkalk," interrupted by evaporitic intercalations and/or ending in thick marls and evaporites, mainly gypsum, called "Keuper." The Middle Triassic part that is developed in the "Muschelkalk" facies is widespread around the western edge of Tethys, where the Eurasian- and Gondwana-Tethys shelves each have developed their particular Muschelkalk faunas. The northern Peri-Tethys shelf encompassing the shallow, widely land-locked Germanic basin of alternating salinity, with faunas at both the generic and specific level of a particular evolutionary deployment, such as ammonoids (*Ceratites*) and conodonts (*Celsigondolella*) and, at the specific level, distinct bivalves. Only few Tethyan forms (*Protrachyceras recubariense*) found their way into this ***Germanic*** basin, providing correlation with the Tethys. The southern Peri-Tethys or Gondwana shelf, extending from Iberia over North Africa to the Middle East, consists of a huge succession of shallow platforms with generally slightly hypersaline waters, called the ***Sephardic*** Bio-province (Hirsch, 1972). Anticipated by Lerman (1960), Huddle (1970), and Parnes (1962), the special nature of the fauna was further evidenced in Parnes (1975, 1986), Hirsch (1977, 1987), Parnes et al. (1985), Goy (1986), Virgili (2008), and Pérez-Valera (2008).

To explain faunal migrations between Tethys and the Germanic basin, many invoked a hypothetical "Burgundian Gate" (Durand, 2011), although others did not (Hirsch, 1977; Márquez-Aliaga et al., 2000; Rein, 2008). Courel (1984) and Lienhardt et al. (1984) were decisive in showing the unlikelihood of a fully marine communication with Tethys via Burgundy during the Anisian and Ladinian, but preferred communication through a "Western Gate" proposed by Narkiewicz and Szulc (2004), or passing the "Allemanic Gate" across the Swiss Alps, as proposed by Feist-Burkhardt et al. (2008). Beyond the barrier of Vindelician Land, the so-called Mediterranean, Alpine, or Tethyan facies appear progressively in a succession of epicontinental belts, Brianconnais (Peninnic) in the Western Alps and are fully developed in the Austro-alpine Northern Limestone Alps. The west-end of the narrowing Paleo-Tethys' Meliata ocean separated the latter from the southern peri-Tethys shelf, extending at present south of the peri-Adriatic tectonic lines. While common 239.3 to 240.91 Ma old "pietra verde" (Brack et al., 2005; Bruehwiler et al., 2007) witness a relative proximity of Northern and Southern Alps, the curionii Zone, base of the Ladinian (GSSP) at Bagolino yields pelagic conodonts such as *Gladigondolella,* and members of the *S. truempyi* group (Brack et al., 2005). The latter links the Southern Alps of Tethyan character to the Sardic Bio-Province, an intermediate between the Tethys and Sephardic domains, as well as to the Brianconnais domain of more northern affinity. The sephardic *Ps. murcianus,* present from the Early Ladinian of the sephardic domain, is found in the Late Ladinian–Early Carnian San Cassian fauna (Mastandrea et al., 1998), and in the Wengen and Pratodonto formations of the Val di Scalve and Val Camonica, where it is accompanied by the assemblage of *S. diebeli, S. mostleri, S. mungoensis, G. malayensis, Pa. inclinata, Metapolygnathus foliatus, M. polygnathiformis* (Balini et al., 2000). This assemblage with *Ps. murcianus* also appears at Bukit Kehil (Malayan Peninsula) in the furthermost Cimmerian terrane of Sibumasu (Ishida and Hirsch, 2001).

11.1 The Germanic Domain

During the Anisian and Ladinian stages, conodont animals followed the Tethyan transgression from East to West and at some point penetrated inside what developed as the Germanic Basin, an interior sea with ephemeral access to the Tethyan oceanic seaway. Thus, for conodont-animals, their adaptation to new environments brought morphological changes, a fate similar to that of ammonites. A number of endemic genera and species are characteristic for the Germanic Province. Transgression during the Anisian reached its height prior to the Compressus ammonite Zone which corresponds to the Tethyan curionii ammonite Zone, at the end of conodont Zone 2 (Kozur, 1968). From then on, the regression that lasted through the Ladinian initiated the typical retrograde evolution of both conodonts and ceratites. Franz et al. (2015, fig. 5) shave schematized the transgressive–regressive trends of the Germanic Basin in time and space.

11.2 Sephardic Domain

No one expressed better than Pérez-López and Pérez-Valera (2007) how "the Sephardic domain includes endemic bivalve, ammonoid and conodont species that span a paleogeographic domain S and SW of Tethys, from Spain and North Africa to the Middle East." The particular Muschelkalk biofacies, other than the Germanic with Alpine influence observed in Catalonia, Valencia, Castellón, and Teruel (Spain) by Sopeña et al. (1983), is what Hirsch (1972, 1975) suggested some years earlier and from this, coined the term "Sephardic"domain. Hirsch (1977) mentioned the evolutionary autonomy of species affecting a vast paleogeographic area from the Iberian Peninsula to the E Mediterranean, independent from that of Tethys but with some influence of its fauna. Finally, Costamagna and Barca (2002) extended the Sephardic domain across the whole N region of Africa and E of Spain as far as western Sardinia. Thus, Sephardic biofacies developed all along the Gondwana shelf south of the Tethyan biofacies. It is characterized at the generic level by Ladinian ammonoids *Gevanites, Israelites,* and *Negebites* and of the conodonts *Pseudofurnishius* and *Sephardiella*; several bivalves are endemic at the specific level. Characteristic brachiopods, ostracods, and reptiles add to the biofacies. Tethyan elements are present as immigrants in the more open marine parts of the realm, such as *Daonella* in Menorca, *Eoprotrachyceras* and *Protrachyceras* in Catalonia and Israel. In the course of the Late Ladinian, shallower parts of the Western Tethys became increasingly hypersaline and "Sephardic" faunal elements migrated into the Southern Alps, Apulia, and the Cimmerian terranes. Problematic are some microcontinents (Tisza). In the course of the Late Triassic the westernmost domains of the Sephardic realm fell under influence of the Keuper regime, while the eastern part of the string of Cimmerian terranes drifted off Gondwana, opening a Meso-Tethys branch and, consuming the Paleo-Tethys, finally collided with Eurasia. These Cimmerian terranes remain identifiable as far East as the Malayan Peninsula (Hirsch, 1994a,b) by the conodont *Ps. murcianus,* that marks the Triassic GTD.

Anisian ammonoids are mostly of Tethyan affinity and bivalves are cosmopolitan (Alpine and Germanic) in Iberia, while in the Levant some Sephardic elements appear earlier (*Gervillia joleaudi*), as some are strictly Levantine.

Different from the endemic taxa developed from Tethyan immigrants in the Germanic realm, Sephardic ammonoids such as *Israelites, Negebites,* and *Gevanites* are endemic (Parnes, 1962, 1975, 1986). Kummel (1960: p. 289) noted that L.P. Spath held the majority of Awad's (1945) ammonites from Araif el Naqa (Sinai) for a subgenus of *Ceratites,* comparable to the group of *C. flexuous,* indicating a lower Ladinian age.

Tethyan ammonoids developed endemic species in the Sephardic realm, such as *Protrachyceras hispanicum* and *P. villanovae* or the subspecies *Eoprotrachyceras curionii ramonensis.*

Among bivalves, the Sephardic species *G. joleaudi, Enantiostreon flabellum,* and *Costatoria goldfussi* (=*C. multicostata*) dominate over mostly cosmopolitan forms from Iberia to the Levant. *Placunopsis teruelensis* and *Costatoria kiliani* are endemic to Iberia, and *Costatoria coxi is* endemic to the Levant.

If oceanic belts often coincide with boundaries between provinces, then a certain biomarker can show the divide between realms. During the Triassic, the fauna of the Gondwanian shelf stands out, and the GTD may coincide with the mayor tectonic lines that separate Gondwana-related terranes from Paleo-Tethyan Eurasia. The divide can be followed from Biscay to Vardar and further East along synchronous ophiolite strings that suggest the axis of ocean branches. However, it remains problematic to estimate the extent of transform components on faults (e.g., Michalik, 1994) and others and that may alter substantially any Triassic reconstruction. The Sephardic bioprovince collapsed during the Carnian "Saharan Salinity Crisis."

11.3 Sardinia and Provence

Costamagna and Barca (2002) extend the Sephardic domain across the whole northern region of Africa and East of Spain as far as western Sardinia. Previously Hirsch (1977) and Bouquet and Stoppel (1981) qualified this area as the "Sardic-Provencal" biofacies. Urlichs (1997) contested these attributions with the finding of *Ceratites* (*Alloceratites*) *toulonensis* and that at

Lagoubran of *C.* cf. *laevigatus*, a ceratite that appears in Germany above the Cycloidesbank in the enodis Zone. In his discussion of the relations of Mediterranean and Germanic ceratites, this author concluded that the ceratite-bearing layers of the Muschelkalk in SE-France (Unit F) have to be correlated with the Central European enodis Zone (Early Ladinian). This is corroborated by the presence of the conodont *S. truempyi* and by the two *Germanonautilus bidorsatus* and *G. suevicus*, in these beds. In the German Muschelkalk, *G. bidorsatus* disappears in the postspinosus Zone, and its descendant, *G. suevicus*, appears in the enodis Zone (Mundlos and Urlichs, 1984). Brocard's (1991) correlation is based on occurrence of *S. truempyi* in Caron's (1967) Formation II, which corresponds to the Early to Late Ladinian *curionii* to archelaus Zones Hirsch (1971). Spores and pollen (Brocard, 1991) already appear in the Early Ladinian.

The Balearic Island of Minorca and Catalonia yield well-diversified Middle Triassic ammonoids that consist of taxa unknown in the Germanic Basin (Goy, 1995). These include *Gevanites*, a taxon formerly assigned to *Ceratites*. Despite some morphologic similarities there is no direct family relationship between Iberic *Gevanites* and Germanic *Ceratites*. Hypothetically a common Anisian ancestor of both may have lived in the eastern parts of the Sephardic province (Rein, 2008).

Anisian ammonoids in both Iberia and the Levant are mostly of Tethyan affinity; Iberian bivalves are cosmopolitan (Alpine and Germanic), while a number of Levantine and Sephardic elements appear in the Levant.

Ladinian ammonoids are dominantly Sephardic, strongly differentiated from the endemic taxa that developed from Tethyan immigrants in the Germanic realm. Tethyan ammonoid genera developed endemic species in the Sephardic realm, such as *P. hispanicum* and *P. villanovae* or subspecies like *P. curionii ramonensis*. Dominantly endemic Sephardic ammonoid genera (*Gevanites, Israelites, Negebites*); *Protrachyceras* of Tethyan affinity developed in the Sephardic Realm (*P. hispanicum, P. villanovae, E. curionii ramonensis*). Bivalves are mostly cosmopolitan, but two sephardic bivalves *G. joleaudi* and *E. flabellum* dominate over mostly cosmopolitan ones from Iberia to the Levant; *P. teruelensis* is endemic to Iberia; *C. coxi* and *C. multicostata* are endemic to the Levant.

Carnian: The Sephardic bioprovince collapsed during the Carnian "Saharan Salinity Crisis."

If oceanic belts often coincide with boundaries between provinces, then a certain biomarker can show the divide between realms. During the

Triassic, the fauna of the Gondwanian shelf stands out and the GTD may coincide with the mayor tectonic lines that separate Gondwana-related terranes and Paleo-Tethyan Eurasia. The divide can be followed from Biscay to Vardar and further east along synchronous ophiolite strings that suggest the axis of ocean branches (Pérez-López and Pérez-Valera, 2007).

11.4 Short History of Conodont Research in the Levantine Region

This chapter was motivated by one particular issue in the Triassic conodont literature that distorts the otherwise well-established conodont and palynomorphs-based biostratigraphy of the Triassic sequence across the Dead Sea Rift. East of the Jordan River, a Ladinian age attribution to conodonts, found in the Mukheiris Formation at Wadi Siyala (Sadeddin, 1990; Sadeddin and Kozur, 1992) falls outside the otherwise sound regional conodont and palynomorphs biostratigraphy (Abu Hamad, 2004; Bandel and Waksmundzki, 1985). Since the 1974 Vienna Symposium (Zapfe), the Anisian identity of the coeval Ra'af and Hisban formations on both sides of the Dead Sea Rift was clarified (Druckman, 1974). Subsequently, the overlying Gevanim and Mukheiris formations, dated by Anisian ammonites (Parnes, 1962) and palynomorphs (Abu Hamad, 2004), respectively, represent a clastic interval between "lower" and "upper Muschelkalk" facies developments in the Middle East. West of the DST, these units are well dated by ammonites (Parnes, 1962, 1986), conodonts (Hirsch and Gerry, 1974) and palynomorphs (Eshet, 1990). East of the DST, conodonts (Bandel and Waksmundzki, 1985) and palynomorphs (Abu Hamad, 2004) are supplemented by additional conodont findings (Sadeddin, 1990). At the locality of Naur, but without illustration nor mention of level or sample number, the conodont *S. truempyi* would precise the Fassanian age of the lower part of the Iraq Al Amir Formation. In the lower half of the Pelsonian or Illyrian Mukheiris Formation at Wadi Siyala, a stream adjacent to Wadi Hisban, Sadeddin and Kozur (1992) found *Ps. priscus* and *Ps. siyalaensis* that they erroneously considered Fassanian in age. These neospathid morphs, important elements in the phylogeny of *Pseudofurnishius,* occur at various levels, including *Ps. shagami* (Benjamini and Chepstow-Lusty), also possibly found in Naur (sample Ab8).

At Wadi Hisban, Wetzel and Morton (1959) described ca. 200 m of mostly limestones in the lower half that represent the Hisban Limestone Formation, followed by sandstones, erroneously attributed to the Liassic Subeihi

Fm. This could only represent the Mukheiris Fm. In Sadeddin and Kozur (1992, fig. 3) 18 m limestone with *Beneckeia* are followed by "ca. 38 m of dolomite and sandstone," still attributed to the Illyrian (Anisian). Erroneously the sudden income of "Fassanian" with *Trigonodus* yields the solace *Ps. priscus*.

11.5 Triassic Conodonts Were First Found in the Levant

In the year 1946, Eicher (1947) discovered conodonts in Araif e Naqa (Sinai) and Wadi Raman (Palestine), which he claimed as Triassic, but lacking illustrations, the discovery was dismissed as reworked from the Paleozoic. The very same conodonts were then extracted by Diebel (1956) from the alleged Turonian of the Mungo River (Cameroon) and were revealed later to be Ladinian from Mora de Ebro (Spain). Next, they were identified as "*Sephardiella*" *mungoensis* (Diebel) and *Ps. murcianus* Van den Boogaard with the latter being restricted to the Sephardic Gondwanian Tethys-shelf (Hirsch, 1972). Huddle (1970) found these taxa in Stratigraphic sequence in Ramon (Negev, S. Israel). Hirsch (1971) added *S. truempyi* from the Fassanian of Provence (S. France). Hirsch and Gerry (1974) proposed the Fassanian Murcianus Zone and Longobardian–Julian Murcianus–Mungoensis Zones.

Other Middle Triassic genera were first recognized in the Muschelkalk of the Germanic Faunal Province (Tatge, 1956). Additional taxa from the wide realm of the Tethys were first found in the Eastern Alps (Huckriede, 1958). Here ends the first phase of Triassic conodont research.

Staesche's (1964; after Hirsch, 1975) Early Triassic conodonts, found in Humrat Ma'in (Jordan) by Huckriede (in Bender, 1968), occur in the Negev subsurface (Hirsch, 1975).

12. PHYLOGENY OR ONTOGENY?

Taxonomy and evolution of the Late Anisian (Pelsonian)–Early Carnian (Cordevolian) conodont genus *Pseudofurnishius* were revealed (Plasencia et al., 2015) to be of gondolellid stock. Its most commonly found species, *Ps. murcianus*, appears at the base of the Ladinian (Fassanian). The Ladinian material of Spain reveals its ontogeny characterized by initial lateral protrusions from the carina that may develop first monoplatform and later biplatform units, finally evolving into elaborated multidenticulated forms. While a phylogenesis of *Ps. priscus* → *Ps. shagami* → *Ps. murcianus* → *Mosherella* is in line with the general view of Plasencia et al. (2015) as well as of Sadeddin and Kozur (1992), large discrepancies exist in the details. *Ps. murcianus* is known for its morphological variability and while Hirsch and Gerry (1974) mention the ratio of mono- and biplatform specimens within the species, biplatforms have been recognized since Van den Boogaard and Simon (1973) as *P. huddlei,* the predecessor of *Ps. murcianus*. Sadeddin and Kozur (1992) saw the evolution of *huddle—murcianus* as a string of subspecies ending in the monoplatform type, although for Bandel and Waksmundzki (1985), the large population of Naur J 34 consisted of all types of mono- and biplatforms, within a single species *Ps. murcianus*, nullifying *P. huddlei* as a species. In his thesis, Plasencia (2009) addressed the variability in the irregular pattern of distribution of denticles on both platforms within *Ps. murcianus* to growth, thus to ontogeny, biplatforms representing adults.

The atavistic *N. kockeli, Ps. priscus, Ps. siyalensis, Ps. shagami,* and *Mosherella* are pacing gondolellid evolution as proteromorphs (sensu Guex, 2001).

13. RESULTS

13.1 Triassic Conodont Lineages Take Place in Three Waves

The first wave starts with Permian–Induan survivors, a few of them producing an enormous wealth of genera that radiate abundantly until the end of the Spathian. Most lineages that take their origin during the Scythian do not persist beyond the Olenekian; the *Neospathodus–Chiosella* lineage continues until the final extinction of conodonts; the Olenekian *Neogondolella*-like forms may have persisted until the Ladinian or risen again from Anisian Chiosella-stock; the *Gladigondolella* lineage dwells in pelagic seas of the Late Spathian until the end of the Julian. The Spathian–Anisian transition represents a first bottle neck through which only *Gladigondolella* and *Kashmirella* could pass. *The second wave* consists of *Gladigondolella, Paragondolella,* and *Neogondolella* lineages, also encompassing the punctuated iterative proteromorphs *Nicoraella, Pseudofurnishius,* and *Mosherella*, one of them generating the lineages of *Pseudofurnishius* and *Sephardiella*. They all range into the Julian, thought only *Paragondolella* survived the second bottle neck of the Early Julian, giving rise to *Metapolygnathus,* 15 Ma into the Triassic. *The third wave* of lineages starts an equally long Julian–Tuvalian time span of 16 Ma, during which the *Paragondolella* lineage produces *Metapolygnathus, Mazzaella, Quadralella, Norigondolella, Carnepigondolella,* and *Ancyrogondolella* that reaches the ultimate extent of the platform widening evolutionary trend by the end of the Lacian. The Alaunian reversal of this trend into platform-narrowing was initiated in the last 20 Ma of the Triassic. *Mockina* dominated the the Late Sevatian fauna, and fewer speciations occurred. *Misikella* was paramount in the Rhaetian along with the atavistic genera *Parvigondolella, Oncodella,* and *Zieglericonus* that terminate the third wave and final extinction of conodonts.

13.2 Triassic Conodonts Response to Environmental Changes

The opening of new niches (Induan, Late Olenekian, Ladinian, and Late Carnian) was accompanied by the appearance of new genera (e.g., *Neospathodus*). In many cases stress conditions first generated neospathid morphs from which the new taxa emerged (Smithian *Borinella* "restoration" of *Neogondolella*-like appearance, Late Spathian ascent of pelagic *Gladigondolella*, Pelsonian appearance of local *Pseudofurnishius*, followed by the unfolding of worldwide Sephardiellinae. Reduction of niches due to the Late Ladinian–Early Carnian global regression and salinity crisis causing extinction (*Neogondolella, Sephardiella,* and *Pseudofurnishius*) and the iteration of the neospathid *Mosherella*. The Julian Pluvial Event may coincide with the ascent of *Mazzaella*. Tuvalian, Lacian, and Alaunian oscillations of sea level coincide with *Quadralella, Carnepigondolella, Ancyrogondolella,* and *Mockina*, respectively. A meteorite impact may have precipitated the final extinction of conodonts (Rhaetian *Misikella*), this last iteration being to no avail.

13.3 Atavism in Conodont Evolution

Rutherford and Lindquist (1998) demonstrated that external stress, such as abnormal temperature or chemical stress could uncover morphological variants for selection to act upon, allowing some of the rapid morphological radiations observed in the fossil record (Guex, 2001). According to Korte and Kozur (2010), strong activity of the Siberian Trap terminates at the end of the Griesbachian. Hermann et al. (2011) report recurrent environmental changes triggering Early Triassic ecosystem instabilities, such as the Middle Smithian spore spike concurring with a negative $\delta^{13}C$ anomaly, preceding a Late Smithian marine extinction event. The major phylogenetic developments in Triassic conodonts seem to coincide with severe sea level changes. Correspondingly, chemical stress and high temperatures may cause inheritable resurgences of atavistic structures. To give some examples, the iteration of Paleozoic morphologies such as *Cornudina* and *Zieglericonus*, and that of Neospathid genera *Nicoraella, Mosherella, Neocavitella,* and *Misikella* are possibly the result of retrogradation in the sense of Guex (2001). For this author atavistic homeomorph forms appear during sublethal stress events, often separated by several millions of years, as in the case of the proteromorphic retrogradation into the genus *Norigondolella*. Hypothetically a simple and economic architecture being a potential source for renewed phylogenetic development explains such sudden appearances (Guex et al., 2014). A Late Tuvalian example may be the appearance of *Acuminatella,* an atavistic retrogradation issued from *Carnepigondolella* that represents the first appearance of the Alaunian trend that characterizes *Mockina*.

13.4 Biological Factors May Have Been Decisive in Conodont Extinction

Twenty-five years ago, Francis D. Por, in a memorable conversation on conodont extinction with the senior author, suggested faunal competition as the cause for their extinction. Martinez-Perez et al. (2013) stressed that the extinction of conodonts was the accumulative result of geologic and biological circumstances, in which biological factors may have been decisive. Thus, since Julian times, conodont taxa were unable to adapt to both stressful environmental conditions and competition from new Mesozoic biota.

14. CONCLUSIONS

1. What is challenging Darwin? Our phylogenetic reconstructions are of anagenetic nature in a Darwinian way. Only the interpretation of some morphs as retrogradations at first seemed challenging, was it not that they integrate atavistic morphologies within a normal evolutionary succession. These apparently "loose" neospathodid morphs appear sporadically, not as "ghosts" of an imaginary phylogenic succession, but as proteromorph links between successive lineages at the very moment of highest environmental stress.
2. The major causes of stress that affected evolutionary trends in Triassic conodont lineages are environmental, such as marine regressions, anoxic episodes, and thermal or trophic (nutritional) stress (Hallam, 1978; Mancini, 1978; Valentine et al., 1994). In comparison with Cope's rule, these constraints cause decreasing sizes and morphological complexity, which have sometimes been explained by elimination of the more complex forms during extinction events (Saunders et al., 1999). While the larger forms are more vulnerable to extinction (r-selection), more decreasingly small forms with a short life span and rapid proliferation are favored during stressful episodes.

3. The atavistic genera *Neospathodus–Kashmirella, Nicoraella, Pseudofurnishius–Mosherella, Acuminatella, Neocavitella,* and *Misikella* are linking the anagenetic lineages of the genera *Clarkina; Scythogondolella; Borinella; Pridaella; Paragondolella; Neogondolella; Pseudofurnishius–Sephardiella; Metapolygnathus–Mazzaella; Quadralella–Carnepigondolella–Ancyrogondolella–Mockina* and *Norigondolella.*
4. Three major waves characterize the evolution of Triassic conodonts:
   - **I.** The 5-Ma long Scythian recovery and intensive speciation (23 S/Ma);
   - **II.** The 10-Ma long Dinarian intensive radiation (8.5 S/Ma);
   - **III.** The 16-Ma long Julian–Lacian survival (2 S/Ma) and the 20 Ma of Alaunian–Rhaetian decline (1 S/Ma or less) until extinction.

The disappearance of conodonts was not caused by a single mass extinction event, but the cumulative effect of stress factors from hostile environments over a long period of time.

ACKNOWLEDGMENTS

We are indebted to our reviewers, who permitted us to improve the text: Dr. E.S. Carter (University of Portland, USA), Prof. H.R. Feldman (American Museum of Natural History, New York, USA), and one anonymous.

REFERENCES

Abu Hamad, A.M.B., 2004. Paläobotanik und Palynostratigraphie der Permo-Trias Jordaniens. Dissertation Hamburg, 330 pp.

Awad, G.H., 1945. On the occurrence of Marine Triassic (Muschelkalk) deposits in Sinai. Bull. Inst. Égypte 27, 397–427.

Balini, M., Germani, D., Nicora, A., Rizzi, E., 2000. Ladinian ammonoids and conodonts from the classic Schilpario-Pizzo Camino Area (Lombardy): revaluation of the biostratigraphic support to chronostratigraphy and paleogeography. Riv. Ital. Paleontol. Stratigr. 106, 19–58.

Bandel, K., Waksmundzki, B., 1985. Triassic conodonts from Jordan. Acta Geol. Pol. 35, 288–304.

Barash, M.S., 2006. Development of marine biota in the Paleozoic in response to abiotic factors. Oceanology. ISSN: 0001-437046 (6), 848–858.

Bender, F., 1968. Geologie von Jordanien. Borntraeger, Berlin. 230 pp.

Bouquet, C., Stoppel, D., 1981. Contribution l'etude du Trias carbonaté des Pyrenees occidentales et centrales. Bull. Bur. Rech. Géol. Min. 1 (2), 119–134.

Brack, P., Rieber, H., Nicora, A., Mundil, R., 2005. The global boundary stratotype section and point of the Ladinian Stage (Middle Triassic) at Bagolino (Southern Alps, Northern Italy) and its implications for the Triassic time scale. Episodes 28, 233–244.

Brocard, C.H., 1991. La plate-forme provençale au Trias moyen: Un modèle de rampe carbonatée en milieu restreint—Dynamique des paléoenvironnements—Evolution diagénétique. PhD thesis, Université de Provence-Marseille. 282 pp. (unpublished).

Bruehwiler, T., Bucher, H., Brayard, A., Goudemand, N., 2007. High-resolution biochronology and diversity dynamics of the Early Triassic ammonoid recovery: the

Smithian faunas of the Northern Indian Margin. Palaeogeogr. Palaeoclimatol. Palaeoecol. 297, 491–501.

Bucher, H., Orchard, M.J., 1995. Intercalibrated ammonoid and conodont succession, Upper Anisian-Lower Ladinian of Nevada. Albertiana 15, 66–71.

Budurov, K., 1976. Structures, evolution and taxonomy of the Triassic platform conodonts. Geol. Balcan. 6 (1), 13–20.

Budurov, K., 1977. Revision of the Late Triassic platform conodonts. Geol. Balcan. 7 (3), 31–48.

Budurov, K., Petrunova, L., 1998. Muschelkalk conodonts as components of the Peri-Tethyan Conodont Fauna. Epicontinental Triassic International Symposium. Hallesches Jahrbuch fur Geowissenschaften, Reihe B, Beiheft 5 Halle, pp. 28–29.

Budurov, K., Sudar, M., 1990. Late Triassic conodont stratigraphy. In: Ziegler, W. (Ed.), 1st International Senckenberg Conference and 5th European Conodont Symposium (ECOS V) Contributions IV: Papers on Conodonts and Ordovician to Triassic Conodont Stratigraphy. Courier Forsch. Inst. Senckenberg, vol. 118, pp. 203–240.

Budurov, K.J., Sudar, M.N., 1998. New conodont taxa from the Middle Triassic. In: Geology of Himalayas—Paleontology, Stratigraphy and Structure (Contribution to Himalayan Geology), vol. 4, pp. 250–254.

Budurov, K.J., Buryi, G.I., Sudar, M.N., 1988. *Smithodus* n. gen. (Conodonta) from the Smithian Stage of the Lower Triassic. Mitt. Österr. Geol. Bergbaustud. 34 (35), 295–299.

Buryi, G.I., 1996. Triassic conodonts from the cherts of the Nadanhada Range, north-east China. Acta Micropalaeontol. Sin. 13, 207–214 (In Russian).

Buryi, G.I., 1997. Evolution of the platform elements of the conodont genus *Metapolygnathus* and their distribution in the Upper Triassic of Sikhote-Alin. In: Dickins, J.M. (Ed.), Late Palaeozoic and Early Mesozoic Circum-Pacific Events and Their Global Correlation. Cambridge University Press, pp. 193–197. World and Regional Geology, UK, 10.

Caron, J.P.H., 1967. Étude stratigraphique du Muschelkalk supérieur calcaire et dolomitique de Basse-Provence occidentale entre Bandol et Hyères (Var). Bull. Soc. Géol. Fr. 9, 670–677.

Carter, E., Orchard, M.J., 2013. Intercalibration of conodont and radiolarian faunas from the Carnian-Norian boundary interval in Haida Gwaii, British Columbia, Canada. In: Tanner, L.H., Spielmann, J.A., Lucas, S.G. (Eds.), The Triassic System, pp. 67–92. New Mexico Museum of Natural History and Science, Bulletin 61.

Clark, D.L., Behnken, F.H., 1971. Conodonts and biostratigraphy of the Permian. In: Sweet, W.C., Bergstrom, S. (Eds.), Symposium on Conodont Biostratigraphy, pp. 415–439. Geological Society of America Memoir 127.

Costamagna, L.C., Barca, S., 2002. The "Germanic" Triassic of Sardinia (Italy): a stratigraphic, depositional and palaeogeographic review. Riv. Ital. Paleontol. Stratigr. 108, 67–100.

Courel, L., 1984. Trias. In: Debrand-Passard, S., Courbouleix, S., Lienhardt, M.J. (Eds.), Synthèse géologique du Sud-Est de la France: Stratigraphie et Paléogéographie, pp. 61–118. Mémoire du Bureau de Recherches Géologiques et Minières 125.

Croft, J.S., 1978. Upper Permian Conodont and Other Microfossils From the Rustler Formation, West Texas. Ohio State University, MA, pp. 103–144. Thesis, Part II.

Dagys, A.A., 1984. Early Triassic Conodonts of Northern Central Siberia. Nauka, Moscow, p. 69 (in Russian).

De Renzi, M., Budurov, K., Sudar, M., 1996. The extinction of conodonts—in terms of discrete elements—at the Triassic-Jurassic boundary. Cuad. Geol. Iber. 20, 347–364.

Diebel, K., 1956. Conodonten in der Oberkreide von Kamerun. Geologie 6425, 424–450.

Dollo, L., 1893. Les lois de l'evolution. Bull. Soc. Belge Geol. Pal. Hydr. 7, 164–166.

Druckman, Y., 1974. The stratigraphy of the Triassic sequence in Southern Israel. Geol. Surv. Israel Bull. 64, 76 pp.

Durand, M., 2011. General outlines of the French South-East Basin. In: Durand, M., Caron, J.P., Hagdorn, H. (Eds.), Triassic of the Southeast France. Pan-European Correlation of the Triassic, 8th International Field Workshop, Guide Book, pp. 2–4.

Dzik, J., 1991. Evolution of oral apparatuses in the conodont chordates. Acta Palaeontol. Pol. 36 (3), 265–323.

Dzik, J., Trammer, J., 1980. Gradual evolution of conodontophorids in the Polish Triassic. Acta Palaeontol. Pol. 25, 55–89.

Eicher, D.B., 1947. Micropaleontology of the Triassic of North Sinai. Bull. Inst. Égypte 28, 87–92.

Embry, A.F., 1997. Global sequence boundaries of the Triassic and their identification in the Western Canada Sedimentary Basin. In: Moslow, T., Wittenberg, J. (Eds.), Triassic of Western Canada Basin, pp. 415–433. Canadian Society of Petroleum Geological Bulletin 45.

Eshet, Y., 1990. Paleozoic-Mesozoic palynology of Israel. 1. Palynological aspects of the Permo-Triassic succession in the subsurface of Israel. Isr. Geol. Surv. Bull. 81, 42 pp.

Feist-Burkhardt, S., Goetz, A.E., Szulc, J., 2008. Triassic. In: McCaen, T. (Ed.), The Geology of Central Europe 2. Geological Society of London, Special Publications, pp. 749–821.

Foerster, R., Weddige, K., 1979. Die Herkunft der Mungo-Kreide Conodonten. Mitteilungen der Bayerischen Staatssammlung für Palaontologie und Historische. Geologie 19, 187–194.

Franz, M., Kaiser, S.I., Fischer, I., Heunisch, C., Kustatscher, E., Luppold, F.W., Berner, U., Röhling, H.G., 2015. Eustatic and climatic control on the Upper Muschelkalk Sea (late Anisian/Ladinian) in the Central European Basin. Global Planet. Change 135, 1–27.

Gaetani, M., Jakobshagen, V., Nicora, A., Kauffmann, G., Tselepidis, V., Fantini Sestini, N., Mertmann, D., Skourtis-Corononeou, V., 1992. The Early-Middle Triassic boundary at Chios (Greece). Riv. Ital. Paleontol. Stratigr. 98 (2), 181–204.

Galton, D.J., 2015. *Man of Science, Man of God*—Gregor Mendel—Discovering the Gene—For His 150th Anniversary. Timaeus Press, Great Britain (227 pp.).

Gazdzicki, A., Planderova, E., Svkora, M., 1979. An Upper Triassic-Lower Jurassic sequence in the Krizna nappe (West Tatra Mts, West Carpathians, Czechoslovakia)—Zdpadne Karpaty. Geologic 5, 119–148.

Gedik, I., 1975. Die Conodonten der Trias auf der Kocaeli-Halbinsel (Turkei). Palaeontogr. Abt. A 150, 99–160.

Gerry, E., Honigstein, A., Rosenfeld, A., Hirsch, F., Eshet, Y., 1990. The Carnian salinity crisis: ostracods and palynomorphs as indicators of palaeoenvironment. In: Whatley, R., Maybury, C. (Eds.), Ostracoda and Global Events. Chapman and Hall, UK, pp. 87–100.

Goudemand, N., Orchard, M., Tafforeau, P., Urdy, S., Bruehwiler, T., Brayard, A., Galfetti, T., Bucher, H., 2012. Early Triassic conodont clusters from South China: revision of the architecture of the 15 element apparatus of the superfamily Gondolelloidea. Palaeontology 55 (5), 99–160.

Gould, J.S., 1977. Ontogeny and Phylogeny. Belknap. Harvard University Press, Cambridge, MA. 490 pp.

Goy, A., 1986. Ammonoideos del Triásico de España. Universidad Complutense de Madrid publ (28 pp.).

Goy, A., 1995. Ammonoideos del Triasico Medio de Espana: Bioestratigrafia y correlaciones. Cuad. Geol. Iber. 19, 21–26.

Guex, J., 1992. Origine des sauts évolutifs chez les ammonites. Bull. Soc. Vaud. Sci. Nat. 82 (2), 117–144.

Guex, J., 1995. Ammonites hettangiennes de la Gabbs Valley Range (Nevada USA). Mém. Géol. (Lausanne) 27, 1–131.

Guex, J., 2001. Environmental stress and atavism in ammonoid evolution. Eclogae Geol. Helv. 94, 321–328.
Guex, J., 2006. Reinitialization of evolutionary clocks during sublethal environmental stress in some invertebrates. Earth Planet. Sci. Lett. 242, 240–253.
Guex, J., 2016. Retrograde evolution during major extinction crises. In: Springer Briefs in Evolutionary Biology. (77 p.).
Guex, J., Caridroit, M., Kuwahara, K., O'Dogherty, L., 2014. Retrograde evolution of *Albaillella* during the Permian-Triassic crisis. Rev. Micropaléontol. 57, 39–43.
Hallam, A., 1978. How rare is phyletic gradualism and what is its evolutionary significance? Evidence from Jurassic bivalves. Palaeobiology 4, 16–25.
Haq, B., Hardenbol, J., Vail, P., 1988. Mesozoic and Cenozoic Chronostratigraphy and Cycles of Sea-level Change in Sea-Level Changes. SEPM Spec. Publ. Ê4Ê2, pp. 71–108.
Hayashi, S., 1968. The Permian conodonts in chert of the Adayama Formation, Ashio Mountains, central Japan. Earth Sci. 22 (2), 63–77.
Hayashi, S., 1987. Conodont arekore. Geosc. Mag. Chigaku Kenkyu 37, 1–16 (in Japanese).
Henderson, C., Mei, S., 2007. Geographical clines in Permian and lower Triassic gondolellids and its role in taxonomy. Palaeoworld 16, 190–201.
Hermann, R., Hochuli, P.A., Bucher, H., Bruhwiler, T., Hatmann, M., Ware, D., Roohi, G., 2011. Terrestrial ecosystems on North Gondwana following the end-Permian extinction. Gondw. Res. 20 (2–3), 630–637.
Hirsch, F., 1971. Conodontes nouvelle du Trias Méditerranéen. C. R. Soc. Phys. Hist. Nat. Genève 6, 65–69.
Hirsch, F., 1972. Middle Triassic conodonts from Israel, southern France and Spain. Mitt. Ges. Geol. Bergbaustud. Wien 21, 811–828.
Hirsch, F., 1975. Lower Triassic conodonts from Israel. Bull. Geol. Surv. Israel 66, 39–48.
Hirsch, F., 1977. Essai de correlation bioestratigraphique des niveaux meso et neotriassiques de facies Muschelkalk du domaine sepharade. Cuad. Geol. Iber. 4, 511–526.
Hirsch, F., 1981. Some Late Eo- and Mesotriassic conodont-multi-elements: notes on their taxonomy, phylogeny and distribution. Arch. Sci. 34 (2), 201–210.
Hirsch, F., 1987. Biostratigraphy and correlation of the marine Triassic of the Sephardic Province. Cuad. Geol. Iber. 4, 815–826.
Hirsch, F., 1994a. Triassic conodont multielements versus eustatic cycles. Lausanne Mémoires de Géologie 22, 35–52.
Hirsch, F., 1994b. Triassic conodonts as ecological and eustatic sensors. In: Embry, A.F., Beauchamp, B., Glass, D.J. (Eds.), Pangea: Global Environments and Resources, pp. 949–959. Memoir of the Canadian Society of Petroleum Geologists 17.
Hirsch, F., Gerry, E., 1974. In: Conodont and ostracode biostratigraphy of the Triassic in Israel. Proceedings of the International Symposium of the Alpin-Mediterranean Triassic Stratigraphy, Vienna, May 1978, pp. 107–118. Schriftenreihe Erdwiss, Komm. Osterr. Akad. Wiss. 2.
Hirsch, F., Ishida, K., 2002. The Izanami Plateau: pre-accretionary origin of Japan's low latitude Triassic pelagic carbonates. Eclogae Geol. Helv. 95, 43–55.
Hoffmann, H., 1869. Untersuchungen zur Bestimmung des Werthes von Species und Varietat. Ein Beitrag zur kritik der Darwinschen hypothese. Ricker, Giessen.
Huckriede, R., 1958. Die Conodonten der Mediterranen Trias und ihr stratigraphischer Wert. Palaontol. Z. 32, 141–175.
Huddle, J.W., 1970. Triassic conodonts from Israel. U. S. Geol. Surv. Res. 700B, 124–130.
Ishida, K., 1981. Fine stratigraphy and conodont biostratigraphy of a bedded-chert member of the Nakagawa Group. J. Sci. Univ. Tokushima 14, 107–137. (in Japanese with English abstract).

Ishida, K., 1984. The order of appearance of radiolarians in Anisian Bedded-chert bodies in the South Zone of the Chichibu Belt, Eastern Shikoku. J. Sci. Univ. Tokushima 17, 15–29. (in Japanese with English abstract).

Ishida, K., Hirsch, F., 2001. Taxonomy and faunal affinity of late Carnian–Rhaetian conodonts in the Southern Chichibu Belt, Shikoku, SW Japan. Riv. Ital. Paleontol. Stratigr. 107 (2), 227–250.

Ishida, K., Hirsch, F., 2011. The Triassic conodonts of the NW Malayan Kodiang limestone revisited: taxonomy and paleogeographic significance. Gondw. Res. 19, 22–36.

Jiang, H., Lai, X., Luo, G., Aldridge, R., Zhang, K., Wignall, P., 2007. Restudy of conodont zonation and evolution across the P/T boundary at Meishan section, Changxing, Zhejiang, China. Global Planet. Change 55, 39–55.

Johnson, A., Simms, M., 1989. The timing and cause of Late Triassic marine invertebrate extinctions: evidence from scallops and crinoids. In: Donovan, S. (Ed.), Mass-Extinctions, Processes and Evidence. Belhaven Press, London, pp. 174–194. 4 figs.

Kiliç, A.M., 2004. Multielement Taxonomy of the Triassic conodonts of the Kocaeli Region. PhD ThesisCumhuriyet University, Sivas, Turkey. 132 pp. (in Turkish with English Abstract Unpublished).

Kiliç A.M., Plasencia P., Debate on the skeletal elements of the Triassic conodont *Cornudina* Hirschmann, Acta Geol. Pol. (unpublished\accepted paper).

Kiliç, A.M., Plasencia, P., Ishida, K., Guex, J., Hirsch, F., 2015a. Conodonts versus Triassic climatic and eustatic changes. Prog. Earth Planet. Sci. 15, 321–324.

Kiliç, A.M., Plasencia, P., Ishida, K., Hirsch, F., 2015b. The case of the Carnian (Triassic) conodont genus *Metapolygnathus* Hayashi. J. Earth Sci. 26 (2), 219–223.

Kiliç, A.M., Plasencia, P., Ishida, K., Guex, J., Hirsch, F., 2016. Proteromorphosis of *Neospathodus* (conodonta) during the Permian–Triassic crisis. Rev. Micropaléontol. 59, 33–39.

Klets, T., Kopylova, S., 2007. The problem of Triassic Gondolellid conodont systematics Conodontophorida, Conodonta). In: Lucas, S.G., Spielmann, J.A. (Eds.), The Global Triassic, pp. 131–133. New Mexico Museum of Natural History and Science Bulletin 41.

Korte, C., Kozur, H.W., 2010. Carbon isotope stratigraphy across the Permian–Triassic boundary: a review. J. Asian Earth Sci. 39, 215–235.

Kovacs, S., Kozur, H., 1980. Stratigraphische Reichweite der wichtigsten Conodonten (ohne Zahnreihen conodonten) der Mittel- und Obertrias. Geol. Pal. Mitt. Innsbruck 10, 47–78.

Kozur, H., 1968. Conodonten aus dem Muschelkalk des germanischen Binnenbeckens und ihr stratigraphischer Wert. Part I: Conodonten vom Plattformtyp und stratigraphische Bedeutung der Conodonten aus dem Oberen Muschelkalk. Geologie 17 (8), 930–946 (Part II: Zahnreihen-Conodonten. Geologie 17(9), 1070–1085).

Kozur, H.W., 1989. In: The taxonomy of the Gondolellids conodonts in the Permian and Triassic. 1st International Senckenberg Conference and 5th European Conodont Symposium (ECOS V): Contributions III, Papers on Ordovician to Triassic Conodonts 117, pp. 409–469.

Kozur, H.W., 2003. Integrated ammonoid-, conodont and radiolarian zonation of the Triassic. Hall Jahrb. Abt. Geowiss. B 25, 49–79.

Kozur, H., Lucas, S.G., 1996. A new conodont from the uppermost Lamar Limestone of the Delaware Basin of West Texas. Tex. J. Sci. 48, 96–101.

Kozur, H., Mock, R., 1974. Zwei neue Conodonten-Arten aus der Trias der Slowakischen Karstes. Casopis pro mineralogii a geologii, roc. 19, 135–139.

Krystyn, L., 1980. Triassic conodont localities of the Salkammergut region (northern Calcareous Alps). ECOS-2 Guidebook and Abstracts. Abh. Geol. Bundesanst 35, 61–98.

Krystyn, L., 1983. Das Epidaurus Profil (Griechenland) Ein Beitrag zur Conodonten-Standard zonierung des Tethyalen Ladin and Unterkam. Österr. Akad. Wiss. Schriften. Erdwiss. Komm. Wien 5, 231–258.

Kummel, B., 1960. New Zealand Triassic ammonoids. N. Z. J. Geol. Geophys. 3, 486–509.

Lerman, A., 1960. Triassic pelecypods from southern Israel and Sinai. Bull. Res. Counc. Israel 9G, 1–51.

Lieberman, D.E., 2011. The Evolution of the Human Head. The Belknap Press of Harvard University Press, USA. ISBN: 978-0674046368, ISBN-10. xii + 756 pp.

Lienhardt, M.J., et al., 1984. Planches Trias. In: Debrand-Passard, S., Courbouleix, S. (Eds.), Synthèse géologique du Sud-Est de la France. Atlas. Mémoire du Bureau de Recherches Géologiques et Minières, 126.

Mancini, E.A., 1978. Origin of micromorph faunas in the geologic record. J. Paleontol. 52, 311–322.

March, M., Budurov, K., Hirsch, F., Marquez-Aliaga, A., 1990. *Sephardiella* nov. gen. (Conodonta), emendation of *Carinella* (Budurov, 1973) from the Ladinian (middle Triassic) type area in Catalonia (NE Spain), Sephardic Province. Courier Forsch. Inst. Senckenberg 118, 197–201.

Márquez-Aliaga, A., Valenzuela-Ríos, J.I., Calvet, F., Budurov, K., 2000. Middle Triassic conodonts from northeastern Spain: biostratigraphic implications. Terra Nova 12, 77–83.

Martinez-Perez, C., Plasencia, P., Cascales-Minana, B., Mazza, M., Botella, H., 2013. New insights into the diversity dynamics of Triassic conodonts. Hist. Biol. 6 (5), 591–602.

Mastandrea, A., Neri, C., Russo, F., 1998. Sella pass sections. In: Perri, M.C., Spalletta, C. (Eds.), Southern Alps Field Trip Guidebook, ECOS VII, pp. 282–291. Giornale di Geologia Special Issue 60.

Mazza, M., Furin, S., Spötl, C., Rigo, M., 2010. Generic turnovers of Carnian/Norian conodonts: climatic control or competition? Palaeogeogr. Palaeoclimatol. Palaeoecol. 290, 120–137.

Mazza, M., Cau, A., Rigo, M., 2012. Application numerical cladistic analyses to the Carnian-Norian conodonts: a new approach for phylogenetic interpretations. J. Syst. Palaeontol. 10 (3), 401–422.

Mei, S., Jin, Y., Wardlaw, B.R., 1994. Succession of Wuchiapingian conodonts from northeast Sichuan Province and its worldwide correlation. Acta Micropalaeontol. Sin. 11, 121–139.

Mendel, G., 1865. Versuche über Pflanzen-Hybriden. In: Verhandlungen des naturforschenden Vereines in Brünn, pp. 3–47. Bd. IV fur das Jahr 1865; Abhandlungen.

Michalik, J., 1994. Notes on the paleogeography and paleotectonics of the Western Carpathian Area during the Mesozoic. Mitt. Österr. Geol. Ges. 86, 101–110.

Mosher, L.C., 1968. Evolution of Triassic platform conodonts. J. Paleontol. 42 (4), 895–946.

Mosher, L.C., 1973. Triassic conodonts from British Columbia and the northern arctic islands. Bull. Geol. Surv. Can. 222, 141–192.

Mundlos, R., Urlichs, M., 1984. Revision von *Germanonautilus* aus dem germanischen Muschelkalk (Oberanis Ladin). Stuttg. Beitr. Naturk. B 99, 1–43.

Muttoni, G., Gaetani, M., Budurov, K., Zagorchev, I., Trifonova, E., Ivanova, D., Petrounova, L., Lowrie, W., 2000. Middle Triassic paleomagnetic data from northern Bulgaria: constraints on Tethyan magnetostratigraphy and paleogeography. Palaeogeogr. Palaeoclimatol. Palaeoecol. 160, 223–237.

Narkiewicz, K., Szulc, J., 2004. Controls on migration of conodont fauna in peripheral oceanic areas. An example from the Middle Triassic of the Northern Peri-Tethys. Geobios 37, 425–436.

Nicora, A., 1976. Conodont-fauna, stratigraphic position and relations to the Tethyan successions of the Shoshonensis Zone (Pelsonian) of Nevada. Riv. Ital. Paleontol. Stratigr. 82, 627–648.

Orchard, M.J., 1983. *Epigondolella* populations and their phylogeny and zonation in the Upper Triassic. Fossils Strata 15, 177–192.

Orchard, M.J., 1991a. Late Triassic conodont biochronology and biostratigraphy of the Kunga Group, Queen Charlotte Islands, British Columbia. In: Woodsworth, G.J. (Ed.), Evolution and Hydrocarbon Potential of the Queen Charlotte Basin, British Columbia, pp. 173–193. Geological Survey of Canada 90-10.

Orchard, M.J., 1991b. Upper Triassic conodont biochronology and new index species from the Canadian cordillera. In: Orchard, M.J., McCracken, A.D. (Eds.), Ordovician to Triassic Conodont Paleontology of the Canadian Cordillera, pp. 299–335. Geological Survey of Canada Bulletin 417.

Orchard, M.J., 2005. Multielement conodont apparatuses of Triassic Gondolelloidea. Spec. Pap. Palaeontol. 73, 73–101.

Orchard, M.J., 2007. New conodonts and zonation, Ladinian-Carnian boundary beds, British Columbia, Canada. In: Lucas, S.G., Spielmann, J.A. (Eds.), The Global Triassic. New Mexico Museum of Natural History and Science Bulletin 41.

Orchard, M.J., 2010. Triassic conodonts and their role in stage boundary definition. In: Lucas, S.G. (Ed.), The Triassic Timescale, pp. 139–161. Geological Society, London, Special Publications, 334.

Orchard, M.J., 2013. Five new genera of conodonts from the Carnian-Norian boundary beds of black Bear ridge, Northeast British Columbia, Canada. In: Tanner, L., Spielmann, J.A., Lucas, S.G. (Eds.), The Triassic System. New Mexico Museum of Natural History and Science, Bulletin 61.

Orchard, M.J., 2014. Conodonts from the Carnian-Norian Boundary (Upper Triassic) of Black Bear Ridge, northeastern British Columbia, Canada. New Mexico Museum of Natural History and Science, Bulletin 64, 1–139, 2014.

Orchard, M.J., Tozer, E.T., 1997. Triassic conodont biochronology, its calibration with the ammonoid standard, and a biostratigraphic summary for the Western Canada Sedimentary Basin. Bull. Can. Petrol. Geol. 45 (4), 675–692.

Orchard, M.J., Zonneveld, J.P., Johns, M.J., McRoberts, C.A., Sandy, M.R., Tozer, E.T., Carrelli, G.G., 2001. Fossil succession and sequence stratigraphy of the Upper Triassic of Black Bear Ridge, northeast British Columbia, a GSSP prospect for the Carnian-Norian boundary. Albertiana 25, 10–22.

Parnes, A., 1962. Triassic ammonites from Israel. Geol. Surv. Isr. Bull. 88, 1–76.

Parnes, A., 1975. Middle Triassic ammonite biostratigraphy in Israel. Geol. Surv. Isr. Bull. 66, 25 pp.

Parnes, A., 1986. Middle Triassic cephalopods from the Negev (Israel) and Sinai (Egypt). Geol. Surv. Isr. Bull. 79, 59 pp.

Parnes, A., Benjamini, C., Hirsch, F., 1985. New aspects of Triassic ammonoid biostratigraphy, paleoenvironments and paleogeography in southern Israel (Sephardic Province). J. Paleontol. 59, 656–666.

Payne, J.L., Lehrmann, D.J., Wei, J., Orchard, M.J., Schrag, D.P., Knoll, A.H., 2004. Large perturbations of the carbon cycle during recovery from the end-Permian extinction. Science 305, 506–509.

Pérez-López, A., Pérez-Valera, F., 2007. Palaeogeography, facies and nomenclature of the Triassic units in the different domains of the Betic cordillera (S Spain). Palaeogeogr. Palaeoclimatol. Palaeoecol. 254, 606–626.

Pérez-Valera, A., 2008. Stratigraphy and sedimentology of Muschelkalk carbonates of the Southern Iberian Continental Palaeomargin (Siles and Cehegín Formations, Southern Spain). Facies 54, 61–87.

Perri, M.C., Andraghetti, M., 1987. Permian-Triassic boundary and Early Triassic conodonts from the Southern Alps, Italy. Riv. Ital. Paleontol. Stratigr. 93 (3), 291–328.
Plasencia, P., 2009. Bioestratigrafia y paleobiologia de conodontos del Triasico Medio del Sector Oriental de la Peninsula Iberica. Servei de Publicacions de la Universitat de Valencia. Tesis doctoral Universitat de Valencia, Spain, 408 pp.
Plasencia, P., Hirsch, F., Márquez-Aliaga, A., 2007. Sephardiellinae, a new Middle Triassic conodont subfamily. J. Iber. Geol. 33 (2), 163–172.
Plasencia, P., Márquez-Aliaga, A., Sha, J., 2013. A data base of Triassic conodonts from a comprehensive revision of literature. Span. J. Palaeontol. 28 (2), 215–226.
Plasencia, P., Hirsch, F., Sha, J., Márquez-Aliaga, A., 2015. Taxonomy and evolution of the Triassic conodont *Pseudofurnishius*. Acta Palaeontol. Pol. 60 (2), 385–394.
Purnell, M., Donoghue, C., Aldridge, R., 2000. Orientation and anatomical notation in conodonts. J. Paleontol. 74 (1), 113–122.
Ramirez, I., 2015. Who First Connected Darwin's Work With Mendel's Work? https://www.quora.com/Who-first-connected-Darwins-work-with-Mendels-work.
Ratterman, N.G., 1976. Conodont Biostratigraphy of the Upper Phosphoria Formation (Upper Permian). Ohio State University Publication, USA, 66 pp.
Rein, S., 2008. *Ceratites atavus* auf Mallorca?—Die Muschelkalkammonoideen des Museu Balear de Sciencies Natural Sòller. Vernate 27, 5–19.
Rigo, M., Preto, N., Roghi, G., Stanley, S.M., 2005. Upper Triassic conodont extinction events. Geophys. Res. Abstr. 7, 02339.
Rutherford, S.L., Lindquist, S., 1998. Hsp90 as a capacitator for morphological evolution. Nature 396, 336–342.
Sadeddin, W., 1990. *Pseudofurnishius priscus* n.sp. (Conodonta) and its stratigraphical significance for the Ladinian (Middle in Jordan). N. Jb. Geol. Palaeontol. Abh. 178, 369–382.
Sadeddin, W., Kozur, H., 1992. s. N. Jb. Geol. Palaeontol. Monatsh. 6, 356–368.
Saunders, W.B., Work, D.M., Nikolaeva, S.V., 1999. Evolution of complexity in Paleozoic ammonoid sutures. Science 286, 760–763.
Schaltegger, U., Guex, J., Bartolini, A., Schoene, B., Ovtcharova, M., 2008. Precise U–Pb age constraints for end-Triassic mass extinction, its correlation to volcanism and Hettangian post-extinction recovery. Earth Planet. Sci. Lett. 267, 266–275.
Schoene, B., Guex, J., Bartolini, A., Schaltegger, U., Blackburn, T.J., 2010. Correlating the end-Triassic mass extinction and flood basalt volcanism at the 100 ka level. Geology 38, 387–390.
Shaviv, N.J., Veizer, J., 2003. Celestial driver of Phanerozoic climate? GSA Today, 4–11.
Shimer, H.W., 1908. Dwarf faunas. Am. Nat. 42, 472–490.
Sopeña, A., Virgili, C., Arche, A., Hernando, S., Ramos, A., 1983. El Triasico. In: Fontbote, J.M. (Ed.), In: Libro Homenaje a J.M. Rios, vol. 2. IGME, Madrid, pp. 47–62.
Staesche, U., 1964. Conodont aus dem Skyth von Südtirol. N. Jb. Geol. Palaontol. Abh. 119, 247–306.
Tatge, U., 1956. Conodonten aus dem germanischen Muschelkalk. Palaöntol. Z. 30, 108–147.
Thom, R., 1972. Stabilité structurelle et morphogenèse. Benyamin, New York.
Trammer, J., 1971. Middle Triassic (Muschelkalk) conodonts from the SW margins of the Holy Cross Mts. Acta Geol. Pol. 21 (3), 379–386.
Trammer, J., 1974. Evolutionary trends and patterns of extinction of Triassic conodonts. Acta Palaeontol. Pol. 19 (2), 251–264.
Trotter, J.A., Williams, I.S., Nicora, A., Mazza, M., Rigo, M., 2015. Long-term cycles of Triassic climate change: a new $\delta^{18}O$ record from conodont apatite. Earth Planet. Sci. Lett. 415, 165–174.
Truempy, R., 1982. Das Phänomen Trias. Geol. Rundsch. 71 (3), 711–723.

Urlichs, M., 1997. Die Gattung Ceratites (Ammonoidea) aus dem Muschelkalk der Provence (Mitteltrias, Sudost-Frankreich). Stuttg. Beitr. Naturkunde B 252, 1–12.
Valentine, J.W., Collins, A.G., Meyer, C.P., 1994. Morphological complexity increase in metazoans. Palaeobiology 20, 131–142.
Van den Boogaard, M., Simon, O.J., 1973. Pseudofurnishius (Conodonta) in the Triassic of the Betic Ranges SE Spain. Scripta Geol. 16, 1–23.
Virgili, C., 2008. The Permian-Triassie transition: historical review of the most important ecological crises with special emphasis on the Iberian Peninsula and Western-Central Europe. J. Iber. Geol. 34 (1), 123–158.
Wardlaw, B.R., Collinson, J.W., 1986. Paleontology and deposition of the Phosphoria Formation: University of Wyoming. Contrib. Geol. 24 (2), 107–142.
Wardlaw, B.R., Mei, S., 1998. A discussion of the early reported species of Clarkina (Permian conodonta) and the possible origin of the genus. In: Jin, et al. (Ed.), Permian Stratigraphy, Environments and Resources, pp. 33–52. Palaeoworld 9(2).
Ware, D., Jenks, J.F., Hautman, M., Bucher, H., 2011. Dienerian (Early Triassic) ammonoids from the Candelaria Hills (Nevada, USA) and their significance for palaeobiogeography and palaeoceanography. Swiss J. Geosci. 104, 161–181.
Wetzel, R., Morton, D.M., 1959. Contribution a la geologie de la Transjordanie. Muszeum National d'Histoire Naturelle, Notes et Memoirs Moyen-Orient, Paris, 7, pp. 95–191.
Yin, H.F., Zhang, K.X., Tong, J.N., Yang, Z.Y., Wu, S.B., 2001. The global stratotype section and point (GSSP) of the Permian–Triassic boundary. Episodes 24, 102–114.
Zawidzka, K., 1975. Conodont stratigraphy and sedimentary environment of the Muschelkalk in Upper Silesia. Acta Geol. Pol. 25.

FURTHER READING

Benjamini, C., Chepstow-Lusty, A., 1986. *Neospathodus* and other Conodonta from the Saharonim Formation (Anisian-Ladinian) at Makhtesh Ramon. J. Micropalaeontol. 5 (1), 67–75.
Hirsch, F., Márquez-Aliaga, A., Santiesteban, C., 1987. Distribución de moluscos y conodontos del tramo superior del Muschelkalk en el sector oriental de la provincia sefardí. Cuad. Geol. Iber. 11, 799–814.
Hirsch, F., Márquez-Aliaga, A., Plasencia, P., Perez-Valera, A., 2014. The Triassic Gondwana margin of the Tethys: outline of the sephardic bioprovince. Triassic Outcrops of the Negev Guide Book for 11th International Workshop Pan-European Correlation Triassic, pp. 22–25. Geological Survey of Israel.
Kovács, S., 1994. Conodonts of stratigraphical importance from the Anisian/Ladinian boundary interval of the Balaton Highland, Hungary. Riv. Ital. Paleontol. Stratigr. 99, 473–514.
Kozur, H., 1974. Biostratigraphie der Germanischen Trias. Part I. Freib. Forsch. C 280 (1), 1–56 (Part II. Freib. Forsch. C 280(2), 1–71).
Kozur, H., 1980. Revision der Conodonten zonierung der Mittel- und Obertrias des Tethyalen Faunenreiches. Mitt. Geol. Paläontol. Innsbruck 10, 79–172.
Kozur, H., Mock, R., 1972. Neue Conodonten aus der Trias der Slowakei und ihre stratigraphische Bedeutung. Mitt. Geol. Paläontol. Innsbruck 2 (4), 3–17.
Lucas, S., 2012. New Triassic time scale. In: Tanner, L.H., Spielmann, J.A., Lucas, S.G. (Eds.), The Triassic System, pp. 366–374. New Mexico Museum of Natural History and Science Bulletin 61.
Mazza, M., Rigo, M., 2008. Taxonomy and phylomorphogenesis of the Carnian/Norian conodonts from Pizzo Mondello section (Sicani Mountains, Sicily). Berichte der Geologischen Bundesanstalt 76, 50–52.

Rigo, M., Preto, N., Roghi, G., Tateo, F., Mietto, P., 2007. A rise in the carbonate compensation depth of western Tethys in the Carnian (Late Triassic): deep-water evidence for the Carnian Pluvial Event. Palaeogeogr. Palaeoclimatol. Palaeoecol. 246, 188–205.
Shigeta, Y., Komatsu, T., Maekawa, T., Tran, H.D., 2014. Olenekian (Early Triassic) stratigraphy and fossil assemblages in Northeastern Vietnam. National Museum of Nature and Science Monographs. ISSN: 1881-9109, 45, 295 p. Tokyo.
Sweet, W.C., 1970. Uppermost Permian and Lower Triassic conodonts of the Salt Range and Trans-Indus ranges, West Pakistan. In: Stratigraphic Boundary Problems Permian and Triassic of West Pakistan, pp. 207–275.
Van den Boogaard, M., Simon, O.J., 1973. *Pseudofurnishius* (Conodonta) in the Triassic of the Betic ranges SE Spain. Scr. Geol. 16, 1–23.

CHAPTER EIGHT

A Facies-Independent Trans-European Anisian–Ladinian Marker Horizon? Significance and Impact for Sequence Stratigraphy and Intra-Tethyan Correlation

Annette E. Götz*,[1] and Michael Montenari[†]
*Earth and Environmental Sciences, University of Portsmouth, Portsmouth, United Kingdom
[†]Geography, Geology and the Environment, Keele University, Staffordshire, United Kingdom
[1]Corresponding author: e-mail address: annette.goetz@port.ac.uk

Contents

Abstract

Transcontinental correlation of European carbonate systems that developed during the onset of the Mesozoic breakup of Pangaea remains difficult due to pronounced lateral facies variations and in many cases facies-dependent fossil groups used for biostratigraphic dating and correlation. In addition, radiometric dating of ash layers associated with Middle Triassic carbonate successions in Europe is limited to certain areas and times of volcanic activity. Thus, Middle Triassic correlation schemes of the Tethyan and Peri-Tethyan realm—a highly dynamic area in terms of plate reorganization and related development of early Mesozoic epeiric seas—need further refinement using new correlation tools. Here, we highlight the Europe-wide occurrence of a silica-rich

Stratigraphy & Timescales, Volume 2
ISSN 2468-5178
http://dx.doi.org/10.1016/bs.sats.2017.07.001

horizon at the Anisian–Ladinian boundary as a potential trans-European correlation horizon. Chert layers and nodules form in different depositional settings, under different conditions and within different facies, thus being well suited for interregional correlation. However, the processes leading to chert formation in Middle Triassic shallow-marine carbonate deposits are still under discussion and require further research in order to make this prominent feature a powerful correlation tool.

1. INTRODUCTION

Descriptions of silicified layers (so-called "Hornsteinlagen") in Muschelkalk-aged limestones of the Central European Basin date back to the 19th century (Sandberger, 1864; Seebach, 1861; Speyer, 1875). These layers mark the stratigraphic boundary between the uppermost Middle Muschelkalk (late Illyrian) and the basal crinoidal limestones (Trochitenkalk) of the Upper Muschelkalk (early Fassanian) in Germany (Busse and Horn, 1982; Dünkel and Vath, 1990; Freudenberger, 1996; Genser, 1930; Geyer et al., 2011; Götz and Keller, 1998; Hagdorn et al., 1987; Hohenstein, 1913; Langbein and Knaust, 1997; Michael, 1940; Radzinski, 2008; Rüger, 1934; Schwahn and Gahrmann, 1976; Seidel, 1965, 2003; Seufert and Schweizer, 1985; Simon, 1988; Voigt et al., 2001; Weber and Kubald, 1947; Wilczewski, 1967). This silicified horizon is widespread and has been described from the Swiss Jura Mountains (Merki, 1961) in the south to the Subhercynian Basin (Schwahn and Gahrmann, 1976) and Brandenburg (Zwenger, 2015) in the north. Chert nodules at the Anisian–Ladinian boundary were also reported from the Polish Muschelkalk (Kwiatkowski, 2005), the Swiss Muschelkalk (Jordan, 2016; Pietsch et al., 2016), the Anisian–Ladinian Felsőörs key section of the Transdanubian Range in Hungary (Budai and Haas, 1997; Haas and Budai, 1999), the Dobrogea of Romania (Atudorei et al., 1997), the Western Balkanides of Bulgaria (Chatalov, 2013), and the Muschelkalk series of Iberia and Minorca (Escudero-Mozo et al., 2014, 2015; López-Gómez et al., 2002) (Fig. 1). In contrast, the alpine Middle Triassic as documented by extensive carbonate platforms of the Northern Calcareous Alps and Dolomites lacks siliceous horizons within the Anisian–Ladinian boundary interval (Feist-Burkhardt et al., 2008a). However, at the GSSP of the Ladinian Stage at Bagolino (Southern Alps, northern Italy) the Anisian–Ladinian transitional interval is characterized by siliceous limestones and shales interbedded with volcaniclastic layers (Brack et al., 2005). From a paleogeographic point of view chert layers seem to be related to shallow ramp settings of the Peri-Tethyan realm, whereas in the Tethyan realm platform drowning

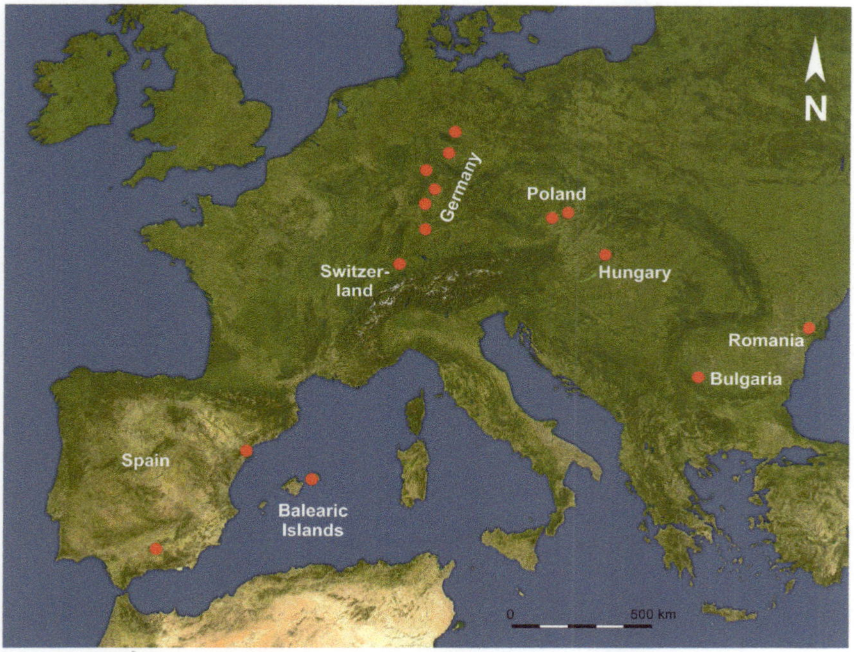

Fig. 1 Location map of Europe with highlighted areas of chert occurrence at the Anisian–Ladinian boundary.

at the Anisian–Ladinian boundary (Rüffer, 1995; Zühlke, 2000) prevented silica precipitation. The Europe-wide occurrence of a silica-rich horizon at the Anisian–Ladinian boundary as documented in the literature thus needs further evaluation with respect to its stratigraphic potential as a trans-European correlation horizon. This study reviews the available data from literature and aims to outline the potential utilization in future refined stratigraphic concepts of the European Triassic.

2. GEOLOGICAL SETTING

The Middle Triassic represents an Epoch in the Phanerozoic that saw the extensive development of carbonate platforms worldwide with the Western Tethyan realm and its adjacent Peri-Tethys basins containing prominent European examples (Burchette and Wright, 1992; Gaetani et al., 2000; Golonka, 2007; Kiessling et al., 2003; Pomar and Hallock, 2008). The sedimentary successions of both realms reflect the long-term transgressive evolution after a global sea-level lowstand at the end of the

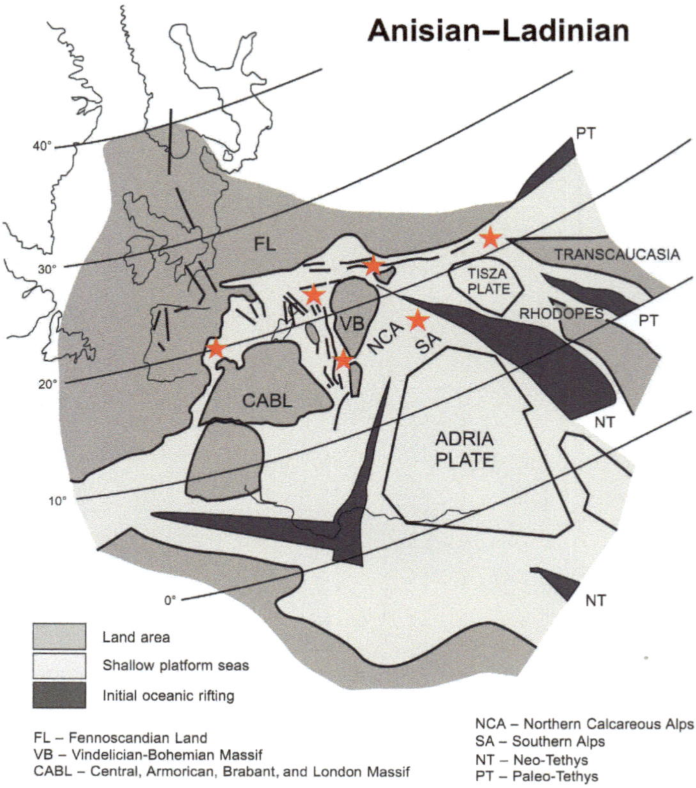

Fig. 2 Paleogeography of the Middle Triassic and position of chert occurrence (*red stars*) at the Anisian–Ladinian boundary. *Map modified from Szulc, J., 2000. Middle Triassic evolution of the northern Peri-Tethys area as influenced by early opening of the Tethys ocean. Ann. Soc. Geol. Pol. 70, 1–48.*

Permian (Hallam, 1992; Haq et al., 1988; Vail et al., 1977), and capture the major recovery phase of marine biota after the most severe mass extinction in Earth's history (Chen and Benton, 2012; Song et al., 2011). Muschelkalk carbonates of the Triassic Peri-Tethys Basin (Germanic Basin) are a characteristic feature of central Europe (Feist-Burkhardt et al., 2008a; Szulc, 1999, 2000; Fig. 2) with a research history dating back to the early 19th century (Alberti, 1834). This basin was bordered by landmasses and open to the Tethyan shelf by three tectonically controlled gates in the south and southeast: the East Carpathian, Silesian-Moravian, and Western gates. The East Carpathian Gate was already active in the late Induan, the Silesian-Moravian Gate opened in the Olenekian (Szulc, 1999, 2000), and the westernmost communication to the Tethys developed during the Anisian

(Feist-Burkhardt et al., 2008b; Götz and Feist-Burkhardt, 2012; Götz and Gast, 2010; Götz et al., 2005). The semi-closed situation of the basin and the diachronous communication with the Tethys Ocean resulted in a distinctive facies differentiation between the western and eastern parts of the basin. While in the Silesian and Carpathian domains the early Anisian is already represented by carbonates, the central and western areas were still dominated by siliciclastic red beds (Röt facies). In the eastern subbasin, open-marine sedimentation continued during almost the entire Anisian, while the western part experienced restricted circulation during the early and late Anisian. The opening of the Burgundy Gate in the Ladinian led again to full marine conditions. However, the diachronous facies pattern prevailed during the entire Ladinian (Szulc, 2000).

The adjacent Tethys shelf area exposes from west to east Anisian–Ladinian carbonate successions of present-day Spain (López-Gómez et al., 1993; Pérez-López and Pérez-Valera, 2007), the Northern Calcareous Alps (Rüffer, 1995), Dolomites (Zühlke, 2000), Hungary (Haas and Budai, 1999; Török, 1998), Bulgaria (Chatalov, 2013, 2016), and Romania (Atudorei et al., 1997). The interaction between an open ocean system and its adjacent epeiric seas during the Middle Triassic is still subject of ongoing research (e.g., Götz et al., 2017), and from this background and based on the need of refined large-scale correlations (e.g., Bernasconi et al., 2017), the assessment of a distinct enrichment of silica at the Anisian–Ladinian boundary across Europe as correlation tool is seen as an innovative step toward Triassic trans-European stratigraphy.

3. PALEOGEOGRAPHIC OCCURRENCE OF CHERT AT THE ANISIAN–LADINIAN BOUNDARY

Silicious horizons occur in ramp settings along the Tethys shelf (Fig. 3) and are described in the literature from Spain, Switzerland, Germany, Poland, Hungary, Romania, and Bulgaria. A brief summary of each area is given below.

3.1 Central European Basin

In Germany (Baden-Wuerttemberg, Franconia, Hesse, Thuringia, Saxony-Anhalt, Brandenburg), chert nodules and layers occur in dolomitic, partly stromatolitic lagoonal limestones of the uppermost Middle Muschelkalk and in oolitic, bioclastic crinoidal limestones (Trochitenkalk), typical ramp shoal deposits, of the lowermost Upper Muschelkalk, representing the stratigraphic interval of the Anisian–Ladinian boundary (Freudenberger, 1996;

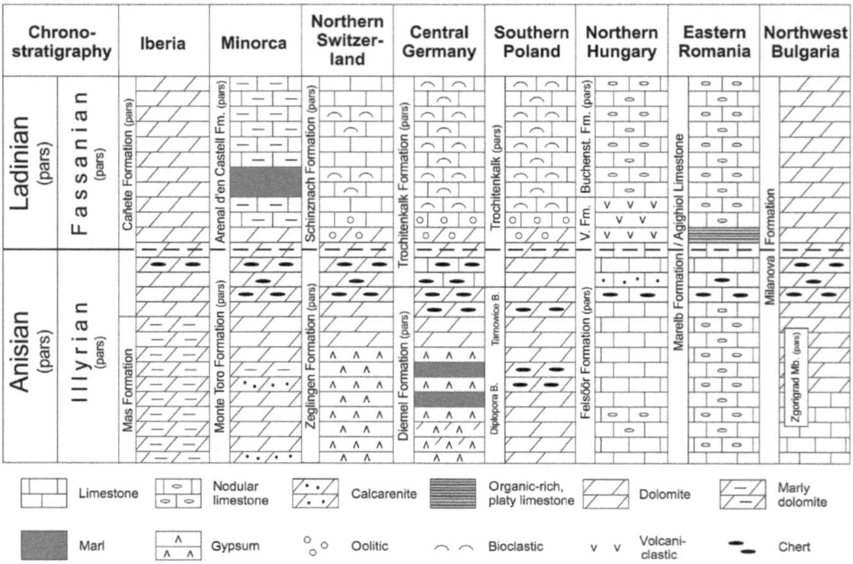

Fig. 3 Stratigraphic position of silicious horizons within the Anisian–Ladinian boundary interval across a trans-European transect from Spain in the west to Bulgaria in the east. Abbreviations used: *B.*, Beds; *Fm.*, Formation; *Mb.*, Member; *V. Fm.*, Vászoly Formation. *Data compiled from Atudorei, V., Baud, A., Crasquin-Soleau, S., Galbrun, B., Grădinaru, E., Mirauta, E., Renard, M., Zerrari, S., 1997. Extended Scientific Report of the Peri-Tethys Project "The Triassic of North-Dobrogea," Lausanne, 55 pp.; Haas, J., Budai, T., 1999. Triassic sequence stratigraphy of the Transdanubian Range (Hungary). Geol. Carpath. 50, 459–475; Kwiatkowski, S., 2005. Origin of chert nodules from the Polish Muschelkalk, Middle Triassic. Ann. Soc. Geol. Pol. 75, 287–308; Feist-Burkhardt, S., Götz, A.E., Szulc, J., Borkhataria, R., Geluk, M., Haas, J., Hornung, J., Jordan, P., Kempf, O., Michalík, J., Nawrocki, J., Reinhardt, L., Ricken, W., Röhling, G.-H., Rüffer, T., Török, Á., Zühlke, R., 2008. Triassic. In: McCann, T. (Ed.), The Geology of Central Europe, vol. 2. Geological Society, London, pp. 749–821; Chatalov, A., 2013. A Triassic homoclinal ramp from the Western Tethyan realm, Western Balkanides, Bulgaria: integrated insight with special emphasis on the Anisian outer to inner ramp facies transition. Palaeogeogr. Palaeoclimatol. Palaeoecol. 386, 34–58; Escudero-Mozo, M.J., Martín-Chivelet, J., Goy, A., López-Gómez, J., 2014. Middle-Upper Triassic carbonate platforms in Minorca (Balearic islands): implications for Western Tethys correlations. Sediment. Geol. 310, 41–58; Escudero-Mozo, M.J., Márquez-Aliaga, A., Goy, A., Martín-Chivelet, J., López-Gómez, J., Márquez, L., Arche, A., Plasencia, P., Pla, C., Marzo, M., Sánchez-Fernández, D., 2015. Middle Triassic carbonate platforms in eastern Iberia: evolution of their fauna and palaeogeographic significance in the western Tethys. Palaeogeogr. Palaeoclimatol. Palaeoecol. 417, 236–260; Jordan, P., 2016. Reorganisation of the Triassic stratigraphic nomenclature of northern Switzerland—overview and the new Dinkelberg, Kaiseraugst and Zeglingen Formations. Swiss J. Geosci. 109, 241–255; and Pietsch, J.S., Wetzel, A., Jordan, P., 2016. A new lithostratigraphic scheme for the Schinznach Formation (upper part of the Muschelkalk Group of northern Switzerland). Swiss J. Geosci. 109: 285–307.*

Geyer et al., 2011; Hagdorn et al., 1987; Radzinski, 2008; Seidel, 2003). In Brandenburg (Rüdersdorf) chert nodules occur in the basal Upper Muschelkalk Transversa-Schichten (Transversa Beds) and Glaukonitischer Kalkstein (Glauconitic Limestone), together representing the basal Ladinian (Fassanian) Trochitenkalk (Zwenger, 2015), and were also documented from this stratigraphic level in boreholes from the Lausitz (Göthel and Grunert, 1996). A study on chert layers in the Thuringian Muschelkalk detailed the different depositional environments (Langbein and Knaust, 1997), whereas in a more recent work on the Central European Basin by Franz et al. (2015) this prominent feature is not included.

3.2 Eastern Gate Area

In Poland (Upper Silesia, Holy Cross Mountains), chert nodules occur in dolomites, dolomitic limestones, cavernous limestones, oolitic and oncoidal limestones, crinoidal limestones, and sponge bioherms in different horizons of the Anisian, which are interpreted as organodetritic limestones or dolomites deposited in lagoonal and evaporitic settings (Kwiatkowski, 2005). The Anisian–Ladinian boundary is marked by chert layers formed under evaporitic conditions where silica precipitation is interpreted to be triggered by the activity of bacteria, causing reduction of sulfates and a decrease of pH (e.g., Birnbaum and Wireman, 1984; Clayton, 1986). Recently, Matysik (2016) provided a detailed facies description and environmental interpretation of the Upper Silesian Muschelkalk deposits, capturing the Anisian–Ladinian interval. Strong lateral facies variability, including peritidal dolomitic, restricted lagoonal and sabkha deposits, reveals a complex paleotopography. Stratigraphically, the transition from restricted to open-marine conditions related to a major transgressive pulse is documented by the deposition of crinoidal shoal deposits (Trochitenkalk).

3.3 Western Gate Area

In Switzerland (northern Switzerland), Muschelkalk deposits of the western gate area are well documented by the drilling program of the NAGRA (National Cooperative for the Disposal of Radioactive Waste) in detailed borehole reports (e.g., Dronkert et al., 1990; Matter et al., 1988; NAGRA, 2001; Peters et al., 1992). Chert nodules occur in dolomites of the uppermost Anisian (Zeglingen Formation) and oolitic, partly stromatolitic dolomites at the Anisian–Ladinian boundary (Schinznach Formation) (Jordan, 2016; Pietsch et al., 2016). The sedimentary features point to lagoonal settings and bacterial activity might have caused silica precipitation. However, the high terrestrial input of organic material (plant debris, wood remains, pollen grains, spores) and permeable oolitic and dolomitic carbonates favor the precipitation of silica by changes in pH caused by the presence of a high amount of sedimentary organic matter, as recently described from Muschelkalk oolites in the southern Peri-Tethys Basin (Götz et al., 2017).

3.4 Eastern Tethys Shelf Area

In Hungary, nodular, cherty limestones are described from the Transdanubian Range in the northern part of the country (Budai and Haas, 1997; Budai and Vörös, 2006; Haas and Budai, 1999). There they represent a characteristic feature of the Anisian Felsőörs Formation (Márton et al., 1997; Vörös et al., 2003). In places, these cherty limestones contain crinoid-brachiopod bioclastic grainstones with tuffitic intercalations and within the boundary interval volcanic tuffs with siliceous limestone lenses or intercalations (Vászoly Formation) become prevailing (Haas, 2012). In contrast, the Middle Triassic ramp deposits of the Mecsek and Villany Mountains of southern Hungary lack silicious horizons within the Anisian–Ladinian boundary interval and are characterized by supra- to peritidal dolomites and ooidal shoals (Török, 1997, 1998).

In Romania, cherty limestones occur at the Anisian–Ladinian boundary in the Agighiol region of the North Dobrogean Orogen. This region is one of the classic study areas for the Triassic System, due to the almost complete succession spanning late Olenekian (Spathian) to late Carnian deposits (Atudorei et al., 1997; Grădinaru, 2000). A succession of alternating nodular limestones and red marly shales (Agighiol Limestone) regarded as Hallstein facies of the outer platform setting contains *Eoprotrachyceras curionii*, *Arcestes* sp., and *Ptychites* sp., indicating a late Anisian (Illyrian) to early Ladinian (Fassanian) age. Within this interval siliceous layers as well as chert nodules are prominent features.

In Bulgaria (Western Balkanides), documenting the easternmost European region, a carbonate ramp system existed from the late Olenekian to early Carnian (Chatalov, 2013, 2016). Dolostones with chert nodules occur in the uppermost part of the late Anisian (Illyrian) Zgorigrad Member (Chatalov, 2013; Chatalov and Stanimirova, 2001; Tronkov, 1968, 1976). This interval is interpreted to illustrate inner to mid ramp deposits with peloidal-ooidal shoals and bioclastic wacke-packstones deposited under normal marine conditions (Chatalov, 2002, 2013).

3.5 Western Tethys Shelf Area

In Spain, the Iberian and Catalan Coastal Ranges and Betic Cordillera of the mainland as well as the Balearic Islands expose Middle Triassic carbonate series of the western Tethyan realm (Calvet et al., 1990; Llompart et al., 1987; López-Gómez et al., 1998, 2002; Péréz-López, 1998). Within the Iberian and Catalan Coastal Ranges three paleogeographic domains (Iberian, Mediterranean, and Levantine–Balearic) are distinguished (Escudero-Mozo et al., 2015), recording Muschelkalk facies and the development of wide carbonate platforms. Chert layers are a prominent feature in the basal dolomitic Cañete Formation (uppermost Anisian) of the Mediterranean and northern Levantine–Balearic domains of the Iberian Ranges, containing a continuous Anisian–Ladinian succession (Escudero-Mozo et al., 2015). In the Catalan Coastal Ranges (Mediterranean domain) chert nodules are reported from this stratigraphic interval to be characteristic of inner ramp stromatolites (Mercedes-Martín et al., 2013, 2014). In Menorca, belonging to the Levantine–Balearic domain (López-Gómez et al., 1998), Muschelkalk facies across the Anisian–Ladinian boundary is characterized by shallow ramp deposits including dolostones and dolomitic marls with stromatolites and intercalated oolitic and bioclastic grainstones (Monte Toro Formation; uppermost Anisian) followed upsection by micritic limestones with an interval of black, organic-rich marls in the basal Arenal d'en Castell Formation (lowermost Ladinian). A siliceous horizon is described from the uppermost dolomitic Monte Toro Formation related to a sequence boundary (Escudero-Mozo et al., 2014). The External Zones of the Betic Cordillera expose Anisian–Ladinian Muschelkalk successions (Majanillos Formation; Péréz-López, 1991) in a complex tectonic setting, corresponding to the folded and faulted Southern Iberian Paleomargin (Vera, 2001). A prominent discontinuity (DM1) marked by a hardground occurs at the Anisian–Ladinian boundary (Goy and Péréz-López, 1996) reflecting a sequence boundary (Péréz-López, 1998), and chert nodules are a common feature in the underlying dolostones.

4. DISCUSSION

Globally, the Middle Triassic documents the prolonged biotic recovery phase of the Permian–Triassic crisis (Erwin, 1998; Fraiser and Bottjer, 2005; Hallam and Wignall, 1997; Payne and van de Schootbrugge, 2007; Pruss and Bottjer, 2005; Stanley, 2003) and a period of major plate tectonic reconfiguration reflected in the onset of the Mesozoic breakup of Pangaea (Golonka, 2004; Golonka and Ford, 2000; Ziegler, 1990). Central Europe witnessed the evolution of shallow epeiric seas along the northern margin of the Tethyan shelf under a fluctuating but ameliorating climate trend with three rapid short-term warming events of decreasing magnitude in the mid–late Anisian, early Ladinian, and latest Ladinian (Trotter et al., 2015). Type and growth of Middle Triassic carbonate platforms might have been related to these short-term climatic pulses; however, other distinctive features, such as the seawater chemistry of an "aragonite sea" with a high Mg/Ca ratio (Stanley, 2008) and high atmospheric pCO_2 (Berner, 2006), are considered to be the controlling factors. In the context of these unique paleogeographic, oceanographic, and climatic conditions, sources of silica in shallow-marine carbonate settings of the Peri-Tethyan realm have to be carefully examined, as they are generally interpreted as biogenic (e.g., skeletal opal produced by radiolarians, diatoms, siliceous sponges), volcano-genetic and hydrothermal (DeMaster, 1981; Flügel, 2004; Packard et al., 2001), or related to continental chemical weathering (Kump et al., 2000; Laschet, 1984). However, extensive ferralitic (humid tropical) weathering is necessary to dissolve silica and can be excluded for the overall arid to warm-temperate climate during the Middle Triassic (Fijałkowska-Mader, 2015; Preto et al., 2010; Sellwood and Valdes, 2006; Stefani et al., 2010). Volcanic activity is not recorded during the Anisian of central Europe and limited to the late Anisian–Ladinian of the western Tethyan realm (Budai and Haas, 1997; Feist-Burkhardt et al., 2008a; Haas and Budai, 1999; Kovács et al., 2011; Szulc, 2000). Biogene silica producers such as diatoms can be excluded, since they first occur in the Lower Cretaceous (Harwood et al., 2007). Radiolarians have not been reported from the shallow epicontinental Peri-Tethyan realm and are still in the recovery phase after the P/T boundary event (De Wever et al., 2006) as are the Tethyan reef communities (Gaetani et al., 1981; Senowbari-Daryan et al., 1993). Rare reef build-ups along the Peri-Tethys margin include *Placunopsis* (bivalve) patch reefs in Germany (Hagdorn et al., 1999) and coral-sponge reefs in Poland (Hagdorn et al., 1999; Szulc, 2000, 2003); it is not before the mid Anisian (Pelsonian) that hexactinellid sponges along with scleractinian corals gave rise to the oldest in situ reefs found in the western Tethyan realm (Szulc, 2007). Further factors to be considered for the presence of silica are the chemistry of the pore fluid, the pH value of the environment, the presence of clay minerals, and the amount of organic material (Flügel, 2004).

From the different settings described above it is obvious that ramp morphology, water depths and water chemistry, as well as proximity to the open Tethys shelf are factors triggering silica precipitation during the Middle Triassic. All these areas are shallow ramp settings close to the hinterland and exposed to high terrestrial input of organic particles, both aquatic through river and delta systems as well as wind-borne. Additionally, bioproductivity of phytoplankton is high, although marine phytoplankton is still in the recovery phase after the P/T boundary event, and it is not before the Late Triassic that acritarch diversity increases and the first dinoflagellates occur in the broad Tethys shelf areas (Fensome et al., 1996; Strother, 1996). However, marginal, shallow lagoonal and inner ramp settings were dominated by prasinophytes with peak abundances during flooding phases (Götz and Feist-Burkhardt, 2012). Recently, silica precipitation and chert formation were studied in the Anisian Muschelkalk of southern Germany (Götz et al., 2017), and it was documented that sedimentary organic matter content in a shallow ramp setting strongly affects pH conditions in the water column, thus triggering *syn*-sedimentary silica precipitation especially in porous and permeable lithologies.

In a sequence stratigraphic context the chert layers occur close to a sequence boundary marking the transgressive surface directly on top of the sequence boundary, since lowstand deposits are not recorded. This initial flooding of Peri-Tethyan ramp systems along the Tethys shelf seems to have provided best conditions to trigger silica precipitation within a very specific pH range of seawater related to high terrestrial input of organic matter from the hinterland as well as high primary bioproductivity in the sea. So far, the effect of organic matter preservation on silicification processes in shallow-marine ramp settings has been overlooked. Most chert accumulations have been interpreted as diagenetic in origin or to be related to volcanic activity, and changes in salinity have been previously discussed to be responsible for the *syn*-sedimentary to early diagenetic formation of chert horizons at the Middle/Upper Muschelkalk boundary in central Germany. This stratigraphic interval is documenting the switch from evaporitic to open-marine conditions (Langbein and Knaust, 1997). Within the semi-enclosed Germanic Basin salinity changes during a major flooding phase across the Anisian–Ladinian boundary are to be expected and well documented in a marked facies change from evaporitic, dolomitic, lagoonal-stromatolitic to bioclastic-ooidal shoal deposits (Szulc, 1999, 2000). For ramp systems open to the Tethyan shelf, the ramp morphology—homoclinal or rimmed—seems to have strongly affected salinity, pH conditions, and organic matter distribution, so that in the case of the southern Hungarian homoclinal system (Török, 1997, 1998) no silica precipitation occurred. On the other hand, rimmed systems with protected inner and shoaly mid ramp areas along the western and central Peri-Tethyan realm provided ideal conditions for silicification processes. Thus, *syn*-sedimentary silica precipitation along the Tethyan–Peri-Tethyan interaction zone becomes plausible considering the specific Middle Triassic paleogeographic situation of epeiric seas during early Pangaean fragmentation as documented by a major marine flooding surface above a sequence boundary developed at the Anisian–Ladinian boundary (An4/La1 in Szulc, 2000; DS-1/DS-2 in Escudero-Mozo et al., 2014). This event seems to be recorded as a prominent and widespread distinctive geochemical signature, that appears to be highly suitable for interregional correlations. Within the Tethyan realm rapid platform drowning is documented, e.g., in the Northern Calcareous Alps and Dolomites (Rüffer, 1995; Stefani et al., 2010; Zühlke, 2000), whereas the Peri-Tethyan realm experienced a major transgression (Szulc, 2000) resulting in broad shallow-marine, partly lagoonal depositional areas with high terrestrial influx of organic matter from the close hinterland as well as high marine phytoplankton productivity across the inner to outer ramp settings. Additionally, porous and permeable lithologies of inner and mid ramp settings favored silica precipitation.

In terms of reservoir geology, Middle Triassic epeiric carbonates of central Europe can serve as outcrop analogues for major Permian–Triassic carbonate reservoirs in the Middle East (Khuff Formation, e.g., Alsharhan, 2006; Götz et al., 2008; Vaslet et al., 2005). The identification of laterally continuous silicious horizons as potential sealing horizons within a hydrocarbon reservoir related to sequence boundaries and transgressive surfaces, respectively, may in the future play a crucial role in the development of new exploration strategies.

5. CONCLUSIONS

Chert layers and nodules form in different depositional settings from evaporitic-lagoonal to normal marine, and a distinct time interval spanning the late Anisian (Illyrian) to early Ladinian (Fassanian) features prominent chert formation across European carbonate systems. However, paleogeographically such chert horizons seem to be restricted to the Peri-Tethyan realm where ramp/platform configuration and morphology, water depths (salinity, primary productivity) as well as the proximity to adjacent land masses (terrestrial influx of organic matter) played a crucial role in silica precipitation. The open Tethys shelf with thick carbonate sequences of the Northern Calcareous and Southern Alps lacks such silica horizons, most likely related to rapid platform drowning at the Anisian–Ladinian boundary.

Evidently, shallow epeiric seas at the boundary interval provided the most favorable conditions for silica precipitation which have to be further investigated in the wider context of the overall development of the Triassic oceans. The stratigraphic potential of silicified layers that developed at the Anisian–Ladinian boundary for larger scale correlations is so far well documented for the Peri-Tethyan realm, and a study of similar settings on a global scale is necessary to unravel the full potential.

In a sequence stratigraphic context, these silicifications occur during the initial transgression (transgressive surfaces) above sequence boundaries and therefore reveal a high potential as transcontinental correlation tool. Furthermore, laterally continuous chert layers are potential sealing horizons within a hydrocarbon reservoir and thus may play a crucial role in reservoir geology. Middle Triassic ramp deposits in Europe represent excellent outcrop analogues for hydrocarbon carbonate reservoirs in the Middle East, and future exploration activities should include the identification of such prominent features for correlation purposes and reservoir characterization.

ACKNOWLEDGMENTS

Joint fieldwork of the "Trans-European Muschelkalk Working Group" initiated the idea to compile the available data on chert formation across the Anisian–Ladinian boundary, and many fruitful discussions with our colleagues are gratefully acknowledged. We thank two anonymous reviewers for their comments, that helped to improve an earlier version of the manuscript.

REFERENCES

Alberti, F.A.v., 1834. Beitrag zu einer Monographie des Bunten Sandsteins, Muschelkalks und Keupers, und die Verbindung dieser Gebilde zu einer Formation. Cotta, Stuttgart. 366 pp.

Alsharhan, A.S., 2006. Sedimentological character and hydrocarbon parameters of the middle Permian to Early Triassic Khuff Formation, United Arab Emirates. GeoArabia 11, 121–158.

Atudorei, V., Baud, A., Crasquin-Soleau, S., Galbrun, B., Grădinaru, E., Mirauta, E., Renard, M., Zerrari, S., 1997. Extended Scientific Report of the Peri-Tethys Project "The Triassic of North-Dobrogea" Lausanne, 55 pp.

Bernasconi, S.M., Meier, I., Wohlwend, S., Brack, P., Hochuli, P.A., Bläsi, H., Wortmann, U.G., Ramseyer, K., 2017. An evaporite-based high-resolution sulfur isotope record of Late Permian and Triassic seawater sulfate. Geochim. Cosmochim. Acta 204, 331–349.

Berner, R.A., 2006. GEOCARBSULF: a combined model for Phanerozoic atmospheric O_2 and CO_2. Geochim. Cosmochim. Acta 70, 5663–5664.

Birnbaum, S.J., Wireman, J.W., 1984. Bacterial sulfate reduction and pH: implications for early diagenesis. Chem. Geol. 43, 143–149.
Brack, P., Rieber, H., Nicora, A., Mundil, R., 2005. The Global Boundary Stratotype Section and Point (GSSP) of the Ladinian Stage (Middle Triassic) at Bagolino (Southern Alps, Northern Italy) and its implications for the Triassic time scale. Episodes 28 (4), 233–244.
Budai, T., Haas, J., 1997. Triassic sequence stratigraphy of the Balaton Highland, Hungary. Acta Geol. Hung. 40 (3), 307–335.
Budai, T., Vörös, A., 2006. Middle Triassic platform and basin evolution of the Southern Bakony Mountains (Transdanubian Range, Hungary). Riv. Ital. Paleontol. Stratigr. 112, 359–371.
Burchette, T.P., Wright, V.P., 1992. Carbonate ramp depositional systems. Sediment. Geol. 79, 3–57.
Busse, E., Horn, M., 1982. Muschelkalk. In: Erläuterungen zur Geologischen Karte Hessen 1:25000, Bl. 4520 Warburg, pp. 55–90.
Calvet, F., Tucker, M.E., Henton, J.M., 1990. Middle Triassic carbonate ramp systems in the Catalan Basin, northeast Spain: facies, systems tracts, sequences and controls. IAS Spec. Publ. 9, 79–108.
Chatalov, A., 2002. Inner ramp carbonate shoals from the Middle Triassic in Northwestern Bulgaria. Rev. Bulg. Geol. Soc. 63, 3–20 (in Bulgarian with English abstract).
Chatalov, A., 2013. A Triassic homoclinal ramp from the Western Tethyan realm, Western Balkanides, Bulgaria: integrated insight with special emphasis on the Anisian outer to inner ramp facies transition. Palaeogeogr. Palaeoclimatol. Palaeoecol. 386, 34–58.
Chatalov, A., 2016. Global, regional and local controls on the development of a Triassic carbonate ramp system, Western Balkanides, Bulgaria. Geol. Mag. http://dx.doi.org/10.1017/S0016756816000923, (in press).
Chatalov, A., Stanimirova, T., 2001. Diagenesis of Spathian and Anisian dolomites from the Western Balkanides, Bulgaria. Neues Jahrb. Geol. Palaeontol. Monatsh. 2001, 301–320.
Chen, Z.Q., Benton, M.J., 2012. The timing and pattern of biotic recovery following the end-Permian mass extinction. Nat. Geosci. 5, 375–383.
Clayton, C.J., 1986. The chemical environment of flint formation in Upper Cretaceous chalks. The scientific study of flint and chert. In: Sieverking, G., Hart, M.B. (Eds.), Proceedings of the Fourth International Flint Symposium. Cambridge University, Cambridge, pp. 43–54.
DeMaster, D.J., 1981. The supply and accumulation of silica in the marine environment. Geochim. Cosmochim. Acta 45, 1715–1732.
De Wever, P., O'Dogherty, L., Goričan, Š., 2006. The plankton turnover at the Permo-Triassic boundary, emphasis on radiolarians. In: Baumgartner, P.O., Aitchison, J.C., De Wever, P., Jackett, S.-J. (Eds.), Radiolaria—Siliceous Plankton Through Time, pp. 49–62. Swiss J. Geosci. 99 (Suppl. 1).
Dronkert, H., Bläsi, H.-R., Matter, A., 1990. Facies and Origin of Triassic Evaporites From the NAGRA Boreholes, Northern Switzerland. Nagra Technischer Bericht, NTB 87-20. 120 pp.
Dünkel, H., Vath, U., 1990. Ein vollständiges Profil des Muschelkalks (Mitteltrias) der Dransfelder Hochfläche, SW Göttingen (Südniedersachsen). Geol. Jahrb. Hessen 188, 87–126.
Erwin, D.H., 1998. The end and the beginning: recoveries from mass extinctions. Trends Ecol. Evol. 13, 344–349.
Escudero-Mozo, M.J., Martín-Chivelet, J., Goy, A., López-Gómez, J., 2014. Middle-Upper Triassic carbonate platforms in Minorca (Balearic islands): implications for Western Tethys correlations. Sediment. Geol. 310, 41–58.
Escudero-Mozo, M.J., Márquez-Aliaga, A., Goy, A., Martín-Chivelet, J., López-Gómez, J., Márquez, L., Arche, A., Plasencia, P., Pla, C., Marzo, M., Sánchez-Fernández, D., 2015.

Middle Triassic carbonate platforms in eastern Iberia: evolution of their fauna and palaeogeographic significance in the western Tethys. Palaeogeogr. Palaeoclimatol. Palaeoecol. 417, 236–260.

Feist-Burkhardt, S., Götz, A.E. & Szulc, J. (coordinators), Borkhataria, R., Geluk, M., Haas, J., Hornung, J., Jordan, P., Kempf, O., Michalík, J., Nawrocki, J., Reinhardt, L., Ricken, W., Röhling, G.-H., Rüffer, T., Török, Á., Zühlke, R., 2008a. Triassic. In: McCann, T., (Ed.), The Geology of Central Europe. vol. 2: 749–821, Geological Society, London.

Feist-Burkhardt, S., Götz, A.E., Ruckwied, K., Russell, J.W., 2008b. Palynofacies patterns, acritarch diversity and stable isotope signatures in the Lower Muschelkalk (Middle Triassic) of N Switzerland: evidence of third-order cyclicity. Swiss J. Geosci. 101, 1–15.

Fensome, R.A., MacRae, R.A., Moldowan, J.M., Taylor, F.J.R., Williams, G.L., 1996. The early Mesozoic radiation of dinoflagellates. Paleobiology 22, 329–338.

Fijałkowska-Mader, A., 2015. A record of climatic changes in the Triassic palynological spectra from Poland. Geol. Q. 59, 615–653.

Flügel, E., 2004. Microfacies of Carbonate Rocks: Analysis, Interpretation and Application. Springer, Berlin. 976 pp.

Fraiser, M.L., Bottjer, D.J., 2005. Restructuring of benthic level-bottom shallow marine communities due to prolonged environmental stress during the aftermath of the end-Permian mass extinction. C. R. Palevol. 4, 583–591.

Franz, M., Kaiser, S.I., Fischer, J., Heunisch, C., Kustatscher, E., Luppold, F.W., Berner, U., Röhling, H.-G., 2015. Eustatic and climatic control on the Upper Muschelkalk Sea (late Anisian/Ladinian) in the Central European Basin. Global Planet. Change 135, 1–27.

Freudenberger, W., 1996. Trias. Bayerisches Geologisches Landesamt, (Ed.), In: Erläuterungen zur Geologischen Karte von Bayern 1:500 000, fourth ed., Bayerisches Geologisches Landesamt: München, pp. 55–64.

Gaetani, M., Fois, E., Jadoul, F., Nicora, A., 1981. Nature and evolution of the Middle Triassic carbonate buildups in the Dolomites (Italy). Mar. Geol. 44, 25–57.

Gaetani, M., et al., 2000. Early Ladinian (238–235 Ma). In: Dercourt, J., et al., (Ed.), Atlas Peri-Tethys, Paleogeographical Maps. CCGM/CGMW, Paris (Map 5).

Genser, C., 1930. Zur Stratigraphie und Chemie des mittleren Muschelkalks in Franken. Geol.-paläontol. Abh. N.F. 17, 381–492.

Geyer, M., Nitsch, E., Simon, T., 2011. Geologie von Baden-Württemberg, fifth ed. Schweizerbart, Stuttgart. 627 pp.

Golonka, J., 2004. Plate tectonic evolution of the southern margin of Eurasia in the Mesozoic and Cenozoic. Tectonophysics 381, 235–273.

Golonka, J., 2007. Phanerozoic paleoenvironment and paleolithofacies maps. Mesozoic. Geologia 33, 211–264.

Golonka, J., Ford, D., 2000. Pangean (Late Carboniferous–Middle Jurassic) paleoenvironment and lithofacies. Palaeogeogr. Palaeoclimatol. Palaeoecol. 161, 1–34.

Göthel, M., Grunert, K., 1996. Neue Erkenntnisse aus der Braunkohlen-Erkundung zur Trias-Stratigraphie der Lausitzer Triasplatte. Freib. Forsch. C464, 175–202.

Götz, A.E., Feist-Burkhardt, S., 2012. Phytoplankton associations of the Anisian Peri-Tethys Basin (Central Europe): evidence of basin evolution and palaeoenvironmental change. Palaeogeogr. Palaeoclimatol. Palaeoecol. 338, 151–158.

Götz, A.E., Gast, S., 2010. Basin evolution of the Anisian Peri-Tethys: implications from conodont assemblages of Lower Muschelkalk key sections (Central Europe). Z. dt. Ges. Geowiss. 161, 39–49.

Götz, A.E., Keller, M., 1998. Muschelkalk in Mainfranken. Kurzfassungen und Exkursionsführer, 13. Sedimentologentreffen. Erlanger Geol. Abh. Sdb. 2, 151–159.

Götz, A.E., Szulc, J., Feist-Burkhardt, S., 2005. Distribution of sedimentary organic matter in Anisian carbonate series of S Poland: evidence of third-order sea-level fluctuations. Int. J. Earth Sci. 94, 267–274.

Götz, A.E., Rameil, N., Borkhataria, R., 2008. Field trip POST5—Epeiric carbonates of the Lower Muschelkalk (Middle Triassic)—outcrop analogue for layer cake reservoirs. 26th IAS Regional Meeting/SEPM-CES Sediment 2008 Bochum, Excursion Guidebook, EDGG, 237, pp. 150–163.

Götz, A.E., Montenari, M., Costin, G., 2017. Silicification and organic matter preservation in the Anisian Muschelkalk: implications for the basin dynamics of the central European Muschelkalk Sea. Cent. Eur. Geol. 60 (1), 35–52.

Goy, A., Peréz-López, A., 1996. Presencia de cefalópodos del tránsito Anisiense-Ladiniense en las facies Muschelkalk de la Zona Subbética (Cordillera Bética). Geogaceta 20, 183–186.

Grădinaru, E., 2000. Introduction to the Triassic geology of North Dobrogea Orogen—an overview of the Triassic System in the Tulcea Unit and the ammonoid biostratigraphy. In: Grădinaru, E. (Ed.), Bucharest Workshop on the Lower–Middle Triassic (Olenekian–Anisian) Boundary, Tulcea, Romania, pp. 5–37. Field Trip Section.

Haas, J., 2012. Geology of Hungary, Regional Geology Reviews. Springer, Berlin. 244 pp.

Haas, J., Budai, T., 1999. Triassic sequence stratigraphy of the Transdanubian Range (Hungary). Geol. Carpath. 50, 459–475.

Hagdorn, H., Hickethier, H., Horn, M., Simon, T., 1987. Profile durch den hessischen, unterfränkischen und baden-württembergischen Muschelkalk. Geol. Jahrb. Hessen 115, 131–160.

Hagdorn, H., Szulc, J., Bodzioch, A., Morycowa, E., 1999. Riffe aus dem Muschelkalk. In: Hauschke, N., Wilde, V. (Eds.), Trias, eine ganz andere Welt: Mitteleuropa im frühen Erdmittelalter. Verlag Dr. Friedrich Pfeil, München, pp. 309–320.

Hallam, A., 1992. Phanerozoic Sea-Level Changes. Columbia University Press, New York. 266 pp.

Hallam, A., Wignall, P.B., 1997. Mass Extinctions and Their Aftermath. Oxford University Press, Oxford. 328 pp.

Haq, B.U., Hardenbol, J., Vail, P.R., 1988. Mesozoic and Cenozoic chronostratigraphy and cycles of sea-level. In: Wilgus, C.K., Hastings, B.S., Kendall, C.G.S.C., Posamentier, H.W., Ross, C.A., Van Wagoner, J.C. (Eds.), In: Sea-Level Changes—An Integrated Approach, vol. 42, pp. 71–108. SEPM Special Publication.

Harwood, D.M., Nikolaev, V.A., Winter, D.M., 2007. Cretaceous records of diatom evolution, radiation, and expansion. In: Starratt, S. (Ed.), Pond Scum to Carbon Sink: Geological and Environmental Applications of the Diatoms. vol. 13, pp. 33–59, The Paleontology Society Papers.

Hohenstein, V., 1913. Beiträge zur Kenntnis des Mittleren Muschelkalks und des unteren Trochitenkalks am östlichen Schwarzwaldrand. Geol.-paläontol. Abh. N.F. 12, 173–272.

Jordan, P., 2016. Reorganisation of the Triassic stratigraphic nomenclature of northern Switzerland—Overview and the new Dinkelberg, Kaiseraugst and Zeglingen formations. Swiss J. Geosci. 109, 241–255.

Kiessling, W., Flügel, E., Golonka, J., 2003. Patterns of Phanerozoic carbonate platform sedimentation. Lethaia 36, 195–225.

Kovács, S., Sudar, M., Grădinaru, E., Gawlick, H.-J., Karamata, S., Haas, J., Péró, C., Gaetani, M., Mello, J., Polák, M., Aljinović, D., Ogorelec, B., Kolar-Jurkovšek, T., Jurkovšek, B., Buser, S., 2011. Triassic evolution of the tectonostratigraphic units of the Circum-Pannonian region. Jahrb. Geol. Bundesanst. 151, 199–280.

Kump, L.R., Brantley, S.L., Arthur, M.A., 2000. Chemical weathering, atmospheric CO_2, and climate. Annu. Rev. Earth Planet. Sci. 28, 611–667.

Kwiatkowski, S., 2005. Origin of chert nodules from the Polish Muschelkalk, Middle Triassic. Ann. Soc. Geol. Pol. 75, 287–308.

Langbein, R., Knaust, D., 1997. Zur Petrographie der Hornsteinlagen im Mittleren Muschelkalk von Thüringen. Beitr. zur Geol. v. Thür. N.F. 4, 37–62.

Laschet, C., 1984. On the origin of cherts. Facies 10, 257–290.

Llompart, C., Rosell, J., Márquez-Aliaga, A., Goy, A., 1987. El Muschelkalk de la isla de Menorca. Cuad. Geol. Iber. 11, 323–335.

López-Gómez, J., Mas, R., Arche, A., 1993. The evolution of the Middle Triassic (Muschelkalk) carbonate ramp in the SE Iberian Ranges, eastern Spain: sequence stratigraphy, dolomitization processes and dynamic controls. Sediment. Geol. 87, 165–193.

López-Gómez, J., Arche, A., Calvet, F., Goy, A., 1998. Epicontinental marine carbonate sediments of the Middle and Upper Triassic in the westernmost part of the Tethys Sea, Iberian Península. Zbl. Geol. Paläontol. 9/10, 1033–1084.

López-Gómez, J., Arche, A., Pérez-López, A., 2002. Permian and Triassic. In: Gibbons, W., Moreno, T. (Eds.), The Geology of Spain, pp. 185–212. Geological Society of London, Special Publication.

Márton, E., Budai, T., Haas, J., Kovács, S., Szabó, I., Vörös, A., 1997. Magnetostratigraphy and biostratigraphy of the Anisian–Ladinian boundary section Felsőörs (Balaton Highland, Hungary). Albertiana 20, 50–57.

Matter, A., Peters, T., Bläsi, H.-R., Meyer, J., Ischi, H., Meyer, C., 1988. Sondierbohrung Weiach—Geologie (Textband). Nagra Technischer Bericht, NTB 86-01. 438 pp.

Matysik, M., 2016. Facies types and depositional environments of a morphologically diverse carbonate platform: a case study from the Muschelkalk (Middle Triassic) of Upper Silesia, Southern Poland. Ann. Soc. Geol. Pol. 86, 119–164.

Mercedes-Martín, R., Salas, R., Arenas, C., 2013. Facies heterogeneity and depositional models of a Ladinian (Middle Triassic) microbial-dominated carbonate ramp system (Catalan Coastal Ranges, NE Spain). Mar. Pet. Geol. 46, 107–128.

Mercedes-Martín, R., Salas, R., Arenas, C., 2014. Microbial-dominated carbonate platforms during the Ladinian rifting: sequence stratigraphy and evolution of accommodation in a fault-controlled setting (Catalan Coastal Ranges, NE Spain). Basin Res. 26, 269–296.

Merki, P., 1961. Der obere Muschelkalk im östlichen Schweizer Jura. Eclogae Geol. Helv. 54, 127–220.

Michael, P., 1940. Weitere geologische Beobachtungen aus der Umgebung Weimars. Beitr. zur Geol. v. Thür. 5, 258–268.

NAGRA, 2001. Sondierbohrung Benken—Untersuchungsbericht. Nagra Technischer Bericht, NTB 00-01. 290 pp.

Packard, J.J., Al-Aasam, I., Samson, I., Berger, Z., Davies, J., 2001. A Devonian hydrothermal chert reservoir: the 225 bcf Parkland Field, British Columbia, Canada. AAPG Bull. 85, 51–84.

Payne, J.L., van de Schootbrugge, B., 2007. Life in Triassic Oceans: links between planktonic and benthonic recovery and radiation. In: Falkowski, P., Knoll, A.H. (Eds.), Evolution of Primary Producers in the Sea. Academic Press, Amsterdam, pp. 165–189.

Péréz-López, A., 1991. El Trías de facies Germánica del sector central de la Cordillera Bética. Unpublished PhD thesis, University of Granada. 400 pp.

Péréz-López, A., 1998. Epicontinental Triassic of the Southern Iberian Continental Margin (Betic Cordillera, Spain). Zentralbl. Geol. Paläontol. Teil 1 9–10, 1009–1031.

Pérez-López, A., Pérez-Valera, F., 2007. Palaeogeography, facies and nomenclature of the Triassic units in the different domains of the Betic Cordillera (S Spain). Palaeogeogr. Palaeoclimatol. Palaeoecol. 254, 606–626.

Peters, T., Matter, A., Meyer, J., Isenschmid, C., Ziegler, H.J., 1992. Sondierbohrung Kaisten—Geologie (Textband). Nagra Technischer Bericht, NTB 86-04, 286 pp.

Pietsch, J.S., Wetzel, A., Jordan, P., 2016. A new lithostratigraphic scheme for the Schinznach Formation (upper part of the Muschelkalk Group of northern Switzerland). Swiss J. Geosci. 109, 285–307.
Pomar, L., Hallock, P., 2008. Carbonate factories: a conundrum in sedimentary geology. Earth Sci. Rev. 87, 134–169.
Preto, N., Kustatscher, E., Wignall, P.B., 2010. Triassic climate—state of the art and perspectives. Palaeogeogr. Palaeoclimatol. Palaeoecol. 290, 1–10.
Pruss, S.B., Bottjer, D.J., 2005. The reorganization of reef communities following the end-Permian mass extinction. C. R. Palevol. 4, 553–568.
Radzinski, K.-H., 2008. Muschelkalk. In: Bachmann, G.H., Ehling, B.C., Eichner, R., Schwab, M. (Eds.), Geologie von Sachsen-Anhalt. Schweitzerbart, Stuttgart, pp. 201–212.
Rüffer, T., 1995. Entwicklung einer Karbonat-Plattform: Fazies, Kontrollfaktoren und Sequenzstratigraphie in der Mitteltrias der westlichen Nördlichen Kalkalpen (Tirol, Bayern). Gaea Heidelbergensis 1, 1–288.
Rüger, L., 1934. Einige Beobachtungen an den Grenzschichten zwischen Mittlerem und Oberem Muschelkalk bei Dallau (Badisches Bauland). Jahresber. Mitt. Oberrhein Geol. Ver. 27, 9–15.
Sandberger, F., 1864. Beobachtungen in der Würzburger Trias. Würzbg. Naturwiss. Z. 5, 201–231.
Schwahn, H.-J., Gahrmann, N., 1976. Der Mittlere Muschelkalk im östlichen Subherzyn. Herzynia N. F. 13 (3), 332–339.
Seebach, K. v., 1861. Die Conchylien-Fauna der Weimarischen Trias. Z. Dtsch. Geol. Ges. 13, 551–666.
Seidel, G., 1965. Zur Ausbildung des Muschelkalkes in NW-Thüringen. Geologie 14 (1), 58–63.
Seidel, G., 2003. Geologie von Thüringen, second ed. Schweizerbart, Stuttgart. 601 pp.
Sellwood, B.W., Valdes, P.J., 2006. Mesozoic climates: general circulation models and the rock record. Sediment. Geol. 190, 269–287.
Senowbari-Daryan, B., Zühlke, R., Bechstädt, T., Flügel, E., 1993. Anisian (Middle Triassic) buildups of the Northern Dolomites (Italy): the recovery of reef communities after the Permian/Triassic crisis. Facies 28, 181–256.
Seufert, G., Schweizer, V., 1985. Stratigraphische und mikrofazielle Untersuchungen im Trochitenkalk (Unterer Hauptmuschelkalk, mo1) des Kraichgaus und angrenzender Gebiete. Jahresber. Mitt. Oberrhein Geol. Ver. 67, 129–171.
Simon, T., 1988. Geologische und hydrogeologische Ergebnisse der neuen Solebohrung Bad Rappenau, Baden-Württemberg. Jahrb. Geol. Landesamt Baden-Württemberg 30, 479–510.
Song, H., Wignall, P.B., Chen, Z.Q., Tong, J., Bond, D.P.G., Lai, X., Zhao, X., Jiang, H., Yan, C., Niu, Z., Chen, J., Yang, H., Wang, Y., 2011. Recovery tempo and pattern of marine ecosystems after the end-Permian mass extinction. Geology 39, 739–742.
Speyer, O., 1875. Die paläontologischen Einschlüsse der Trias in der Umgebung Fulda's. II. Bericht des Vereins für Naturkunde zu Fulda über die Vereinsjahre vom 13. März 1869 bis dahin 1874. pp. 43–86.
Stanley, G.D.J., 2003. The evolution of modern corals and their early history. Earth Sci. Rev. 60, 195–225.
Stanley, S.M., 2008. Effects of global seawater chemistry on biomineralization: past, present, and future. Chem. Rev. 108, 4483–4498.
Stefani, M., Furin, S., Gianolla, P., 2010. The changing climate framework and depositional dynamics of Triassic carbonate platforms from the Dolomites. Palaeogeogr. Palaeoclimatol. Palaeoecol. 290, 43–57.
Strother, P.K., 1996. Acritarchs. In: Jansonius, J., McGregor, D.C. (Eds.), In: Palynology: Principles and Applications, vol. 1. American Association of Stratigraphic Palynologists Foundation, Publisher Press, Salt Lake City, UT, pp. 81–106.

Szulc, J., 1999. Anisian-Carnian evolution of the Germanic basin and its eustatic, tectonic and climatic controls. In: Bachmann, G.H., Lerche, I. (Eds.), Epicontinental Triassic, pp. 813–852. Zentralbl. Geol. Paläontol. Teil I 1998 (7–8).

Szulc, J., 2000. Middle Triassic evolution of the northern Peri-Tethys area as influenced by early opening of the Tethys Ocean. Ann. Soc. Geol. Pol. 70, 1–48.

Szulc, J., 2003. Sponge-microbial stromatolites and coral-sponge reefs recovery in the Triassic of the Western Tethys and northern Peri-Tethys basins. Berichte des Institutes für Geologie und Paläontologie der Karl-Franzens-Universität Graz 7, 108.

Szulc, J., 2007. Sponge-microbial stromatolites and coral-sponge reef recovery in the Triassic of the Western Tethys Domain. In: Lucas, S.G., Spielmann, J.A. (Eds.), The Global Triassic p. 402. N. M. Mus. Nat. Hist. Sci. Bull. 41.

Török, Á., 1997. Triassic ramp evolution in Southern Hungary and its similarities to the Germano-type Triassic. Acta Geol. Hung. 40, 367–390.

Török, Á., 1998. Controls on development of Mid-Triassic ramps: examples from southern Hungary. In: Wright, V.P., Burchette, T.P. (Eds.), Carbonate Ramps, pp. 339–367. Geological Society of London, Special Publication, 149.

Tronkov, D., 1968. Die Grenze untere Trias—mittlere Trias in Bulgarien. Bull. Geol. Inst. Ser. Palaeontol. 17, 113–131(in Bulgarian with German abstract).

Tronkov, D., 1976. Triassische Ammoniten-Sukzessionen im Westlichen Balkangebirge in Bulgarien. C. R. Acad. Bulg. Sci. 29, 1325–1328.

Trotter, J.A., Williams, I.S., Nicora, A., Mazza, M., Rigo, M., 2015. Long-term cycles of Triassic climate change: a new $\delta^{18}O$ record from conodont apatite. Earth Planet. Sci. Lett. 415, 165–174.

Vail, P.R., Mitchum, R.M., Thompson III, S., 1977. Seismic stratigraphy and global changes of sea level. Part 4: global cycles of relative changes of sea level. In: Payton, C.E. (Ed.), Seismic Stratigraphy—Applications to Hydrocarbon Exploration, pp. 83–97. Am. Assoc. Petr. Geol. Mem. 26.

Vaslet, D., Le Nindre, Y.-M., Vachard, D., Broutin, J., Crasqiun-Soleau, S., Berthelin, M., Gaillot, J., Halawani, M., Al-Husseini, M., 2005. The Permian-Triassic Khuff Formation of central Saudi Arabia. GeoArabia 10, 77–134.

Vera, J.A., 2001. Evolution of the South Iberian Continental Margin. In: Ziegler, P.A., Cavazza, W., Robertson, A.H.F., Crasquin-Soleau, S. (Eds.), Peri-Tethys Memoir 6: Peri-Tethyan Rift/Wrench Basins and Passive Margins, pp. 109–143. Mém. Mus. Natl. Hist. Nat. 186.

Voigt, T., Bachmann, G.H., Gaupp, R., 2001. Buntsandstein und Muschelkalk im Thüringer Becken. In: Gaup, R., Van der Klauw, S. (Eds.), Sediment 2001, pp. 137–169. Schriftenreihe der deutschen geologischen Gesellschaft, 13.

Vörös, A., Budai, T., Haas, J., Kovács, S., Kozur, H., Pálfy, J., 2003. GSSP (Global Boundary Stratotype Section and Point) proposal for the base of Ladinian (Triassic)—a proposal for the GSSP at the base of the Reitzi Zone (sensu stricto) at Bed 105 in the Felsőörs section, Balaton Highland, Hungary. Albertiana 28, 35–47.

Weber, H., Kubald, P., 1947. Der Mittlere Muschelkalk an der Reichsautobahn bei Eisenach. Beitr. zur Geol. v. Thür. 8, 167–189.

Wilczewski, N., 1967. Mikropaläontologische Untersuchungen im Muschelkalk Unterfrankens. Inaugural-Dissertation, Würzburg, 125 pp.

Ziegler, P.A., 1990. Geological Atlas of Western and Central Europe, second ed. Shell Internationale Petroleum Maatschappij B.V. and Geological Society, London. 239 pp.

Zühlke, R., 2000. Fazies, hochauflösende Sequenzstratigraphie und Beckenentwicklung im Anis (mittlere Trias) der Dolomiten (Südalpin, N-Italien). Gaea Heidelbergensis 6, 1–368.

Zwenger, W., 2015. Muschelkalk. In: Stackebrandt, W., Franke, D. (Eds.), Geologie von Brandenburg. Schweizerbart, Stuttgart, pp. 182–193.

CHAPTER NINE

Integrated Stratigraphy of Triassic and Jurassic Formations in the Central Oman Mountains: A Synopsis

Markus Schlaich*, Michael Obermaier†, Daniel Bendias‡, Thomas Aigner*,[1]

*University of Tübingen, Tübingen, BW, Germany
†Shell Projects & Technology, The Hague, The Netherlands
‡ENI, London, United Kingdom
[1]Corresponding author: e-mail address: t.aigner@uni-tuebingen.de

Contents

Abstract

The Triassic Mahil Formation and the Lower Jurassic Lower Mafraq Member are part of the AP6 (Arabian Plate) tectonostratigraphic mega-sequence. The overlying Mid- to Upper Jurassic Upper Mafraq Member, Dhruma, Tuwaiq Mountain, Hanifa and Jubaila formations form the AP7 mega-sequence. All formations were studied in outcrops and in borehole cores and cuttings from the subsurface of Oman. Biostratigraphic and chemostratigraphic analyses were carried out in order to determine the age of each formation and to identify major unconformities. Biostratigraphic analyses for the Triassic Mahil Formation revealed an Olenekian to Early Anisian to Carnian age.

This paper expands on the sequence stratigraphic interpretations that have been established for each individual interval from the Lower Triassic to the Upper Jurassic by Pöppelreiter et al. (2011), Obermaier et al. (2012), Bendias and Aigner (2015), Bendias

Stratigraphy & Timescales, Volume 2
ISSN 2468-5178
http://dx.doi.org/10.1016/bs.sats.2017.07.004

(2016), Schlaich (2016) and Schlaich and Aigner (2017). It aims to integrate these previous studies in a consistent sequence-stratigraphic framework based on observations in outcrops of the Central Oman Mountains.

1. INTRODUCTION

In the Sultanate of Oman, the Triassic and Jurassic periods are represented by a sediment package of up to 3 km thickness of varying lithologies which range from dolomites, limestones, and subordinately anhydrite to mixed carbonate-siliciclastics of various depositional environments. The succession contains several unconformities and discontinuities which show evidence of erosion or nondeposition caused by regional tectonic events in an overall passive margin setting.

The Triassic–Jurassic stratigraphy is well exposed in outcrops of the Oman Mountains where it has been studied in detail by various authors who established stratigraphic type sections for each individual formation and member (Bendias, 2016; Bendias and Aigner, 2015; Obermaier et al., 2012; Pöppelreiter et al., 2011; Schlaich, 2016; Schlaich and Aigner, 2017).

This chapter integrates cycle interpretations, bio- and chemostratigraphic evidence and facies analyses from each individual formation with the aim to provide a synopsis of the Triassic–Jurassic stratigraphy in Oman. Since most of the data presented in this study are based on outcrop data, formation names commonly refer to the nomenclature used in outcrops with additional references to equivalent subsurface formation names and observed differences in facies or thickness in core data or cuttings (Table 1).

Sharland et al. (2001) subdivided the sediment succession on the AP6 into tectonostratigraphic megasequences (TMS) and distinctive maximum flooding surfaces (MFS) which are assumed to be regionally extensive. In order to put the Triassic–Jurassic sequence stratigraphic interpretations from Oman into a plate-wide context, tentative correlations to Sharland et al.'s (2001) MFS are presented in respective paragraphs below.

The Triassic succession in Oman is part of Sharland et al.'s (2001) AP6 TMS which began in the Mid-Permian and lasted until the Lower Jurassic. The Triassic part of the AP6 TMS in outcrops of the Oman Mountains

Table 1 Relative Age and Range of All Investigated Formations in the Subsurface of Oman (Forbes et al., 2010) and Outcrops of the Central Oman Mountains (Bendias, 2016; Obermaier, 2013; Schlaich and Aigner, 2017)

	Subsurface		Outcrop			
Period	**Relative Age**	**Formation**	**Relative Age**	**Formation**	**Group**	**TMS**
Jurassic	L. Kimmeridgian to E. Tithonian	Jubaila	*Not present*		Sahtan	AP7
	Kimmeridgian	Hanifa	*Not present*			
	L. Bathonian to Callovian	Tuwaiq Mountain	*Not present*			
	L. Bajocian to Bathonian	Dhruma	E. Bajocian to L. Bathonian	Dhruma		
	Toarcian to L. Bajocian	Upper Mafraq Member	L. Aalenian to Bajocian	Upper Mafraq Member		
	L. Pliensbachian to E. Toarcian	Lower Mafraq Member	L. Hettangian to L. Pliensbachian	Lower Mafraq Member		AP6
Triassic	Rhaetian	Minjur	*Not present*		Akhdar	
	Anisian to Norian	Jilh	M. Anisian to L. Norian	Upper Mahil		
	Olenekian	Sudair Khuff	M. Induan to M. Anisian	Middle Mahil		
Permian	Wordian to Induan (Olenekian?)	Khuff	Induan?	Lower Mahil Saiq		
	Roadian to Wordian	Upper Gharif Member	*Not present*			

TMS: Tectonostratigraphic megasequence.

is called Mahil Formation which involves the Lower Mahil (Upper Khuff in the subsurface), Middle Mahil (Sudair in the subsurface), and Upper Mahil (Jilh in the subsurface) members (Table 1). The unconformably overlying Lower Jurassic mixed carbonate-siliciclastic Lower Mafraq Member (same name in outcrops and subsurface) represents the youngest sediments of the AP6 TMS (Fig. 1). According to Sharland et al. (2001), the dating of the upper boundary of the AP6 megasequence is poorly constrained, but probably coincides in Oman with a hiatus during the Toarcian.

The following sediment accumulation in the Jurassic took place during Sharland et al.'s (2001) AP7 TMS and began after the unconformity in the Toarcian and ended with an unconformity in the Tithonian (e.g., De Matos and Hulstrand, 1995). In the Oman Mountains, the AP7 TMS comprises the Lower Jurassic Upper Mafraq Member (Bendias, 2016; Bendias and Aigner, 2015) and the Middle Jurassic Dhruma Formation (Schlaich, 2016; Schlaich and Aigner, 2017). The Jurassic formations, including the Lower Mafraq Member of the AP6 TMS, form the Sahtan Group (Table 1, Fig. 1) which was originally defined by Glennie et al. (1974) in the Al Hajar Mountains outcrops.

Some 210 km west of the Triassic–Jurassic outcrops in the Oman Mountains, time-equivalent strata in the subsurface of Oman have been investigated in borehole cores and cuttings (14 wells for Triassic, 13 wells for Lower Jurassic, and 20 wells for Middle Jurassic strata). Therefore, the succession exhibits some differences in depositional facies and in the duration of hiatuses. The subsurface Triassic stratigraphy contains an additional formation which is absent in outcrops of the Oman Mountains: the Triassic Minjur Formation (Forbes et al., 2010). The siliciclastic deposits are of Rhaetian age and overly unconformably the Jilh Formation, a subsurface equivalent to the Upper Mahil Member. The top of the Minjur Formation is marked by an unconformity which is of Hettangian to Sinemurian age before the deposition of the Lower Mafraq Member began in the Early Sinemurian.

Owing to the diachronous nature of unconformities across the AP6, also the Jurassic stratigraphy contains some additional formations in subsurface wells located farther west which are absent in outcrops. Whereas the Dhruma Formation in outcrops is followed by a time of nondeposition and erosion, it is overlain by the Middle Jurassic Tuwaiq Mountain and the Upper Jurassic Hanifa/Jubaila formations in the subsurface of central Oman (Table 1).

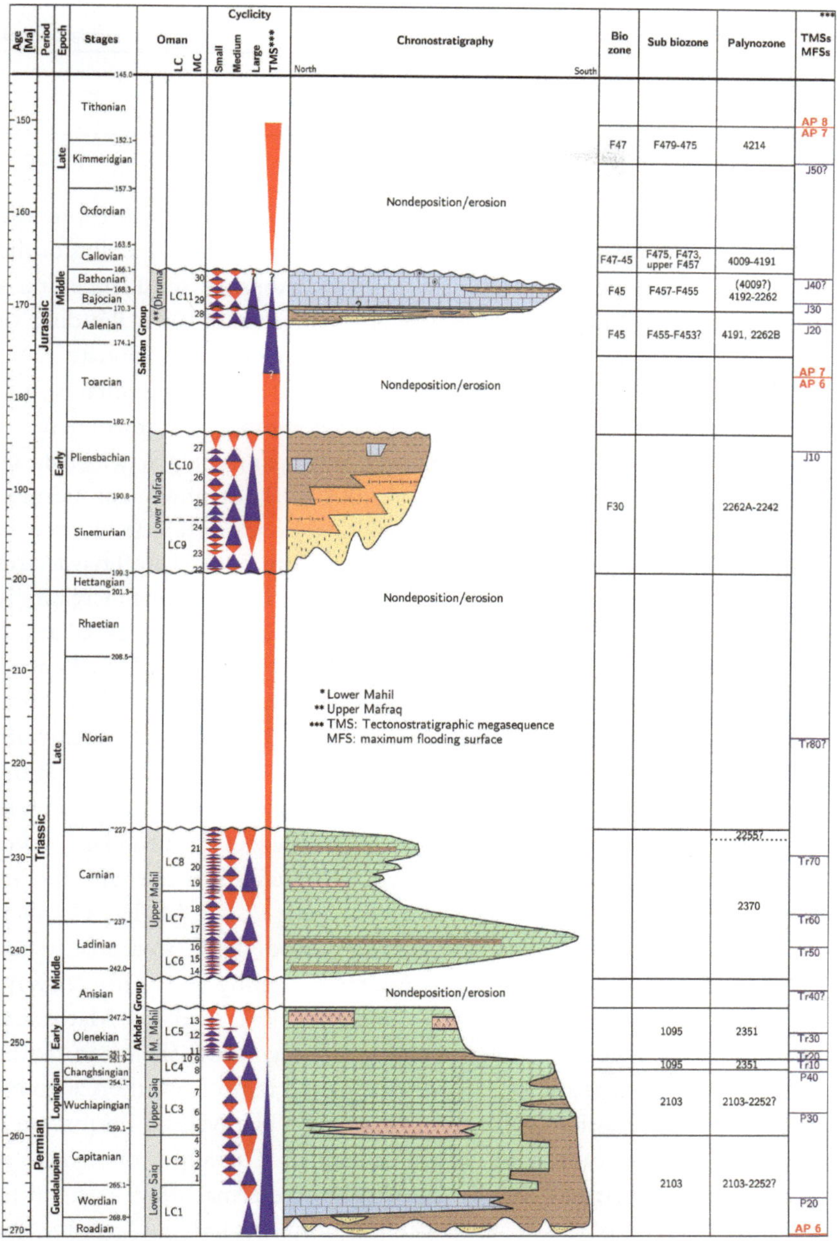

Fig. 1 Chronostratigraphic chart for outcrops of the Oman Mountains based on biostratigraphy from Wadi Sahtan. *Modified after Bendias, D., Aigner, T., 2015. Facies, sequence stratigraphy, reservoir and seal potential of the Mafraq Formation, Sultanate of Oman: an integrated outcrop analogue study. GeoArabia 20 (3), 17–94; Obermaier, M., 2013. The Geology of the Triassic Sudair and Jilh Formations in Outcrop and Subsurface (Sultanate of Oman). (PhD thesis) University of Tübingen; Schlaich, M., Aigner, T., 2017. Facies and Integrated Sequence Stratigraphy of an Epeiric Carbonate Ramp Succession: Dhruma Formation, Sultanate of Oman. The Depositional Record, 41 p.; based on Forbes, G., Jansen, H., Schreurs, J., 2010. Lexicon of Oman subsurface stratigraphy: reference guide to the stratigraphy of Oman's hydrocarbon basins. GeoArabia Special Publication 5, Gulf PetroLink, Bahrain. 371 p.*

2. GEOLOGICAL SETTING

The initial deposition of AP6 strata in North Oman started in the Mid-Permian, following the break-up of Gondwana and the opening of the Neo-Tethys Ocean (Konert et al., 2001). Rapid thermal subsidence and a gradual northward drift of the Arabian platform resulted in a flooding of the passive margin and the development of a flat epeiric carbonate ramp (Sharland et al., 2001). In outcrops of North Oman (Oman Mountains), the coastal encroachment is reflected by onlapping geometries of Permian deposits onto Precambrian and Paleozoic sediments forming in some places a prominent angular unconformity (Bendias et al., 2013). The subsequent Permo-Triassic period is marked by stable depositional conditions that resulted in the deposition of kilometer thick carbonate successions. In the Late Triassic, this stable carbonate platform of Late Permian to Triassic times became increasingly influenced by regional tectonic events causing subaerial exposure (Bendias and Aigner, 2015; Forbes et al., 2010) and a major hiatus during the Late Triassic.

In the Late Triassic to Early Jurassic, the break-up of Pangea led to rifting events and subsidence in the northern part of the AP6 resulting in a northward tilt and an uplift in the south (Grabowski et al., 1994). While deposits of the Jilh Formation are followed in the subsurface of Oman by siliciclastic facies of the Rhaetian Minjur Formation and the Pliensbachian to Toarcian Lower Mafraq Member, in outcrops of the Oman Mountains it is overlain directly by the shallow marine, mixed carbonate-siliciclastic Lower Mafraq Member.

In the following Middle to Upper Jurassic period, the depositional system in Oman evolved into a low-angle, epeiric ramp dipping in a northwestward direction (Schlaich and Aigner, 2017). Murris (1981) and De Matos and Hulstrand (1995) suggest a low-angle ramp setting in the Toarcian to the Bathonian, which turned into a slightly steepened ramp from the Callovian to the end of the Jurassic for the rest of the AP6. The Mafraq Formation is followed by marine carbonates of the Dhruma, Tuwaiq Mountain, and Hanifa/Jubaila Formations. In the Late Jurassic, thermal uplift and rifting marked the opening of the Proto-Owen Ocean (early Indian Ocean) in the southeast resulting in the truncation of the Middle and Upper Jurassic formations in the Late Jurassic throughout Oman.

3. BIO- AND CHEMOSTRATIGRAPHIC DATA

Biostratigraphic analyses on Triassic Mahil strata revealed three Foraminiferal Assemblage Zones (FAZ; Fig. 2)—FAZ 1: *Meandrospira* sp.,

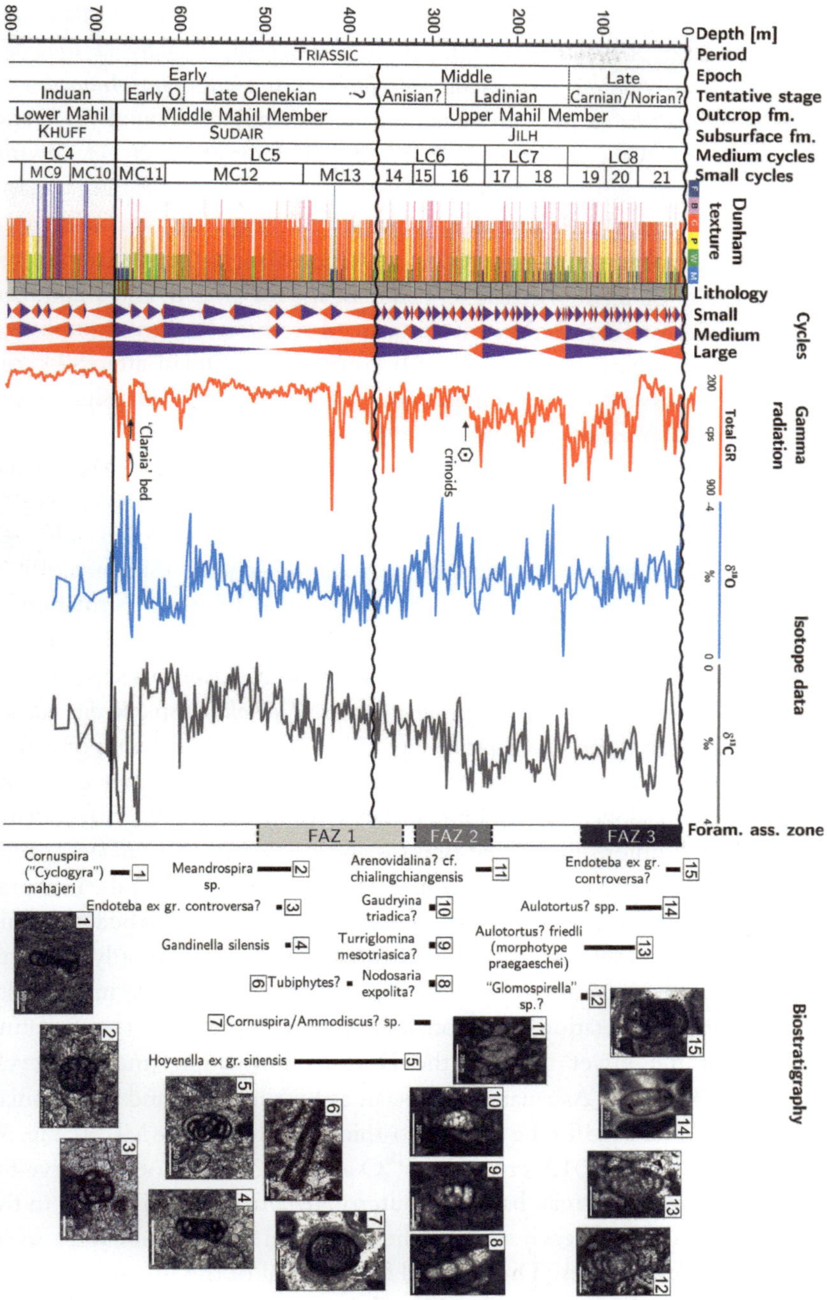

Fig. 2 Overview of the Triassic succession of Wadi Sahtan in the Oman Mountains including age, Dunham texture, lithology, cyclicity, outcrop gamma-ray, isotope data, and biostratigraphy. *Modified after Obermaier, M., 2013. The Geology of the Triassic Sudair and Jilh Formations in Outcrop and Subsurface (Sultanate of Oman). (PhD thesis) University of Tübingen.*

Endoteba ex gr. *controversa*, *Gandinella silensis*, Tubiphytes and *Hoyenella* ex gr. *sinensis*; FAZ 2: *Arenovidalina?* Cf. *chialingchiangenesis*, *Gaudryina triadica*?, *Turriglomina mesotriasica*?, *Nodosaria expolita*? and *Cornuspira/Ammodiscus*? sp.; FAZ 3: *Endoteba* ex gr. *controversa*?, *Aulotortus*? spp., *Aulotortus*? *friedli* and *Glomospirella* sp.? (Obermaier et al., 2012; Pöppelreiter et al., 2011). Owing to a generally low faunal diversity as well as pervasive dolomitization, a taxonomic differentiation of foraminifera in the Triassic succession is challenging. Yet, biostratigraphic analyses revealed an Olenekian to Early Anisian age for FAZ 1 (Fig. 2), a Late Anisian to Ladinian age for FAZ 2, and a Ladinian to Carnian age for FAZ 3. Age interpretations are based on an unpublished Petroleum Development Oman (PDO) zonation for Oman after Sikkema (1992) and Forbes et al. (2010) and thus refer to biozones and their assigned ages which are still in need of comprehensive revision.

More than 100 thin sections were analyzed from the Jurassic Mafraq and Dhruma Formations in outcrops of Wadi Sahtan in the Oman Mountains to establish a biostratigraphic framework (Bendias and Aigner, 2015). The age was determined based on an unpublished zonation scheme (Sikkema, 1992) from PDO. Biostratigraphic data of the Mafraq and Dhruma Formation in the Oman Mountains are mainly based on calcareous and agglutinating benthic foraminifera like *Haurania* spp., *Textularia* spp., *Pseudocyclammina* spp., *Trochamijiella* spp., *Pfenderina* spp., *Nautiloculina oolitica*, *Trocholina* spp., *Redmondoides* spp. *Involutina* spp., and *Orbitopsella* spp. (Fig. 3). Their analysis point to a Hettangian to Late Pliensbachian age for the Lower Mafraq Member and a Late Aalenian to Late Bajocian age for Upper Mafraq Member. The interval contains also so-called "Lithiotis beds" (see Bendias and Aigner, 2015); *Lithiotis* species are technically restricted to the Lower Jurassic which conflicts with the interpreted time range by benthic foraminifera. However, the "Lithiothis beds" contain *Lithiothis* and/or other large bivalves such as Megalodonts mostly as clasts. A detailed subdivision into the different species was not possible in the field, therefore the interpretation based on foraminifera is preferred as age dating method. For the lower third of the Dhruma Formation (medium cycle MC29; Fig. 3), a Late Aalenian to Bajocian age is suggested and a Bathonian to Middle Callovian age for the upper two-thirds (medium cycle MC30; Fig. 3).

Obermaier et al. (2012) created a $\delta^{18}O$ and $\delta^{13}C$ bulk isotope curve for the entire Triassic in Oman based on outcrop data from Wadi Sahtan in the Oman Mountains. For each sample, the $\delta^{18}O$ and $\delta^{13}C$ values were measured on bulk carbonates. The original $\delta^{18}O$ signal of the investigated succession however, is probably altered and reset by the recrystallization and dolomitization processes in the Oman Mountains (Coy, 1997). Hence,

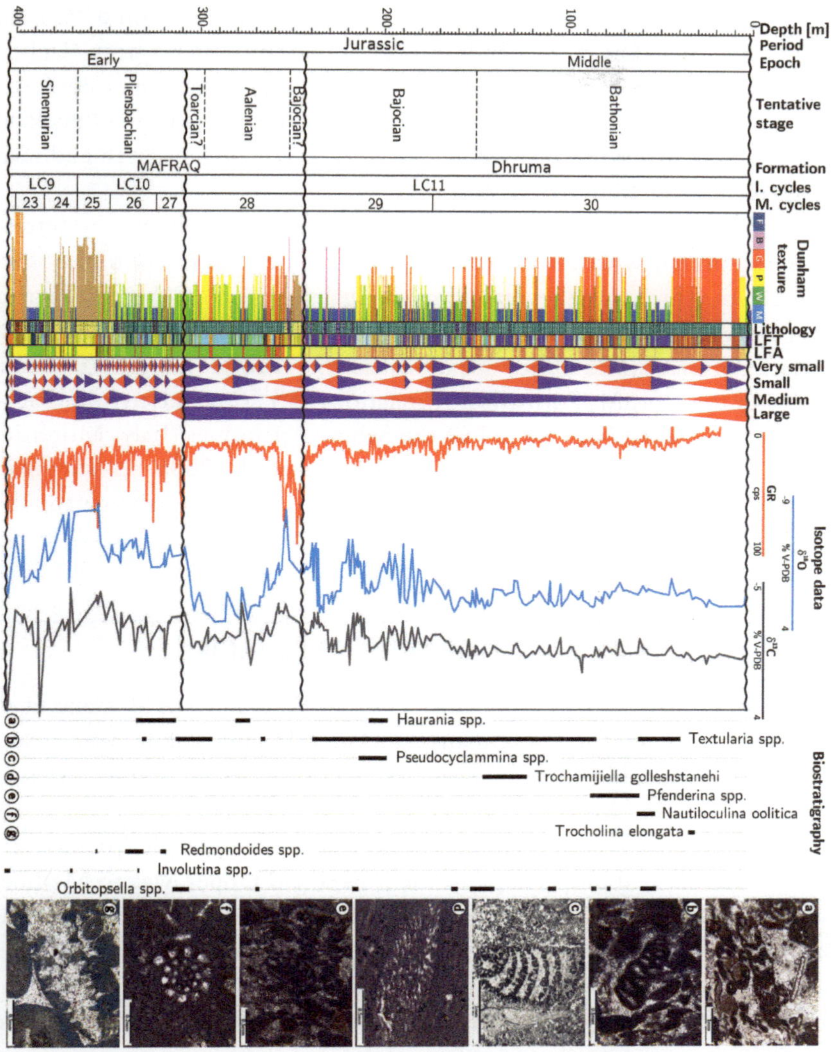

Fig. 3 Overview of the Jurassic succession of Wadi Sahtan in the Oman Mountains including age, Dunham texture, lithology, cyclicity, outcrop gamma-ray, isotope data, and biostratigraphy. *Modified after Schlaich, M., Aigner, T., 2017. Facies and Integrated Sequence Stratigraphy of an Epeiric Carbonate Ramp Succession: Dhruma Formation, Sultanate of Oman. The Depositional Record, 41 p.; Bendias, D., Aigner, T., 2015. Facies, sequence stratigraphy, reservoir and seal potential of the Mafraq Formation, Sultanate of Oman: an integrated outcrop analogue study. GeoArabia 20 (3), 17–94.*

isotope correlations to other Triassic type sections were based on $\delta^{13}C$ bulk isotope data only. The $\delta^{13}C$ trends of over 300 measurements from Richoz (2006) and Obermaier et al. (2012) were tentatively correlated to other locations like South China or Italy (Obermaier, 2013) and point to an Olenekian to Lower Anisian age for the Middle Mahil and a Lower Anisian to Ladinian age for the Upper Mahil (Fig. 2).

The Mafraq Formation represents deposits of a shallow marine to terrestrial environment with a number of unconformities and discontinuity surfaces present which cause several significant gaps in the $\delta^{13}C$ trend. This and rather imprecise chronostratigraphic data for the section prevent a straight forward correlation to other locations at the Tethys margin (Bendias, 2016). For the Dhruma Formation correlations of $\delta^{13}C$ trends to other locations around the Tethys Ocean like the Sabina Apennines (Italy, Morettini et al., 2002) and southern Spain (O'Dogherty et al., 2006) as shown in Schlaich and Aigner (2017) indicate a Bajocian to Bathonian age for the Dhruma Formation, matching the biostratigraphic dating.

4. SEQUENCE STRATIGRAPHY AND FACIES

In outcrops of the Oman Mountains, the AP6 TMS of Sharland et al. (2001) is represented by the Permian Saiq, Triassic Mahil, and Jurassic Lower Mafraq Formation. The Triassic Mahil Formation consists of three different members: Lower, Middle, and Upper Mahil (Koehrer et al., 2010; Obermaier et al., 2012; Pöppelreiter et al., 2011). The Jurassic Mafraq Formation is subdivided into two parts: the Lower and Upper Mafraq Member (Forbes et al., 2010; Figs. 1 and 3). While the Lower Mafraq is still considered as part of the AP6 megasequence, the Upper Mafraq forms together with the overlying Dhruma, Tuwaiq Mountain, and Hanifa/Jubaila formations the AP7 TMS of Sharland et al. (2001). The Dhruma and Tuwaiq Mountain formations are not further subdivided in Oman.

Cycle boundaries were determined in outcrops for example by the presence of paleo-soils, karst features, input of clastics, the occurrence of evaporites, or dependent on variations in Dunham texture, where the most proximal facies mark the cycle boundary and the most distal facies the zone of maximum flooding (Bendias, 2016; Bendias and Aigner, 2015; Obermaier et al., 2012; Schlaich, 2016; Schlaich and Aigner, 2017).

5. LOWER MAHIL MEMBER (UPPER KHUFF FORMATION)

The oldest Triassic sediments in outcrops of Oman are assigned to the Lower Mahil Member. The shallow marine carbonate deposits of Early Triassic age (Induan stage) are time equivalent to the Upper Khuff Formation which can be subdivided into two medium cycles: MC9 and MC10 from bottom to top (KS2 and KS1, Koehrer et al., 2010). The maximum flooding of MC9 most likely corresponds to the regional Tr10 MFS of Sharland et al. (2001), whereas the maximum flooding of MC10 is probably equivalent to Sharland et al.'s (2001) Tr20 MFS (Al-Husseini and Koehrer, 2013).

The Lower Mahil Member which is completely dolomitized in outcrops of the Oman Mountains is dominated by abiotic carbonate components such as ooids, peloids, and microbially coated lithoclasts which form thick carbonate grainstones of several meters in thickness (Koehrer et al., 2010). They were deposited in a storm-influenced shoal to foreshoal setting of a paleogeographically more distal portion of the Khuff carbonate ramp (Koehrer et al., 2010). The boundary between the Lower Mahil and the overlying Middle Mahil Member is marked by the occurrence of shales which are accompanied by a distinctive increase in gamma radiation.

6. MIDDLE MAHIL MEMBER (SUDAIR FORMATION)

The Lower Triassic Middle Mahil Member (Fig. 4) conformably overlies the Lower Mahil and is assumed to represent an outcrop equivalent of the subsurface Sudair Formation (Pöppelreiter et al., 2011). It spans a time interval of approximately 6 million years and constitutes one large cycle (Fig. 1) which is subdivided into three medium cycles (MC11–13; Fig. 2). The maximum flooding of the Middle Mahil large cycle LC5 probably coincides with the Tr30 MFS of Sharland et al. (2001). It is assumed that the Middle Mahil is followed by a hiatus of several million years involving most of the Anisian stage and the MFS Tr40.

Owing to a restricted marine environment and the preceding Permian–Triassic mass extinction event, fauna in the Middle Mahil is generally scarce. The succession is mainly composed of an alternation of dolomitized shoal-associated grainstones and backshoal-related muddy carbonate deposits

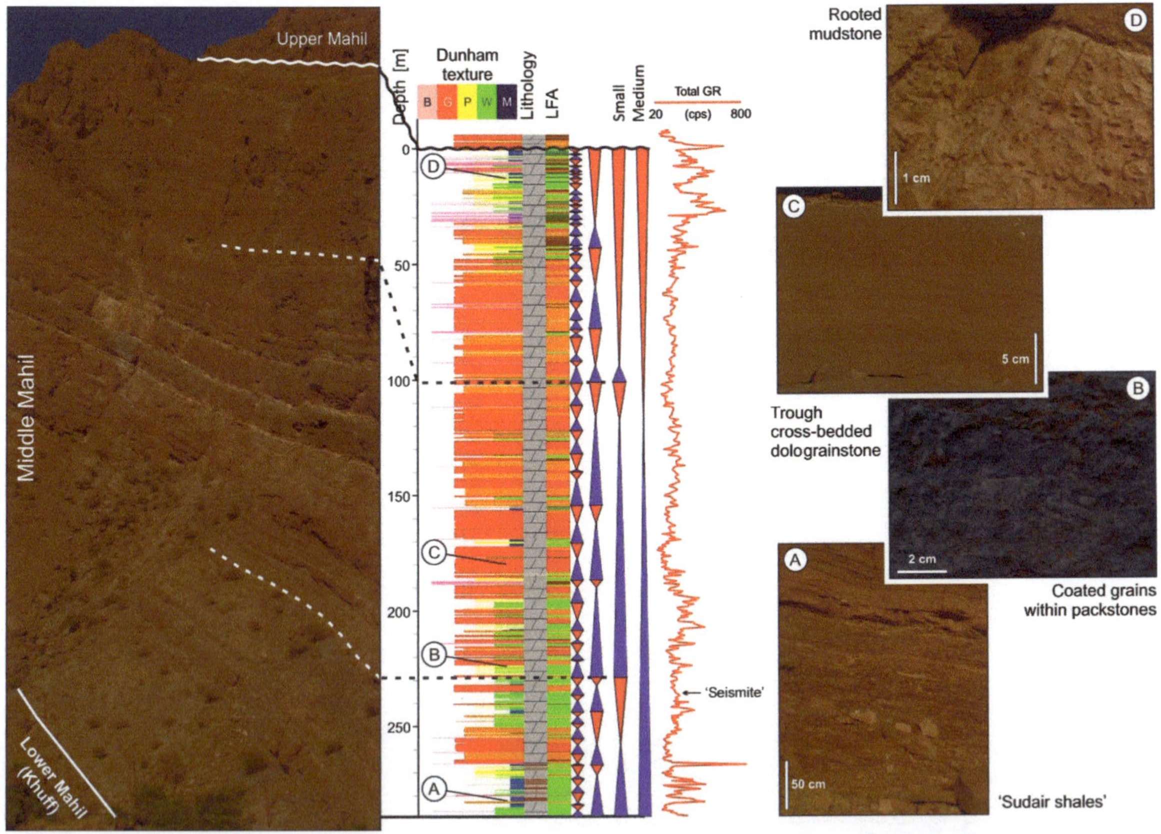

Fig. 4 The c. 289-m thick Middle Mahil Member (Sudair) in Wadi Bani Awf (N 2475259, E 548041). *Modified after Obermaier, M., 2013. The Geology of the Triassic Sudair and Jilh Formations in Outcrop and Subsurface (Sultanate of Oman). (PhD thesis) University of Tübingen.*

(Pöppelreiter et al., 2011). The lowermost Sudair comprises meter-thick claystones which are highlighted by high natural gamma radiation (Fig. 2) and illustrates a crucial marker unit to determine the Khuff/Sudair boundary in the subsurface of Oman. Backshoal facies in the lower part turn during transgression into thick grainstone deposits in the middle part of the succession. Thickest grainstone deposits with the lowest gamma-ray values are interpreted as the maximum flooding interval of medium cycle MC12 within large cycle LC5. Since facies of the overlying Upper Mahil, as well as the Middle Mahil Member indicate a more proximal environment, the maximum flooding in the Lower Mahil (LC4) is assumed to represent also the zone of maximum flooding of the TMS AP6. During the regressive phase of the Middle Mahil large cycle LC5 (Fig. 2), grainstones become less abundant, while muddy carbonates and microbial laminites with higher natural gamma radiation occur more frequently. The upper boundary of the Middle Mahil Member was picked at an exposure-related carbonate bed of a presumably supratidal marsh environment. Biostratigraphic data, outcrop, and subsurface facies suggest that the boundary between the Middle and Upper Mahil is disconformable and comprises a hiatus of a few million years (Forbes et al., 2010).

In a subsurface (well A) to outcrop (Wadi Sahtan) correlation clear facies changes toward a more proximal depositional setting in the west–southwest can be observed. The subsurface Sudair Formation in well A features mostly wacke-, pack- and occasionally grainstone beds which are interbedded by thin shale or anhydrite layers. In lateral equivalent strata in the outcrops of Wadi Sahtan, dolomites consist mainly of shoal grainstone facies with microbial boundstone interlayers (Fig. 5). In the upper Sudair Formation, a sharp facies change from dolo-wackestones to fine sandstones can be observed in the subsurface which is considered as time equivalent to exposure-related supratidal marsh facies in Wadi Sahtan (Fig. 5).

7. UPPER MAHIL MEMBER (JILH FORMATION)

Bio- and chemostratigraphic analyses imply that the Upper Mahil Member (Figs. 5 and 6) is time equivalent to the subsurface Jilh Formation (Obermaier et al., 2012). The Upper Mahil comprises in total 15–21 million years and consists of three large cycles (LC6, LC7, and LC8; Fig. 2), each spanning a time interval of several million years. Each large cycle consists of 2–3 medium cycles. Zones of maximum flooding of the large cycles might

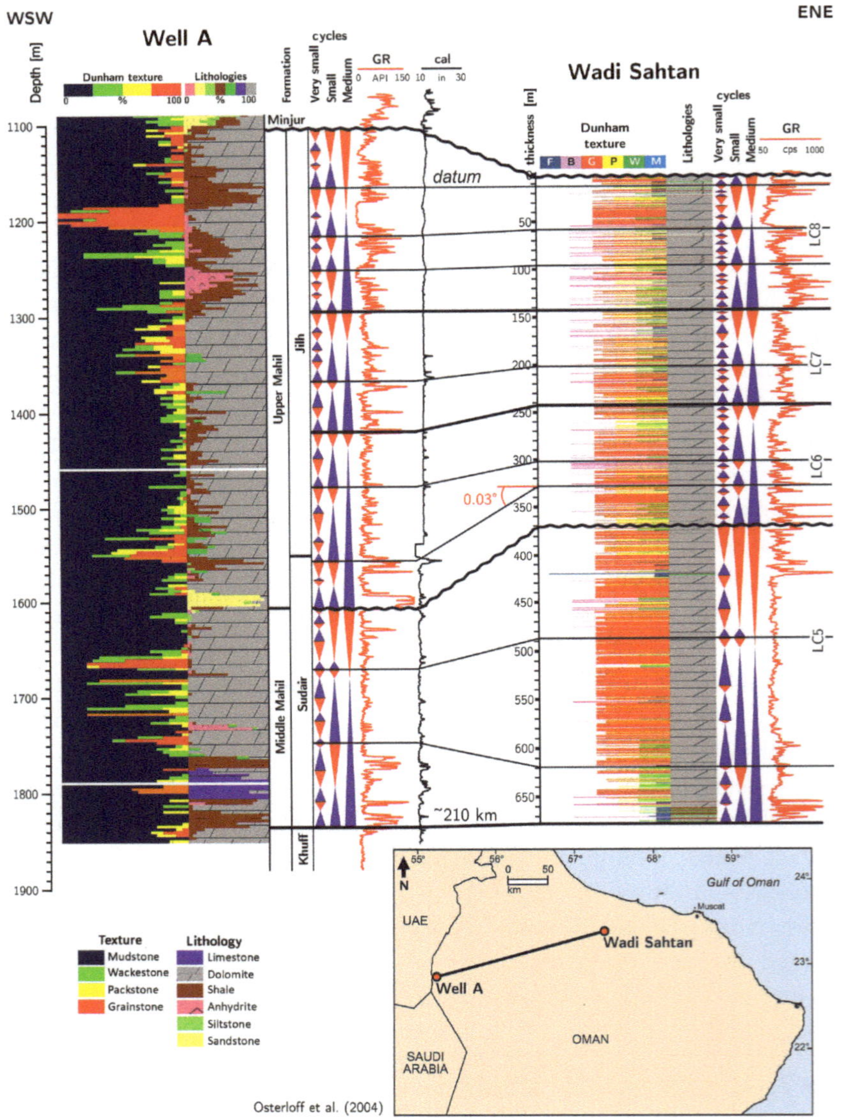

Fig. 5 Correlation of the Middle and Upper Mahil from subsurface well A to the outcrop in Wadi Sahtan (Central Oman Mountains), illustrating facies and thickness trends (Obermaier, 2013). *Modified from Osterloff, P., Al-Harthy, A., Penney, R., Spaak, P., Williams, G., Al-Zadjali, F., Jones, N., Knox, R., Stephenson, M., Oliver, G., Al-Husseini, M., 2004. Depositional sequences of the Gharif and Khuff Formations, subsurface interior Oman. GeoArabia, Special Publication 3, 83–147.*

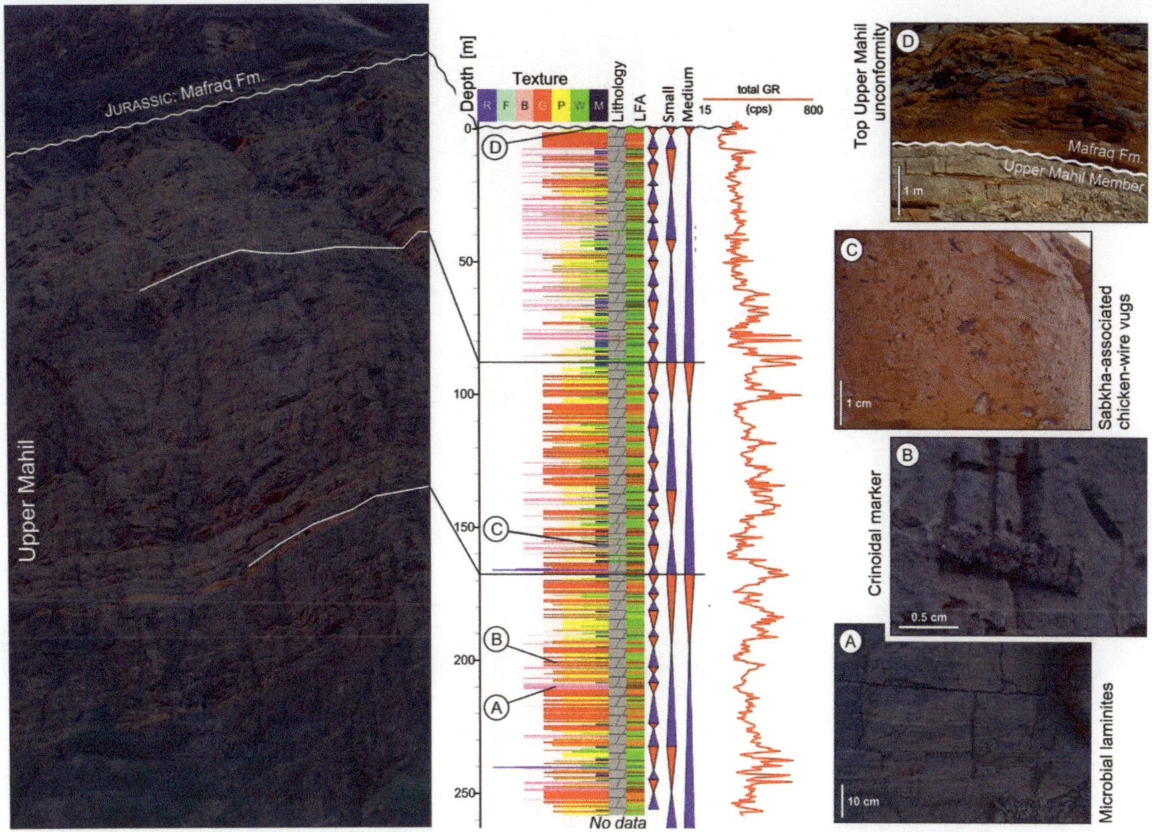

Fig. 6 The c. 258-m thick Upper Mahil Member in Wadi Bani Kharus (N 2572851, E 553266). *Modified after Obermaier, M., 2013. The Geology of the Triassic Sudair and Jilh Formations in Outcrop and Subsurface (Sultanate of Oman). (PhD thesis) University of Tübingen.*

correlate with MFS Tr50, Tr60, and Tr70 of Sharland et al. (2001) (Fig. 1). The upper boundary of the Jilh equivalent is marked by a distinctive erosional unconformity which is related to a "renewed uplift of Neo-Tethys rift shoulders" (Forbes et al., 2010) resulting in a northwestward tilting of the region.

Supratidal marsh, peritidal, and backshoal environments predominate in this Jilh time-equivalent formation (Fig. 2). The lowermost large cycle LC6 comprises mainly peritidal intraclastic grainstones and microbial laminites. A thin packstone unit with more open marine fauna (crinoids) determines the LC6 zone of maximum flooding (Fig. 2). The short regressive phase of this large cycle is dominated by muddy marsh facies (Obermaier et al., 2012). Three shifts in depositional environments within LC6 are interpreted as medium cycles (MC14–16). The overlying large cycle LC7 is marked by two backshoal–shoal–peritidal facies successions interpreted as medium cycles MC17 and MC18 which are well reflected in the gamma-ray pattern. Proximal lithofacies types show typically, due to an increased clay content, higher natural gamma radiation while distal facies exhibit a cleaner gamma-ray pattern. The uppermost large cycle LC8 contains in the lower part many muddy carbonate deposits and karst-related dissolution surfaces. The interbedding of abundant thin shale beds results in a characteristic serrated gamma-ray pattern (Fig. 2) which can also be found in the subsurface Jilh Formation (Fig. 5). A thick grainstone interval in the upper part of LC8 (MC21) of distinctive low gamma-ray defines the zone of maximum flooding of this large cycle. A similar gamma-ray response with a low gamma radiation in the upper Jilh can be observed in well A about 210 km farther WSW of Wadi Sahtan (Fig. 5). The upper boundary of the Upper Mahil is highlighted in outcrops by a lithology change from dolomites to sandstones and limestones of the overlying Jurassic Mafraq Formation. The lithological change indicates an erosional unconformity and depicts the top of the Triassic stratigraphy in outcrops.

The lowermost Jilh Formation (LC6) exhibits generally higher shale proportions and fewer grainstone deposits in the subsurface than in outcrops (Obermaier, 2013; Fig. 5). The thickness of LC6 decreases from well A to Wadi Sahtan by 60 m. Facies in LC7 appear similar with little shale content and some grainstone deposits. LC8 contains shale and anhydrite in well A which are probably equivalent to dissolution breccias and rooted carbonate facies in Wadi Sahtan (Fig. 5). Overall, the comparison between subsurface and outcrop facies from the Upper Mahil (Jilh) supports the regional trend of more proximal depositional environments in a southwestward direction (Ziegler, 2001) (Fig. 5).

8. MAFRAQ FORMATION

The Mafraq Formation is subdivided into a mixed carbonate-siliciclastic Lower Mafraq Member and a carbonate dominated Upper Mafraq Member (Fig. 7). The Lower Mafraq Member reflects the initial clastic pulse at the onset of marine transgression at the base of the Sahtan Group and onlaps on the tilted and eroded Akhdar Group from the NW to the SE (Forbes et al., 2010). It represents two large cycles (LC9 and LC10) composed each of three medium cycles which are subdivided into two to three small cycles. The zone of maximum flooding of the upper large cycle (LC10) might coincide with the MFS J10 of Sharland et al. (2001). According to Forbes et al. (2010), the Lower Mafraq Member ranges in age from the Pliensbachian to the Early Toarcian in the NW of Oman. Palynological data from Bendias and Aigner (2015) from the Oman Mountains (Wadi Sahtan) however suggest a longer time span ranging from the Late Hettangian to Late Pliensbachian, thus covering a time span of approximately 13–17 Ma (Fig. 1).

The depositional environment is characterized by terrestrial to shallow-marine clastics with paleosoils and other exposure features and marine carbonates. The boundary between the lower and upper Mafraq member is marked by a rooted sandstone bed of high gamma radiation in outcrops, separating the AP6 and AP7 megasequences. Biostratigraphic data from Wadi Sahtan indicates a time gap of approximately 10 Ma (Late Pliensbachian to early Aalenian) for this stratal surface (Bendias and Aigner, 2015; Fig. 1).

The Upper Mafraq Member was interpreted by Bendias and Aigner (2015) as one medium cycle (MC28) composed of two small cycles. Its zone of maximum flooding might coincide with the MFS J20 of Sharland et al. (2001). The Upper Mafraq ranges in age from Late Toarcian to Bajocian in the northwest of Oman (Forbes et al., 2010) and in outcrops of Wadi Sahtan from Late Aalenian to Bajocian (Bendias and Aigner, 2015). It thus covers a time span of 2–5 Ma representing one medium cycle.

The Upper Mafraq Member extends toward the southeast of North Oman, where it eventually thins out (Al Huqf area, Bendias and Aigner, 2015). Most parts of the Upper Mafraq Member consist of shallow-marine carbonates from a lagoonal to shoal-associated setting. Ferruginous ooid intervals and clastic input with distinctive high gamma-ray values occur at the top of the Mafraq Fm. and likely represent extensive soil development (Forbes et al., 2010). Outcrop to subsurface correlations indicate a thinning from WSW to ENE from approximately 300 m in well A to 160 m in the outcrop section at Wadi Sahtan for the whole Mafraq Fm. and its members.

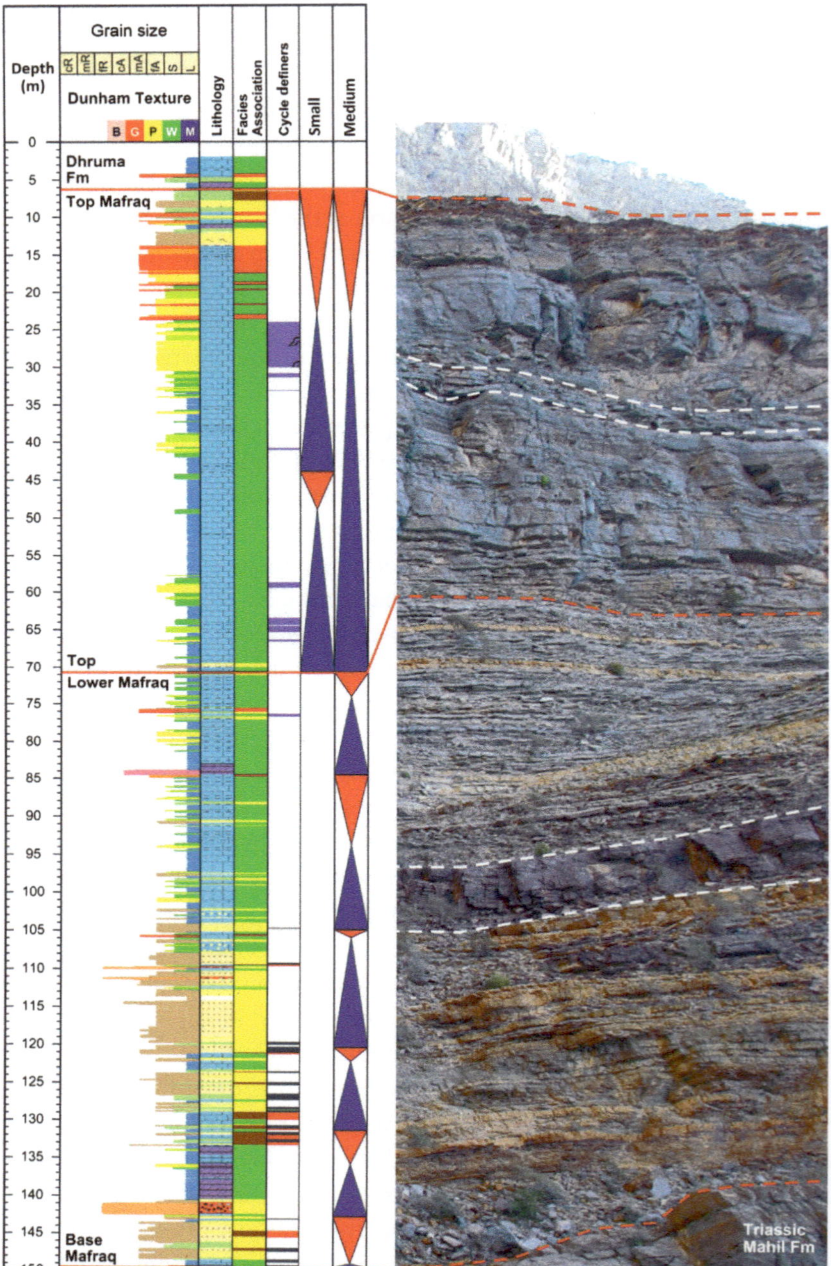

Fig. 7 Overview of the outcrop of the Mafraq Formation in Wadi Hail Bint (N 2583886, E 536000). *Modified after Bendias, D., 2016. The Minjur and Mafraq Formations in Oman, Facies—Sequence Stratigraphy—Reservoir and Seal Potential, from Outcrop to Subsurface. (PhD thesis) University of Tübingen, Germany.*

9. DHRUMA FORMATION

The Dhruma Formation (Fig. 8) consists of two medium cycles (MC29 and MC30) which can be further subdivided (Fig. 3): MC29 into three and MC30 into four small cycles (Fig. 9). While the zone of maximum flooding of MC29 might coincide with the MFS J30 of Sharland et al. (2001), the zone of maximum flooding of MC30 probably corresponds to MFS J40. Biostratigraphic data from Wadi Sahtan, as well as chemostratigraphic correlations suggest an Early Bajocian to Late Bathonian age for the Dhruma Fm. and thus covering a time span of 2.5 Ma (Forbes et al., 2010) to 6.5 Ma (Schlaich and Aigner, 2017). Small cycles appear to be very extensive and might be correlatable over a distance of 213 km from well A to the outcrops in Wadi Sahtan (Fig. 9; same number of small cycles per HFS in outcrops and subsurface).

Similar to the Upper Mafraq Member, the Dhruma Fm. represents deposits of a shallow-marine environment. It is dominated by carbonates and onlaps onto the Mafraq Fm. from the northwest to the southeast in MC29. It is characterized by peritidal, lagoonal, and high-energy "shoal-like" deposits and the occurrence of several discontinuity surfaces such as hard- and firmgrounds (Schlaich and Aigner, 2017). MC30 covers the whole of North and Central Oman but thins toward the south and southeast. Low-energy lagoonal facies are upward increasingly replaced by high-energy facies as well as laterally toward the northwest. Correlations indicate a very low relief as already suggested by Rousseau et al. (2006), thus supporting the existence of a vastly extensive, epeiric carbonate ramp (Schlaich, 2016).

The whole formation as well as its smaller building blocks show a distinct deepening-up, transgressive dominated cyclicity of "lagoonal" mudstones grading into grainstones in Oman. Toward Saudi Arabia and the UAE, the presence of an intrashelf basin becomes increasingly influential on the facies distribution in the Dhruma Formation. According to Forbes et al. (2010), deposits in adjacent Saudi Arabia and the UAE represent regressive dominated cycles ranging from mudstones of an intrashelf basin to shoal-associated grainstones. Regressive sequences at times of global sea level rise are justified by a continued continental-margin uplift (Forbes et al., 2010). However, the aforementioned uplift in southeast Oman appears to result in a different depositional environment. Correlations reveal facies trends ranging from peritidal deposits in the southeast of the study area to low-energy mudstones and increasingly high-energy deposits in the form of wacke-, pack-, and grainstone deposits toward the northwest of Oman. Hence, a foreshoal

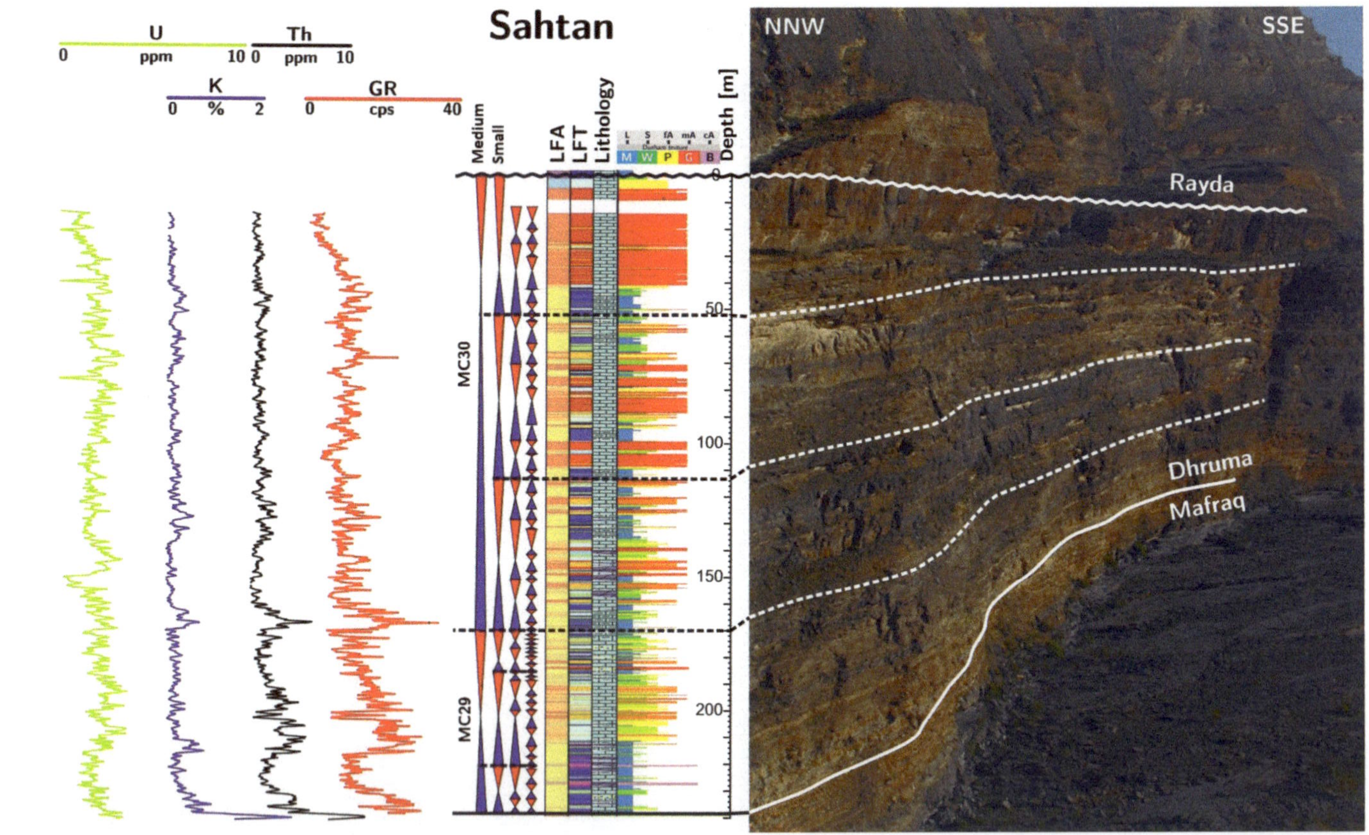

Fig. 8 Overview of the c. 238-m thick Dhruma Formation in outcrops of Wadi Sahtan (N 2583012, E 531597). *Modified after Schlaich, M., 2016. Geology of the Dhruma to Top Jurassic Formations in Outcrops and Subsurface, Sultanate of Oman. (PhD thesis) University of Tübingen, Germany.*

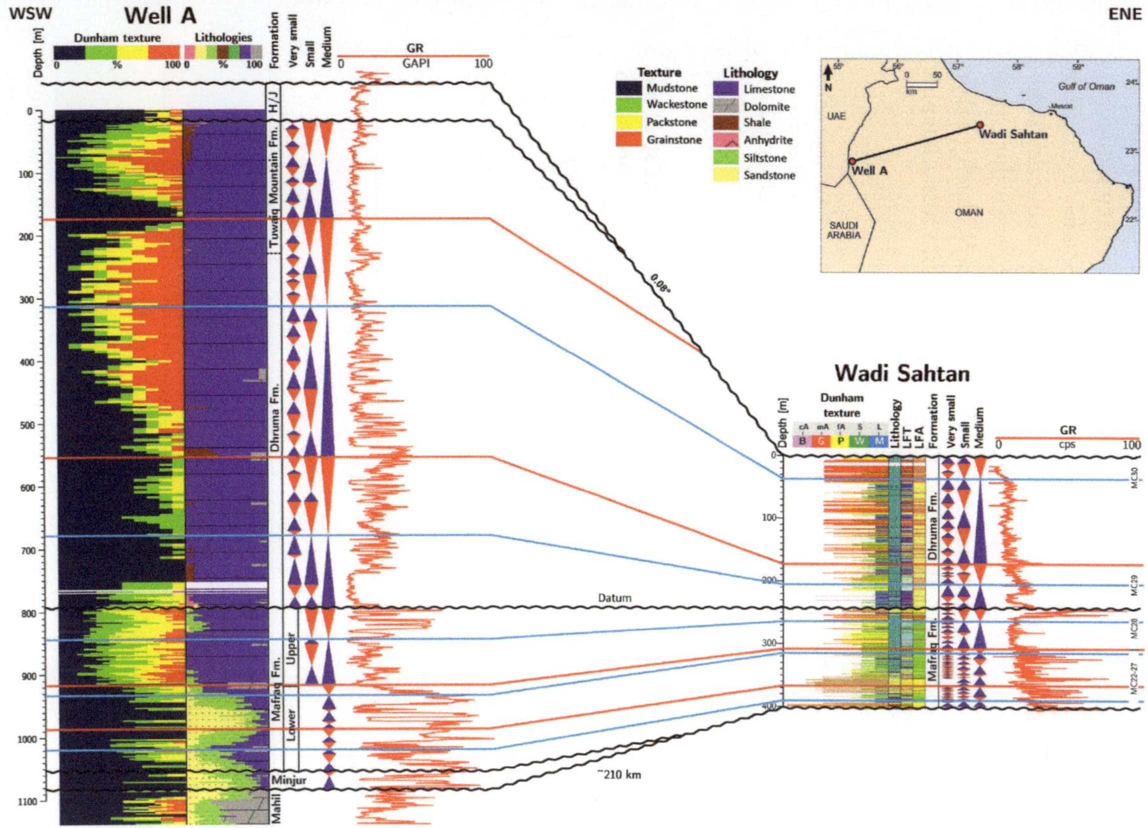

Fig. 9 Correlation of the Minjur, Mafraq, Dhruma, and Tuwaiq Mountain Formations from subsurface well A to the outcrop in Wadi Sahtan (Central Oman Mountains), illustrating facies and thickness trends (Bendias and Aigner, 2015; Schlaich, 2016). *Modified from Osterloff, P., Al-Harthy, A., Penney, R., Spaak, P., Williams, G., Al-Zadjali, F., Jones, N., Knox, R., Stephenson, M., Oliver, G., Al-Husseini, M., 2004. Depositional sequences of the Gharif and Khuff Formations, subsurface interior Oman. GeoArabia, Special Publication 3, 83–147.*

depositional environment is unlikely for mudstones of the Dhruma Fm. in Oman. Discrepancies in cycle hierarchies of the studies about the Mahil, Mafraq, Dhruma, and Tuwaiq Mountain Formations became apparent when reviewing and comparing the cycle duration of all three studies in this chapter: SC correspond to cycle sets, MC to high-frequency sequences, and LC to composite sequences in the Mahil and Lower Mafraq formations from Obermaier, 2013) and Bendias (2016) and SC to high-frequency sequences, MC to composite sequences, and LC to super sequences for the Upper Mafraq and Dhruma Formations from Bendias (2016) and Schlaich (2016). Timespans were calculated after the "International Chronostratigraphic Chart" from 2016 after Cohen et al. (2016; Table 2).

Table 2 Timespan for Tectonostratigraphic Megasequences (TMS) and Average Large (LC), Medium (MC), and Small Scale Cycles (SC) of the Mahil, Mafraq, and Dhruma Formations Based on Various Studies (Bendias, 2016; Obermaier, 2013; Schlaich, 2016; Schlaich and Aigner, 2017)

	Middle Mahil	**Upper Mahil**	**Lower Mafraq**	**Upper Mafraq +LC29 of the Dhrum Fm.**	**LC30 of the Dhruma Fm.**
Timespan [Ma]	M. Induan to M. Anisian	M. Anisian to L. Norian	U. Hettangian to U. Pliensbach.	U. Aalenian to Baj./Bath. boundary	Baj./Bath. boundary to U. Bathonian
	6.1	20.6	16.0	4.4	
TMS [Ma]	~94			~27	
Large cycles [Ma]	>6.10	6.87	8.00	>4.4	
Medium cycles [Ma]	2.03	2.58	2.67	1.35	1.70
Small cycles [Ma]	0.68	0.52	1.07	0.45	0.57

SC correspond to cycle sets, MC to high-frequency sequences, and LC to composite sequences in the Mahil and Lower Mafraq formations from Obermaier (2013) and Bendias (2016) and SC to high-frequency sequences, MC to composite sequences, and LC to super sequences for the Upper Mafraq and Dhruma Formations from Bendias (2016) and Schlaich (2016). This initial discrepancy in cycle hierarchies became apparent when reviewing and comparing the cycle duration of all three previous studies in this chapter. Timespans were calculated after the "International Chronostratigraphic Chart" from 2016 after Cohen et al. (2016).

10. SUBSIDENCE HISTORY: MID-PERMIAN TO LATE JURASSIC

In order to obtain a first-pass indication of changes in subsidence and/or sedimentation rates, a simple burial graph (sediment thickness vs time plot) was created for Mid-Permian to Jurassic times (Fig. 10). For simplification, changes in water depth, eustatic sea level, or differences in sediment compaction were disregarded in this graph.

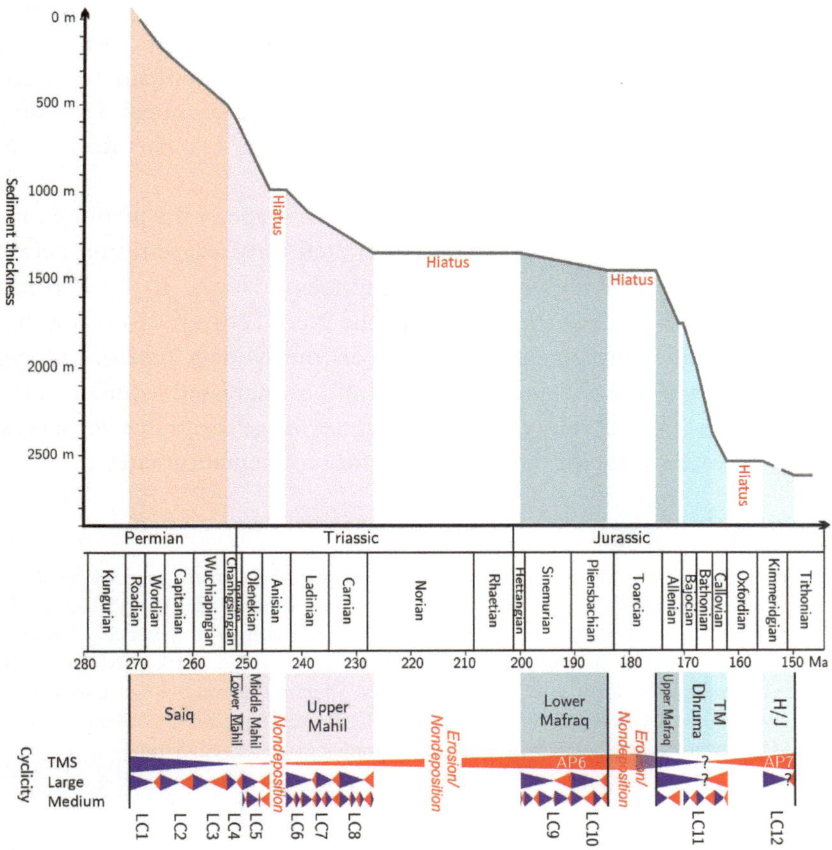

Fig. 10 Sediment thickness vs stratigraphic age for the Permian to Jurassic of North Oman reflects two pulses of rifting followed by thermal subsidence according to Sclater and Christie (1980). *Data from Permian to Triassic from Obermaier, M., 2013. The Geology of the Triassic Sudair and Jilh Formations in Outcrop and Subsurface (Sultanate of Oman). (PhD thesis) University of Tübingen.*

In Mid-Permian to Early Triassic times (Saiq, Lower and Middle Mahil Fm.), a steep increase of sediment thickness suggests high subsidence rates. According to Le Nindre et al. (2003), the increased sediment accumulation in the Early Triassic might be related to a new pulse of tectonic subsidence. After a hiatus in the Anisian, slightly lower accumulation rates are observed for the Upper Mahil. A Norian to Late Hettangian hiatus is succeeded by low subsidence rates in the Lower Mafraq Member in the Late Hettangian to Pliensbachian (Bendias and Aigner, 2015).

A hiatus of Toarcian times separates the Lower from the Upper Mafraq Member and the AP6 from the AP7 TMS. From the Late Toarcian to the Callovian, high rates of subsidence were present, representing the Upper Mafraq Member, the Dhruma and the Tuwaiq Mountain Formations. A hiatus of Oxfordian times separates the Tuwaiq Mountain from the Hanifa/Jubaila Fm. (Forbes et al., 2010). Due to the erosional Tithonian unconformity sediment accumulation rates for the Hanifa/Jubaila Fm. might be higher than shown in the graph.

According to Sclater and Christie (1980), the shape of the graph resembles a general thermal subsidence curve with an early active rifting period represented by initial high sediment accumulation rates in the Permian and Triassic which reflect the opening of the Neo-Tethys Ocean. Another pulse of tectonic subsidence is observed in the Middle Jurassic (Upper Mafraq, Dhruma, and Tuwaiq Mountain Fm.), by increased sediment accumulation rates (Fig. 10). However, uncertainties in age constraints for several formations cause a significant uncertainty for sedimentation rates.

ACKNOWLEDGMENTS

The studies that lead to the present synopsis were supported for many years by Shell and PDO, which is gratefully acknowledged. PDO and the Ministry of Oil and Gas of the Sultanate of Oman are thanked for permission to publish this chapter. P. Jeisecke (University of Tübingen) is thanked for the preparation of thin sections and S. Packer (Millennia SC Ltd.) for biostratigraphic analyses of the thin sections. Shuram Oil and Gas is acknowledged for fieldwork logistics. We are very grateful to ALT Luxemburg for providing access to WellCAD software packages and Schumberger for academic licenses of Petrel.

REFERENCES

Al-Husseini, M.I., Koehrer, B., 2013. Chrono- and sequence-stratigraphy of the Mid-Permian to Early Triassic Khuff sequences of the Arabian Plate. GeoArabia 18 (3), 103–130.

Bendias, D., 2016. The Minjur and Mafraq Formations in Oman, Facies—Sequence Stratigraphy—Reservoir and Seal Potential, from Outcrop to Subsurface (PhD thesis). University of Tübingen, Germany.

Bendias, D., Aigner, T., 2015. Facies, sequence stratigraphy, reservoir and seal potential of the Mafraq formation, Sultanate of Oman: an integrated outcrop analogue study. GeoArabia 20 (3), 17–94.

Bendias, D., Koehrer, B., Obermaier, M., Aigner, T., 2013. Mid-Permian Khuff Sequence KS6: paleorelief-influenced facies and sequence patterns in the Lower Khuff time-equivalent strata, Oman Mountains, Sultanate of Oman. GeoArabia 18 (3), 135–178.

Cohen, K., Finney, S., Gibbard, P., Fan, J., 2016. ICS international chronostratigraphic chart 2016. Episodes 36, 199–204.

Coy, G.A., 1997. Dolomitisation of the Akhdar Group: The Arabian Platform of Oman (PhD thesis). Darwin College, Cambridge.

De Matos, J.E., Hulstrand, R., 1995. In: Al-Husseini, M.I. (Ed.), Regional characteristics and depositional sequences of the Oxfordian and Kimmeridgian, Abu Dhabi. Middle East Geosciences Conference, GEO'94. Gulf PetroLink, Bahrain, 1, pp. 346–356.

Forbes, G., Jansen, H., Schreurs, J., 2010. Lexicon of Oman Subsurface Stratigraphy: reference Guide to The Stratigraphy of Oman's Hydrocarbon Basins. GeoArabia Special Publication 5, Gulf PetroLink, Bahrain. 371 p.

Glennie, K., Boeuf, M., Hughes-Clark, M., Moody-Stuart, M., Pilar, W., Reinhardt, B., 1974. Geology of the Oman Mountains: Enclosures. Royal Dutch Geological and Mining Society.

Grabowski, G.J., Norton, I.O., et al., 1994. In: Al-Husseini, M.I. (Ed.), Tectonic controls on the stratigraphic architecture and hydrocarbon systems of the Arabian Plate. Middle East Petroleum Geosciences Conference, GEO'94. Gulf PetroLink, Bahrain, 1, pp. 413–430.

Koehrer, B., Zeller, M., Aigner, T., Pöppelreiter, M., Milroy, P., Forke, H., Al-Kindi, S., 2010. Facies and stratigraphic framework of a Khuff outcrop equivalent: Saiq and Mahil Formations, Al Jabal al Akhdar, Sultanate of Oman. GeoArabia 15 (2), 91–156.

Konert, G., Afifi, A., Al-Hajri, S., de Groot, K., Al Naim, A., Droste, H., 2001. Paleozoic stratigraphy and hydrocarbon habitat of the Arabian Plate. AAPG Special Volumes 74, 483–515.

Le Nindre, Y.-M., Vaslet, D., Mécour, J., Bertrand, J., Halawani, M., 2003. Subsidence modelling of the Arabian Platform from Permian to Paleogene outcrops. Sediment. Geol. 156 (1–4), 263–285.

Morettini, E., Santantonio, M., Bartolini, A., Cecca, F., Baumgartner, P., Hunziker, J., 2002. Carbon isotope stratigraphy and carbonate production during the Early-Middle Jurassic: examples from the Umbria-Marche-Sabina Apennines (central Italy). Palaeogeogr. Palaeoclimatol. Palaeoecol. 184 (3–4), 251–273.

Murris, R.J., 1981. Middle east: stratigraphic evolution and oil habitat. AAPG Bull. 65 (7), 1358.

O'Dogherty, L., Sandoval, J., Bartolini, A., Bruchez, S., Bill, M., Guex, J., 2006. Carbon-isotope stratigraphy and ammonite faunal turnover for the Middle Jurassic in the Southern Iberian palaeomargin. Palaeogeogr. Palaeoclimatol. Palaeoecol. 239 (3–4), 311–333.

Obermaier, M., 2013. The Geology of the Triassic Sudair and Jilh Formations in Outcrop and Subsurface (Sultanate of Oman) (PhD thesis). University of Tübingen.

Obermaier, M., Aigner, T., Forke, H., 2012. Facies, sequence stratigraphy and reservoir/seal potential of a Jilh Formation outcrop equivalent (Wadi Sahtan, Triassic, Upper Mahil Member, Sultanate of Oman). GeoArabia 17 (3), 85–128.

Pöppelreiter, M., Schneider, C., Obermaier, M., Forke, H.C., Koehrer, B., Aigner, T., 2011. Seal turns into reservoir: Sudair equivalent in outcrops, Al Jabal al-Akhdar, Sultanate of Oman. GeoArabia 16 (1), 69–108.

Richoz, S., 2006. Stratigraphie et variations isotopiques du carbone dans le Permien supérieur et le Trias inférieur de quelques localités de la Néotéthys (Turquie, Oman et Iran) (PhD thesis). Université de Lausanne, Suisse.

Rousseau, M., Dromart, G., Droste, H., Homewood, P., 2006. Stratigraphic organization of the Jurassic sequence in Interior Oman, Arabian Peninsula. GeoArabia 11 (1), 17–50.
Schlaich, M., 2016. Geology of the Dhruma to top Jurassic Formations in Outcrops and Subsurface, Sultanate of Oman (PhD thesis). University of Tübingen, Germany.
Schlaich, M., Aigner, T., 2017. Facies and Integrated Sequence Stratigraphy of an Epeiric Carbonate Ramp Succession: Dhruma Formation, Sultanate of Oman. The Depositional Record, Wiley Online Library, 41 pp.
Sclater, J.G., Christie, P., 1980. Continental stretching; an explanation of the post-mid-cretaceous subsidence of the central North Sea basin. J. Geophys. Res. 85 (B7), 3711–3739.
Sharland, P.R., Archer, R., Casey, D.M., Davies, R.B., Hall, S.H., Heward, A.P., Horbury, A.D., Simmons, M.D., 2001. Arabian Plate Sequence Stratigraphy. GeoArabia Special Publication 2, Gulf PetroLink, Bahrain. 371 p.
Sikkema, W., 1992. Micropalaeontological Zonation and Range Chart of the Jurassic in Oman. Unpublished PDO Exploration Report No. 305, Muscat, Oman.
Ziegler, M.A., 2001. Late Permian to Holocene paleofacies evolution of the Arabian Plate and its hydrocarbon occurrences. GeoArabia 6 (3), 455–504.

MIX
Papier | Fördert gute Waldnutzung
FSC® C083411

Kolibri 360 GmbH
Ferdinand-Jühlke-Straße 7
99095 Erfurt, Deutschland
produktsicherheit@kolibri360.de